Dhanjoo N. Ghista (Ed.)

Biomedical and Life Physics

Dhanjoo N. Ghista (Ed.)

Biomedical and Life Physics

Proceedings of the Second Gauss Symposium, 2–8th August 1993, Munich

vieweg

Die Deutsche Bibliothek – CIP-Einheitsaufnahme

Printing and binding: W. Langelüddecke, Braunschweig
Printed on acid-free paper
Printed in Germany

ISBN 3-528-06877-9

Preface

D. N. Ghista

Biomedical & Life Physics Yesterday, Today & Tomorrow

(Based on the Opening Address at the International Conference on Biomedical Physics & Mathematics of the Gauss Symposium, August 2, 1993 at Ludwig Maximillians Universität, München, Germany)

The traditional practice of Biomedical physics has now expanded to involve multiple aspects of medical practice: development of systems and technology in medical monitoring (e.g., PET visualization of brain receptors to identify neuronal dysfunction), diagnosis (e.g., computer-aided echocardiographic texture analysis to detect myocardial infarcts), organ-support (e.g., peritoneal dialysis), and therapeutic function (e.g., encapsulation of insulin-producing pancreatic islet cells for treatment of diabetes). However, is Biomedical Physics a relatively new field? Not really, although we may have opened up new vistas of it, as presented in this book.

Let us recall some early and well-known physician-cum-biomedical physicists. Both physical and physiological scientists will know of Jean Poiseuille (1799-1869), physician and physiologist; he measured blood pressure with a mercury manometer while being a medical student in Paris, received his medical degree in 1928, and then went on to describe the law of viscous flow (applicable to arteriolar flow). Hermann Von Helmholtz (1821-1894), who was the first to measure the speed of a nerve impulse, was a physician and physiologist, physicist and mathematician; in 1858, he became professor of anatomy and physiology at University of Bonn, and in 1871 he was appointed to the Chair of physics at University of Berlin; he had a remarkable insight into the physics of biological systems, notably in physiological optics and acoustics, with his masterpiece book on 'Sensations of Tone' providing the physiological basis of the theory of music.

William Harvey (1578-1658), who is credited with the discovery of blood circulation, first studied medicine at Cambridge, then received his doctor-of-medicine diploma at the University of Paduo, and thereafter worked as a physician at St. Bartholomem's & St. Thomas'; he reasoned that if the heart is pumping 5 litres of blood every minute, where can this come from and where can it go, except go around! However, it seems that the Indian sage and physician, *Charaka*, noted for his encyclopaedic works on *Ayurveda* (the knowledge of life) dating back to 1000 BC, has clearly described the systemic circulation: "from the great centre (the heart) emanate vessels which carry blood into all parts of the body – an element that nourishes life of all animals, and without which life would be extinct; it is that element which goes to nourish the fetus in the uterus, and which flowing into its body returns to the mother's heart."

On the German 10 DM note is depicted (along with the familiar bell-shaped probability distribution curve) the portrait of **Carl Friedrich Gauss** (1777-1855), who this Symposium commemorates. Ranking as one of the greatest mathematicians of all times, he also made important contributions to astronomy, geodesy and electromagnetism. It is this spirit that constituted the theme of this Symposium: to explore mathematical, physical, biological, and medical sciences in a unified fashion.

A more recent and biomedically intrinsic endeavour of Biomedical Physics, dealt with in the Conference is that of **Organ system physics,** entailing analyses of organ functional mechanisms and processes. This can yield new concepts and theories to deal with the intricacies and complexities of organ processes. The development of biomedical physics of macro and micro body processes can eventually lead to a wider and more rational diagnostic methodology. For instance, in this book, in the section on metabolic pathways regulation (by Reinhart Heinrich), Systems and control theory has been employed to, for instance, model the adrenal cortex system and the dynamics of cortisol levels in healthy and diseased states. On the topic of Glucose metabolism and insulin delivery (developed herein by Eugenio Sarti) the pathophysiology of non-insulin dependent diabetes can be depicted in terms of a glucose-insulin metabolism model, whose parametric simulation of glucose tolerance tests can yield information regarding β-cell sensitivity for glucose and hepatic glucose production; the optimal control theory application to diabetes therapy then entails minimizing fluctuations of blood glucose from a preset level while accounting for large rates of insulin delivery.

Yet another case of organ systems physics is the mechanism of coordinated motility of the small intestine (by Rustem Miftakhov), entailing the following sequence of events: triggered propagation of electrical signals in the enteric nervous system network in the intestinal wall, activation of L-type Ca ion channels located in the smooth-muscle syncytim membrane, propagation of depolarization waves in the syncytium, stress and deformation waves in the intestinal wall; this model can be employed to simulate drug action on motility and also the synaptic neurotransmitter-receptor disorders causing pathological motility. The examples, treated in this book, illustrate an approach to organ system process modelling and parametric estimation, for disease diagnosis. Another aspect of disease, discussed in this Conference, is genetic disease, featuring a new DNA sequencing technology, recognition of DNA sequences by multi-site ligands, and design of drugs capable of inhibition of retrovirus infection.

A relatively new realm of **consciousness** (the faculty of mental awareness and experiencing) was introduced in the Conference by Nobel-laureate Brian Josephson. Consciousness cannot be regarded as just an epiphenomenon of complex brain processes, as asserted by Beverly Rubik. In fact how consciousness also affects the body and disease processes leads us to the concept of Mind-Body medicine. This then requires a new science paradigm for Holistic medicine.

The aspects of Biomedical Physics, presented and discussed in this Conference, may in fact be categorized as follows:

  ❑   **Biomedical Technological Physics,** as exemplified herein (by Masatoshi Itoh) with: Invivo measurements of neurotransmitter functional changes in dimentia brain by PET, and cerebral metabolic rates for glucose to detect multi-focal damage in the brain after stroke using ^{18}FDG & PET; Targeted drug delivery systems (by Arend Hoekstra).

  ❑   **Organ Functional Physics,** involving: Metabolic pathways and regulation in diabetes and cancer, Remodelling of nerve synapses and neuromuscular junctions in growth and aging (by Mohamed Fahim), Simulation of small-bowel motility under the action of cholinergic and adrenergic agonists and antagonists, and Reconstruction of cardiac electrical excitation maps from body surface data for more precise electrocardiographic diagnosis (by Milan Horacek).

☐   A hitherto relatively unexplored aspect, **Life & Behavioural
     Physics,** involving Cosmic or Absolute Consciousness as the
     fundamental entity behind the physical universe, vitalization of
     matter into mind, concept of evolution based on unfolding of
     consciousness, and the role of mind in health and disease; this
     approach has in fact been a long-standing tradition in Eastern
     systems of medicine.

The concept of mind–body interaction owes its origins in Ayurveda,
practiced earlier than 1000 BC, wherein life is defined as the union of body,
senses, mind and soul.   According to Ayurveda, a human being is a
microcosm within the Macrocosm; so the human body is considered to be a
matrix of the five fundamental elements forming the basic tissues and the
basic humors.   A healthy person is one in whom there is an equilibrium of
body humors and tissues, as well as control over the senses by the mind.
The somatic, autonomic and mental activities are controlled by plexii, the
activation of which can not only effect healthy function of associated organ
systems but also enable liberation from psychic ailments and propensities.
This concept, overlooked in Newtonian medicine, indeed constitutes the
basis of mind–body interaction for the mechanisms, diagnosis and
treatment of psycho-somatic diseases in Behavioural Medicine and
Psychiatry. It is now time for us enunciate the science basis for Mind–Body
medicine.

The basic concept of the science paradigm for holistic medicine is that
Absolute Consciousness is indeed the fundamental entity, with matter as its
derivative, as all along asserted in the ancient Science of Yoga Cosmology.
Absolute Consciousness incorporates two principles, the Cognitive principle
and the Operative principle.   It is the Operative principle that qualifies
absolute consciousness into the fine fundamental factors (etherial, aerial,
luminous, liquid and solid) which provide the constituents of the physical
universe. Thus matter is metamorphosed consciousness.

Part and parcel of this science paradigm is the recognition of the existence
of *microvita*, revealed by the modern-day sage Prabhat R. Sarkar (1921-
1990) to be responsible for synthetic reactions infusing energy into matter
(derived from Consciousness), and structuring primitive mental forms; the
emergence of mind is when life originates. Thereafter, augmentation of the
biopsychic field of the mind results in increased complexity of organismic

structures. Development in lower life forms as well as in human beings entails increased unfolding of consciousness and higher states of mind. In this process (according to Richard Gauthier), *microvita* are the information carriers, causing genetic modifications of living organisms. These new concepts of Life Physics, presented at the Conference, were embellished by Nobel laureate Manfred Eigen's narration on the principles and algorithms of the creation and organization of genetic information.

Since Behavioural Medicine deals with psychosomatic diseases, it requires the understanding of the mechanism of mind-body interaction associated with psychosomatic disease process, prevention and treatment. Linking the mind and the physical body are subtle psychic glands (*Chakras*), through which the propensities and behavioural traits are expressed. The *Chakras* control the endocrine glands, and hence regulate organ function through the hormonal secretions of the endocrine glands. Because the mind and its emotions affect the *chakras*, which in turn influence the endocrine glands' secretions and hence organ functions, mental stress can also cause organ dysfunction. Now when the mind gets dilated by the ideation of Cosmic Consciousness carried out at the site of these glands, these sentimental instincts are also controlled, and this also contributes to physical well-being and health. This is the basis of psychic therapy for mental ailments and psychosomatic diseases.

In this setting, and in a spirit of exploration and forward vision, it was also appropriate for this Conference to revisit Medical education. In a problem-based curriculum, biomedical sciences are taught in relation to clinical taxonomy. However (as pointed out herein by Vimla Patel), the key role of the study of basic medical sciences principles, in defining chains of causal mechanisms, could well be in developing the ability of coherent medical reasoning in diagnosis and intervention. Further, the above enunciated paradigm of Consciousness could provide an impetus to a holistic approach to health and disease in Medical education. It is hoped that this book can not only provide impetus for new methodologies in internal medicine and rethinking in medical education, but also help lead to new approaches in psychiatry involving mental rejuvenation as an aspect of the integration of physical and psychic aspects of health and disease.

Authors' adresses

M.A. Andrade, Dpto. de Bioquímica y Biología Molecular I, Facultad de CC. Químicas, Universidad Complutense, E-28040 Madrid

R. Bafunno, Dept. of Electronics, Computer and System Science, University of Bologna, Italy

D.J. Benor, 19 Fare Street, Bishopsteignton, South Devon TQ14 9QR, U.K.

I. Bernhardt, Humboldt-Universität Berlin, Institute of Biology, Experimental Biophysics, Invalidenstr. 42, 10115 Berlin

E. Biermann, Institut für Diabetesforschung, Kohnerplatz 1, 80804 München, Germany

M. Bier, Section of Plastic and Reconstructive Surgery, Dept. of Surgery, MC 6035, University of Chicago, 5841 South Maryland Av., Chicago, IL 60637, USA

G. Borst, Section of Neurophysiology, Dept. of Biology, Vrije Universiteit, de Boelelaan 1087, NL-1081 HV Amsterdam

R. Carroll, Dept. of Mathematics, Statistics and Computig Science, Dalhousie University, Halifax, N.S., Canada

M. Casolaro, Dept. of Chemistry, University of Siena, Italy

P. Ciaccia, Dept. of Electronics, Computer and System Science, University of Bologna, Italy

J.C. Clements, Dept. of Mathematics, Statistics and Computing Science, Dalhousie University, Halifax, N.S., Canada

C. Coltelli, Dept. of Electronics, Computer and System Science, University of Bologna, Italy

D.L. Delanoy, Dept. of Psychology, University of Edinburgh, 7 George Square, Edinburgh EH8 9JZ, U.K.

J. Delgado, Texas A&M University, Dept. of Chemical Engineering, College Station, Texas, 77843-3122, USA

K. Denner, Humboldt-Universität Berlin, Intitute of Biology, Experimental Biophysics, Invalidenstr. 42, 10115 Berlin, Germany

W. Düchting, Universität Siegen, Dept. of Electrical Engineering and Computer Science, Hölderlinstr. 3, 57068 Siegen, Germany

J.A. Edens, Institut de Génie Biomédical, Université de Montréal, Montréal, Qué., Canada

M.A. Fahim, Dept. of Physiology, Faculty of Medicine and Health Sciences, U.A.E. University, P.O.Box 17666, Al-Ain, United Arab Emirates

D.A. Fell, School of Biological and Molecular Sciences, Oxford Brookes University, Haedington, Oxford OX3 OBP, U.K.

P. Colli Franzone, Instituto di Analisi Numerica, Consiglio Nazionale Ricerche, Corso Carlo Alberto 5, I-27100 Pavia

R.F. Gauthier, Microvita Research Institute, Weisenauer Weg 4, 55129 Mainz, Germany

D. N. Ghista, Dept. of Bio Medical Engineering, University College of Engineering, Osmania University, Hyderabad-500 007, India

T. Ginsberg, Universität Siegen, Dept. of Electrical Engineering and Computer Science, Hölderlinstr. 3, 57068 Siegen, Germany

L. Guerri, Faculty of Engineering, University of Trento

R. Heinrich, Humboldt-Universität Berlin, Institute of Biology, Theoretical Biophysics, Invalidenstraße 42, 10115 Berlin, Germany

R. Heinrich, Freie Universität Berlin, Institute of

Biochemistry, Dept. of Chemistry, Thielallee 63, 14195 Berlin, Germany

C. Hilgetag, Humboldt-Universität Berlin, Institute of Biology, Theoretical Biophysics, Invalidenstr. 42, 10115 Berlin, Germany

A. Hoekstra, Cordis Europa N.V. P.O. Box 38, NL-9300 AA Roden

H.-G. Holzhütter, Humboldt-Universität Berlin, Institute of Biochemistry, Hessische Str. 3-4, 10115 Berlin, Germany

B.M. Horáček, Dept. of Physiology and Biophysics, Dalhousie University, Sir Charles Tupper Medical Building, Halifax, Nova Scotia, Canada B3H 4H7

G. Huiskamp, Dept. of Medical Physics ad Biophysics, University of Nijmegen, Geert Grooteplein noord 21, NL-6525 EZ Nijmegen

F. Greensite, Dept. of Radiology, UCI Medical Center, 101 City Drive, Orange, CA 92668, USA

P.R. Jensen, Dept. of Microbiology, Technical University of Denmark,DK-2800 Lyngby

B.D. Josephson, Cavendish Laboratory, Madingley Road, Cambridge CB3 0HE, U.K.

B.N. Kholodenko, E.C. Slater Institute, BioCentrum University of Amsterdam, PlantageMuidergracht 12, NL-1018 TV Amsterdam and A.N. Belozersky Institute of Physico-Chemical Biology, Moscow State University

K. Kits, Section of Neurophysiology, Dept. of Biology, Vrije Universiteit, de Boilelaan 1087, NL-1081 HV Amsterdam

E. Klipp, Humboldt-Universität Berlin, Institute of Biology, Theoretical Biophysics, Invalidenstr. 42, 10115 Berlin, Germany

V.A. Kolombet, The Institute of Theoretical and Experimental Biophysics, Russian Academy of Sciences, Pushchino, Moscow Region, Russia

J. Lankelma, Dept. of Oncology, FreeUniversity Hospital, NL-Amsterdam

A.P.M. Jongsma, The Netherlands Cancer Institute, H5, Plesmanlaan 121, NL-1066 CX Amsterdam

Y. Lenbury, Mahidol University, Dept. of Mathematics, Faculty of Science, Rama 6 Rd., Bankok 10400, Thailand

L.J. Leon, Institut de Génie Biomédical, Université de Montréal, Montréal, Qué., Canada

J.C. Liao, Texas A&M University, Dept. of Chemical Engineering, College Station, Texas,

77843-3122, USA

D. Lorimer, Scientific and Medical Network, Lesser Halings, Tilehouse Lane, Denham, Nr. Uxbridge, Middlesex UB9 5DG, U.K.

E. Meléndez-Hevia, Dpto. de Bioquímica, Facultad de Biología, Universidad de La Laguna, E-38206-Tenerife, Canary Islands

P.E. Micevych, Dept. of Anatomy and Cell Biology and Laboratory of Neuroendocrinology of the Brain Research, Institute UCLA, School of Medicine, Los Angeles, CA 90024, USA

P. Moniz-Barretto, School of Biological and Molecular Sciences, Oxford Brookes University, Haedington, Oxford OX3 0BP, U.K.

F. Montero, Dpto. de Bioquímica y Biología Molecular I, Facultad de CC. Químicas, Universidad Complutense, E-28040 Madrid

F. Morán, Dpto. de Bioquímica y Biología Molecular I, Facultad de CC. Químicas, Universidad Complutense, E-28040 Madrid

J. Nenonen, Dept. of Technical Physics, Helsinki University of Technology, Rakentajanaukio 2C, SF-02150 Espoo

J.C. Nuño, Dpto. de Bioquímica y Biología Molecular I, Facultad de CC. Químicas, Universidad Complutense, E-28040 Madrid

A. van Oosterom, Dept. of Medical Physics and Biophysics, University of Nijmegen, Geert Grooteplein Noord 21, NL-6525 EZ Nijmegen

M. Pennacchio, Instituto di Analisi Numerica, Consiglio Nazionale Ricerche, Corso Carlo Alberto 5, I-27100 Pavia

C. Pérez-Iratxeta, Dpto. de Bioquímica y Biología Molecular I, Facultad de CC. Químicas, Universidad Complutense, E-28040 Madrid

J.-J. Qian, Dept. of Radiology, UCI Medical Center, 101 City Drive, Orange, CA 92668, USA

P. Richard, DBCM, Section of Bioenergetique, Bft. 532, CE. Saclay, F-91191 Gif-sur-Yvette Cedex

S. Richter, Humboldt-Universität Berlin, Intitute of Biology, Experimental Biophysics, Invalidenstr. 42, 10115 Berlin, Germany

B. Rubik, Center for Frontier Sciences, Temple University, Ritter Hall 003-00, Philadelphia, PA 19122, USA

E. Sarti, Dipartimento di Elettronica Informatica e Sistematica, Università degli studi di Bologna, Viale

Risorgimento, 2, I-40136 Bologna

R. Schuster, Humboldt-Universität Berlin, Institute of Biochemistry, Hessische Str. 3/4, 10115 Berlin, Germany

S. Schuster, Humboldt-Universität Berlin, Institute of Biology, Theoretical Biophysics, Invalidenstr. 42, 10115 Berlin, Germany

J. Sicilia, Universidad de La Laguna, Departamento de Estadística, e I.O., Facultad de Matemáticas, E-38206 Tenerife, Canary Islands

E.C. Spoelstra, Dept. of Oncology, Free University-Hospital, NL-Amsterdam

Y. Takahashi, Mikuni R &D Corp., Berkeley, CA, USA

J. Tomas, Unitat d'Histologia i Neurobiologia (UHN) Facultad de Medicina i Ciències de la Salut, Universitat de Catalunia, Rovica i Virgilii (URV),

Sant Leorenc 21, E-43201 Reus (Tarragona)

M. Towsey, Biophysics Dept., Faculty of Medicine and Health Science, United Arab Emirates University, P.O.Box 17666, Al Ain, United Arab Emirates

W. Ulmer, Max-Planck-Institute of Biophysical Chemistry, 37073 Göttingen, Germany

N.N. Vasilyev, Dept. of Solid State Physics, Scientific Research Institute of Physics, St. Petersburg University, St. Petersburg, Russia

T.G. Waddell, University of Tennessee at Chattanooga, Dept. of Chemistry, 615 McCallie Avenue, Chattanooga, TN 37403, USA

H.V. Westerhoff, The Netherlands Cancer Institute, H5, Plesmanlaan 121, NL-1066 CX Amsterdam and E.C. Slater Institute, BioCentrum University of Amsterdam, Plantage Muidergracht 12,NL-1018 TV Amsterdam

Contents

Metabolic Control, Gowth, and Optimization

Introduction to Session Metabolic Regulatory Systems 1
R. Heinrich

Metabolic Control Analysis Using Transient State Data 7
J.C. Liao and J. Delgado

Control of Dynamics and Steady State: Applications to Multidrug Resistance 25
H.V. Westerhoff, M. Bier, D. Molenaar, E.C. Spoelstra, J. Lankelma, A.P.M. Jongsma, P.R. Jensen, P. Richard, and and B.N. Kholodenko

Parameter Space Classification of Solutions to a Mathematical Model for the Cortisol
Secretion System in Normal Man 33
Y. Lenbury

Optimization of Glycogen Design in the Evolution of Metabolism 49
E. Meléndez-Hevia, T.G. Waddell, and J. Sicilia

Tumor Growth and Treatment Planning: A Novel Simulation Approach 61
W. Düchting, T. Ginsberg, and W. Ulmer

Maximization of Enzymic Activity under Consideration of Various Constraints 71
E. Klipp

Metabolic Control Theory 85
R. Heinrich

Determining Elementary Modes of Functioning in Biochemical Reaction Networks at
Steady State 101
S. Schuster, C. Hilgetag, and R. Schuster

On Inferring the Kinetic Scheme of an Ion Channel from Postsynaptic Currents 115
M. Bier, G. Borst, K. Kits

Mathematical Modelling of Red Blood Cell Enzyme Deficiencies as an Example for
Large-Scale Parameter Changes in Biochemical Reaction Systems 125
R. Schuster and H.-G. Holzhütter

Simulation of Dioxygen Free Radical Reactions: Their Importance in the Initiation of
Lipid Peroxidation 137
P. Moniz-Barreto and D.A. Fell

Residual Sodium and Potassium Fluxes Through Red Blood Cell Membranes 145
I. Bernhardt, S. Richter, K. Denner, and R. Heinrich

Electrocardiography: Processes and Analyses

The Role of Natural Selection and Evolution in the Game of the Pentose Phosphate
Cycle 155
F. Montero, J.C. Nuño, M.A. Andrade, C. Pérez-Iratxeta, F. Morán, E. Meléndez-Hevia

The Forward and Inverse Problems of Electrocardiography 169
B.M. Horáček

Simulation of the Spread of Excitation in the Myocardial Tissue 173
P. Colli Franzone, L. Guerri, and M. Pennacchio

A Hybrid Model of Propagated Excitation in the Ventricular Myocardium 181
B.M. Horáček, J. Nenonen, J.A. Edens, and L.J. Leon

Extracardiac Field due to Propagated Excitation in the Ventricular Myocardium 191
J. Nenonen and B.M. Horáček

Spatio-Temporal Constraints in Inverse Electrocardiography 203
A. van Oosterom and G. Huiskamp

When is the Inverse Problem of Electrocardiography Well-Posed 215
F. Greensite and J.-J. Qian

Local Regularization and Adaptive Methods for the Inverse Problem 223
C.R. Johnson and R.S. MacLeod

On Regularization Parameters for Inverse Problems in Electrocardiography 235
J.C. Clements, R. Carroll, and B.M. Horáček

Gastrointestinal Mechanisms and Disorders

The Forces Exerted on the Duodenum and Sphincter of Oddi During ERCP, and Sphincterotomy 247
R. Ratani and P. Swain

Small Bowel Propulsion: Transport of a Solid Bolus 253
R. Miftakhov and G. Abdusheva

Regulatory Role of the Adrenergic Synapse Within a Neural Circuit 261
R. Miftakhov, G. Abdusheva, A. Mougalli

Duodenal Bioelectrical Waxing and Waning Activity 269
S. Salinari and R. Mancinelli

Drug Delivery Systems
Drug Delivery Systems: Technology and Clinical Applications   281
A. Hoekstra

Recent Developments in Drug Delivery Systems   283
A. Hoekstra

Port-Catheter Systems: Design, Advantages, and Applications   297
A. Hoekstra

Hydrocephalus Management With a Flow-Control Shunt: Overdrainage and Proximal
Obstruction: Controllable Complications of Shunting   311
A. Hoekstra and M. Sussman

Subtle Energies
Introduction to the Microvita Section   333
R.F. Gauthier

The Origins of Mind   335
M. Towsey and D.N. Ghista

Microvita: A New Approach to Matter, Life and Health   347
R.F. Gauthier

The Quantum Field of the Healing Force   359
N.N. Vasilyev

Is Another Physics Needed to Explain Fundamental Fluctuations in Physical Measurements?   367
V.A. Kolombet

Consciousness
Scientific Research into Consciousness   373
B.D. Josephson

The Challenge of Consciousness Study   375
B. Rubik

First-Person Experience and the Scientific Exploration of Consciousness   383
B.D. Josephson

The Near-Death Experience and the Nature of Consciousness   391
D. Lorimer

Experimental Evidence Suggestive of Anomalous Consciousness Interactions   397
D.L. Delanoy

Theoretical Speculations and Energy Phenomena Associated with Spiritual Healing ..   411
D.J. Benor

Towards a Science of Consciousness 417
M. Towsey and D.N. Ghista

Glucose Metabolism and Insulin Delivery
Glucose Metabolism and Insulin Delivery....................................... 429
E. Sarti

Recent Developments and Open Problems in Feedback Control of the Glucose System
in Diabetes 431
E. Sarti and C. Cobelli

Autonomic Dysfunction in Diabetes Mellitus.................................... 445
I.T. El Mugamer, K.D. Desai, M. Towsey, and D.N. Ghista

Physiologic Modelling of Type-2-Diabetes for Medical Education 457
E. Biermann

Rechargeable Glucose Biosensors 469
E. Wilkins and P. Atanasov

The Influence of Temperature on pH-Responsive Chemical Valve Systems: Thermodynamic Aspects for Drug Delivery 477
M. Casolaro

Control Algorithms for a Wearable Artificial Pancreas........................... 483
P.G. Fabietti, L. Tega, and S. Allegrezza

Adaptive Glycaemia Control Using Neural Networks 493
R. Bafunno, C. Coltelli, P. Ciaccia, and Y. Takahashi

Neuromuscular Remodelling in Development and Ageing
Neuromuscular Remodelling During Development and Aging..................... 505
M.A. Fahim

Synaptic Plasticity During Aging of the Neuromuscular Junction 507
M.A. Fahim

Neuromuscular Remodelling in the Adult Induced by Small Physiologic Changes in the
Locomotor Activity 521
M.R. Fenoll-Brunet, J. Tomas, M. Santafé, and M.A. Lanuza

Anabolic-Androgenic Steroid Regulation of Gene Expression in Spinal Motoneurons . 533
P.E. Micevych, P. Popper, and C.E. Blanco

Introduction to Session Metabolic Regulatory Systems

R. Heinrich

In recent years mathematical modelling of biochemical reaction networks as found in living cells became an important subject of theoretical biology. Most studies concern the development of models for the simulation of stationary and time dependent states, the analysis of complex dynamic phenomena such as oscillations, the quantitative characterization of control properties of metabolic pathways and the explanation of the structural design of enzymic networks on the basis of evolutionary optimization principles. This is also reflected by the contributions to the session *Metabolic Regulatory Systems* which cover a broad spectrum of the given field.

A necessary prerequisite of the modelling of complex cellular processes is the mathematical description of their elements or subsystems using, for example, principles of enzyme kinetics or methods allowing a quantification of membrane transport processes. This aim is followed by two contributions to the present section. Bernhardt *et al.* present a model for the *passive sodium and potassium fluxes* across the erythrocyte membrane. The mathematical description is based on a carrier mechanism taking into account a rapid competitive binding of both ions to the transport protein and subsequent slow translocation steps. The model could be fitted to experimental data for the rates of the inward and outward transport of sodium and potassium in dependence on their extracellular concentrations. Other individual transport processes, namely *neurotransmitter gated ion channels*, are studied by Bier *et al.* They present a simple reaction scheme for transitions between open and closed states of the channel and treat the system by linear differential equations.

Processes which involve many individual reactions steps are simulated by P. Moniz-Barreto and D. Fell as well as by Y. Lenbury. Both models are based on systems of nonlinear differential equations and concern systems of *free radical reactions* and the *cortisol secretion system* in man, respectively. In the process studied by Moniz-Barreto and Fell the problem arises that the individual reactions take place on very different time scales and that the concentrations of some intermediate compounds are extremely low. Accordingly, the differential equation system is very stiff. Y. Lenbury demonstrates that hormonal systems involving many regulatory loops may show very complicate dynamics, including chaotic behaviour. She makes extensive use of bifurcation analysis and presents simulations for the time dependence of hormone concentrations which may be compared with clinical data.

At present a great number of papers in the field of metabolic regulation is devoted to the extension of *Metabolic Control Analysis* including applications to various real biochemical pathways. This theory which dates back to the early seventies elucidates in quantitative terms to what extent the various reactions of metabolic systems determine the fluxes and metabolite concentrations. For that, two types of coefficients have been introduced, the *control coefficents* characterizing the systemic response of metabolic variables (fluxes, concentrations etc.) after parameter perturbations and *elasticity coefficients* which quantify the changes of reaction rates after perturbation of the substrate concentrations or kinetic parameters under isolated conditions. Up to now a comprehensive theory has only developed for the control of stationary states. There are, however, various attempts to extend metabolic control analysis to time-dependent processes (relaxation processes or oscillations) and by taking into account the hierarchical structure of cellular processes using a *modular approach*. A main result of metabolic control analysis is that the flux within a given metabolic system is generally not determined by one *rate limiting step*. In contrast, flux control is in most cases distributed over many reactions.

This session contains three papers on metabolic control analysis. In the paper of J.C. Liao and J. Delgado a new method for the determination of control coefficients is presented. It is based on an analysis of *transient states of metabolism* and makes use of the fact that in the close neighbourhood of steady states the control coefficients as well as the changes of system variables are mathematically determined by linear relationships resulting from a Taylor expansion of the system equations. The methods are applied to analyze the control properties of glycolysis under *in vitro* conditions.

The paper of H.V. Westerhoff *et al.* draws attention to the fact that there are various phenomena in metabolic systems which are not covered by traditional metabolic control analysis. These include the control properties of pathways involving *enzyme-enzyme interactions* (e.g. metabolic channeling) or *metabolic oscillations*. In the latter case some relevant estimates concerning the importance of specific reactions in determining the frequency of oscillations are obtained by analyzing the parameter dependence of the imaginary parts of eigenvalues close to Hopf-bifurcations, i. e. for situations where limit-cycle oscillations with small amplitudes may arise. Up to now there exists, however, no general theoretical basis for characterizing the parameter dependence of the frequency of limit cycles with large amplitudes.

The basic priciples of metabolic control analysis are outlined in a chapter by R. Heinrich. It is shown that for infinitesimally small perturbations of the kinetic parameters the systemic response is determined by linear equation systems relating the control coefficients of steady state variables to the elasticity coefficients of the individual enzymes. These equation systems may be transformed into a set of equations known as *summation theorems* and *connectivity theorems*. Since regulation of enzymes by effectors can cause large changes of their activities attempts have been made to consider the effect of *finite parameter perturbations*. As shown in the paper of R. Heinrich this is possible by expanding the steady state equations into a Taylor series and by taking into account not only linear but also second-order terms. In this way one arrives at first order and second order control coefficients which together allow a more accurate description of the system behaviour. However, the resulting nonlinear equations are difficult to handle and require a very detailed characterization of the kinetic properties of isolated reactions in terms of second order elasticity coefficients.

An alternative method for the consideration of finite parameter perturbations is presented in the chapter of R. Schuster and H.G. Holzhütter. These authors study the full nonlinear equation system of a given pathway by numerical methods. More specifically, they address the question to what extent finite alterations of maximal activities of the enzymes are tolerated by a metabolic system. A corresponding analysis is carried out for pathways involved in the *energy metabolism of erythrocytes*. It sheds a new light on the response of metabolic systems if they are severely impaired by enzymopathies or toxic drugs.

Metabolic systems are characterized by two distinct groups of experimental data. One set is composed of the variables (essentially concentrations and fluxes), while the other set comprises the system parameters (stoichiometric coefficients, kinetic constants, etc.). Simulation models serve to compute the system variables on the basis of given values for parameters. The question arises whether the latter quantities are also amenable to theoretical explanation. To answer this question, one should consider time-scales on which the kinetic properties and stoichiometry of enzymatic properties have changed, that is the dimension of biological evolution. In contrast to chemical reactions of inanimate nature, all the enzyme-catalyzed processes in the living cell are the outcome of *natural selection* which has acted over billions of years. A certain degree of understanding the structure of metabolic systems may be gained by considering evolution as an optimization process. This view implies that metabolic systems found in living cells show some *fitness properties*, which may be described by *extremum principles*. In the present section three chapters are devoted to the evolutionary optimization on different levels of biological organization.

The paper of E. Meléndez-Hevia *et al.* deals with the optimization of the structure of biological macromolecules. In particular, it is shown that the *structural design of glycogen*, in particular the number of tiers in the molecule, the degree of branching and the length of the chains, may be explained on the basis of various optimization principles related to the number of glucose molecules which may stored in a certain volume and to the velocity of the release of glucose by the phosphorylase.

The chapter of E. Klipp concerns the problem whether the *kinetic design of enzymes* given by the values of elementary rate constants may be explained on the assumption of maximal catalytic power. It is shown that such an analysis necessitates the consideration of two types of constraints, i.e. upper limits for the values of the rate constants and a fixed thermodynamic equilibrium constant. The corresponding nonlinear optimization problem is solved for enzymic uni-uni reactions with ordered mechanisms involving two or three reaction steps.

F. Montero *et al.* apply evolutionary optimization principles on the level of multienzyme systems. They consider different chemical possibilities for the interconversion of hexoses and pentoses and try to find out whether the actual *design of the pentose phosphate cycle* represents an optimal solution with respect to a minimal number of reaction steps and a maximum of the steady state flux. Their mathematical optimization procedure is based on a *quasi-species model* considering different stiochiometric designs and different kinetic constants of the participating reactions.

It is worth mentioning that, from the methodological point of view, evolutionary optimization is related to optimization in biotechnology. Also here, relevant objectives concern the maximization of metabolic yield or other criteria. It may be supposed, therefore, that the optimization methods presented in the above-mentioned chapters may become relevant also in the design and improvement of bioreactors.

Traditionally, most mathematical models of metabolic systems are based on systems of differential equations whose solutions generally require numerical values for the kinetic parameters of the participating reactions. However, these parameters are often unknown and are subject to frequent changes even in short time periods. S. Schuster *et al.* draw attention to the fact that many interesting results may be derived from a mere algebraic analysis of the topology of the metabolic networks without taking into account the kinetic details of the reactions. They develop the concept of *elementary flux modes* which represent the simplest routes leading from certain substrates to some product. These routes may be represented by specific vectors in the so-called null-space of the stoichiometry matrix, i.e. the space of all conceivable steady-state fluxes. It is suggested that the concept of elementary flux modes may be useful for the detection of optimal biosynthetic routes in biotechnology and for understanding the effects of enzyme deficiencies.

The development of models being of practical use for clinical therapies often necessitates to consider the cell as a whole instead of investigating individual metabolic systems. This is, for example, done in the paper of Düchting *et al.* There a *cell cycle model* is presented which allows to simulate the three-dimensional growth of a tumor spheroid in a nutrient medium. Using this model these authors are able to describe in quantitative terms radiation treatments and to derive some conclusions for the optimization of radiation strategies.

Metabolic Control Analysis Using Transient State Data

J.C. Liao and J. Delgado

1. Introduction

It is well recognized that metabolic control is complex: it involves the regulation at the gene expression level as well as the enzyme kinetic level. Extensive investigation over the past few decades has unveiled several mechanisms for regulation at the transcriptional and translational levels, which regulate the amount of enzymes in the cell with time scales typically longer than 10 minutes. Regulation of enzyme activity at the kinetic level can occur at time scales less than a minute by ligand-binding (competitive or non-competitive) and protein modification such as phosphorylation, adenylylation, and urydylylation. In higher animals, these mechanisms are also complicated by the involvement of hormones, membrane potential, and secondary messengers. In some cases, the metabolic pathways are fully integrated with signal transduction pathways, and with gene expression at a longer time scale. Despite encouraging progress, signal transduction and gene expression in higher organisms are still poorly understood. Even in well studied microorganisms such as *Escherichia coli* and *Saccharomyces cerevisiae*, many unknowns still exist in the seemingly straightforward central metabolism[1]. One striking lesson was the 1980 discovery of 2,6-fructose bisphosphate[2], the most important effector for phosphofructokinase and fructose 1,6-bisphosphatase in many eucaryotic organisms. Therefore, it is possible that new metabolites or new roles of known metabolites will be found in the future. However, this does not invalidate the value of theoretical analysis and mathematical modelling. Rather, it calls for more rigorous statement of assumptions and careful interpretation of results.

This chapter aims to provide a rigorous theoretical basis for the experimental determination of control coefficients. The emphasis here is the theoretical relationships that exist in metabolic systems under well defined transient states. Although *in vivo* metabolite measurements in the transient state is difficult, these relationships may provide directions for the development of experimental techniques, in addition to direct applications in *in vitro* systems.

2. The system

The system considered here requires a postulated stoichiometry. If the detailed stoichiometry is unknown, lumped stoichiometry can be used assuming that lumping does not lose significant regulatory features. Unfortunately, it is difficult to verify the validity of lumping without extensive experimentation. Therefore, systems with well defined stoichiometry and little interaction with connecting pathways outside of the system are the best candidates for the analysis here. If the interaction with adjacent pathways is the key regulatory feature, all these pathways must be considered in the system. If intracellular compartmentation exists, the same metabolite in different compartments must be considered individually, and the transport between compartments must be modelled as a reaction. Spatial distribution is not considered within the same compartment. Under these conditions, a system with m metabolites connected by r reactions can be modelled by the following set of ordinary differential equations

$$\frac{d\mathbf{x}}{dt} = \mathbf{Nv} \tag{1}$$

where $\mathbf{x}$ is a $m \times 1$ vector whose elements are metabolite concentrations, $\mathbf{v}$ is a $r \times 1$ reaction rate vector, and $\mathbf{N}$ is the $m \times r$ stoichiometric matrix, in which the element of ith row and jth column is the stoichiometric coefficient of metabolite i in reaction j. The stoichiometric coefficients are positive for products and negative for reactants (or substrates).

The kinetic rate laws in v are typically derived from purified enzymes *in vitro*, and each can be generally described as a function of **x** and **p**:

$$v_i = f(\mathbf{x}, \mathbf{p}) \tag{2}$$

where v_i is the kinetic rate law of enzyme *i*, and **p** is a parameter vector whose elements include kinetic parameters such as K_m, K_i and V_{max}, physical parameters such as temperature and pH, chemical environment such as concentrations of metal ions or metabolites that are not included in **x**, and the concentration of the active enzyme which is regulated by gene expression and protein modification. If all the kinetic rate laws in the system are experimentally determined, the system can be completely modelled by a set of ordinary equations represented by Eq. (1), assuming that the parameters are time-invariant.

Note that the definition of metabolites can be generalized to include any variables that interact with the system through specific rate laws. For example, in the regulation of glutamine synthetase in *Escherichia coli*, glutamine synthetase (GS) is adenylylated by a cascade of proteins-mediated events which in turn respond to the concentration of intracellular α-ketoglutarate and glutamine. Since the concentration of active GS may be a function of time, it can be included in the **x** vector and be considered as a "metabolite" from a mathematical viewpoint. If the rate of GS modification can be specified, the whole metabolic system can be modeled by solving Eq. (1) with proper initial conditions.

3. Which enzymes or parameters are important?

Typically, the biochemist would like to know which parameter is the most important in the control of the metabolic pathway. To answer this question, one has to define the "response" or "objective" of the control under analysis. For example, the steady-state flux through the pathway, the steady-state concentration of a given metabolite, or any combination of these quantities can be defined as a system response, if an asymptotically stable steady state exists. If the complete rate laws (Eq. 2) and the stoichiometry of the system are known, the system response can be directly calculated from the model (Eq. 1).

To determine the most important parameters in these cases, a simulated experiment would be to change the parameters, one at a time, by a given fraction, and compare the change in the system response. Mathematically, the sensitivity of the system response, R_i, with respect to the parameter, p_j, can be defined as

$$\frac{\partial R_i}{\partial p_j} \qquad \text{or in a normalized form as} \qquad \frac{\partial \ln R_i}{\partial \ln p_j}$$

These sensitivity coefficients are defined as the response coefficients in Metabolic Control Analysis 3-7. If one wishes to know which enzyme is most important to the regulation of the system response, then the partial derivative with respect to the rate function or the active enzyme concentration are taken:

$$\frac{\partial R_i}{\partial v_j} \quad , \quad \frac{\partial \ln R_i}{\partial \ln v_j} \quad \text{or} \quad \frac{\partial R_i}{\partial e_j} \quad , \quad \frac{\partial \ln R_i}{\partial \ln e_j}$$

these partial derivatives are defined as the control coefficients. Specifically, the first two are defined as the v-type control coefficients and the latter two as the e-type control coefficients[6,8]. Strictly speaking, the e-type control coefficient is a response coefficient, because enzyme concentration is treated as a parameter here. The distinction between the response coefficients and the v-type control coefficients is important particularly at the steady state. The existence of the steady state gives the control coefficients a set of properties that are useful in the discussion of metabolic regulation and in the determination of control coefficients experimentally. Fortunately, most living systems have a steady state at least under a defined condition. For example, the rate of substrate uptake remains constant under balanced growth conditions in microorganisms. This situation can occur if the intracellular kinetics are insensitive to the relatively small changes in environmental conditions during the time frame of batch growth. The system remains at a quasi-steady state before the depletion of nutrient or the change in environmental condition upsets the intracellular kinetics.

At the steady state, Eq. (1) becomes

$$\mathbf{Nv} = \mathbf{0} \qquad\qquad\qquad\qquad\qquad\qquad\qquad\qquad\qquad\qquad (3)$$

The equality becomes an approximation in the quasi-steady state. The steady-state flux function, $\mathbf{J}$, and the steady-state metabolite concentration function, σ, can be expressed as the following composite functions:

$$\mathbf{J} = \mathbf{F}(\mathbf{v}, \mathbf{x_0}) \qquad (4)$$

$$\sigma = \mathbf{F}(\mathbf{v}, \mathbf{x_0}) \qquad (5)$$

where $\mathbf{v}$ is in turn a function of $\mathbf{x}$ and $\mathbf{p}$ (Eq. 2) and $\mathbf{x_0}$ is the vector containing the initial metabolite concentrations. Thus, the control coefficients and the response coefficients can be related by the chain rule:

$$\frac{\partial J_i}{\partial p_j} = \sum_{k=1}^{r} \left(\frac{\partial J_i}{\partial v_k} \right) \left(\frac{\partial v_k}{\partial p_j} \right) \qquad (6)$$

$$\frac{\partial \sigma_i}{\partial p_j} = \sum_{k=1}^{r} \left(\frac{\partial \sigma_i}{\partial v_k} \right) \left(\frac{\partial v_k}{\partial p_j} \right) \qquad (7)$$

where $\dfrac{\partial J_i}{\partial v_k} \equiv Z_k^i$ and $\dfrac{\partial \sigma_i}{\partial v_k} \equiv \Gamma_k^i$ are defined as the flux and metabolite concentration control coefficients, respectively, if the partial derivatives are evaluated at the steady state.

4. Conservation equations and properties of control coefficients

Biochemical systems usually contain conservation relationships among the metabolites. Common examples are adenine nucleotide and nicotinamide phosphate pools, which are conserved within typical time frames of interest. These relationships determine, in part, the structural properties of the control coefficients. Because of these conservation relationships, Eq. (1) is linearly dependent. If there are no independent equations, Reder[9] proposed to decompose the stoichiometric matrix as

$$N = LN_R \tag{8}$$

where N_R is a $m_0 \times r$ matrix formed by the first m_0 rows of N that constitute a basis for its row space. L is a $m \times m_0$ matrix that has the form

$$L = \begin{bmatrix} I_{m_0} \\ L_0 \end{bmatrix} \tag{9}$$

where I_{m_0} is the identity matrix of rank m_0 and L_0 is the $(m - m_0) \times m_0$ matrix that satisfies Eq. (8). Similarly, the concentration vector can be decomposed as

$$x = \begin{bmatrix} x_R \\ x'_R \end{bmatrix} \tag{10}$$

where x_R is $m_0 \times 1$ and x'_R is $(m - m_0) \times 1$. Using the above definitions, one can partition Eq. (1) into two equations:

$$\frac{d x_R}{dt} = N_R v \tag{11}$$

and

$$\frac{d}{dt}(x'_R) L_0 x_R = 0 \tag{12}$$

Integrating Eq. (12) between two arbitrary time points one gets

$$\Delta x'_R = L_0 \Delta x_R \tag{13}$$

This representation allows us to remove the linear dependencies due to conservation relationships from the system shown in Eq. (1). Using Eqs. (10), (13) and (9), in that order, one obtains

$$\Delta x'_R = L \Delta x_R \tag{14}$$

Therefore, to obtain $\mathbf{x}$ as a function of time, it is only necessary to solve the differential system given by Eq. (11), obtain $\mathbf{x_R}$, and then calculate $\mathbf{x}$ from Eq.(14) and any known reference point.

Having identified the conservation relationships among the metabolites, one can show that the control coefficients are[9]

$$\mathbf{Z} = \mathbf{I_r} - \frac{\partial \mathbf{v}}{\partial \mathbf{x}} \mathbf{L} \left(\mathbf{N_R} \frac{\partial \mathbf{v}}{\partial \mathbf{x}} \mathbf{L} \right)^{-1} \mathbf{N_R} \tag{15}$$

$$\mathbf{\Gamma} = -\mathbf{L} \left(\mathbf{N_R} \frac{\partial \mathbf{v}}{\partial \mathbf{x}} \mathbf{L} \right)^{-1} \mathbf{N_R} \tag{16}$$

where $\mathbf{Z}$ is the matrix of flux control coefficients whose elements are Z_k^i, $\mathbf{\Gamma}$ is the matrix of metabolite control coefficients whose elements are Γ_k^i, $\mathbf{I_r}$ is the identity matrix of rank r, $\partial \mathbf{v}/\partial \mathbf{x}$ is the $r \times m$ matrix of partial derivatives of $\mathbf{v}$ with respect to $\mathbf{x}$, and $\mathbf{N_R} \, \partial \mathbf{v}/\partial \mathbf{x} \mathbf{L}$ is the Jacobian matrix of the system of Eq. (11). All the terms in Eqs (15) and (16) are evaluated at the steady state. These two equations are derived based on the chain rule of differentiation, the conservation relationships, and the assumption that the Jacobian matrix is invertible (i.e., a steady state exists). Although difficult to interpret, these two equations are useful in mathematical derivations.

Matrices $\mathbf{Z}$ and $\mathbf{\Gamma}$ show some interesting properties that are dependent on the stoichiometry of the system. The first of these relationships are the so-called summation relationships, which read[9]

$$\mathbf{ZK} = \mathbf{K} \tag{17}$$
$$\mathbf{\Gamma K} = \mathbf{0} \tag{18}$$

In Eqs. (17) and (18), $\mathbf{K}$ is a $r \times (r - m_0)$ matrix with columns that constitute a basis for the kernel of $\mathbf{N}$, that is, $\mathbf{K}$ contains a basis for the solutions for the system $\mathbf{NK} = \mathbf{0}$. Note that the summation properties depend only on the structure (stoichiometry) of the pathway: they do not depend on the functional form of the rate functions, nor on the values assumed by the parameters. Other structural relationships among these coefficients are

$$N_R Z = 0 \tag{19}$$

$$\begin{bmatrix} -L_0 & I_{m-m_0} \end{bmatrix} \Gamma = 0 \tag{20}$$

and

$$\text{rank}(\Gamma) = m_0 \tag{21}$$

We will use some of these relationships in the following derivations.

5. The flux control coefficients and the transient fluxes

Let us continue to consider a reaction system formed by m internal intermediate linked by r reactions described by Eq. (1) and the flux control coefficients given by Eq.(15). Multiplying both sides of Eq. (15) by $v - J^0$, where J^0 is the steady-state reaction rate through the pathway and v is the reaction rate vector at any *transient time*, we obtain

$$Z(v - J^0) = I_r(v - J^0) - \frac{\partial v}{\partial x} L \left(N_R \frac{\partial v}{\partial x} L \right)^{-1} N_R (v - J^0) \tag{22}$$

But, from the reduced system of differential equations that describe the system (Eq. (11)), we can write

$$N_R(v - J^0) = \frac{d(x_r - \sigma_R^0)}{dt} \tag{23}$$

where σ_R^0 is the steady-state concentration vector for the reduced system. Because the first-order approximation of the reaction rates around the steady state is

$$v - J^0 = \frac{\partial v}{\partial x}(x - \sigma^0) = \frac{\partial v}{\partial x} L (x_R - \sigma_R^0) \tag{24}$$

we have

$$\frac{d(x_R - \sigma_R^0)}{dt} = N_R(v - J^0) \approx N_R \frac{\partial v}{\partial x} L (x_R - \sigma_R^0) \tag{25}$$

Therefore,

$$\left(N_R \frac{\partial v}{\partial x} L\right)^{-1} N_R \left(v - J^0\right) = \left(x_R - \sigma_R^0\right) \tag{26}$$

Substituting Eq. (26) into Eq. (22) gives

$$Z\left(v - J^0\right) = \left(v - J^0\right) - \frac{\partial v}{\partial x} L\left(x_R - \sigma_R^0\right) \tag{27}$$

Using the linear approximation for $v - J^0$ (Eq. (24)) one more time to eliminate $\partial v / \partial x$, we obtain

$$Z\left(v - J^0\right) = \left(v - J^0\right) - \left(v - J^0\right) = 0 \tag{28}$$

Finally, because J^0 belongs to the kernel of N_R, we have $ZJ^0 = J^0$ (from Eq. 21) and thus

$$Zv = J^0 \tag{29}$$

This equation relates the transient reaction fluxes, v, to the steady-state fluxes and the flux control coefficients. This equation has been derived using a slightly different approach[10]. There are two assumptions involved in the derivation of Eq. (29). First, it is assumed that the Jacobian matrix $N_R \, \partial v / \partial x L$ of the differential equations that describe the system (Eq. 1) is invertible. If the system has an asymptotically stable steady or quasi-steady state, this condition is satisfied. Second, by using a linear approximation for the reaction rates, we assumed that the kinetic rate laws are sufficiently linear around the steady state of interest (see Eq. 28). Although this assumption may seem very restrictive because reaction rates are usually non-linear, the linear approximation is often satisfactory for practical purposes. Note that in the linear approximation of the reaction rates we did not include the external reactants and products to ensure the existence of a steady state and invertibility of the Jacobian matrix.

Eq. (29) is applicable to any pathway stoichiometry as long as the assumptions stated above are valid. Note that although the flux control coefficients are defined for the steady state, we use transient reaction rates in the above equation. This equation states that the transient and the steady-state fluxes through each reaction are related to each other by the flux control coefficients. The reactions with larger flux control coefficients have a more significant effect in controlling the transient fluxes than the reactions with small flux control coefficients. This implies that during the transient state, the reaction rates of the steps with small flux control coefficients can fluctuate significantly, whereas reaction rates of the steps with large flux control coefficients must remain relatively constant. In pathways with arbitrary stoichiometry, the equation $N_R Z = 0$ (see Eq. 19) states that there will be m_0 relationships among the rows of Z. Therefore, there will be only $r - m_0$ independent rows in Z and Eq. (29) will consist of $r - m_0$ independent equations.

6. The flux control coefficients and the transient concentrations

One can easily express Eq. (29) in terms of the transient intermediate concentrations by considering the equation

$$\frac{dy}{dt} = Av \tag{30}$$

where y is an $n \times 1$ vector containing the concentrations of the *internal and external* species, and A is the full stoichiometric matrix, which includes the external substrates and products, and n is the total number of species. The inclusion of external metabolites is necessary in closed systems because they allow the determination of input and output fluxes. It is unnecessary in open systems where input and output flux can be determined. If the system is closed, the species in y are linearly dependent, and the system can be reduced to

$$\frac{dy_R}{dt} = A_R v \tag{31}$$

where y_R and A_R are defined according to Eqs. (10) and (8), respectively. One can express the transient reaction rates in terms of the transient intermediate concentrations by solving Eq. (31) for v. This is known as solving v by least-squares, and the solution is

$$v = A_R^+ \frac{dy_R}{dt} \tag{32}$$

The matrix A_R^+ is known as the pseudo-inverse or Moore-Penrose generalized inverse of the full stoichiometric matrix A_R, also denoted as $\left(A_R{}^T A_R\right)^{-1} A_R{}^T$ [11]. Substituting Eq. (32) into Eq. (29) we obtain

$$ZA_R^+ \frac{dy_r}{dt} = J^o \tag{33}$$

Integrating Eq. (33) between two arbitrary time points yields

$$ZA_R^+\left[y_r(t_2) - y_r(t_1)\right] = (t_2 - t_1)J^o \tag{34}$$

If one defines α as a $r \times n$ matrix such that

$$\alpha = ZA_R^+ \tag{35}$$

then, Eq. (34) becomes

$$\alpha\left[y_r(t_2) - y_r(t_1)\right] = (t_2 - t_1)J^o \tag{36}$$

This equation is valid before and during the quasi-steady state. It fails when the external species start to influence the intracellular kinetics. In cases where the reaction rates depend on the external species, one can buffer or keep them at a constant level so that changes in their concentrations will not affect the reaction rates within the time scale of the experiment. In these cases, the system will reach a quasi-steady state and the flux control coefficients will be evaluated at this point. Within the valid range, one can measure the transient concentrations and the quasi-steady state flux, which can be used to determine α by use of linear regression. The flux control coefficients can then be calculated as

$$Z = \alpha A_R \tag{37}$$

Because the derivation of Eq. (36) is based on Eq. (29), the assumptions involved are the same, namely, the Jacobian matrix must be invertible and the reaction rates are linear functions of the intermediate concentrations. Besides these, the columns of A_R must be linearly independent. This condition is necessary to ensure the invertibility of the matrix $A_R{}^T A_R$ and existence of the pseudo-inverse of A_R[11]. From a practical point of view, this condition means that one has to be able to calculate the transient reaction rates from the measured transient intermediate concentrations. A similar equation relating the transient concentrations to metabolite concentration control coefficients has been derived elsewhere[12].

7. Experimental applications to in vitro glycolysis

We now apply the above approach to a partial glycolytic pathway reconstituted *in vitro* with commercial enzymes, which convert glucose into glycerol-3-phosphate (G3P) (Fig. 1). The detailed experimental conditions were described elsewhere[13]. The ATP supply was maintained constant with an auxiliary system consisting of phosphocreatine and creatine phosphokinase. Moreover, glucose and NADH were supplied in sufficient quantities to ensure the saturation of hexokinase and glycerol-3-phosphate dehydrogenase (G3PD), respectively. The temperature was 25 °C in all experiments. The dynamic response was initiated by the addition of hexokinase to complete the final reaction mixture. The flux through G3PD can be directly monitored spectrophotometrically by following NADH consumption. Samples of the reaction mixture were withdrawn periodically. The reactions were stopped immediately by the addition of perchloric acid and the metabolite concentrations were measured using NADH/NADPH-coupled assays. After initiating the transient, the metabolites and NADH were measured as a function of time. The transient metabolite concentrations are shown in Fig. 2.

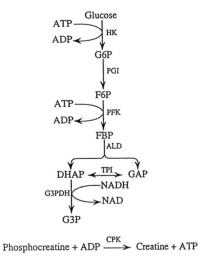

Phosphocreatine + ADP $\xrightarrow{\text{CPK}}$ Creatine + ATP

Figure 1 *In vitro* reconstituted glycolytic pathway. G6P: glucose-6-phosphate; F6P: fructose-6-phosphate; FBP: fructose-1,6-bisphosphate; DHAP: dihydroxyacetone phosphate; GAP: glyceraldehyde-3-phosphate; G3P: glycerol-3-phosphate. Reactions are catalyzed by hexokinase (HK), phosphoglucose isomerase (PGI), phosphofructokinase (PFK), fructose-bisphosphate aldolase (ALD), triose-phosphate isomerase (TPI), glycerol-3-phosphate dehydrogenase (G3PDH), and creatine phosphokinase (CPK).

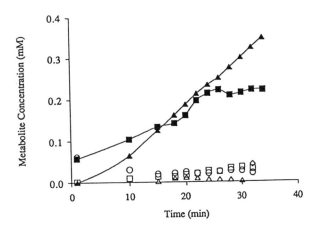

Figure 2 Time courses of metabolite concentrations after the initiation of the transient. Symbols are: glucose-6-phosphate (■); fructose-6-phosphate (□); fructose-1,6-bisphosphate (Δ); dihydroxy-acetone phosphate (O); glyceraldehyde-3-phosphate (◊) and glycerol-3-phosphate (▲).

The concentrations of fructose-6-phosphate (F6P), fructose-1,6-bisphosphate (FBP), dihydroxyacetone phosphate (DHAP), and glyceraldehyde-3-phosphate (GAP) reached a quasi-steady state early in the experiment and their concentrations were much smaller compared with those of glucose-6-phosphate (G6P) and G3P. Therefore, the fluxes between G6P and G3P were roughly the same because of the negligible accumulations of the intermediate metabolites. Hence, one can consider the reactions between G6P and G3P as a single lumped step. The system can then be described by the following lumped reaction scheme:

$$\text{Glucose} \xrightarrow{v_1} \text{G6P} \rightarrow \rightarrow \rightarrow \rightarrow \rightarrow \underbrace{\hspace{4cm}}_{v_2} 2\text{G3P}$$

Reaction v_1 corresponds to hexokinase, and v_2 corresponds to the lumped reaction of phosphoglucose isomerase, phosphofructokinase, aldolase, triosephosphate isomerase, and glycerol-3-phosphate dehydrogenase.

Now, $\mathbf{y_R}$ and $\mathbf{A_R}$ in Eq. (31) are selected as

$$\mathbf{y_R} = \begin{bmatrix} y_{G6P} \\ y_{G3P} \end{bmatrix} \tag{38}$$

$$\mathbf{A_R} = \begin{bmatrix} 1 & -1 \\ 0 & 2 \end{bmatrix} \tag{39}$$

Eq. (36) becomes

$$\alpha_{G6P}\, y_{G6P}(t) + \alpha_{G3P}\, y_{G3P}(t) = t\, J_2^0 \tag{40}$$

Here, J_2^0 is the steady-state flux through v_2 in the lumped pathway and the coefficients α_{G6P} and α_{G3P} were estimated using linear regression with $t\, J_2^0$ as the dependent variable. With the data of G6P and G3P in Fig. 2, and the measured value of J_2^0 (6.74 μ moles/(min • L)). α_{G6P} and α_{G3P} were estimated to be 35.05 and 74.62, respectively. The flux control coefficients were then determined from

$$[Z_1 \quad Z_2] = [\alpha_{G6P} \quad \alpha_{G3P}]A_R \qquad\qquad (41)$$

which gives $Z_1 = 0.2361$ [0.05021] and $Z_2 = 0.7701$ [0.1293]. The numbers in square brackets are the standard errors of the estimates, which can be determined from the standard errors of the regression coefficients[13]. The sum of the estimated flux control coefficients is 1.0062. The close agreement of the experimental result with the theoretical value of one from the summation theorem, suggests the consistency and validity of the estimated flux control coefficients. These values were also confirmed by enzyme titration[13].

8. Concluding remarks

We have shown the validity of using the transient concentration profiles to determine the flux control coefficients, and illustrated its experimental implementation in a reconstituted multi-enzyme system. In this case, only two transient metabolites are used to determine two flux control coefficients. Although the dynamic approach is derived based on a linear approximation of enzyme kinetics, this approximation is accurate enough for this experimental system. Because the control coefficients should be used to detect large differences in the flux control capacity of different enzymes, the error caused by linear approximation should be acceptable for practical purposes. The major source of error arises in the estimation of α by linear regression, which is greatly corrupted by collinearity. The problem is partially solved by the use of the reduced system, y_R, whose elements are linearly independent. However, if some of the metabolites reach quasi-equilibrium or the characteristic reaction path[14] (generalized quasi-equilibrium) within a short time, these metabolites will show collinearity. Therefore, metabolites in y_R must be carefully chosen so that they are not in a quasi-equilibrium state within the time scales of interest. However, if quasi-equilibrium still exits among some species in the best-chosen set of y_R, these metabolites must be lumped together to eliminate the net reactions inter-converting these species. We have shown that reactions that reach fast quasi-equilibrium have a small flux control coefficients. Similar results have been reached by assuming Michaelis-Menten rate laws[3,4].

Erroneous results can be detected either by significant deviations from the summation theorem or by large standard errors associated with the estimated control coefficients. In the glycolysis example, lumping was performed based on the quasi-steady state of metabolites, and was verified by the enzyme titration experiments. Several methods can be used to generate transient responses. Here, we initiated the dynamics by the addition of the first enzyme, hexokinase. This choice is arbitrary and is equivalent to the addition of the first substrate (glucose) or the perturbation of the system from an existing steady state. If some of the intermediate concentrations are too low to be measured accurately, one can start from high initial concentrations of these intermediates and observe their evolution to the steady state. All these methods of initiating the dynamics of the system are mathematically equivalent for systems with little non-linearity, and can be chosen based on practical considerations.

The dynamic approach presented above for determining the control coefficients can be applied to any system which is described by a set of ordinary equations (Eq. 1) with a defined stoichiometry, and which has a (quasi-)steady state. For example, chemostat bioreactors, immobilized multi-enzyme reactors, and many pharmacokinetic systems are amenable to this treatment. If the system involves physical rate processes such as flow into a compartment or diffusion across the compartment membrane, these processes can be treated as pseudo-reactions and their stoichiometry can be defined accordingly.

References

(1) Fraenkel, D.G. Genetics and Intermediary Metabolism. *Annu. Rev. Genet.* 1992, **26**, 159-177.
(2) van Schaftingen, E.; Hue L.; Hers, H. Fructose 2,6-bisphosphate, the probable structure of the glucose- and glucagon-sensitive stimulator of phosphofructokinase, 1980, *Biochem J.* **192**, 897-901.
(3) Kacser, H.; Burns, J. A. The control of flux. In: *Rate Control of Biological Processes*; Davies, D. D., Ed.; Cambridge University Press: Cambridge, MA, 1973; pp 65-104.
(4) Heinrich, R.; Rapoport, T. A. A linear steady-state treatment of enzymatic chains. General properties, control and effector strength. *Eur. J. Biochem.* 1974, **42**, 89-95.
(5) Fell, D. A. Metabolic control analysis: a survey of its theoretical and experimental development. *Biochem. J.* 1992, **286**, 313-330.

(6) Liao, J. C., Delgado, J. Advances in metabolic control analysis. *Biotechnol. Prog.* 1993, **9**, 221-233.

(7) Cornish-Bowden, A.; Cçrdenas, M. L., Eds. *Control of Metabolic Processes*, Plenum Press: New York, 1990.

(8) Schuster, S.; Heinrich, R. The definitions of metabolic control analysis revisited. *BioSystems*, 1992, **27**, 1-15.

(9) Reder, C. Metabolic control theory: A structural approach. *J. Theor. Biol.* 1988, **135**, 175-201.

(10) Delgado, J.; Liao, J. C. Determination of flux control coefficients using transient metabolite concentrations. *Biochem. J.* 1992, **282**, 919-927.

(11) Stewart, G. W. *Introduction to Matrix Computations*; Academic Press: New York, 1973.

(12) Delgado, J.; Liao, J. C. Metabolic control analysis using transient metabolite concentrations. Determination of metabolite concentration control coefficients. *Biochem. J.* 1992, **285**, 965-972.

(13) Delgado, J.; Meruane, J.; Liao, J. C. Experimental determination of flux control coefficients. Application to an in vitro reconstituted glycolysis. *Biotechnol. Bioeng.*, 1993, **41**, 1121-1128.

(14) Liao, J. C., Lightfoot, E. N. Characteristic reaction paths of biochemical reaction systems with time scale separation. *Biotechnol. Bioeng.* 1988, **31**, 847-854.

Control of Dynamics and Steady State: Applications to Multidrug Resistance

H.V. Westerhoff, M. Bier, D. Molenaar, E.C. Spoelstra, J. Lankelma, A.P.M. Jongsma, P.R. Jensen, P. Richard, and and B.N. Kholodenko

1. Introduction

The Control Theory of steady states in "ideal" cells is largely complete. We shall discuss some recent advances in the development of control theory for a more realistic cellular physiology, which includes channelling of metabolites, group relay and signal transduction pathways, as well as hierarchical control systems involving regulated gene expression. With respect to dynamic systems we shall introduce some definitions relevant for the characterization of the control of dynamic systems such as glycolytic oscillations in yeast. These include a control coefficient quantifying the control of the distance from the Hopf bifurcation point, as well as the control exerted by parameters on the amplitudes of the various Fourier components of the sustained oscillations we observe experimentally.

We also apply control analysis to the control of the drug resistance in cells with amplified P-glycoprotein.

2. Group transfer

In ideal metabolic pathways, there is no direct interaction between enzymes and enzymes do not significantly reduce the total concentration of coenzymes. In realistic pathways of metabolism, gene expression and signal transduction direct, enzyme-enzyme interactions may occur and in fact be important. An example is any relay (or group-transfer) pathway. Here a phosphoryl group is transferred sequentially between a set of enzymes. Enzyme-enzyme interactions are essential for the transfer of the phosphoryl group.

In ideal metabolic pathways, the sum of the control exerted by all the enzymes on any flux is 100%. In nonideal pathways, the enzymes have extra control because they participate in more than one process (Westerhoff and Kell, 1988). Recently, exact expressions for the total control have been derived both for group-transfer pathways (Van Dam *et al.*, 1993) and for metabolite channelling (Kholodenko and Westerhoff, 1993, 1994).

3. Sustained metabolic oscillations

Control analysis of time dependent systems is a more recent development. We focus on the control of metabolite systems that exhibit transient or sustained oscillations. Intact yeast cells is a long-known experimental system exhibiting transient oscillations of glycolysis under some conditions. Only recently, we have been able to define the conditions that lead to sustained oscillations (Richard *et al.*, 1993, 1994).

Standard metabolic control analysis focuses on the control of steady-state fluxes and concentrations by system parameters such as enzyme activities (Kacser and Burns, 1973; Heinrich and Rapoport, 1974; Fell, 1992). Of which properties should one study the control, in the case of oscillations? The properties that have come to mind include the frequency and amplitude of the oscillations. One may, however, also ask what determines whether the system exhibits steady state, transient oscillations or sustained oscillations.

For systems close to the Hopf bifurcation, and probably for other systems as well, it may be useful to inspect how the eigenvalues of the Jacobian at the fixed point, are controlled by the activities of the enzymes in the system. For a two variable system the control of the frequency of the oscillations becomes equal to the control exerted on the imaginary part of the eigenvalues:

$$C_{e_i}^{\omega} = C_{e_i}^{Im\lambda} = \frac{d\ln(Im\lambda)}{d\ln e_i} \tag{1}$$

The control of the real part of these eigenvalues describes the extent to which an enzyme determines the distance of the fixed point from the Hopf bifurcation, and has been defined by:

$$C_{e_i}^{Re\lambda} = \frac{d\ln(Re\lambda)}{d\ln e_i} \tag{2}$$

At the Hopf bifurcation point itself, this control coefficient becomes infinite.

4. Multidrug resistance

These more elaborate aspects of metabolic control theory, but also its standard aspects, may be applied to various experimental systems of interest. One of the latter is the multidrug resistance that often develops in human tumors treated with anticancer drugs (Gottesman, 1993). Here the concentration of a plasma membrane protein, such as P-glycoprotein, may increase. The P-glycoprotein extrudes many hydrophobic molecules rather indiscriminately from the cells. Some inhibitors of the P-glycoprotein exist.

The intracellular drug concentration (X in Fig. 1) is a function of the plasma membrane permeability for that drug, as well as of the activity of the pump. One may ask what is the limiting step for drug toxicity.

Metabolic Control Analysis of various systems has led to the recognition that asking for "the limiting step" may not always be the right thing to do; control may be distributed over various steps (Groen *et al.*, 1982). For P-glycoprotein mediated multidrug resistance the following quantitative model has been shown to be in accordance with experimental data (Spoelstra *et al.*, 1992, 1994) (we here simplify the model somewhat, see Fig. 1):

$$v_p = \frac{V \cdot \left(\dfrac{X}{K_X}\right)^h}{1 + \left(\dfrac{X}{K_X}\right)^h}$$ (3)

$$v_l = k \cdot (X - D)$$ (4)

Here v_p and v_l are the pump and leak rates respectively. D and X are the extracellular and intracellular drug concentrations, respectively. h is the Hill coefficient measuring the positive cooperativity between substrate molecules (Spoelstra *et al.*, 1992, 1994; Guiral *et al.*, 1994). k is proportional to the passive membrane permeability for the drug. V is the maximum rate of the pump. The complete model (Spoelstra *et al.*, 1994) describes the pump as reversible, but in practice the reverse rates seem negligible.

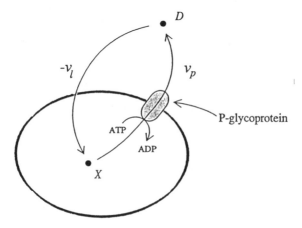

Figure 1 Scheme of passive and active drug fluxes in a multidrug resistant cell.

Accordingly, Eq. 3 assumes Michaelis-Menten kinetics for an irreversible, and even product uninhibited pump reaction. Eq. 4 assumes a parallel passive membrane permeation for the drug. We shall make a simplification that is realistic for most drugs (Spoelstra *et al.*, 1992): the intracellular drug concentration is far below its Michaelis constant (i.e., $X << K_X$). Pending the steady state requirement that leak rate plus pump rate must equal zero, D then depends on X by:

$$D = X + \frac{V}{k} \cdot \left(\frac{X}{K_X}\right)^h \tag{5}$$

In our models the toxicity of the drug is determined by its intracellular concentration. One may now compare the extracellular drug doses that are equal ("iso-") toxic for two cell lines by requiring that their intracellular drug concentrations be equal (Westerhoff *et al.*, in preparation). The question what limits drug toxicity then translates into: What controls the extracellular isotoxic drug concentration? The control coefficient for P-glycoprotein activity with respect to the isotoxic drug dose is defined by:

$$C^d_{P-gp} = \frac{d\ln D}{d\ln[P-gp]} \tag{6}$$

where [P-gp] represents the concentration of P-glycoprotein. Since V is proportional to [P-gp] one finds by differentiation of Eq. 5:

$$\frac{1}{C^D_{P-gp}} = \frac{1}{\left(\frac{d\ln D}{d\ln V}\right)_X} = 1 + \frac{K_X \cdot k}{V} \cdot \left(\frac{K_X}{X}\right)^{h-1} \tag{7}$$

In line with the summation theorem for concentration control coefficients (Heinrich and Rapoport, 1974), the control by the passive drug permeability is negative and precisely the opposite of the control by the drug efflux pump activity.

For the control by the passive peremability, k, one finds:

$$\frac{1}{C_k^d} = -\left(1 + \frac{K_X \cdot k}{V} \cdot \left(\frac{K_X}{X}\right)^{h-1}\right) \tag{8}$$

Using data for 2780^{AD} cells from Spoelstra et al., (1994) and Jongsma et al., in preparation ($V = 10\text{-}15$ mol min^{-1}cell^{-1}; $K_X = 1$ μM; h = 1.5; $X = 3$ nM; $k = 10^{-11}$ l min^{-1}cell^{-1}), the above equation predicts that the coefficient for the control by the pump activity of the isotoxic drug dose should amount to 0.82. The passive permeability should control the isotoxic drug concentration for minus 85%. If the Hill coefficient amounts to 2, then the control by pump and leak are estimated at +23 and -23% respectively. These estimations suggest that there is no single limiting step for the toxic drug dose. Even in this simple model, control is necessarily distributed among pump and leak. Most likely, the control of either is significant.

In this paper we have discussed extensions to the theory of metabolic control. Importantly, these extensions allow metabolic control analysis to proceed beyond the limits of ideal, text book metabolic chemistry, into the realms of signal transduction and metabolic dynamics. These limits removed, much of the interest may shift to experimental applications. The application to multidrug resistance of tumor cells, initiated here, may ultimately help in assessing optimum strategies of its management.

Acknowledgements

This study was supported by the Netherlands Organization of Scientific Research (NWO).

References

Fell DA. Biochem. J. **286** (1992) 313-330
Gottesman MM. Cancer Res **53** (1993) 747-754.
Groen AK, Van der Meer R, Westerhoff HV, Wanders RJA Akerboom TPM and Tager JM . in:Metabolic comparmentation, ed. H Sies (Academic Press, New York 1982) pp 9-37.
Guiral M, Viratelle O, Westerhoff HV and Lankelma J. FEBS Lett **346** (1994) 141-145.
Heinrich R and Rapoport TA. Eur J Biochem **42** (1974) 31-35.
Kacser H and Burns JA. Symp.Soc.Exp.Biol. **27** (1973) 65-104.
Kholodenko BN and Westerhoff HV. FEBS Lett. **320** (1993) 71-74.
Kholodenko BN and Westerhoff HV. Trends Biochem. Sci. (1994) in press.

Richard P, Teusink B, Westerhoff HV and Van Dam K. *FEBS Lett* **318** (1993) 80-82.

Richard P, Diderich JA, Bakker BM, Teusink B, Van Dam K and Westerhoff HV. *FEBS Lett* **341** (1994) 223-226.

Spoelstra EC, Westerhoff HV, Dekker H and Lankelma J. *Eur J Biochem* **207** (1992)567-579.

Spoelstra EC, Westerhoff HV, Pinedo HM, Dekker H and Lankelma J. *Eur J Biochem* **221** (1994) 363-373.

Van Dam K, Van der Vlag J, Kholodenko BN and Westerhoff HV. *Eur J Biochem* **212** (1993) 791-799.

Westerhoff HV and Kell DB. *Comm Molec Cellul Biophys* **5** (1988) 57-107.

Parameter Space Classification of Solutions to a Mathematical Model for the Cortisol Secretion System in Normal Man

Y. Lenbury

1. Introduction

There have been numerous studies describing the fluctuation of plasma corticosteroid levels in the human adult [1-3]. According to Krieger *et al.* [2], it has not been possible to derive from these studies statistically valid definition of normal circadian periodicity to serve as a basis for characterizing deviations observed in various patient populations. The main reason was that methodology utilized, especially with regard to frequency of sampling, has not been comparable in these studies. It has however become evident that the circadian cycles of the hormone levels in man is not a smooth curve when analyzed at short time intervals [2, 3]. Efforts to remove these peaks and valleys by averaging and weighting manipulations of the data still shows the plasma levels plotted against clock time remains a jagged line. The majority of authors considered experimental errors, sampling or dilution errors to be the likely cause but still cannot adequately explain these departure from a smooth line. It has been suggested by E. D. Weitzman, on the other hand, that the secretory bursts observed in the hormone patterns are associated with the central nervous system activity [3].

The study of such transient and oscillatory phenomenon is seen to be not only a necessity for understanding and efficient monitoring of various highly complexed systems, especially those self-regulative processes in the physiological activities of the human life, but also a powerful research tool for the study of regulatory mechanisms in a normal human body.

2. Formulation of the Model

The corticotropin-releasing hormone (CRH) is synthesized in the hypothalamus and reaches ACTH-producing cells of the anterior pituitary via the hyposeal portal system [4]. In response to CRH, corticotrophic cells of the pituitary synthesize and secrete the adrenocorticotropic hormone (ACTH), which circulates on the surface of adrenocortical cells to stimulate synthesis and secretion of cortisol (F). Cortisol exerts negative feedback control on ACTH gene in the pituitary and by suppressing formulation of CRH in the hypothalamus. In addition, there is a similar feedback regulation between the plasma ACTH level and the process of secreting CRH. Therefore, the entire process can be considered as a self-regulating process, which may be summerized by the diagram in Fig. 1.

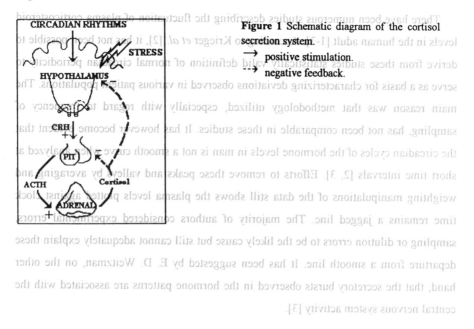

Figure 1 Schematic diagram of the cortisol secretion system.
→   positive stimulation.
- -→ negative feedback.

In considering this diagram a mathematical model can be formulated in order to study the dynamics of the system and to explain the clinical observations. To keep our model mathematically tractable and not too complicated, we do not consider other hormones secreted by this system since there are no correlations between them and the hormones CRH and ACTH. Also, we shall consider the control capacity of the nerve center as an external factor to the subsystem under discussion.

Letting $x(t)$ be the concentration of the plasma CRH at time t; $y(t)$ the concentration of the plasma ACTH at time t; $z(t)$ the concentration of the plasma F at time t, all divided by their respective daily averaged values. the self-regulative system can be described by the following system of ordinary differential equations [5]

$$\frac{dx}{dt} = \alpha \, \exp\left[a\left(1-z^2\right)+b\left(1-y^2\right)\right] - \alpha x + D \cos(\omega t) \tag{1.1}$$

$$\frac{dy}{dt} = \beta x \, \exp\left[b\left(1-z^2\right)\right] - \beta y \tag{1.2}$$

$$\frac{dz}{dt} = \gamma y - \gamma z. \tag{1.3}$$

Here, the parameters α, β, and γ indicate the capacity of secreting corresponding hormones. The term $D \cos(\omega t)$ incorporates periodic response signals of the nervous system, following E. D. Weitzman's suggestion [3]. Such sinusoidal response is frequently encountered in studies of biological feedback control systems as in the work of R. W. Jones [6] on the principles of biological regulations, where it is suggested that the postural control system behavior is dominated by a pair of complex poles, i. e. the nervous response may be given by an expression of the form

$$\exp(-\theta t) \cos(2\pi f t + \delta)$$

In our model, we assume $\omega = 2\pi f$ and that the system receives undisssipating stimulation from the nerve center (undamped), i.e. $\theta = 0$. We also put $\delta = 0$ without loss of generality. The value of parameter D thus indicates the control capacity of the nerve center.

Studying the diagram in Fig. 1, we can conceive that, with the change in hormone z level, the secretion rate of hormone x should be suppressed by z at a specific rate which is directly proportional to the amount of hormone z at any time t, that is

$$\frac{1}{x_t} \frac{\partial x_t}{\partial z} = -kz \tag{2}$$

for some positive constant k. This leads, upon integrating Eq. (2), to the exponential term $\exp\left[a\left(1 - z^2\right)\right]$, where $a = k/2$, in Eq. (1.1), and similarly for the other feedback terms in the model equations. Such decaying exponential terms are frequently encountered in biological regulation systems [6].

Moreover, the rate of secretion of each of the three hormones is indirectly regulated by the amount of that hormone itself at any time t, and hence the terms $-\alpha x, -\beta y$, and $-\gamma z$ in Eqs. (1.1), (1.2), and (1.3), respectively. In particular, if $a = 0$, $b = 0$ and $D = 0$, Eq. (1.1) reduces to

$$\frac{dx}{dt} = \alpha(1 - x)$$

which means that, disregarding the direct inhibitory effect of hormones y, and z, and without influence from the nerve center, the rate of secretion of hormone x should decrease with the increase in hormone x level.

Theoretical study of the above system of Eqs. (1.1) - (1.3) has shown that the model exhibits bifurcation and chaotic patterns for a certain range of parametric values [5]. The results helped the authors in the attempt to classify different dynamic patterns exhibited by the model and relate them to the circadian activities of these hormones in the normal being or the diseased population.

According to Krieger et al. [2], hitherto lacking clinically were suitable conditions for determination and definition of the normal circadian patterns of plasma corticosteroid levels.

With this in mind, we attempt in this paper to theoretically delineate more precisely the temporal patterns of adrenal secretory activity in a normal human and a patient by deriving boundary conditions to divide the parameter space of different dynamic behavior. Simulated time course of hormone levels is then compared to available clinical data in order to check the reliability of our theoretical conclusions.

3. Bifurcation Analysis

To conduct a theoretical study [7] of the proposed model, we remove the explicit dependence of the oscillatory term in Eq. (1.1) by letting $u = \cos(\omega t)$ and $v = \sin(\omega t)$ which transforms Eqs. (1.1) - (1.3) into

$$\frac{dx}{dt} = \alpha \exp\left[a\left(1 - z^2\right) + b\left(1 - y^2\right)\right] - \alpha x + Du \tag{3.1}$$

$$\frac{dy}{dt} = \beta x \exp\left[b\left(1 - z^2\right)\right] - \beta y \tag{3.2}$$

$$\frac{dz}{dt} = \gamma y - \gamma z \tag{3.3}$$

$$\frac{du}{dt} = -\omega v \tag{3.4}$$

$$\frac{dv}{dt} = \omega u \tag{3.5}$$

A steady state solution (or critical point) $(\bar{x}, \bar{y}, \bar{z}, \bar{u}, \bar{v})$ of the above system can be found by equating the right sides of Eqs. (3.1) - (3.5) to zero, from which we find that the model system (3) has a critical point $(\bar{x}, \bar{y}, \bar{z}, \bar{u}, \bar{v}) = (1,1,1,0,0)$. The Jacobian matrix J of the model system evaluated at this critical point is

$$J = \begin{bmatrix} -\alpha & -2\alpha b & -2\alpha a & D & 0 \\ \beta & -\beta & -2\beta b & 0 & 0 \\ 0 & \gamma & -\gamma & 0 & 0 \\ 0 & 0 & 0 & 0 & -\omega \\ 0 & 0 & 0 & \omega & 0 \end{bmatrix} \tag{4}$$

If we let

$$\mu = \alpha + \beta + \gamma \tag{5}$$

$$\delta = \alpha\beta + \beta\gamma + \alpha\gamma + 2b\beta(\alpha + \gamma) \tag{6}$$

$$\rho = (1 + 4b + 2a)\alpha\beta\gamma \tag{7}$$

$$\phi = \frac{\delta}{3} - \frac{\mu^2}{9} \tag{8}$$

$$\psi = \frac{1}{6}(\delta\mu - 3\rho) - \frac{\mu^2}{27} \tag{9}$$

then the 5 eigenvalues of the Jacobian matrix J are

$$\lambda_{1,2} = -\frac{1}{2}(s_1 + s_2) - \frac{\mu}{3} \pm \frac{\sqrt{3}}{2}(s_1 - s_2)i \tag{10}$$

$$\lambda_3 = (s_1 + s_2) - \frac{\mu}{3} \tag{11}$$

$$\lambda_{4,5} = \pm i\omega \tag{12}$$

where

$$s_{1,2} = \left[\psi \pm (\phi^3 + \psi^2)^{1/2}\right]^{1/3} \tag{13}$$

We note that if

$$\phi^3 + \psi^2 > 0 \tag{14}$$

then $\lambda_1(a)$ and $\lambda_2(a)$ are complex conjugates, in which case the real part of $\lambda_{1,2}$ vanishes when

$$2\mu = -3(s_1 + s_2) \tag{15}$$

Raising (15) to the third power and using the fact that

$$s_1^3 + s_2^3 = 2\psi \tag{16}$$

$$s_1 s_2 = -\phi \tag{17}$$

Eq. (15) simplifies to

$$8\mu^3 + 54\phi\mu + 54\psi = 0 \tag{18}$$

On using (8) and (9) in (18), we find that the real part of $\lambda_{1,2}$ vanishes when

$$\rho = \delta\mu \tag{19}$$

Using condition (19) together with Eqs. (5) - (7), we can conclude, therefore, that if (14) holds and

$$a < \frac{(\alpha+\beta+\gamma)\left[\alpha\beta+\beta\gamma+\alpha\gamma+2b\beta(\alpha+\gamma)\right]}{2\alpha\beta\gamma} - 4b - 1$$

then all 3 eigenvalues $\lambda_{1,2}$ and λ_3 have negative real parts. Since $\lambda_{4,5}= \pm i\omega$, the solution trajectories in the phase space will tend to a closed curve around the critical point $(1,1,1,0,0)$. Letting

$$a_c \equiv \frac{(\alpha+\beta+\gamma)\left[\alpha\beta+\beta\gamma+\alpha\gamma+2b\beta(\alpha+\gamma)\right]}{2\alpha\beta\gamma} - 4b - 1 \tag{20}$$

it can be shown that for values of a in an open interval (a_c , $a_c+\varepsilon$) the system of Eqs. (3.1) - (3.3) with $D = 0$ has a family of periodic solutions bifurcating from the critical point $(\bar{x},\bar{y},\bar{z})$, $= (1,1,1)$. This result is stated in the following theorem.

Theorem There exists an $\varepsilon > 0$ such that the system of equations (3.1) - (3.3) with $D = 0$ will have a family of periodic solutions for values of a in the open interval $(a_c, a_c +\varepsilon)$.

Proof We first note that Re $\lambda_1(a_c) = 0$ since a_c is defined so that

$$a = a_c \tag{21}$$

is equivalent to Eq. (19) which is the necessary condition for Re $\lambda_1(a_c)$ vanishing at a_c .
We then show that Re $\lambda_1'(a_c)\neq 0$. By differentiating

$$\text{Re } \lambda_1 = -\frac{1}{2}(s_1 + s_2) - \frac{\mu}{3} \tag{22}$$

with respect to a, we find that

$$\text{Re } \lambda_1'(a_c) = -\alpha\beta\gamma\left[\frac{1 + \psi\left(\psi^2 + \phi^3\right)^{-1/2}}{s_1^2} + \frac{1 - \psi\left(\psi^2 + \phi^3\right)^{-1/2}}{s_2^2}\right]$$

Suppose, by contradiction, that $\text{Re } \lambda_1'(a_c) = 0$, that is

$$\frac{\left[\psi + \left(\psi^2 + \phi^3\right)^{1/2}\right]s_2^3}{s_2\left(\psi^2 + \phi^3\right)^{1/2}} = -\frac{\left[\psi - \left(\psi^2 + \phi^3\right)^{1/2}\right]s_1^3}{s_1\left(\psi^2 + \phi^3\right)^{1/2}}$$

Using Eq. (14), we arrive at $s_1 = -s_2$ which means that we must have $\psi = 0$.

Using Eq. (9), the above equation states that

$$\frac{\delta\mu - 3\rho}{6} - \frac{\mu^3}{27} = 0 \tag{23}$$

When $a = a_c$, we have $\rho = \delta\mu$, in which case (23) reduces to $9\delta\mu + \mu^3 = 0$.

Since both μ and δ are positive, the above equation cannot hold and thus $\text{Re }\lambda_1'(a_c) \neq 0$.

Now, in order to show that $\text{Im } \lambda_1(a_c) \neq 0$, we need $s_1 - s_2$ to be real and $s_1 \neq s_2$. This is equivalent to showing that, at $a = a_c$, the inequality (14) holds. Using Eqs. (8) and (9), we find that at $a = a_c$ ($\rho = \delta\mu$)

$$\phi^3 + \psi^2 = \left(\delta + \mu^2\right)^2 > 0 \text{ and so } \text{Im } \lambda_1(a_c) \neq 0.$$

Finally, the remaining eigenvalue is $\lambda_3 = (s_1 + s_2) - \frac{\mu}{3}$. At the point when $\text{Re } \lambda_1 = 0$,

$s_1 + s_2 = -\frac{2}{3}\mu$ which implies that $\lambda_3 = -\mu$. Hence, λ_3 is negative when $a = a_c$.

Thus, all requirements for Hopf bifurcation are met and the system of equations (3.1) - (3.3) with $D = 0$ will have periodic solutions bifurcating from the critical point $(\bar{x}, \bar{y}, \bar{z})$, $= (1,1,1)$ for values of a in some open interval $(a_c, a_c + \varepsilon)$.

For the system of equations (3.1) - (3.5) with $D \neq 0$, this means that as a increases above the critical value a_c, Hopf bifurcation occurs, on top of the existing closed orbit about the point $(1,1,1,0,0)$ in the (x, y, z, u, v) space due to the eigenvalues $\lambda_{4,5} = \pm i\omega$. The resulting solution trajectory then lies on a 2-torus in the five dimensional phase space.

4. Stability Condition

After having arrived at one boundary condition, namely Eq. (21), we now attempt to derive another, which determines the stability of the system solution on the torus, in the hope that this will also delineate the parameter space into a region of different dynamic behavior indicative of certain other states of health, such as that of a person suffering from the Cushing's syndrome.

With this goal in mind, we find the stability condition for the limit cycle as it bifurcates from the steady state $(\bar{x},\bar{y},\bar{z})$ of the system of Eqs. (3.1) - (3.3) with $D = 0$. Since the occurrence of this limit cycle gives rise to the solution on the 2-torus in the 5-dimensional phase space, the stability of the torus in the (x,y,z,u,v)-space is dependent upon the stability of the limit cycle in the (x, y, z)-space bifurcating from the steady state $(1,1,1)$.

Applying the stability condition derived in [8] to the system of Eqs. (3.1) - (3.3) with $D = 0$, we find that the stability of the bifurcated periodic solution of our model system can be determined by the sign of the following function:

$$f(\alpha,\beta,\gamma,b) \equiv 2\delta(\mu^2 + \delta)\left[\frac{L_1}{\beta\gamma^2} - \frac{2L_3}{\mu^3 + \rho}\right] + L_2 + \frac{2\delta}{\nabla^2 + 4\delta}(L_4 + 2L_5) \tag{24}$$

where $\nabla = -\dfrac{\mu^3 + \rho}{\mu^2 + \delta}$,

$$L_1 = \frac{\alpha}{\gamma}\left(\theta_3 + \theta_4 + 6a_c\theta_2 + 6b\theta_1 + \frac{\delta(\theta_4 + 2a_c\theta_2)}{\gamma^2}\right)$$

$$-\frac{(\beta+\gamma)}{\beta\gamma^2}(2\alpha\theta_2\theta_6 + \beta\gamma\theta_4 - (\delta + \alpha\mu)\theta_2) - \frac{\alpha\mu}{\gamma^2}\left(\theta_4 + 4a_c\theta_2 + 2b\theta_1 + \frac{\delta\theta_4}{\gamma^2}\right)$$

$$L_2 = \left(\frac{\alpha\delta}{\gamma^2}(\theta_2 + \theta_5) + \frac{2\delta b(\beta+\gamma)^2}{\beta\gamma^2}\right)\left(-\frac{\alpha}{\gamma}\left(\theta_6 + \frac{\delta\theta_2}{\gamma^2}\right) + \frac{(\beta+\gamma)}{\beta\gamma}\left(\frac{4\alpha b\theta_6}{\gamma} + \beta\theta_2\right)\right)$$

$$\left(\frac{\alpha\mu}{\gamma}\left(\theta_8 + \frac{\delta\theta_2}{\gamma^2}\right) + \frac{(\delta+\alpha\mu)}{\beta\gamma}\left(\frac{4\alpha b\theta_6}{\gamma} + \beta\theta_2\right)\right)\left(\frac{\alpha\mu(\theta_2 + \theta_5)}{\gamma^2} - \frac{2b(\beta+\gamma)(\delta + a_c\mu)}{\beta\gamma^2}\right)$$

$$+\left(-\frac{\alpha\theta_8}{\gamma} + \frac{(\beta+\gamma)}{\beta\gamma}\left(\frac{4\alpha b\theta_6}{\gamma} + \beta\theta_2\right)\right)\left(\frac{\alpha\mu\theta_8}{\gamma} + \frac{(\delta+\alpha\mu)}{\beta\gamma}\left(\frac{4\alpha b\theta_6}{\gamma} + \beta\theta_2\right)\right)$$

$$+\frac{\alpha^2\delta^2\mu\theta_2^2}{\gamma^4}$$

$$L_3 = \left[-\frac{\alpha}{\gamma}\left(\theta_1 + \theta_5 - \frac{(\alpha+\beta)}{\gamma}(\theta_2 + \theta_5) \right) - \frac{\beta+\gamma}{\beta\gamma}\left(\frac{2b}{\gamma}(\theta_7 + \alpha\mu - \alpha\theta_6) - \beta\theta_2 \right) \right.$$

$$-\frac{\alpha\mu}{\gamma^2}\left(\frac{(\alpha+\beta)\theta_2}{\gamma} - \theta_5 \right) - 2b(\delta+\alpha\mu)\frac{(\alpha+\beta)}{\beta\gamma^2} \left] \left[-\frac{\alpha}{\gamma}\left(\theta_8 + \frac{\delta\theta_2}{\gamma^2} \right) \right.$$

$$\left. + \frac{\beta+\gamma}{\beta\gamma}\left(\frac{4\alpha b\theta_6}{\gamma} + \beta\theta_2 \right) \right]$$

$$L_4 = \left[-\frac{\alpha}{\gamma}\left(\theta_1 + \theta_5 - \frac{(\alpha+\beta)}{\gamma}(\theta_2 + \theta_5) \right) - \frac{(\beta+\gamma)}{\beta\gamma}\left(\frac{2b}{\gamma}(\theta_7 + \alpha\mu - \alpha\theta_6) - \beta\theta_2 \right) \right.$$

$$\left. + \frac{\alpha\mu}{\gamma^2}\left(\frac{(\alpha+\beta)\theta_2}{\gamma} - \theta_5 \right) + (\delta+\alpha\mu)\frac{2b}{\beta\gamma^2} \right]\left[\nabla\left(-\frac{\alpha}{\gamma}\left(\theta_8 - \frac{\delta\theta_2}{\gamma^2} \right) \right.\right.$$

$$\left.\left. + \frac{(\beta+\gamma)}{\beta\gamma}\left(\frac{4b\alpha\theta_6}{\gamma} + \beta\theta_2 \right) \right) + 4\left(\frac{\alpha\delta}{\gamma^2}(\theta_2 + \theta_5) + \frac{2b\delta}{\beta\gamma^2}(\beta+\gamma)^2 \right) \right]$$

$$L_5 = \left[-\frac{\alpha\delta}{\gamma^2}\left(\frac{(\alpha+\beta)\theta_2}{\gamma} - \theta_5 \right) + \frac{2b\delta}{\beta\gamma^2}(\beta+\gamma)^2 - \frac{\alpha\mu}{\gamma}\left(\theta_1 + \theta_5 - \frac{(\alpha+\beta)}{\gamma}(\theta_2 + \theta_5) \right) \right.$$

$$\left. - \frac{(\delta+\alpha\mu)}{\beta\gamma}\left(-\frac{2b}{\gamma}(\theta_7 + \alpha\mu - \alpha\theta_6) + \beta\theta_2 \right) \right]\left[\frac{\nabla}{\beta\gamma^2}\left(\frac{\alpha}{\gamma^2}(\theta_2 + \theta_5) + \frac{2b}{\beta\gamma^2}(\beta+\gamma)^2 \right) \right.$$

$$\left. + \frac{\alpha}{\gamma}\left(\theta_8 - \frac{\delta\theta_2}{\gamma^2} \right) - \frac{\alpha\delta\theta_2}{\beta\gamma^3}(\beta+\gamma) \right]$$

while $\theta_1 = 2a_c(2a_c - 1)$, $\theta_2 = 2b(2b - 1)$, $\theta_3 = 4a_c^2(2a_c - 3)$, $\theta_4 = 4b^2(2b - 3)$,

$\theta_5 = 4a_c b$, $\theta_6 = \beta + \gamma + 2b\beta$, $\theta_7 = \beta\gamma(1+2b)$, and $\theta_8 = \theta_1 + \theta_2 + 2\theta_5$.

The stability of the bifurcated periodic solution will be lost at the point where $f(\alpha,\beta,\gamma,b)$

changes its sign, namely at the point where

$$f(\alpha,\beta,\gamma,b) = 0. \tag{25}$$

5. Numerical Solution and Discussion

In order to check our theoretical results, we plot the graph of Eq. (25) in the (β,α)-plane for fixed values of γ and b. This together with the graph of Eq. (21) for a fixed value of $a = a_c$ are shown in Fig. 2. Here, $\gamma = 0.5$, $b = 0.7$ and the graph of Eq. (21) is plotted for $a = 12.4$. Thus, the graph of Eq. (21) divides the plane into 2 regions of different dynamic behavior.

One is the region where $a_c < 12.4$, while the other is where $a_c > 12.4$. The two boundary conditions divide the (β, α)-plane, then, into 3 regions as seen in Fig. 2.

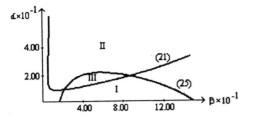

Figure 2 The graphs of equations (21) and (25) divide the parameter space into 3 regions of different dynamic behavior. Here, $\gamma = 0.5$, b = 0.7.

Fig. 3 shows numerical simulation of (3.1) - (3.5) with (β, α) in Region I of the parameter space and using $a = 12.4$. Since (β, α) is in Region I, $a = 12.4 < a_c$, so that the critical point $(1,1,1)$ of the system (3.1) - (3.3) with $D = 0$ is stable and no bifurcation occurs. Due to the eigenvalues $\pm \omega i$, therefore, we see in Fig. 3 the projection onto the (x,y)-plane of the space trajectory of the system (3) approaching a limit cycle surrounding the vortex point $(1,1,1,0,0)$ in the (x, y, z, u, v)-space.

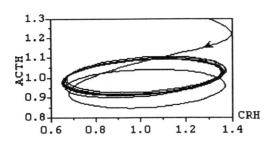

Figure 3 Numerical simulation of equations (3.1)-(3.5) in Region I with $\alpha = 0.01$, $\beta = 0.5$, $\gamma = 0.5$, $a = 12.4$, $b = 0.7$, $D = 0.8228$, and $\omega = 2$.

Fig. 4 shows numerical simulation of (3.1) - (3.5) with $a = 12.4$ and (β, α) in Region II. Here, $a > a_c$ and $f(\alpha, \beta, 0.5, 0.7) < 0$, and Hopf bifurcation occurs on top of the existing closed curve. We see the projection onto the (x,y)-plane of the solution trajectory on the apparently stable torus.

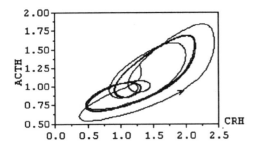

Figure 4 Numerical simulation of Eqs. (3.1)-(3.5) in Region II with $\alpha = 0.2$, $\beta = 1.2$, $\gamma = 0.5$, a = 12.4, b = 0.7, D = 0.8228, and $\omega = 2$.

Fig. 5 shows the solution trajectory for $a = 12.4$ and (β,α) is in Region III, where $a > a_c$ and $f(\alpha,\beta,0.5,0.7) > 0$. Here, it is expected that the solution on the surface of the torus is no longer stable. We obtain a solution trajectory on the surface of the torus only if the initial point is on the torus exactly. If the starting point is only slightly perturbed away from this position the solution trajectory may tend away from the torus toward a limit cycle, becoming periodic, or develope into a space trajectory which has the appearance of a higher dimensional attractor as seen in Fig. 5.

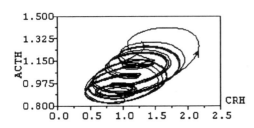

Figure 5 Numerical simulation of equations (3.1)-(3.5) in Region III with $\alpha = 0.15$, $\beta = 0.35$, $\gamma = 0.5$, a = 12.4, b = 0.7, D = 0.8228, and $\omega = 2$.

Figs. 6 and 7 demonstrate the simulated time courses of plasma ACTH (y) with parametric values in Region I of Fig. 3, and Region II of Fig. 4, respectively. Also shown in Fig. 6 is a sample of clinical data of hormone level in a sick person with phychotic depression, taken from the work of Sachar *et al.* [9]. In Fig. 7, we also include a sample of the clinical data of a normal person, taken from the work of Krieger *et al.* [2].

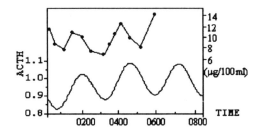

Figure 6 Simulated time course, ———— , of hormone ACTH level of the case seen in Fig. 3, compared with the clinical data, •————•.

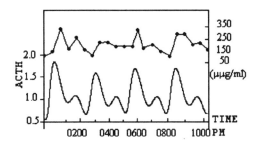

Figure 7 Simulated time course, ———— , of hormone ACTH level of the case seen in Fig. 4, compared with the clinical data, •————•.

We observe that the time course of the periodic closed curve of Fig. 6 seems to compare well with the circadian pattern of a sick person while that of the temporal pattern of the curve on the torus of Fig. 7 appears to compare well with the more irregular pattern of a normal person. Our theoretical analysis indicates that this transition from health to sickness can take place at the point when the feed back capacity constant a decreases below the critical value a_c given by Eq. (21).

It needs to be noted that all clinical data depicted here shows only measurements taken every half hour, following Krieger et $al.$'s example. Other measurements taken at a more frequent intervals are not included in order that data from independent sources can be compared on a common time scale.

Fig. 8 demonstrates the simulated time course of plasma cortisol level (z) with parametric values in Region III of Fig. 5. Difficulty arises, however, when we attempt to compare this with clinical data since available to us on patients suffering from the Cushing's syndrome is the time course of plasma 11-OHCS levels presented by Krieger et $al.$ [2], in their search of criteria for the definition of circadian periodicity of plasma corticosteroid levels.

Nontheless, on comparing the time course (in Fig. 8) of Cushing's syndrome plasma 11-
OHCS level from the work of Krieger *et al.* with the simulated curve of plasma 17-
OHCS, we find close resemblance between them.

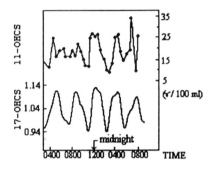

Figure 8 Simulated time
course, ———— , of hormone ACTH
level of the case seen in Fig. 5,
compared with the clinical
data, •————•.

In particular, one common characteristics is the broard columns appearing in both
patterns. This is dynamically different from the sharp rise and fall- a high peak followed
by a low one- we observe in the temporal pattern of a normal person. According to
Krieger *et al.* [2], in their studies of patients with active Cushing's syndrome, "the
patterns observed in these patients are best described as a seemingly random pattern of
oscillations". This seems to indicate, then, that if the ecretion capacity constants α and
β have values in this third region, we have a patient in an unstable condition who may
fall prey to a developing sickness under a slight disturbance.

6. Conclusion

We have derived the two boundary conditions, namely Eqs. (21) and (25) which
divide the (β, α)-plane into essentially 3 regions of dynamically different behavior.
Choosing the feedback capacity constant a as bifurcation parameter, we find that if the
value of a falls below the critical value a_c given by Eq. (20), a change in the circadian
pattern of the hormone levels may be expected. For fixed values of γ, a, and b, this event
occurs if (β, α) falls in the Region I in the parameter space (Fig. 3). Here $a_c > a$ and
accentuated periodicity in the solution trajectory of the system model is observed
indicitive of people suffering from diseases perhaps due to endocrinopathy [9]. In Region

II, on the other hand, the system model exhibits more complicated solution trajectories comparable to oscillatory patterns of a normal human.

However, if the secretion capacity constants (β, α) have values that fall in Region III where the stability expression $f(\alpha, \beta, 0.5, 0.7)$ becomes positive, a dynamically different pattern is observed. The solution on the 2-torus loses its stability resulting in more regular or more chaotic pattern, depending on the initial point, indicative again of a different state of health.

We note that in order to keep our model mathematically tractable and not too complicated, we consider the control capacity of the nerve center as an external factor to the subsystem under discussion. The response signals of the nervous system is incorperated into the model by the use of a simple single frequency sinusoidal expression. It is possible that a response expression with varying frequencies be incorperated which would, however, render the model much more untractable mathematically. With our simple model consisting of 3 Eqs. (1.1) - (1.3) it is possible to derive the boundary condition together with the complicated stability condition yielding helpful information about the system under study which may have been lost otherwise. Theoretical investigation carried out here coupled with simultaneous clinical study of various plasma levels and other influencing criteria or variables should give insight into the dynamics of the altered periodicity observed and also in parts serve as a basis for the determination and definition of the normal or abnormal fluctuation patterns observed in various patient populations.

References

[1] J. Simpkins and J. I. Williams. *Advanced Biology*. Unwin Hyman Ltd., London, 1989.
[2] D. T. Krieger, W. Allen, F. Rizzo, and H. P. Krieger. Characterization of the Normal Temporal Pattern of Plasma Corticosteroid Levels. *J. Clin. Endocrinol.* 32 : 266-284, 1971.
[3] E.D. Weitzman, D. Fukushima, C. Nogeire, H. Roffwarg, T. F. Gallagher, and L. Hellman. Twenty-four Hour Pattern of the Episodic Secretion of Cortisol in Normal Subjects. *J. Clin. Endocrinol.* 33 :14-22, 1971.
[4] W. Foster. *Williams Textbook of Endocrinology*. W. B. Saunders, Philadelphia, 1985.

[5] Y. Lenbury and P. Pacheenburawana. Modelling Fluctuation Phenomena in Plasma Cortisol Secretion System in Normal Man. *BioSystems.* 26 : 117-125, 1991.

[6] R. W. Jones. *Principles of Biological Regulation.* Academic Press, New York, 1973.

[7] D. Ruelle. *Elements of Differentiable Dynamics and Bifurcation Theory.* Academic Press, San Diego, 1989.

[8] J. E. Marsden and M. McCracken. *Theory and Applications of Hopf Bifurcation.* Springer-Verlag, New York , 1976.

[9] E. J. Sachar, L. Hellman, H. P. Roffwarg, F. S. Halpern, D. K. Fukushima, and T. F. Gallagher. Disrupted 24-Hour Patterns of Cortisol Secretion in Psychotic Depression. *Arch. Gen. Psychiatry.* 28 : 19-24, 1973.

Optimization of Glycogen Design in the Evolution of Metabolism

E. Meléndez-Hevia, T.G. Waddell, and J. Sicilia

1. Introduction

The animal glycogen molecule has to be designed in accordance with its metabolic function as a very effective fuel store which allows the quick release of large amounts of glucose.

One well known feature of biological evolution is its role as an optimization process. An example of this is the study on the optimization of the pentose phosphate cycle (Meléndez-Hevia, 1990; Meléndez-Hevia & Isidoro, 1985; Meléndez-Hevia, Waddell & Montero, 1994). The aim of the present work was to investigate the evolutionary optimization of molecular design in the structure of animal glycogen, a molecule of fundamental importance for energy metabolism and survival (see also Meléndez-Hevia, Waddell & Shelton, 1994).

The main function of skeletal muscle glycogen is to be the source of glucose for anaerobic glycolysis, a very quick metabolic source of ATP, which makes rapid macroscopic motion possible. The efficiency of the features of animal macroscopic behavior related with glycogen metabolism depends on a good design of the glycogen molecule. We present here a mathematical model which describes the structure of glycogen and which allows us to calculate the values of its parameters which optimize the variables mentioned above. Our results demonstrate that the structure of the glycogen molecule has an optimized design for (a) maximizing the total stored glucose in the smallest volume, (b) maximizing the amount of it which can be directly released by phosphorylase, before any debranching, and (c) maximizing the number of non-reducing ends (points of attack for phosphorylase), which maximizes the quickness of fuel release.

2. Glycogen structure and mathematical model

Glycogen structure

The model of Whelan (Gunja-Smith *et al.*, 1970; see also Gunja-Smith *et al.*, 1971) derived from glycogen enzymatic degradation data is generally accepted for describing glycogen structure (Goldsmith *et al.*, 1982; Bullivant *et al.*, 1983). According to these results, the main features of glycogen structure can be described as follows:

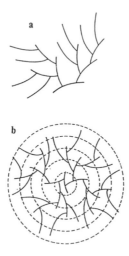

Figure 1 *Structure of glycogen.* Scheme showing the structure of the glycogen molecule. (a) Extended structure to show the branching structure; (b) A more realistic drawing showing the disposition of the successive branches forming concentric tiers (numbered circles). Both schemes show a simplified molecule with only 4 tiers in (a), and 5 in (b).

The glycogen molecule is formed by two different kinds of chains: B chains, branched, and A chains, not branched. The branching of the B chains is uniformly distributed (with branching degree equal 2), so every B chain has two branches on it, creating further A- or B-chains. There are four glucose residues between branches and a tail after the second branch in the B chains. Both A- and B-chains have a uniform length, the same for both kinds of chains; this length has a mean value of 13 glucose residues. Phosphorylase can only work on the A chains since the length of the tail of the B chains is too short -about 4 glucose residues- which is at the limit of phosphorylase action. The glycogen molecule is spherical and is structurally organized in concentric tiers; every tier has the same length (1.9 nm). There are 12 tiers in a ß-particle, with a total radius of 21 nm. Every A-chain is in the most external tier (number 12). As a consequence of the degree of branching, $r = 2$ (every B-chain gives 2 chains), the number of chains in any tier is twice that of the previous one, and the same number of chains as the summation of all other previous tiers. From these data it can be derived that there are the same amount of A-chains (all them in the last tier) as B-chains, and that the amount of glucose directly available to be released by phosphorylase is 34.6% of the total molecule. Assuming a general kind of organization for glycogen as described above, the values of certain parameters of design can give glycogens with different structural features whose properties condition different functional possibilities. These parameters are: the chain length (g_C); the branching degree (r); and the number of tiers (t). Any value for these parameters different from those of cellular glycogen would give a molecule with a similar shape, but different properties. We analyze here these questions with a mathematical model which describes the glycogen structure, in order to find the values of the three parameters mentioned above, for obtaining a molecular design which accounts for maximization of all stored glucose, glucose available for phosphorylase, number of A-chains (points of attack for phosphorylase), and minimization of the molecular volume; then we compare these results with the values of these parameters in cellular glycogen.

Mathematical model

Our model is made under a set of hypotheses according to the general features of the molecular structure given by Gunja-Smith *et al.* (1970) and Goldsmith *et al.* (1982); see also Bullivant *et al.* (1983). The structure of glycogen molecule can be described by the following set of equations: let r be the branching degree (number of branching points on each B-chain), and thus, the factor which multiplies the number of chains in a tier giving the number of chains in the next one. Let t be the number of tiers in the molecule. Then, the total number of chains (C_T) is:

$$C_T = \sum_{i=1}^{t} r^{(i-1)} = \frac{1-r^t}{1-r} \tag{1}$$

and the number of chains in a given tier t_i

$$\mathbf{C}_{t_i} = r^{(t_i-1)} \tag{2}$$

Every A-chain is in the most external tier. Thus, Eq.(2) gives us the number of A-chains:

$$C_A = r^{(t-1)} \tag{3}$$

Let g_c be the number of glucose residues in any A-chain. Not all of these units can be released by phosphorylase. A physical limit exists in the progress of phosphorylase to digest any A-chain.

This limit was empiricaly determined by Walker & Whelan (1960) as 4. There are clear steric reasons which can justify the value of this parameter, so we take it fixed in our reasoning. So, the number of glucose residues available for phosphorylase in each A-chain (G_{PC}) is:

$$G_{PC} = g_c - 4 \tag{4}$$

The total amount of glucose available for phosphorylase in the whole molecule (G_T) is written as:

$$G_{PT} = C_A \cdot G_{PC} = C_A \cdot (g_c - 4) \tag{5}$$

and the total glucose in the whole molecule (G_T):

$$G_T = C_T \cdot g_c \tag{6}$$

which by Eq.(1) gives:

$$C_T = g_c \sum_{i=1}^{t} r^{(i-1)} = g_c \cdot \frac{1 - r^t}{1 - r} \tag{7}$$

According to Goldsmith et al. (1982), on average, a branch starts halfway up a chain, and it gains about 0.35 nm because of the (1-6) bond; in the same paper they give a length of 0.24 nm per glucose in a chain. This gives an effective length per tier (L) of

$$L_t = 0.12 \, g_c + 0.35 \ \text{nm}^3 \tag{8}$$

The molecule has a spherical shape; this is consistent with Whelan's model and experimental data from other groups (see, e.g., Madsen & Cori, 1958), including electron micrographs. The radius of the sphere with t tiers is:

$$R_s = L_t \cdot t$$

which applying Eq.(8) is:

$$R_s = t(0.12 \, g_c + 0.35) \ \text{nm}^3$$

and the volume of the sphere:

$$V_s = \left(\tfrac{4}{3}\right) \pi t^3 (0.12 \, g_c + 0.35)^3 \ \text{nm}^3 \tag{9}$$

Thus, our mathematical model consisting of Eqs. (3), (5), (7) and (9) (see Fig. 2) will allow us to calculate any of the variables of the glycogen molecule mentioned above as a function of r, t and g_c parameters. We shall use them to find the values for these parameters which optimize the variables.

$$C_A = r^{(t-1)} \tag{3}$$

$$G_{PT} = C_A \cdot G_{PC} = C_A \cdot (g_c - 4) \tag{5}$$

$$C_T = g_c \sum_{i=1}^{t} r^{(i-1)} = g_c \cdot \frac{1 - r^t}{1 - r} \tag{7}$$

$$V_S = \left(\frac{4}{3}\right)\pi t^3 (0.12\, g_c + 0.35)^3 \text{ nm}^3 \tag{9}$$

Figure 2 *Mathematical model of glycogen structure.* This set of equations describe all structural relationships among the differents parameters and variables of the glycogen molecule, and can therefore be used to study any structural feature of this molecule. (see Fig.1). Symbols are: (a) parameters: r, branching degree, i.e., number of new branches started from each former branch; t, numer of tiers; glycogen molecule is organized in concentric tiers as a consequence of the branched structure (see Fig.1b); g_c, chain length in glucose residues; (b) variables: C_A, number of A-chains, or non-reducing ends of the molecule; G_T, total amount of glucose stored in the glycogen molecule; G_{PT}, total amount of glucose available directly for phosphorylase, in the most external tier, with no debranching; V_S, volume of the molecule.

3. Optimization of glycogen molecular design

(a) The number of tiers in the glycogen molecule

The cellular glycogen molecule has 12 tiers. Goldsmith *et al.* (1982) have calculated that the volume of the hypothetical 13rd tier in the glycogen molecule would be 10,000 nm^3, and there would be about 55,000 glucose residues in such a tier. Assuming an approximate volume of 0.113 nm^3 for the glucose molecule (van der Waals volume) this would give 6,215 nm^3 , which means that in that tier 62% of the space would be strictly occupied by glucose, leaving practically no space for phosphorylase. The same reasoning can be applied to glycogen synthesizing enzymes. This is a very efficient way for controlling the molecule size. Madsen & Cori (1959) calculated the surface area of the glycogen molecule according to the different number of tiers; they noted that the glucose residues became more and more crowded as the structure grows in size, and suggested that a structure of this type could be self-limiting in size. This conclusion is right. The physical limit for the size of the glycogen molecule, at 12 tiers is not only empirically known. The size of the enzymes involved in glycogen metabolism can explain it. It is interesting to note that in glycogen storage disease type II (Pompe's disease) where a lack in lysosomal amylo 1-4 glucosidase occurs, there is a significant increase in the amount of glycogen in liver and muscle cells. However, that is because of a great increase of glycogen particles accumulated in lysosomal vacuoles, not because of a larger molecule (Baudhuim *et al.*, 1964; Garancis, 1968).

(b) The branching degree

If the branching degree in the glycogen molecule was very high, then it would lead to an extremely dense molecule useless both for phosphorylase action and for the primary purpose of fuel storage.

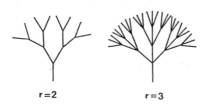

$r=2$ $r=3$

Figure 3 *Branching degree in glycogen structure.* A very simplified scheme for the structure of glycogen, showing the properties which derive from the branching degree $(r=2$ in cellular glycogen). Greater values of r give a molecule extremely dense with a poor capacity for storing glucose since it can contain few tiers.

Data derived from our mathematical model demonstrate that if the branching degree were $r = 3$, then a molecule with the same density as cellular glycogen could only have 7 tiers; such a molecule would be much less efficient than the cellular one: its capacity for storing glucose would be 27% of cellular glycogen, and glucose available for phosphorylase only 36% of cellular glycogen, these values decreasing dramatically for larger values of r. (e.g., for $r = 4$, a molecule with the same density would only have five tiers, 8% of stored glucose, and 13% of glucose available for phosphorylase). We therefore conclude that a degree of branching of $r = 2$ maximizes both the capacity of glucose storage, and the amount of glucose available for phosphorylase. This is precisely the degree of branching in cellular glycogen in Whelan's model, as stated by Gunja-Smith *et al.* (1970, 1971), and later confirmed by Goldsmith *et al.* (1982), and Bullivant *et al.* (1983) among others. It is interesting to note that when this branching degree is $r = 2$, then the number of chains in the most external tier (12) approximately equals the number of chains in all the inside tiers, since

$$2^{t} \approx \sum_{i=1}^{t-1} 2^{i} = 2^{t} - 1 \tag{10}$$

(c) The chain length

Fig. 4 shows two glycogen molecules which have the same amount of stored glucose, but with different designs, the chain length being the only difference between them; they are short in Design *a*, and long in Design *b*. It can be seen that a number of properties is derived from the value of this parameter; for example, for a given total amount of glucose Design *a* occupies less space and has more A-chains (more tiers), but they are longer in Design *b*, and there are more glucose units available for phosphorylase. The four properties which an optimized molecule should have are: (a) the maximum points for phosphorylase attack; (b) the maximum stored glucose; (c) the maximum glucose residues directly available for phosphorylase with no previous debranching; and (d) the minimum volume of the molecule. All these properties must be optimized, so this is a typical case of multi-objective optimization. Each of these properties can be 'expressed by a given variable, all of them being related by the equations of the model here described; in effect, A-chains (C_A) are the points for phosphorylase attack, G_T is the total stored glucose in the molecule, G_{PT} is the total glucose available for phosphorylase, and V_S is the volume of the molecule.

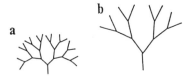

Figure 4 *Chain length in glycogen structure.* Scheme which illustrates the problem of optimization for determining the most appropriate chain length in the design of the glycogen molecule. The different properties of two glycogen designs both with the same branching degree, $r = 2$, but with different chain lengths are shown as *Design a* and *Design b*, respectively. *Design a* has short chains and *Design b* has long chains. The pictures show two glycogens with the same amount of stored glucose (the same amount of ink was spent in each one).

The relationships among these variables are given by Eqs. (3), (5), (7) and (9), see Fig. 2. They describe the properties of the glycogen molecule whose variables have to be optimized. In accordance with the previous reasoning the aim now is to find the values of t and g_C which maximize C_A, G_T and G_{PT}, and which minimizes V_S ; i.e., to maximize the function

$$f = \frac{G_T \cdot C_A \cdot G_{PT}}{V_S} \tag{11}$$

C_A, G_T, G_{PT} and V_S being related by Eqs.(3), (4), (5) and (7). According to the result of the previous section it must be $r = 2$. Therefore, Eq.(14) can be written as:

$$f = K \cdot \frac{g_c \cdot (g_c - 4)}{(0.12\, g_c + 0.35)^2} \tag{12}$$

with

$$K = \frac{(6C_A - 3)C_A{}^2}{4\pi t^3} = \frac{\left[6 \cdot 2^{(3t-3)}\right] - \left[3 \cdot 2^{(2t-2)}\right]}{4\pi t^3} \tag{13}$$

Eqs.(12) and (13) show that the value of g_C which maximizes f is independent of the value of t (and consequently also of C_A), since t and g_C are independent variables. On the other hand, as discussed below, there is a physical limit for t. The value of g_C that maximizes f is calculated by making the derivative of f with respect to g_C, in Eq.(12) and making it equal to zero. This is:

$$f = K \cdot \frac{(2g_c - 4)(0.12\, g_c + 0.35)^3 - 3 \cdot (0.12\, g_c + 0.35)^2 0.12\, g_c(g_c - 4)}{(0.12\, g_c + 0.35)^6}$$

Making it equal to zero, we obtain:

$$0.12\, g_c{}^2 - 1.66\, g_c + 1.4 = 0$$

whose two solutions are: $g_C = 12.93$, and $g_C = 0.90$. The first is the value of the maximum; this can be demonstrated by seeing that the second derivative is positive, but it is not necessary, since $g_C = 0.90$ has no physical meaning, because it cannot be lesser than 4. Therefore, we found that the optimum value of g_C, which maximizes f from the root of

$df/dg_C= 0$ is $g_C = 12.93$. This is well illustrated in Fig. 5, where the optimization function is plotted; thus, $g_C = 13$ is the value of the chain length which optimizes the structure of glycogen. Empirical data obtained by several groups are in good agreement with this theoretical result: Cori's group (Illingworth *et al.*, 1952) reported data of twelve glycogens analyzed from different sources, which gave an average chain length of between 10.8 and 15.4 glucose residues, with a mean value of 12.91 (for example: it was 13 in fetal guinea pig liver and Sprague-Dawley rat liver, and 12.5 in Busch strain rat liver and cat liver). Manners (1957) has noted that of 84 different glycogens examined, 62 had chain lengths of 11-13 residues. Manners & Wright (1962) reported a chain length of 13 glucose residues in rabbit muscle and rat liver. Gunja-Smith *et al.* (1971) have reported chain lengths of 11.5 in human muscle, 12 in skate liver, 12.5 in *Ascaris*, 13 in rabbit muscle, 14 in cat liver and rabbit liver, 14.5 in *Trichomonas foetus*, and 15 in horse diaphragm. Bullivant *et al.* (1983) have reported a value of 12-14 for white rabbit liver. An overall chain length of 12-14 glucose residues was stated by Ryman & Whelan (1971) for the majority of glycogens (see also Smith, 1968).

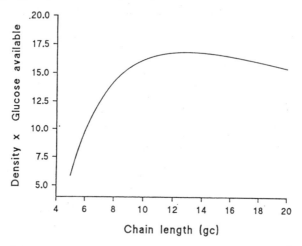

Figure 5 *Plot of the optimization function.* Optimization of the chain lenght (g_c) in the glycogen molecule in order to maximize the capacity for storing glucose in the least possible volume, the total amount of glucose which can be released by phosphorylase before any debranching occurs, and the points of attack for the enzyme. A value of 13 glucose residues as the chain length optimizes the glycogen structure. The optimization function has dimensions of glucose available for phosphorylase multiplied by density, and it represents the maximum available glucose stored in a molecule which has the maximum possible density.

By examining the optimization function in Fig. 5 it is clear that these small ranges aroud 13 do not indicate a significant deviation from the optimum value, and we can conclude that the chain length is a well optimized parameter in glycogen structure.

Acknowledgements This work was supported by grants from Dirección General de Investigación Científica y Técnica, Ministerio de Educación y Ciencia (Spain), Ref.No. PB90-0846, and Consejería de Educación del Gobierno de Canarias (Spain), Ref. 91/010. E.M.H. is grateful to Consejería de Educación del Gobierno de Canarias for the additional funds for a three month stay at the University of Tennessee at Chattanooga, where part of this work was done.

References

Baudhuim,P., Hers, H.G. & Loeb, H. (1964) *Lab. Invest.* **13**, 1139-1152

Bullivant, H. M., Geddes, R. & Wills, P. R. (1983) *Biochem. Int.* **6**, 497-506

Garancis, J. C. (1968) *Amer. J. Med.* **44**, 289-300

Goldsmith, E., Sprang, S. & Fletterick, R. (1982) *J. Mol. Biol.* **156**, 411-427

Gunja-Smith, Z., Marshall, J. J., Mercier, C., Smith, E. E. & Whelan, W. J. (1970) *FEBS lett.* **12**, 101-104

Gunja-Smith, Z., Marshall, J.J. & Smith, E. E. (1971) *FEBS lett.* **13**, 309-311

Helmreich, E. & Cori, C. F. (1965) *Adv. Enz. Regul.* **3**, 91-107

Illingworth, B., Larner, J. & Cori, G. T. (1952) *J. Biol. Chem.* **199**, 631-640

Madsen, N. B. & Cori, C. F. (1958) *J. Biol. Chem.* **233**, 1251-1254

Manners, D. J. (1957) *Adv. Carbohydrate Chem.* **12**, 261-298

Manners, D. J. & Wright, A. (1962) *J. Chem. Soc.* pp. 1597-1602

Meléndez-Hevia, E. (1990) *Biomed. Biochim. Acta,* **49**, 903-916

Meléndez-Hevia, E. & Isidoro, A. (1985) *J. Theor. Biol.* **117**, 251-263

Meléndez-Hevia, E., Waddell, T. G. & Shelton,E. (1994) *Biochem. J.* **295**, 477-483.

ap Res,T. (1974) in *Plant Biochemistry* (Northcote, D.H., ed.), *MTP Int. Rev. Sci.* **11**, 89-127

Ryman, B. E. & Whelan, W. J. (1971) *Advan. Enzymol. Relat. Areas Mol. Biol.* **34**, 285-443

Smith, E. E. (1968) in *Control of glycogen metabolism* (Whelan, W. J., ed.), pp. 203-213, Universitetsforlaget, Oslo.

Walker, G. J. & Whelan, W. J. (1960) *Biochem. J.* **76**, 264-270

Tumor Growth and Treatment Planning: A Novel Simulation Approach

W. Düchting, T. Ginsberg, and W. Ulmer

The aim of this contribution is to outline the way how methods of systems analysis, control theory and computer science can be applied to simulate cell growth and to optimize cancer treatment. Based on biological observations and cellkinetic data, our group has constructed an oversimplified cell-cycle model describing the spatial (3D) and temporal growth of a tumor spheroid in a nutrient medium. This growth model has been extended by an irradiation model based on the linear-quadratic survival function, which enables the study of the radiation effect on tumor cells. Different clinical fractionation schemes (standard, super-, hyperfractionation, irradiation with a weekly high single dose) have been applied to different in-vitro growing tumor spheroids. As a result, an individual optimal treatment schedule was gained. In order to minimize the radiation effect on normal cells, we are at present constructing compartment models describing the growth and renewal of rapidly-proliferating as well as of slowly-proliferating normal tissues. The ultimate goal of our work is to test different clinical therapy schedules by computer experiments and to develop new optimal treatment strategies.

1. Introduction

Our research group is intensively involved in the mathematical modeling of benign and malignant cell growth and applies methods of control theory [1] to it. This approach and other aspects have brought about the hypothesis to associate tumor growth with a cellular division control circuit which has become structure-unstable. The scope of these investigations was to stepwise develop numerous models describing the chronological and spatial growth of tumors [2]. In one of our latest papers ([3] and references therein) the in-vitro tumor growth model was extended by a radiobiological dose-response model basing on the linear-quadratic approach. Thus, clinical irradiation schemes can be tested by computer experiments at in-vitro tumor spheroids providing a contribution to the optimization of therapy planning. Regarding an overall optimal therapy planning, it is not only necessary to reach a maximal tumor killing but also simultaneously to get a minimal killing of the normal cells. Therefore, compartment models are under construction considering the irradiation of rapidly and slowly proliferating normal tissues [4] in order to study the acute and late radiation effects.

2. Modeling and simulation of spheroidal tumor growth

Model approach

The basic idea of our work was in a first step to substitute in-vitro experiments of spheroidal tumor growth by computer models. One of these approaches is described in a previous paper [5]. It is important to note that spheroidal growth stops at a diameter of 3...4 mm, that means there is a maximum volume of the spheroid beyond which no further expansion occurs [6, 7]. The cell-cycle process of a tumor cell with its different discrete phases (G1, S, G2, M, G0, N) is condensed in Fig. 1. The G1, S, G2, M-cells are called proliferating cells (P-Cells) and the noncycling "G0" and N-cells are named resting (dormant) cells and necrotic cells, respectively.

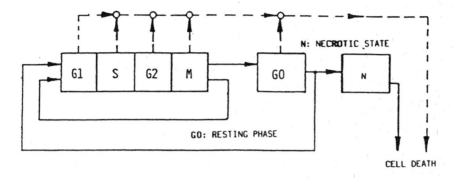

Figure 1 Simplified cytokinetic model of a tumor cell [5]

The construction of a model describing in-vitro tumor growth requires:

1. A cytokinetic model [5] describing the division of a tumor cell (Fig. 1)

2. Experimentally gained data of the cell-cycle phase durations

3. Cell production and interaction rules describing the cell-to-cell communication.

For instance some rules of the catalogue may say:

(i) The multiplication of a tumor cell is possible only if the distance between a dividing tumor cell and the nutrient medium is less than three cell layers. All tumor cells residing in a distance of more than three cell layers enter the resting phase GO because of the lack of oxygen and nutrient supply.

(ii) If there is no free position around the dividing tumor cell, it divides into the direction which has the shortest distance between the tumor cell and the nutrient medium. If there are equal minimum distances in various directions, a pseudo-random number generator determines the direction in which the tumor cell will divide.

The cell space is limited to 42 x 42 x 42 cell units and is - excluding the tumor cells - symbolically filled up with a nutrient medium. Every unit is described by a cell code which includes the typical parameters (cell species, remaining life span of the cell, cell cycle phase duration).

After transforming these statements, rules and equations into algorithms, computer programs were written in FORTRAN IV. The input data to the simulation model are the cell-cycle phase durations of a specific tumor cell and its initial configuration.

Simulation results

Based on this model, the dynamics of the tumor cell growth for tumors with different cell-cycle-phase durations (Table 1) are calculated. The simulation runs start with one single tumor cell which resides in the state of mitosis at $t = 0$ hours. The different growth curves can mathematically be described by the so-called Gompertz-function. Moreover, the tumor growth model constructed by our group in [5] makes it possible to illustrate the spatial three-dimensional growth of tumors. Fig. 2 represents the spatial growth of a rapidly growing tumor spheroid (A) with the cell-cycle time $T_C = 10$ h at three different time moments.

Table 1 Cell cycle phase durations (cell cycle time $T_C = T_{G1} + T_S + T_{G2} + T_M$) of different growing tumors

Kind of tumor	Notation of phase duration	T_{G1}	T_S	T_{G2}	T_M	T_{G0}	T_N
Rapidly growing tumor (e.g. small cell lung carcinoma, $T_c = 10$ h) (A)	Phase duration in h	3	5	1	1	25	40
	Deviation in h	0	1	0	0	5	2
Moderately fast growing tumor (e.g. squamous carcinoma of the lung, $T_c = 15$ h) (B)	Phase duration in h	5	7	2	1	25	40
	Deviation in h	0	1	0	0	5	2
Slowly growing tumor (e.g. brain tumor, $T_c = 30$ h) (C)	Phase duration in h	11	13	4	2	25	40
	Deviation in h	1	2	1	0	5	2

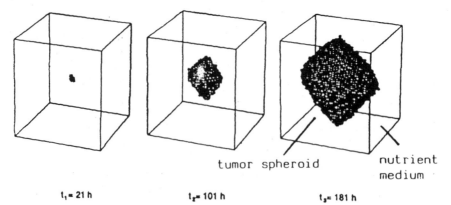

tumor spheroid   nutrient medium

$t_1 = 21$ h $t_2 = 101$ h $t_3 = 181$ h

Figure 2 3D illustration of a rapidly growing tumor spheroid (A) in a nutrient medium (data see Table 1)

3 Modeling and simulation of fractionated radiotherapy

Model approach

In order to construct a model describing radiation treatment, it is necessary to know the number of tumor cells hit by radiation. The computation of the percentage of the tumor cells killed by irradiation is based on the "linear-quadratic model" with the survival function [8]

$$S(D) = e^{-\alpha D} * e^{-\beta D^2} \quad .$$ (1)

In equation 1, D stands for dose, and α, β are symbolizing parameters depending on the kind of tumor and on the type of radiation. For different growing tumors, the radiation parameters α and β as well as the total dose D_{TOTAL} used in this paper are listed in Table 2. These parameters are related to the application of γ-rays and fast electrons. The number of the tumor cells to be killed can be determined via equation 1. Subsequently, the killing is performed by means of pseudo-random-number generators in the computer model. To become more realistic, the simplified model was enriched by implementing the following additional assumptions:

1. 30% of the hit tumor cells are repaired within 15 hours and proceed growing.
2. 15 % of the dormant GO-tumor cells still existing after irradiation are recruited into the cell cycle (consideration of the effect of reoxygenation).
3. The lysis and transportation of the lethally damaged tumor cells need about 120 hours (=five days), that means only after this time the position of a lethally injured tumor cell may be occupied by another tumor cell or by the nutrient medium.

It is well known in radiobiology and radiotherapy that, contrary to a single tumor irradiation, a fractionated application offers many advantages [9...13]. Going into further detail, advanced results of radiobiology provide many rationales for modifications of the daily applied irradiation dose of 1 x 2 Gy (standard fractionation scheme) to be usually found in clinical routine. Thus, the task is to develop a treatment scheme which leads to a maximum cell kill of tumor cells and to a minimum damage of normal cells.

Because of the complexity and heterogenity of biological systems, at present it seems to be impossible to give a general optimized treatment scheme. Therefore, we start with very simple simulation experiments to study clinical treatment schedules applied to in-vitro tumor growth under different oversimplified assumptions and restrictions.

Table 2 Irradiation parameters of different growing tumors

Irradiation parameters	α in 1/Gy	β in 1/Gy^2	D_{TOTAL} in Gy
Rapidly growing tumor (e.g. small cell lung carcinoma, $T_c = 10$ h) Ⓐ	0.42	0.02	42
Moderately fast growing tumor (e.g. squamous carcinoma of the lung, $T_c = 15$ h) Ⓑ	0.38	0.02	60
Slowly growing tumor (e.g. brain tumor, $T_c = 30$ h) Ⓒ	0.09	0.02	60

Simulation studies

Five different fractionation schemes (Table 3) are simulated for different growing tumors (**Table 1**) with irradiation parameters according to Table 2. The systematical investigation of their therapeutic effectiveness on in-vitro tumors via computer simulation is extremely useful, because, so far, the novel fractionation regimens are only sporadically tested in the clinical practice [14, 15]. The simulation runs are performed on a DEC-computer (VAXstation 3100 M 38). The computing time takes about four hours for simulating a single therapy course comprising 1500 hours (≈ 63 days).

The constructed computer models allow the calculation and representation of the spatial configuration and of the time behavior of the irradiated tumor spheroid. To demonstrate the power of the model, Fig. 3 shows the spatial configuration of the hyperfractionated irradiation (scheme 3) of a moderately fast growing tumor spheroid (B) (data see Table 1) at three different time moments starting at t = 500 h.

Table 3 Different fractionation schemes

Fractionation scheme		Dose in Gy
Standardfractionation	(1)	1 x 2 Gy per day
Superfractionation	(2)	3 single doses per day in an interval of 4 h: 0.7 / 0.6 / 0.7 Gy
Hyperfractionation I	(3)	3 single doses per day in an interval of 4 h: 1 / 1 / 1 Gy
Hyperfractionation II	(4)	3 single doses per day in an interval of 4 h: 1.5 / 1.5 / 1.5 Gy
Weekly high single dose	(5)	1 x 6 Gy per week

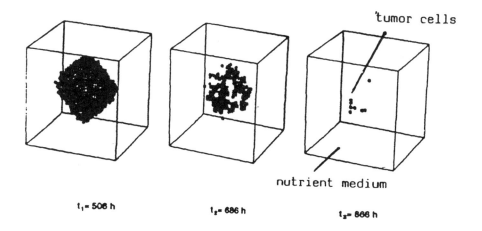

'tumor cells

nutrient medium

t_1 = 506 h t_2 = 686 h t_3 = 866 h

Figure 3 Illustration of a hyperfractionated irradiation of a moderately fast growing tumor spheroid (B)

The time course of the number of tumor cells of a rapidly growing tumor spheroid (A) is plotted with five different fractionation schemes in Fig. 4. In this case the hyperfractionation (scheme 3 or 4) will lead, above all, to a particularly good tumor effectiveness (see Fig. 4). These model-predicted results are in excellent agreement with clinical observations of several authors [8, 14]. In the case of moderately fast growing tumor spheroids the simulation outcome shows [3] that the hyperfractionation proves to be the optimal radiation scheme, too, provided that the criteria are a high tumor effectivity and simultaneously the least possible overall treatment time. In contrast to the preceding results, in the case of the slowly growing tumor spheroid (C) a clear statement in favor of the treatment with a weekly high dose (scheme 5) can be made in accordance to the model results [3]. This fractionation scheme has been proposed and tested in clinical practice [15]. However, the clinical evaluation has led to a controverse discussion so that a further clinical confirmation of the model predictions presented by our group has not yet been reached.

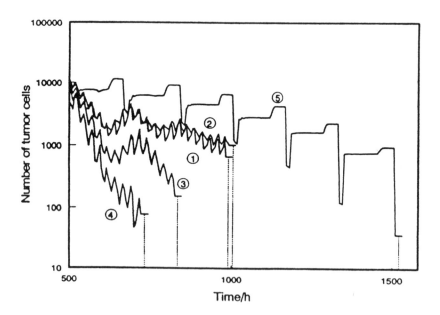

Figure 4 Radiotherapy of a rapidly growing tumor spheroid (A) with different fractionation schemes (data see Tables 1 to 3)

References

1. Düchting,W.: Tumor growth simulation. Comput. & Graphics 14 (1990), 505-508.
2. Düchting, W.: Simulation of malignant cell growth. In: Fractal Geometry and Computer Graphics, J.L. Encarnacao, H.-O. Peitgen, G. Sakas, G. Englert (Eds.), Springer-Verlag, Berlin 1992, 135-143.
3. Düchting, W., Ulmer, W., Lehrig, R., Ginsberg, T. and Dedeleit, E.: Computer simulation and modelling of tumor spheroid growth and their relevance for optimization of fractionated radiotherapy. Strahlenther. Onkol. 168 (1992), 354-360.
4. Saile, C.: Simulation unterschiedlicher Radiotherapien von Tumorsphäroiden sowie von Akutwirkungen der Bestrahlung auf gesunde Zellen. Diplomarbeit am "Institut für Regelungs- und Steuerungstechnik", Universität Siegen 1993.
5. Düchting, W. and Vogelsaenger, Th.: Three-dimensional pattern generation applied to spheroidal tumor growth in a nutrient medium. Int. J. Bio-Med. Comput. 12 (1981), 377-392.
6. Acker, H., Carlsson, J., Durand, R., Sutherland, R.M.: Spheroids in cancer research. Springer, Berlin 1984.
7. Mueller-Klieser, W.: Multicellular-spheroids. J. Cancer Res. Clin. Oncol. 113 (1987), 101-122.
8. Fowler, J.F.: The linear-quadratic formula and progress in fractionated radiotherapy. Brit. J. Radiol. 62 (1989), 676-694.
9. Thames, H.D. and Hendry, J.H.: Fractionation in radiotherapy. Taylor & Francis, London 1987.
10. Withers, H.R.: Some changes in concepts of dose fractionation over 20 years. Front. Radiat. Ther. Onc. 22 (1988), 1-13.
11. Scherer, E., Sack, H.: Strahlentherapie. Thieme, Stuttgart 1989.
12. Cohen, L.: Biophysical models in radiation oncology. CRC Press, Boca Raton 1983.
13. Peres, C.A., Brady, L.W.: Principles of radiation oncology. J.B. Lippincott Comp., Philadelphia 1989.
14. Lesche, A. und Kulpe, T.: Ergebnisse hyperfraktionierter Bestrahlung des Bronchialkarzinoms. Radiobiol. Radiother. 31 (1990), 391-394.
15. Schuhmacher, W.: Nutzbarmachung neuer Erkenntnisse über die Fraktionierung bei der Bestrahlung bösartiger Tumoren für die Praxis. Röntgen-Ber. 1 (1972), 91-100.

1. Optimization of kinetic parameters of ordered mechanisms - basic aspects

An enzymic reaction with ordered mechanism in n elementary steps can be described with Scheme I where X_0, X_1,..., X_i,..., X_{n-1} are the different enzyme species, and $\tilde{k}_i$ and $\tilde{k}_{-i}$ are the rate constants of the forward and backward steps. Conservation of enzyme species can be expressed as

$$E_t = \sum_{i=0}^{n-1} X_i .$$

(1.1)

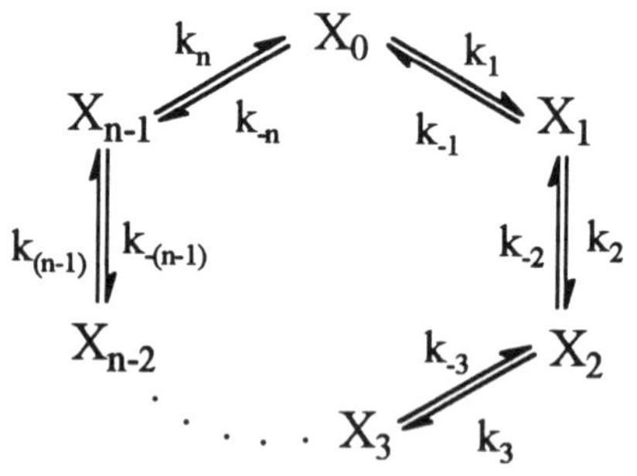

Scheme I

The parameters $\tilde{k}_{\pm i}$ are real or apparent first order rate constants. In the case of binding of a reactant $R_{\pm i}$ in step i the apparent rate constant equals the product of the real second order rate constant $k_{\pm i}$ and the concentration $R_{\pm i}$ of the reactant ($\tilde{k}_{\pm i} = R_{\pm i} k_{\pm i}$). The real and apparent thermodynamical equilibrium constants of the enzymic reaction are

$$q = \prod_{i=1}^{n} \frac{k_i}{k_{-i}} \quad , \qquad\qquad \tilde{q} = \prod_{i=1}^{n} \frac{\tilde{k}_i}{\tilde{k}_{-i}} \quad ,$$

(1.2a,b)

resp. The steady state velocity $V = v_i = \tilde{k}_i X_{i-1} - \tilde{k}_{-i} X_i$ is the same in each step. One finds an expression for the reaction rate V according to the rules of King and Altman (1956):

$$V = \frac{E_t(\tilde{q}-1)}{\sum_{i=1}^{n} \frac{1}{\tilde{k}_{-i}} + \sum_{i=1}^{n} \sum_{j=1}^{n-1} \frac{1}{\tilde{k}_{-(i-1)}} \prod_{r=i}^{i+j-1} \frac{\tilde{k}_r}{\tilde{k}_{-r}}} = \frac{E_t(\tilde{q}-1)}{N}$$

(1.3)

with the cyclic notation $\tilde{k}_{\pm i} = \tilde{k}_{\pm(n+i)}$ At fixed equilibrium constant the numerator in expression (1.3) does not depend on the values of the elementary rate constants. Therefore, maximization of $|V|$ has mathematically the same meaning as minimization of the denominator $N = N(\tilde{k}_{\pm i})$.

Here only the case $q \geq 1$ shall be considered. For symmetry reasons optimal solutions for $q < 1$ can be deduced from the solutions for $q > 1$ via transformation $q \leftrightarrow 1/q$, $R_{\pm i} \leftrightarrow R_{\mp(n-i+1)}$, $k_{\pm i} \leftrightarrow k_{\mp(n-i+1)}$ and $V \leftrightarrow -V$, resp.

Uni-uni-reactions. For reactions with one substrate $S = R_1$ of concentration S and one product $P = R_{-n}$ of concentration P the reaction rate can be expressed in the following form

$$V = E_t \frac{Sq - P}{N} \tag{1.4a}$$

with

$$N = D_1 + SD_2 + PD_3 \tag{1.4b}$$

$$D_1 = \frac{1}{k_{-n}} + \sum_{r=1}^{n-1} \frac{1}{k_{-r}} \prod_{j=r+1}^{n} \frac{k_j}{k_{-j}} \tag{1.4c}$$

$$D_2 = \frac{1}{k_{-n}} \sum_{i=1}^{n-1} \prod_{j=1}^{i} \frac{k_j}{k_{-j}} + \sum_{i=1}^{n-2} \prod_{s=1}^{i} \frac{k_s}{k_{-s}} \sum_{r=i+1}^{n-1} \frac{1}{k_{-r}} \prod_{j=r+1}^{n} \frac{k_j}{k_{-j}} \tag{1.4d}$$

$$D_3 = \sum_{j=1}^{n-1} \frac{1}{k_{-j}} + \sum_{i=2}^{n-1} \sum_{r=1}^{i-1} \frac{1}{k_{-r}} \prod_{j=r+1}^{i} \frac{k_j}{k_{-j}} \ . \tag{1.4e}$$

Phenomenological parameters. The rate equation can be rewritten using Michaelis constants K_{mS} and K_{mP} of substrate and product, resp., and maximal velocities V^+ and V^- of the forward and backward reaction, resp.

$$V = \frac{S \cdot \dfrac{V^+}{K_{mS}} - P \dfrac{V^-}{K_{mP}}}{1 + \dfrac{S}{K_{mS}} + \dfrac{P}{K_{mP}}} \quad , \tag{1.5}$$

A comparison of formulae (1.4) and (1.5) yields

$$K_{mS} = \frac{D_1}{D_2} \ , \quad K_{mP} = \frac{D_1}{D_3} \ , \quad V^+ = \frac{E_t q}{D_2} \ , \quad V^- = \frac{E_t}{D_3} \ . \tag{1.6a-d}$$

Upper limits of individual rate constants: The reaction rate (1.3) is a homogeneous function of first degree of the individual rate constants $k_{\pm i}$ with

$$\tau V (k_{\pm i}) = V (\tau k_{\pm i}) \tag{1.7}$$

for arbitrary $\tau > 0$. Therefore, an increase of all kinetic constants $k_{\pm i}$ by a common factor leads to an increase of V by that factor; maximal values of the reaction rate can be found only under consideration of upper limits for the values of the elementary rate constants

$$0 < k_{\pm i} \le k_{\pm i, max} \ . \tag{1.8}$$

Concerning the upper limits two possibilities will be considered: separate upper limits for all elementary rate constants (*separate limit model, SLM* [1;4]) and further a coupling of the upper limits by means of a cost function expressing the least evolutionary effort nessecary to increase the rate constants from very low values to the recent values (*overall limit model, OLM* [3]).

2. Separate limit model

To take into account the different physico-chemical meaning and the different units of first and second order rate constants different upper limits shall be considered:

$$0 < k_{bimol} \leq k_d \qquad (2.1a)$$

$$0 < k_{monomol} \leq k_m . \qquad (2.1b)$$

Normalization. For sake of simplicity and comparability the model parameters will be normalized using the corresponding maximal values the rate constants can assume. In order to avoid new symbols the previous quantities will simply be redefined.

$$\frac{k_{bimol}}{k_d} \to k_{bimol} \quad , \qquad \frac{k_{monomol}}{k_m} \to k_{monomol} \quad , \qquad (2.2a\text{-}c)$$

which implies for all i

$$0 < k_{\pm i} \leq 1 . \qquad (2.3)$$

Concentrations of reactants and Michaelis constants, resp., as well as the reaction rate and maximal velocities, resp., are normalized, too

$$\frac{R_{\pm i} k_d}{k_m} \to R_{\pm i} \quad , \qquad \frac{V}{k_m E_t} \to V . \qquad (2.4a,b)$$

It is worth mentioning that after normalization the form of the rate equation is retained exept that in Eqs.(1.3) and (1.4) one has to omit the factor E_t.

Optimization strategy. The fact that the rate is a homogeneous function of first degree of the rate constants (Eq. 1.7) implies that in optimal states at least one of the rate constants has to attain its maximal value. The number $K(n)$ of combinations of maximal and non-maximal rate constants counts $K(n) = 2^{2n} - 2^n$. In addition, from the structure of the denominator N in Eq. (1.3) follows that in optimal states from the couples (k_i, k_{-i}) and

$(k_i, k_{-(i-1)})$, resp., not both constants can assume non-maximal values (note the cyclic notation for the indices) [4]. This exclusion rule reduces the number $Z(n)$ of possible combinations of maximal and non-maximal rate constants drastically. $K(n)$ and $Z(n)$ are compared in Table 2.1. for different numbers n of mechanism steps. All combinations which are in accordance with the exclusion rule lead to solutions of the optimization problem as long as the space of the concentrations of the reactants has enough dimensions (at least $n-1$). These solutions can be classified into types $T_{\alpha,\beta}$ according to the number α of submaximal forward and the number β of submaximal backward rate constants they include.

Table 2.1. Number of possible combinations of submaximal rate constants $K(n)$ compared to the number of optimal solutions $Z(n)$ in the separate limit model depending on the number n of steps.

number n of steps	1	2	3	4	5
combinations of rate constants $K(n)$	2	12	56	240	992
number of optimal solutions $Z(n)$	1	3	10	31	91

Two-step mechanism. For a reversible uni-uni-reaction with two steps as depicted in the following scheme

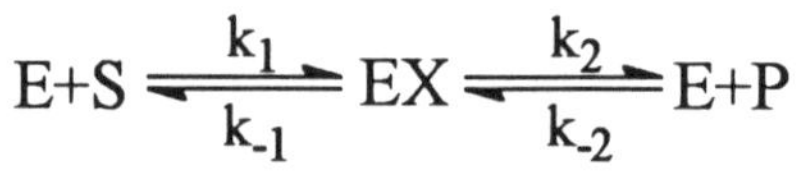

$$E+S \underset{k_{-1}}{\overset{k_1}{\rightleftharpoons}} EX \underset{k_{-2}}{\overset{k_2}{\rightleftharpoons}} E+P$$

Scheme 2

the denominator of the rate equation (1.1) reads

$$N = \frac{1}{k_{-2}} + \frac{k_2}{k_{-1}k_{-2}} + \frac{Sk_1}{k_{-1}k_{-2}} + \frac{P}{k_{-1}} \tag{2.5}$$

According to Table 2.1. one finds three solutions of the optimization problem with the optimal rate constants listed in Table 2.2. The optimal denominators N are

$$N_1 = 1 + q + Sq + Pq, \tag{2.6a}$$

$$N_2 = 2q + Sq + P, \tag{2.6b}$$

$$N_3 = q + Sq + 2(Pq)^{1/2}. \tag{2.6c}$$

A comparison of N_1, N_2 and N_3 shows that solution L_1 is valid for $P < 1/q$, solution L_2 for $P > q$ and solution L_3 for $1/q \leq P \leq q$ [4].

Table 2.2. Optimal rate constants of the three solutions for the two-step mechanism in the *separate limit model* depending on the concentration of the product P.

solution	Typ	k_1	k_{-1}	k_2	k_{-2}
L_1	$T_{0,1}$	1	$1/q$	1	1
L_2		1	1	1	$1/q$
L_3	$T_{0,2}$	1	$(P/q)^{1/2}$	1	$(Pq)^{-1/2}$

Three-step mechanism

$$E+S \underset{k_{-1}}{\overset{k_1}{\rightleftharpoons}} ES \underset{k_{-2}}{\overset{k_2}{\rightleftharpoons}} EP \underset{k_{-3}}{\overset{k_3}{\rightleftharpoons}} E+P$$

Scheme 3

For $n=3$ the denominator of the rate equation reads

$$N = \frac{1}{k_{-3}} + \frac{k_3}{k_{-2}k_{-3}} + \frac{k_2 k_3}{k_{-1}k_{-2}k_{-3}} + S\frac{k_1}{k_{-1}k_{-3}}\left(1 + \frac{k_3}{k_{-2}} + \frac{k_2}{k_{-2}}\right) + P\left(\frac{1}{k_{-1}} + \frac{1}{k_{-2}} + \frac{k_2}{k_{-1}k_{-2}}\right) \tag{2.7}$$

According to the optimization strategy explained above one finds ten solutions depending on the reactant concentrations. The optimal rate constants are listed in Table 2.3. The conditions for the validity of the solutions divide the (S,P)-plane into ten regions depicted in Fig. 2.1 [1]. The solutions may be classified into four types of optimal parameter combinations: a) three solutions with in each case one non-maximal backward rate constant, b) three solutions with in each case two non-maximal backward rate constants, c) three solutions with in each case one non-maximal forward and one non-maximal backward rate constant, and d) one *central solution* with maximal values of all forward rate constants and non-maximal values of the backward rate constants. The *central solution* (L_{10}) is valid for intermediate reactant concentrations. For high values of the equilibrium constant it covers the main part of the (S,P)-space. For very low reactant concentrations one finds a solution (L_9) with submaximal values of the reactant release constants which implies improved binding. For high reactant concentrations (L_7, L_8) low constants of the binding steps favour the isomerisation of enzyme bound states.

Table 2.3. Optimal rate constants of the ten solutions for the three-step mechanism in the *separate limit model* depending on the concentrations of the substrate S and the product P.

solution	Typ	k_1	k_{-1}	k_2	k_{-2}	k_3	k_{-3}
L_1	$T_{0,1}$	1	$1/q$	1	1	1	1
L_2		1	1	1	$1/q$	1	1
L_3		1	1	1	1	1	$1/q$
L_4	$T_{0,2}$	1	$(P/q)^{1/2}$	1	1	1	$(Pq)^{-1/2}$
L_5		1	$\left(\dfrac{S+P}{q(1+P)}\right)^{1/2}$	1	$\left(\dfrac{1+P}{q(S+P)}\right)^{1/2}$	1	1
L_6		1	1	1	$\left(\dfrac{2P}{q(1+S)}\right)^{1/2}$	1	$\left(\dfrac{1+S}{2Pq}\right)^{1/2}$
L_7	$T_{1,1}$	$\left(\dfrac{2q(1+P)}{S}\right)^{1/2}$	1	1	$\left(\dfrac{2(1+P)}{Sq}\right)^{1/2}$	1	1
L_8		1	1	$\left(\dfrac{2q(1+S)}{P}\right)^{1/2}$	1	1	$\left(\dfrac{2(1+S)}{Pq}\right)^{1/2}$
L_9		1	$\left(\dfrac{2(S+P)}{q}\right)^{1/2}$	1	1	$(2q(S+P))^{1/2}$	1
L_{10}	$T_{0,3}$	1	$k_{-1}^4 + k_{-1}^3 - \dfrac{Pk_{-1}}{q} - \dfrac{SP}{q} = 0$	1	$\dfrac{P}{qk_{-1}^2}$	1	$\dfrac{1}{qk_{-1}k_{-2}}$

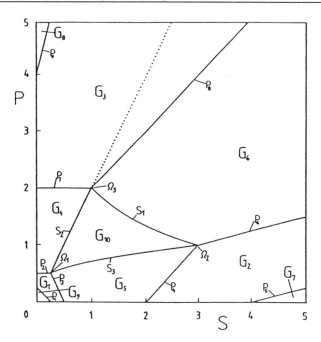

Figure 2.1
Subdivision of the (S,P)-plane into subregions corresponding to the solutions of the three-step mechanism in the *separate limit model* for $q=2$.
Corners of the *central region* $\Omega_1\left(q^{-1}, q^{-2}\right)$, $\Omega_2(2q-1,1)$, $\Omega_3(1,q)$

3. Overall limit model

To take into account further dependencies between upper limits of the rate constants of the elementary steps an overall upper limit for the whole of the elementary rate constants is considered

$$\sum_{i=1}^{n}\left((\beta_i k_i)^\nu + (\beta_{-i} k_{-i})^\nu\right) \le C^\nu \tag{3.1}$$

which resembles the constraint of constant "evolutionary effort" introduced in [2]. With $\beta_{\pm i} = \alpha$ for rate constants of bimolecular steps and $\beta_{\pm i} = 1$ for rate constants of monomolecular steps and under consideration of the fact that Eq.(1.3) is a homogeneous function of the rate constants it follows

$$(\alpha k_1)^\nu + \sum_{i=2}^{n} k_i^\nu + \sum_{i=1}^{n-1} k_{-i}^\nu + (\alpha k_{-n})^\nu = C^\nu \tag{3.2}$$

and for the rate constants $k_{1,-n} \leq C/\alpha$, $k_{i,-i} \leq C$ (else) holds true.

Normalization. Rate constants will be normalized with respect to their maximal values

$$k_{1,-n} \frac{\alpha}{C} \to k_{1,-n} \quad , \qquad \frac{k_{\pm i}}{C} \to k_{\pm i} \ \text{(else)} \tag{3.3a,b}$$

which yields

$$\sum_{i=1}^{n} \left(k_i^v + k_{-i}^v \right) = 1. \tag{3.4}$$

Normalization with respect to parameter α yields dimensionsless values of reactant concentrations and Michaelis constants and with respect to $C \cdot E_t$ dimensionless values of rates.

Optimization strategy. We want to minimize the denominator N in Eq.(1.3) under two constraints, i.e. a fixed value of the equilibrium constant q and condition (3.4) for the upper limits of the rate constants of the individual steps. Using the method of Lagrange multipliers we construct the function

$$\Psi(k_{\pm i}) = N(k_{\pm i}) + \lambda_1 \varphi_1(k_{\pm i}) + \lambda_2 \varphi_2(k_{\pm i}) \tag{3.5}$$

where the two constraints are taken into account by

$$\varphi_1 = \prod_{i=1}^{n} \frac{k_i}{k_{-i}} - q = 0 \tag{3.6}$$

$$\varphi_2 = (\alpha k_1)^v + (\alpha k_{-n})^v + k_{-1}^v + k_n^v + \sum_{i=2}^{n-1} \left(k_i^v + k_{-i}^v \right) - C^v = 0 \tag{3.7}$$

The necessary conditions for a minimum of N under the given constraints read

$$\frac{\partial \Psi(k_{\pm i})}{\partial k_{\pm i}} = \frac{\partial N(k_{\pm i})}{\partial k_{\pm i}} + \lambda_1 \frac{\partial \varphi_1}{\partial k_{\pm i}} + \lambda_2 \frac{\partial \varphi_2}{\partial k_{\pm i}} = 0 \tag{3.8}$$

which represent together with Eqs.(1.2a) and (3.4) $2n+2$ equations for the unknown kinetic parameters as well as for the multipliers λ_1 and λ_2. One finds $\lambda_2 = N/v$.

Due to the nonlinear nature of equation system (3.8) the optimal solutions have to be found in general numerically. The special case of normalized reactant concentrations $S=P=1$ permits an analytical treatment. Here Eqs.(3.8) are solved with equal values for all forward rate constants and equal values for all backward rate constants, i.e. $k_i = k_+$ and $k_{-i} = k_-$, resp., and one gets

$$k_+ = q^{1/n}\left(n\left(1+q^{v/n}\right)\right)^{-1/v} , \quad k_- = \left(n\left(1+q^{v/n}\right)\right)^{-1/v} \qquad (3.9a,b)$$

Using the method of Lagrange multipliers one can show that for $v\to\infty$ the solutions of the *overall limit model* tend to those of the *separate limit model*.

Two-step mechanism. (see Scheme 2, Eq. 2.5) For the case $v=1$ the condition (3.4) allows to represent the feasible region of the kinetic constants as a 3-dimensional simplex (k-simplex, Fig. 3.1). At the corners $\Omega_{\pm i}$ one of the rate constants $k_{\pm i}$ is equal to unity while all other rate constants are vanishing. On the edges $\omega_{\pm i,\pm j}$ the sum of two constants $k_{\pm i}$ and $k_{\pm j}$ is equal to unity at vanishing values of the other rate constants .

The relation (1.2a) defines a 2-dimensional surface ζ of all possibles parameter values within the k-simplex. Except of the edges $\omega_{1,2}$ and $\omega_{-1,-2}$ all edges of the k-simplex belong to the surface ζ. The shape of surface ζ depends on the value of the equilibrium constant q. Its mathematical form is a saddle with the saddle-point at $k_1 = k_2$, $k_{-1} = k_{-2}$. All optimal solutions depending on S and P are located on this surface ζ. Using Eqs.(1.2a),(3.4) one may express k_1 and k_{-2} for $v=1$ as functions of k_{-1} and k_2. The equations resulting from $\partial N/\partial k_{-1} = \partial N/\partial k_2 = 0$ are used for numerical solution of the optimization problem. Fig. 3.1 gives a simplex-representation of the optimal rate constants for $q=1$. The lines connect solutions within the surface ζ obtained for equal values of S or P, resp. All optimal solutions for $0\le S < \infty$ and $0\le P < \infty$ are contained within a subset ζ' of surface ζ.

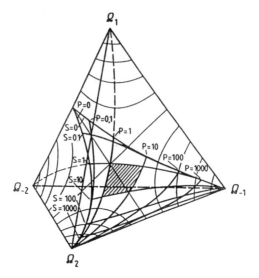

Figure 3.1
Simplex representation for the rate constants of the two-step mechanism with the corners $\Omega_{\pm i}\{k_{\pm i}=1\}$ and the edges $\omega_{\pm i,\pm j}\{k_{\pm i}+k_{\pm j}=1\}$ for $\nu=1$ and $q=1$. The accessible region for rate constants is represented by thin lines for the curves $k_1+k_2=$const. on the surface ζ (eq. (1.2)) and broken lines for $k_1=k_2$ or $k_{-1}=k_{-2}$, resp. The optimal rate constants are functions of the reactant concentrations. The thick lines represent in each case optimal rate constants for constant values of S $(0\le P<\infty)$ or P $(0\le S<\infty)$ calculated by numerical solution of the equations resulting from conditions $\partial N/\partial k_{-1}=0$ and $\partial N/\partial k_2=0$. Hatched area: solutions for $1\le S,P\le10$.

The set ζ of possible solutions is bounded by the lines for $S=0$ and $P=0$ as well as by the line for $S+P\to\infty$ (edge $\omega_{2,-1}$). The following conclusions are derived for $q=1$:

a) In the limit $S,P\to0$ the optimal solution approaches the point $k_1+k_{-2}=1$, $k_1=k_{-2}$, $k_{-1},k_2\to0$ which is located on the edge $\omega_{1,-2}$. This means that for low substrate and product concentrations maximal reaction rates are achieved by a strong binding of both reactants (*high reactant-affinity solution*).

b) Increasing concentration of substrate S is characterized by decreasing optimal values of k_1 and increasing optimal values of k_2 (*low S-affinity solutions*).

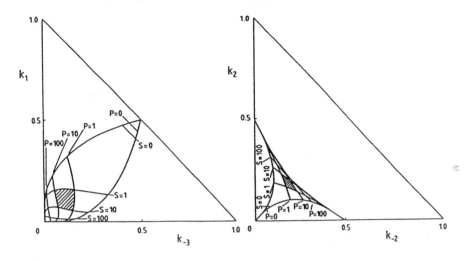

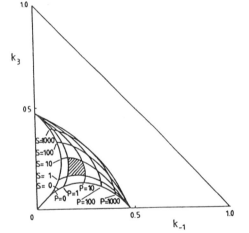

Figure 3.2
Optimal solutions for the three-step mechanism represented within the accessible space of the rate constants for the parameter values $v=1$ and $q=1$. Curves indicate solutions for constant values of S $(0 \le P < \infty)$ or P $(0 \le S < \infty)$. A) Second order rate constants, B) first order rate constants of the isomerization steps, C) first order rate constants for the steps of reactant release. Hatched area: solutions for $1 \le S, P \le 10$.

c) Increasing concentration of product P is characterized by decreasing optimal values of k_{-2} and increasing values of k_{-1} (*low P-affinity solutions*).

d) In the limits $S \to \infty$ and $P \to \infty$ one obtains as optimal solutions the points $k_2 \to 1$, $k_1, k_{-1}, k_{-2} \to 0$ (Ω_2) and $k_{-1} \to 1, k_1, k_2, k_{-2} \to 0$ (Ω_{-1}), resp.

e) For $S + P \to \infty$ the solutions are located on the edge $\omega_{2,-1}$ $(k_{-1} + k_2 = 1)$, i.e. both second order rate constants approach zero and a maximal reaction rate is achieved by a weak binding of the reactants.

Three-step mechanism. With Eqs. (1.2a) and (3.4) for $n=3$ two rate constants are expressed in terms of the four others which in turn are calculated numerically by a *Gradient method* from the conditions N=min. In Fig. 3.2 the optimal rate constants are represented for $q=1$ and $v=1$ in diagrams showing in each case the optimal rate constants of binding and dissociation steps and the rate constants k_2 and k_{-2} of the isomerization step. The plots are symmetrical with respect to the transformation $S \leftrightarrow P$, $k_1 \leftrightarrow k_{-3}$, $k_2 \leftrightarrow k_{-2}$, and $k_{-1} \leftrightarrow k_3$. The results obtained for the second order rate constants as well as for the first order rate constants of reactant release resemble those of the two-step mechanism. In particular, low concentrations of S and P imply high rate constants of binding processes and low values of the rate constants of the release processes. For $q>1$ one obtains diagrams (not shown) which resemble those of Fig. 3.2 but the regions of optimal parameter values are shifted to higher values of the forward rate constants.

References

[1]   Heinrich, R. & Hoffmann, E. 1991. Kinetic parameters of enzymatic reactions in states of maximal activity; an evolutionary approach. *J. theor. Biol.* **151**, 249-283.

[2]   Heinrich, R., Holzhütter, H.-G. & Schuster, S. 1987. A theoretical approach to the evolution and structural design of enzymatic networks; linear enzymatic chains, branched pathways and glycolysis of erythrocytes. *Bull. Math. Biol.* **49**, 539-595.

[3] Klipp, E. & Heinrich, R. 1994. Evolutionary optimization of enzyme kinetic parameters; Effect of constraints. *J. theor. Biol.*, in press.

[4] Wilhelm, T., Hoffmann-Klipp, E. & Heinrich, R. 1994. An evolutionary approach to enzyme kinetics: Optimization of ordered mechanisms. *Bull. Math. Biol.* **56**, 65-106.

Metabolic Control Theory

R. Heinrich

1. Introduction

Metabolic control theory investigates the effects of infinitesimally small parameter perturbations on the variables of metabolite systems. Originally, it was designed to quantify the concept of rate limitation in biochemical pathways [5,6,10,12] (and cf. [3,4]). Later on the analysis was extended by considering the parameter dependence of steady state concentrations of metabolites [6,19] and of other systemic variables (transient times, volumes, transmembrane potential etc. [2,8,14]). Further extensions concern relaxation processes [7,9], finite parameter perturbations [11,18] (cf. sections 4. and 5.), oscillations [1] and the consideration of metabolic "modules" [13,17].

Metabolic control theory can be applied to systems which are described by the following system of differential equations

$$\frac{dS_i}{dt} = \sum_{i=1}^{r} n_{ij} v_j \quad , \quad \frac{dS}{dt} = N v \quad , \quad v = v(S,p) \tag{1.1}$$

where $N = \{n_{ij}\}$ denotes the matrix of stoichiometric coefficients, $S = (S_1,...,S_n)$ the vector of metabolite concentrations, $v = (v_1,...,v_r)^T$ the vector of reaction rates, and $p = (p_1,...,p_m)$ kinetic parameters of the enzymes.

The equations

$$\mathbf{N} v(S, p) = 0 \tag{1.2}$$

define in an implicite manner the parameter dependence of the concentrations and the fluxes, i. e. the functions

$$S = S(p) \quad , \quad J = v\big(S(p), p\big) \tag{1.3a,b}$$

where J denotes the vector of steady state fluxes.

2. Definitions

For flux control coefficients the following definition has been proposed

$$C_{jk}^{J} = \left(\frac{v_k}{J_j} \frac{\Delta J_j}{\Delta v_k} \right)_{\Delta v_k \to 0} = \frac{v_k}{J_j} \frac{\partial J_j}{\partial v_k} \tag{2.1}$$

[6] where Δv_k denotes the change of the activity of a reaction k under isolated conditions, i.e. substrates, products as well as internal modifier concentrations are considered to be constant. Since mathematically the fluxes J_j cannot directly be expressed as functions of the rates v_k Eq. (2.1) has to be regarded as an abbreviated notation of

$$C_{jk}^{J} = \frac{v_k}{J_j} \frac{\partial J_j / \partial p_k}{\partial v_k / \partial p_k} \tag{2.2}$$

where p_k is a kinetic parameter which affects only reaction k directly, i. e.

$$\frac{\partial v_k}{\partial p_k} \neq 0 \quad , \quad \frac{\partial v_l}{\partial p_i} = 0 \text{ for any } l \neq k \quad . \tag{2.3}$$

Control coefficients calculated on the basis of formula (1.3) are independent of the special choice of the parameter p_k as long as condition (2.3) is fulfilled (cf. section 3.). This means that the coefficients (2.2) can be interpreted as the extent to which reaction k (rather than some parameter) controls a given steady state flux.

Concentration control coefficients have defined as follows

$$C_{ik}^{S} = \left(\frac{v_k}{S_i} \frac{\Delta S_i}{\Delta v_k} \right)_{\Delta v_k \to 0} = \frac{v_k}{S_i} \frac{\partial S_i / \partial p_k}{\partial v_k / \partial p_k} \tag{2.4}$$

[6]. Using for an enzymatic network the total enzyme concentrations E_k as perturbation parameters one obtains under the assumption

$$\frac{E_j}{v_j} \frac{\partial v_j}{\partial E_j} = \frac{\partial \ln v_j}{\partial \ln E_j} = 1 \tag{2.5}$$

which means that the reaction rates are linearly dependent on the enzyme concentrations for the control coefficients

$$C_{kj}^{J} = \frac{E_j}{J_k} \frac{\partial J_k}{\partial E_j} \quad , \qquad C_{ij}^{S} = \frac{E_j}{S_i} \frac{\partial S_i}{\partial E_j} \quad . \tag{2.6a,b}$$

(2.6a) corresponds to the definition of flux control coefficients proposed by Kacser and Burns [12].

Besides the coefficients (2.2, 2.4) non-normalized expressions have been introduced into metabolic control theory which are defined as follows

$$C_{jk}^{J} = \frac{\partial J_j / \partial p_k}{\partial v_k / \partial p_k} \quad , \qquad C_{ik}^{S} = \frac{\partial S_i / \partial p_k}{\partial v_k / \partial p_k} \tag{2.7a,b}$$

The control coefficients may be considered as elements of control matrices $\mathbf{C}^J = \{C_{jk}^J\}$ and $\mathbf{C}^S = \{C_{ik}^S\}$.

3. A general approach

The response of the system variables toward small parameter perturbations can be systematically analyzed in the following way. Implicit differentiation of Eq. (1.2) with respect to p yields

$$\mathbf{N}\frac{\partial v}{\partial S}\frac{\partial S}{\partial p}+\mathbf{N}\frac{\partial v}{\partial p}=\mathbf{0} \tag{3.1}$$

[7]. In case that the system does not contain conservation quantities and the steady state is asymptotically stable, the Jacobian $\mathbf{M}=\mathbf{N}\,\partial v/\partial S$ is non-singular. Therefore, one derives from Eq. (3.1) for the metabolite concentrations

$$\frac{\partial S}{\partial p}=-\left(\mathbf{N}\frac{\partial v}{\partial S}\right)^{-1}\mathbf{N}\frac{\partial v}{\partial p}=-\mathbf{M}^{-1}\mathbf{N}\frac{\partial v}{\partial p}\equiv\mathbf{R}^{S} \tag{3.2}$$

and for the steady state fluxes using Eq. (1.3b)

$$\frac{\partial J}{\partial p}=\frac{\partial v}{\partial p}+\frac{\partial v}{\partial S}\frac{\partial S}{\partial p}=\left[\mathbf{I}-\frac{\partial v}{\partial S}\left(\mathbf{N}\frac{\partial v}{\partial S}\right)^{-1}\mathbf{N}\right]\frac{\partial v}{\partial p}\equiv\mathbf{R}^{J} \quad . \tag{3.3}$$

(cf.[7]). It is seen that the response coefficients $\mathbf{R}^{S}$ and $\mathbf{R}^{J}$ can be split into two terms. The terms

$$\mathbf{C}^{S}=-\left(\mathbf{N}\frac{\partial v}{\partial S}\right)^{-1}\mathbf{N} \quad , \qquad \mathbf{C}^{J}=\mathbf{I}-\frac{\partial v}{\partial S}\left(\mathbf{N}\frac{\partial v}{\partial S}\right)^{-1}\mathbf{N}=\mathbf{I}+\frac{\partial v}{\partial S}\mathbf{C}^{S} \tag{3.4a,b}$$

depend via the stoichiometric coefficients on the systemic properties of the network but are independent of the special choice of the perturbation parameters. In contrast to that the term $\partial v / \partial p$ is independent of the systemic properties of the network and characterizes the effect of parameter changes on the individual reactions at fixed concentrations of the metabolites. If the parameter perturbations are infinitesimally small $(\Delta p = \delta p)$ one may use for Δv as well as for ΔS and ΔJ linear approximations, i. e.

$$\Delta v \approx \delta v = \left(\partial v / \partial p\right)\delta p \quad , \qquad \Delta S \approx \delta S = \left(\partial S / \partial p\right)\delta p \quad , \qquad \Delta J \approx \delta J = \left(\partial J / \partial p\right)\delta p \quad . \quad (3.5)$$

By definition, the elements of the vector δv are the immediate changes of the reaction rates after parameter perturbation at $t = t_0$ while the vectors δS and δJ contain as elements the final changes of the concentrations which are attained after adjustment of the system to the parameter pertubations for $t \to \infty$. With (3.4) it follows from Eqs. (3.2, 3.3)

$$\delta S = \mathbf{C}^S \, \delta v \quad , \quad \delta J = \mathbf{C}^J \, \delta v \qquad\qquad\qquad\qquad\qquad (3.6a,b)$$

i. e. the matrices $\mathbf{C}^S$ and $\mathbf{C}^J$ transform the vector δv into the vectors δS and δJ, respectively.

The partial derivatives of reaction rates v with respect to substrate concentrations or kinetic parameters, i.e. the elements of the matrices $\partial v / \partial S$ and $\partial v / \partial p$, respectively, are called elasticity coefficients. Using the notations

$$\varepsilon_{ij} = \frac{\partial v_i}{\partial S_j} \quad : \quad \varepsilon\text{-elasticities} \quad ,$$

$$\pi_{ij} = \frac{\partial v_i}{\partial p_j} \quad : \quad \pi\text{-elasticities} \qquad\qquad\qquad\qquad (3.7a,b)$$

Eqs. (3.2, 3.3) may be written as follows, respectively

$$R_{ik}^S = \sum_m C_{im}^S \, \pi_{mk} \qquad \text{or} \qquad \mathbf{R}^S = \mathbf{C}^S \boldsymbol{\pi} \qquad\qquad (3.8)$$

and

$$R_{ik}^J = \sum_m C_{im}^J \, \pi_{mk} \qquad \text{or} \qquad \mathbf{R}^J = \mathbf{C}^J \boldsymbol{\pi}. \qquad\qquad (3.9)$$

Reaction systems with conservation equations: Formulae (3.2) and (3.3) for the calculation of control coefficients have to be modified if metabolic systems with conservation equations are considered [15]. In this case the stoichiometric matrix has not full rank and therefore the Jacobian $\mathbf{M} = \mathbf{N}(\partial v / \partial S)$ is singular. By rearranging the rows of $\mathbf{N}$ so that the upper ρ rows are linearly independent, one may write

$$\mathbf{N} = \mathbf{L}\mathbf{N}^0 = \begin{pmatrix} \mathbf{I} \\ \mathbf{L}' \end{pmatrix} \mathbf{N}^0 = \mathbf{L}\mathbf{N}^0 \tag{3.10}$$

where $\mathbf{N}^0$ is a submatrix of $\mathbf{N}$ with a maximal number of linearly independent rows. $\mathbf{L}$ is called the link matrix [15]. The system equations of the reaction network assume the form

$$\frac{d}{dt} \begin{pmatrix} S_a \\ S_b \end{pmatrix} = \begin{pmatrix} \mathbf{I} \\ \mathbf{L}' \end{pmatrix} \mathbf{N}^0 v \tag{3.11}$$

where the vector of concentrations is split into two vectors $S_a = (S_1, \dots, S_\rho)^T$ and $S_b = (S_{\rho+1}, \dots, S_n)^T$. From Eq. (3.11) one derives the relations

$$\frac{dS_b}{dt} = \mathbf{L}' \frac{dS_a}{dt} \quad , \qquad S_b = \mathbf{L}' S_a + const. \tag{3.12a,b}$$

Accordingly, the metabolite concentrations S_i with $i > \rho$ may be expressed as linear combinations of the concentrations S_i with $i \leq \rho$.

Implicit differentiation of the linearly independent steady state equations with respect to the kinetic parameters yields

$$\mathbf{N}^0 \frac{\partial v}{\partial S_a} \frac{\partial S_a}{\partial p} + \mathbf{N}^0 \frac{\partial v}{\partial S_b} \frac{\partial S_b}{\partial S_a} \frac{\partial S_a}{\partial p} + \mathbf{N}^0 \frac{\partial v}{\partial p} = 0 \tag{3.13}$$

and from that with Eqs. (3.10 - 3.12)

$$\mathbf{N}^0 \frac{\partial v}{\partial S} \mathbf{L} \frac{\partial S_a}{\partial p} + \mathbf{N}^0 \frac{\partial v}{\partial p} = 0 \tag{3.14}$$

with $\partial v / \partial S = (\partial v / \partial S_a , \partial v / \partial S_b)$. From (3.14) one gets

$$\frac{\partial S_a}{\partial p} = -\left(\mathbf{M}^0\right)^{-1} \mathbf{N}^0 \frac{\partial v}{\partial p} \quad , \qquad \mathbf{M}^0 = \mathbf{N}^0 \frac{\partial v}{\partial S} \mathbf{L} \tag{3.15a,b}$$

where $\mathbf{M}^0$ denotes the Jacobian of the reduced system, which is a non-singular matrix. Taking into account (3.12) one gets from Eq. (3.17)

$$\frac{\partial S}{\partial p}=\mathbf{C}^S\frac{\partial v}{\partial p} \quad , \qquad \mathbf{C}^S=-\mathbf{L}\left(\mathbf{M}^0\right)^{-1}\mathbf{N}^0 \; . \tag{3.16a,b}$$

The matrix of flux control coefficients can be expressed as follows

$$\mathbf{C}^J=\mathbf{I}-\frac{\partial v}{\partial S}\mathbf{L}\left(\mathbf{M}^0\right)^{-1}\mathbf{N}^0 \; . \tag{3.17}$$

For systems without conservation equations $(\mathbf{L}=\mathbf{I})$ Eqs. (3.16b, 3.17) simplify to Eqs. (3.4a,b).

Normalized coefficients: It is often useful to transform the unscaled elasticity and control coefficients into a normalized form

$$(\mathrm{dg}J)^{-1}\varepsilon(\mathrm{dg}S)\rightarrow\varepsilon \quad , \qquad (\mathrm{dg}v)^{-1}\pi(\mathrm{dg}p)\rightarrow\pi \tag{3.18a,b}$$

$$(\mathrm{dg}J)^{-1}\,\mathbf{C}^J\,(\mathrm{dg}J)\rightarrow\mathbf{C}^J \quad , \qquad (\mathrm{dg}S)^{-1}\,\mathbf{C}^S(\mathrm{dg}J)\rightarrow\mathbf{C}^S \tag{3.19a,b}$$

where the reaction rates $(v = J)$ and substrate concentrations S in the reference state are used for normalization. The matrices ε and π contain the elements, respectively

$$\varepsilon_{ij}=\frac{S_j}{v_i}\frac{\partial v_i}{\partial S_j} \quad , \qquad \pi_{ik}=\frac{p_k}{v_i}\frac{\partial v_i}{\partial p_k} \tag{3.20a,b}$$

while the normalized control coefficients assume the form given in Eqs. (2.2, 2.4). Using normalized control and elasticity coefficients Eqs. (3.4a,b) are replaced by

$$\mathbf{C}^S=-\left(\mathbf{N}(\mathrm{dg}J)\varepsilon\right)^{-1}\left(\mathbf{N}(\mathrm{dg}J)\right) \quad , \qquad \mathbf{C}^J=\mathbf{I}+\varepsilon\mathbf{C}^S \; . \tag{3.21a,b}$$

The various control coefficients are not fully independent of each other. Two types of relationships between concentration control coefficients as well as flux control coefficients can be derived which are generally valid irrespective of the complexity of the considered reaction network. Some of the relationships named *summation theorems* reflect the structural properties of the reaction network and are independent of the kinetic parameters of the individual enzymes. In contrast to that the *connectivity theorems* relate the properties of the enzymes to the systemic behaviour.

Summation theorems: We consider first the normalized control coefficients defined in Eqs. (3.21a,b). Postmultiplication of Eqs. (3.21a,b) with the r-dimensional vector $\mathbf{1}=(1,...,1)^T$ yields under consideration of the steady state condition $\mathbf{N}\boldsymbol{J}=\boldsymbol{0}$

$$C^S\boldsymbol{1}=\boldsymbol{0} \ , \quad \text{or} \quad \sum_{j=1}^{r}C_{ij}^S=0 \tag{3.22}$$

and

$$C^J\boldsymbol{1}=\boldsymbol{1} \ , \quad \text{or} \quad \sum_{j=1}^{r}C_{kj}^J=1 \tag{3.23}$$

i. e. for each metabolic compound the sum of the concentration control coefficients is equal to zero while the control coefficients of a given steady state flux sum up to unity. Relation (3.22) represents the summation theorem for the concentration control coefficients [6] and relations (3.23) the summation theorem for the flux control coefficients [6, 12, 15].

It has been shown by Reder [15] that relations (3.22, 3.23) are special cases of *generalized summation theorems*. This may be best seen by using the matrices of unscaled control coefficients $\mathbf{C}^S$ and $\mathbf{C}^J$ which fulfil in the general case Eqs.(3.16b) and (3.17), respectively. Postmultiplication of these equations by the nullspace matrix $\mathbf{K}$ yields under consideration of $\mathbf{N}\mathbf{K}=\boldsymbol{0}$

$$\mathbf{C}^S\mathbf{K}=\boldsymbol{0} \ , \quad \mathbf{C}^J\mathbf{K}=\mathbf{K} \quad . \tag{3.24a,b}$$

The number of generalized summation relations equals the number of linearly independent k-vectors.

Connectivity theorems: Postmultiplication of Eqs.(3.16b, 3.17) by $(\partial v/\partial S)\mathbf{L}$ yields under consideration of (3.15b)

$$\mathbf{C}^S\frac{\partial v}{\partial S}\mathbf{L}=-\mathbf{L} \ , \quad \mathbf{C}^J\frac{\partial v}{\partial S}\mathbf{L}=\boldsymbol{0} \tag{3.25a,b}$$

respectively, which are the connectivity theorems of metabolic control theory [12,19,15]. Using normalized control and elasticity coefficients they may be rewritten in the following form

$$\mathbf{C}^S \, \varepsilon (\mathbf{dg}S)^{-1} \mathbf{L} = (\mathbf{dg}S)^{-1} \mathbf{L} \quad , \qquad \mathbf{C}^J \, \varepsilon (\mathbf{dg}S)^{-1} \mathbf{L} = 0 \ . \tag{3.26a,b}$$

For systems without conservation relations $(\mathbf{L} = \mathbf{I})$ Eqs. (3.26a,b) simplify to the relations

$$\sum_{j=1}^{r} C_{ij}^S \, \varepsilon_{jk} = -\delta_{ik} \quad , \qquad \sum_{j=1}^{r} C_{ij}^J \, \varepsilon_{jk} = 0 \tag{3.27a,b}$$

originally derived by Kacser & Burns [12] and Westerhoff & Chen [19], respectively.

It has been shown that the connectivity and summation theorems can be used to calculate the flux and concentration control coefficients in terms of elasticities and stoichiometry [7,8, 15].

4. Time dependent control coefficients

Hitherto, the analysis of parameter perturbations was confined to steady states, i.e. the time dependence of the system behaviour during the relaxation period was not studied. Obviously, this restriction can be misleading at the interpretation of experimental results. In particular, it can be practically impossible to approach a steady state in reasonable times. In the present section control theory is extended to time dependent states in the neighbourhood of a stable steady state (cf. [9]).

Suppose that for $t < 0$ S^0 is a stable steady-state solution of system (1.1) for a given parameter vector p^0. At time zero the parameter is perturbed and takes the value p for all positive times. For p close to p^0, the solution $S(t, p)$ of Eq. (1.1) can be approximated by

$$S(t, p) = S^0 + \frac{\partial S}{\partial p} (t, p^0)(p - p^0). \tag{4.1}$$

The time-dependent flux vector $J(t,p)$ is defined by

$$J(t,p) = v[S(t,p),p],$$ (4.2)

for p close to p^0, it can be approximated as

$$J(t,p) = J^0 + \frac{\partial J}{\partial p}(t,p^0)(p-p^0)$$ (4.3)

with $J^0 = v(S^0,p)$. It follows from (1.1) that $\partial S / \partial p$ is the matrix solution of

$$\frac{d}{dt}\left(\frac{\partial S}{\partial p}\right) = \left(N\frac{\partial v}{\partial S}\right)\frac{\partial S}{\partial p} + N\frac{\partial v}{\partial p}$$ (4.4)

with $\partial S / \partial p(t=0,p^0) = 0$. From (4.2) one derives

$$\frac{\partial J}{\partial p} = \frac{\partial v}{\partial S}\frac{\partial S}{\partial p} + \frac{\partial v}{\partial p}.$$ (4.5)

Let us assume that the rows of the stoichiometric matrix N are linearly independent, i.e. that there are no conservation relationships for the metabolite concentrations. Since the references state is assumed to be stable, all the eigenvalues of the Jacobian $M = N(\partial v / \partial S)$ have negative real parts; in particular, M is invertible. The formal solution of the linear differential equation system (4.4) reads

$$\frac{\partial S}{\partial p}(t,p^0) = C^S(t)\frac{\partial v}{\partial p} \quad , \qquad C^S(t) = [\exp(t M) - I]M^{-1}N$$ (4.6a,b)

where I denotes the $n \times n$ identity matrix. From (4.5, 4.6b) one gets

$$\frac{\partial J}{\partial p}(t,p^0) = C^J(t)\frac{\partial v}{\partial p} \quad , \qquad C^J(t) = I + \frac{\partial v}{\partial S}C^S(t) = I + \frac{\partial v}{\partial S}[\exp(t M) - I]M^{-1}N.$$

(4.7a,b)

In the present case the control matrices $\mathbf{C}^S(t)$ and $\mathbf{C}^J(t)$ are time dependent operators which transform the initial perturbations δv of the reaction rates into the concentration and flux variations δS and δJ at time t.

Since the eigenvalues of $\mathbf{M}$ have negative real parts, the matrix $\left[\exp(t\mathbf{M})\right]$ approaches zero when t tends to infinity. Therefore, one gets for $t \to \infty$ from (4.6b) and (4.7b) the usual unscaled time-independent concentration and flux control matrices $\mathbf{C}^S$ and $\mathbf{C}^J$.

If the network is such that some metabolite concentration sums are conserved the rows of the stoichiometric matrix $\mathbf{N}$ are not linearly independent. In this case one derives that the time-dependent control matrices may be expressed as

$$\mathbf{C}^S(t) = \mathbf{L}\left[\exp(t\mathbf{M}^0) - \mathbf{I}\right](\mathbf{M}^0)^{-1}\mathbf{N}^0 \tag{4.8}$$

$$\mathbf{C}^J(t) = \mathbf{I} + \frac{\partial v}{\partial S}\mathbf{L}\left[\exp(t\mathbf{M}^0) - \mathbf{I}\right](\mathbf{M}^0)^{-1}\mathbf{N}^0 \ . \tag{4.9}$$

It follows immediately from Eqs (4.8, 4.9) that the matrices of control coefficients fulfil the following relationships

$$\mathbf{C}^J(t_2)\mathbf{C}^J(t_1) = \mathbf{C}^J(t_1 + t_2) \ , \quad \mathbf{C}^S(t_2)\varepsilon\mathbf{C}^S(t_1) + \mathbf{C}^S(t_1) + \mathbf{C}^S(t_2) = \mathbf{C}^S(t_1 + t_2)$$

(4.10a,b)

[16]. Summation and connectivity theorems can also be derived for the time dependent control coefficients. One derives

$$\mathbf{C}^S(t)\mathbf{K} = \mathbf{0} \ , \qquad \mathbf{C}^J(t)\mathbf{K} = \mathbf{K} \tag{4.11a,b}$$

and

$$\mathbf{C}^S(t)\frac{\partial v}{\partial S}\mathbf{L} = \mathbf{L}\left[\exp(t\mathbf{M}^0) - \mathbf{I}\right] \ , \qquad \mathbf{C}^J(t)\frac{\partial v}{\partial S}\mathbf{L} = \frac{\partial v}{\partial S}\mathbf{L}\exp(t\mathbf{M}^0) \tag{4.12a,b}$$

Eqs. (4.11, 4.12) reveal the interesting fact that time enters explicitly only the connectivity theorems but not the summation theorems.

5. A second order approach

Owing to relation (3.5) control coefficients describe the response of the system variables to infinitesimally small rate perturbations. In this sense they characterize local properties of a biochemical network in the vicinity of a stable steady state. Since regulation of enzymes by effectors can cause substantial changes of their activities it may be questioned to what extent the effects of relevant parameter perturbations can be described by the linear approximation [11,18]. In the present section it is analyzed how metabolic control theory can be extended to give more accurate predictions for the changes of the system variables than the simple linear approximation.

The analysis is based on the Taylor expansion

$$\Delta S = \frac{\partial S}{\partial p}(p)\Delta p + \frac{1}{2}\sum_{\alpha,\beta}\frac{\partial^2 S}{\partial p_\alpha \,\partial p_\beta}(p)\Delta p_\alpha \Delta p_\beta \tag{5.1a}$$

$$\Delta J = \frac{\partial J}{\partial p}(p)\Delta p + \frac{1}{2}\sum_{\alpha,\beta}\frac{\partial^2 J}{\partial p_\alpha \,\partial p_\beta}(p)\Delta p_\alpha \Delta p_\beta \quad . \tag{5.1b}$$

It has been shown in section 3. that the first order terms $\partial S/\partial p$ and $\partial J/\partial p$ can be obtained by differentiation of the steady state equation (1.2, 1.3b). In a similar way the second order terms are obtained by differentiating Eqs. (1.2, 1.3b) twice with respect to p resulting in

$$\frac{\partial^2 S}{\partial p_\alpha \,\partial p_\beta} = \sum_{i,l,m=1}^{r}\sum_{j,k=1}^{n} C_{ai}^S \frac{\partial^2 v_i}{\partial S_j \,\partial S_k} C_{jl}^S C_{km}^S \frac{\partial v_l}{\partial p_\alpha}\frac{\partial v_m}{\partial p_\beta}$$
$$+ \sum_{i,k=1}^{r}\sum_{j=1}^{n} C_{ai}^S \left(\frac{\partial^2 v_i}{\partial S_j \,\partial p_\beta} C_{jk}^S \frac{\partial v_k}{\partial p_\alpha} + \frac{\partial^2 v_i}{\partial S_j \,\partial p_\alpha} C_{jk}^S \frac{\partial v_k}{\partial p_\beta} \right)$$
$$+ \sum_{i=1}^{r} C_{ai}^S \frac{\partial^2 v_i}{\partial p_\alpha \,\partial p_\beta} \tag{5.2}$$

and

$$
\frac{\partial^2 J_b}{\partial p_\alpha \partial p_\beta} = \sum_{i,l,m=1}^{r} \sum_{j,k=1}^{n} C_{bi}^{J} \frac{\partial^2 v_i}{\partial S_j \partial S_k} C_{jl}^{S} C_{km}^{S} \frac{\partial v_l}{\partial p_\alpha} \frac{\partial v_m}{\partial p_\beta}
$$

$$
+ \sum_{i,k=1}^{r} \sum_{j=1}^{n} C_{bi}^{J} \left(\frac{\partial^2 v_i}{\partial S_j \partial p_\beta} C_{jk}^{S} \frac{\partial v_k}{\partial p_\alpha} + \frac{\partial^2 v_i}{\partial S_j \partial p_\alpha} C_{jk}^{S} \frac{\partial v_k}{\partial p_\beta} \right)
$$

$$
+ \sum_{i=1}^{r} C_{bi}^{J} \frac{\partial^2 v_i}{\partial p_\alpha \partial p_\beta} \tag{5.3}
$$

[11]. In addition to the quantities of the linear theory (control coefficients, ε - and π - elasticities) the second order terms contain the following second derivatives of the individual rates

$$
\varepsilon_{ijk} = \frac{\partial^2 v_i}{\partial S_j \partial S_k} \quad : \quad \text{second-order } \varepsilon \text{-elasticities} \tag{5.4a}
$$

$$
\pi_{i\alpha\beta} = \frac{\partial^2 v_i}{\partial p_\alpha \partial p_\beta} \quad : \quad \text{second-order } \pi \text{-elasticities} \tag{5.4b}
$$

and

$$
(\varepsilon - \pi)_{ij\alpha} = \frac{\partial^2 v_i}{\partial S_j \partial p_\alpha} \quad : \quad \text{mixed second-order} (\varepsilon - \pi) \text{-elasticities} \tag{5.4c}
$$

Hence, the local characterization of the individual rates has to be extended to the second-order elasticity coefficients in order to determine the response of the system variables to parameter perturbations in the quadratic approximation. Owing to the occurrence of mixed derivatives of the reaction rates with respect to metabolite concentrations, a general definiton of parameter-independent second-order control coefficients is not possible. In particular, the parameter perturbations cannot be replaced by the rate perturbations as independent variables in Eqs. (5.1a,b). Therefore, the perturbation parameters do not merely play a technical role as in the linear theory. Another interesting feature of the second-order terms is that they contain, besides derivatives characterizing the influence of a single reaction on a steady state variable, also mixed derivatives, e. g. $\partial^2 S_a / \partial p_\alpha \partial p_\beta$

where p_α and p_β may belong to different rate equations. The effects of simultaneous perturbations of several rates are thus not simply approximated as the sum of the individual effects as in the case of the linear theory. A first discussion of Eqs. (5.2, 5.3) becomes easier if reaction-specific perturbation parameters are considered which enter the rate laws linearly. It turns out that, under this condition, the following definition of second-order flux control coefficients is appropriate

$$D^J_{b\alpha\beta} = \frac{1}{2}\frac{\partial^2 J_b}{\partial p_\alpha p_\beta} \bigg/ \frac{\partial v_\alpha}{\partial p_\beta}\frac{\partial v_\beta}{\partial p_\beta} \tag{5.5}$$

and that one can introduce the rate perturbations in place of the parameter perturbations as independent variables. One gets

$$\Delta J_j = \sum_{\alpha=1}^{r} C^J_{j\alpha}\Delta v_\alpha + \sum_{\alpha,\beta}^{r} D^J_{j\alpha\beta}\Delta v_\alpha \Delta v_\beta \tag{5.6}$$

with

$$D^J_{b\alpha\beta} = \frac{1}{2}\left\{\sum_{i=1}^{r}\sum_{j,k=1}^{n}C^J_{bi}\varepsilon_{ijk}C^S_{j\alpha}C^S_{k\beta}\right.$$
$$\left. +\frac{1}{v_\alpha}C^J_{b\alpha}C^J_{\alpha\beta}+\frac{1}{v_\beta}C^J_{b\beta}C^J_{\beta\alpha}-\delta_{\alpha\beta}\left(C^J_{b\alpha}/v_\alpha+C^J_{b\beta}/v_\beta\right)\right) \tag{5.7}$$

Similar equations are obtained for the second order concentration control coefficients.

For the second-order control coefficients *summation theroerms* exist similar to those of the linear theory. Denoting by k_γ and k_δ two column vectors of **K** one obtains

$$\sum_{\alpha,\beta}^{r}D^J_{b\alpha\beta}k_{\alpha\gamma}k_{\beta\delta}=0 \quad , \quad \sum_{\alpha,\beta}^{r}D^S_{a\alpha\beta}k_{\alpha\gamma}k_{\beta\delta}=0 \tag{5.8}$$

In contrast to the coefficients of the linear theory the summation relationships for the second-order control coefficients for metabolite concentrations and fluxes have the same form.

References

[1]   Baconnier PF, Pachot P and Demongeot J, An attempt to generalize the control
        coefficient concept. *Journal of Biological Systems* **1** (1993) 335-347.
[2]   Brumen M and Heinrich R, A metabolic osmotic model of human erythrocytes.
        BioSystems **17** (1984) 155-169.
[3]   Cornish-Bowden A & Cardenas ML (eds.), *Control of Metabolic Processes* (1990)
        Plenum Press (New York).
[4]   Fell DA, Metabolic control analysis: a survey of its theoretical and experimental
        development. *Biochem. J.* **286** (1992), 313-330.
[5]   Heinrich R and Rapoport TA, Linear theory of enzymatic chains:
        Acta biol.med.germ. **31** (1973) 479-494.
[6]   Heinrich R and Rapoport TA, A linear steady-state treatment of enzymatic chains.
        General properties, control and effector strength. *Eur. J. Biochem.* **42** (1974) 89-95.
[7]   Heinrich R and Rapoport TA.The regulatory principles of the glycolysis of
        erythrocytes in vivo and in vitro, *Symp. Biol. Hung.* **18** (1974) 173-212.
[8]   Heinrich R and Rapoport TA, Mathematical analysis of multienzyme systems.
        II. Steady state and transient control. *BioSystems* **7** (1975) 130-136.
[9]   Heinrich R and Reder C, Metabolic control analysis of relaxation processes.
        J. theor. Biol. **151** (1991) 343-350.
[10]   Higgins JJ, Dynamics and control in cellular reactions. In *Control of Energy
        metabolism*, ed. by Chance B, Estabrook RW and Williamson JR (Academic Press,
        New York and London, 1965).
[11]   Höfer T and Heinrich R, A second-order approach to metabolic control analysis.
        J. theor. Biol. **162** (1993) 85-102.
[12]   Kacser H and Burns JA, The control of flux. *Symp. Soc. Exp. Biol.* **27** (1973)
        65-105.
[13]   Kahn D and Westerhoff HV, Control theory of regulatory cascades.
        J. theor. Biol. **159** (1991) 255-285.
[14]   Meléndez-Hevia E, Torres NV, Sicilia J and Kacser H, Control analysis
        of transition times in metabolic systems. *Biochem J.* **265** (1990) 195-202.
[15]   Reder C, Metabolic control theory: a structural approach. *J. theor. Biol.* **135**
        (1988) 175-201.
[16]   Schuster S and Heinrich R, The definitions of metabolic control analysis revisited.
        BioSystems **27** (1992) 1-15.
[17]   Schuster S, Kahn D and Westerhoff HV, Modular analysis of the control of
        complex metabolic pathways. *Biophys. Chem.* **48** (1993) 1-17.
[18]   Small RJ and Kacser H, Response of metabolic systems to large changes in enzyme
        activities and effectors, 1.The linear treatment of unbranched chains.
        Eur. J. Biochem. **213** (1993) 613-624.
[19]   Westerhoff HV and Chen YD, How do enzyme activities control metabolite
        concentrations? *Eur. J. Biochem.* **142** (1984) 425-430.

Determining Elementary Modes of Functioning in Biochemical Reaction Networks at Steady State

S. Schuster, C. Hilgetag, and R. Schuster

1. Introduction

In the modelling of biochemical reaction systems, analysis of steady states plays an important role, because virtually stationary regimes are frequently encountered in experimental settings and under *in-vivo* conditions. Since in steady states, all intermediates have to be balanced with respect to inputs and outputs, the vector, V, of stationary reaction rates (called fluxes below) has to be situated in the null-space (kernel) of the stoichiometry matrix, N (cf. Clarke, 1980, 1988; Thomas and Fell, 1993),

$$N\,V = 0 \quad \Leftrightarrow \quad V \in \ker(N). \tag{1}$$

Here, 0 denotes the null matrix. For basic concepts of linear algebra, the reader is referred to Groetsch and King (1988). Consider, for example, the branched reaction system depicted in Fig. 1. It has the stoichiometry matrix

$$N = (1 \ \ -1 \ \ -1). \tag{2}$$

Accordingly, for this system Eq. (1) means that $v_1 - v_2 - v_3 = 0$. This equation determines a plane in the three-dimensional space spanned by v_1, v_2, and v_3. This plane is the nullspace of N for this example.

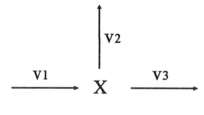

Figure 1 Branched reaction system composed of three mono-molecular reactions. X stands for an intermediate. The v_i denote reaction rates. The substrate of the first reaction and the products of reactions 2 and 3 are not shown, because they are considered as external metabolites (i.e., their concentrations are assumed to be constant).

The null-space of the stoichiometry matrix can be conceived of as spanned by basis vectors. It is of interest to find those basis vectors that can be interpreted in terms of functions of the particular biochemical system (for example, synthesis of ATP or a particular amino acid). This amounts to a decomposition of the biochemical system into the principal routes involved, i.e., the simplest biochemically meaningful steady-state flux vectors possible. All other admissible flux patterns should then be a superposition of these elementary modes.

Leiser and Blum (1987) proposed to identify "fundamental modes" of systems containing substrate cycles, by invoking that any steady-state flux pattern could be decomposed as a linear superposition of cyclic and non-cyclic modes and that these modes are all thermodynamically "realizable". Sign constraints for fluxes are frequent in modelling studies, in particular if the reaction rates are defined as unidirectional rates (Clarke, 1980, 1988). As for virtually irreversible reactions, such as many kinase reactions, also the net rates can be assumed to be non-negative. Fell (1990, 1993) proposed to define fundamental modes by a proper choice of basis vectors of the null-space. He observed that this method meets with the difficulties that irreversibility constraints may be violated and that there may be more relevant ways how to connect the inputs to the outputs of the system than vectors are needed to form a basis.

In the present contribution, we give a mathematical definition of the concept of elementary mode and outline an algorithm for detecting these modes in systems of any complexity. This analysis is tangential to methods developed by Seressiotis and Bailey (1988) and Mavrovouniotis *et al.* (1990, 1992) for constructing, by computer, metabolic routes leading from a given substrate to a given product. In those methods, a distinction is made between required substrate, required product, allowed substrates, allowed products and allowed intermediates. In determining the elementary modes, we only distinguish between internal and external metabolites.

2. Definition of elementary modes

Consider again the system shown in Fig. 1. Given that all reactions are irreversible in the direction indicated by the arrows, it is plausible to attach to the system two elementary modes represented by the flux vectors

$$\mathbf{V}^{(1)} = (1 \quad 1 \quad 0)^T, \quad \mathbf{V}^{(2)} = (1 \quad 0 \quad 1)^T, \tag{3a,b}$$

because each vector of steady-state fluxes not violating the irreversibility conditions can be written as non-negative linear combination of $\mathbf{V}^{(1)}$ and $\mathbf{V}^{(2)}$. The units of the components of $\mathbf{V}^{(i)}$ are arbitrary. Each of the two elementary modes represents the biosynthetic route from the substrate to one of the products.

As the term "elementary mode" should be so general that two vectors that differ by a positive factor are not considered different, elementary modes are classes of flux vectors that convert into one another by multiplication by a positive scalar. Two elementary modes that differ by a negative factor are to be considered different, because opposite directions of flux correspond to different functions, *e.g.* ATP production and proton transport.

By decomposing the flux vector into a subvector, $\mathbf{V}^{irr}$, corresponding to the (physiologically) irreversible reactions and a subvector, $\mathbf{V}^{rev}$, corresponding to the reversible reactions, we can write the irreversibility constraint as

$$\mathbf{V}^{irr} \geq \mathbf{0}. \tag{4}$$

By reversible reactions, we here mean reactions that can proceed in either direction under physiological conditions, such as the reactions shared by glycolysis and gluconeogenesis.

DEFINITION 1. A flux mode, $\mathbf{M}$, is defined as the set

$$\mathbf{M} = \{\mathbf{V} \in \mathbf{R}^r \mid \mathbf{V} = \lambda \mathbf{V}^*, \lambda > 0\}, \tag{5}$$

where $\mathbf{V}^*$ is an r-vector (r denotes the number of reactions in the system) that is unequal to the null vector and fulfils the following two conditions,

(C1) Steady-state condition. $\mathbf{V}^*$ satisfies Eq. (1).

(C2) Sign restriction. The subvector, $\mathbf{V}^{irr}$, of $\mathbf{V}^*$ fulfils inequality (4).

Remark: In what follows, we shall often denote a flux mode by simply giving a representative of $\mathbf{M}$, the elements of which are reduced to integers the absolute values of which are as small as possible.

DEFINITION 2. A flux mode $\mathbf{M}$ with a representative $\mathbf{V}^*$ is called elementary (flux) mode if, and only if, $\mathbf{V}^*$ fulfils the condition

(C3) Simplicity. There exists no couple of vectors $\mathbf{V}'$, $\mathbf{V}''$ (unequal to the null vector) with the following properties:

(i) $\mathbf{V}^*$ is a non-negative linear combination of $\mathbf{V}'$ and $\mathbf{V}''$,

$$\mathbf{V}^* = \lambda_1 \mathbf{V}' + \lambda_2 \mathbf{V}'', \quad \lambda_1, \lambda_2 > 0. \tag{6}$$

(ii) $\mathbf{V}'$, $\mathbf{V}''$ obey restrictions (C1) and (C2),

(iii) both $\mathbf{V}'$ and $\mathbf{V}''$ contain zero elements wherever $\mathbf{V}^*$ does so, and they include at least one additional zero component each.

Remark: Condition (C3, iii) is a formalization of the concept of genetic independence used by Seressiotis and Bailey (1988). If this condition were phrased by just saying that both $\mathbf{V}'$ and $\mathbf{V}''$ are to have more zero components than $\mathbf{V}^*$, any flux mode that can be decomposed into two simpler ones would not be elementary, even if these simpler modes involve a reaction that is not involved in $\mathbf{V}^*$. In this case, however, $\mathbf{V}'$ and $\mathbf{V}''$ would be genetically independent of $\mathbf{V}^*$ since an extra enzyme needs to be expressed.

DEFINITION 3. A flux mode $\mathbf{M}$ is called reversible flux mode if, and only if, $\mathbf{M}' = \{-\mathbf{V} \mid \mathbf{V} \in \mathbf{M}\}$ is a flux mode as well. Otherwise, $\mathbf{M}$ is called irreversible flux mode.

The same distinction can then be made for elementary modes.

If we assume that in the example depicted in Fig. 1, only reaction 2 is irreversible, we obtain the elementary flux vectors $(1\ 1\ 0)^T$, $(0\ 1\ -1)^T$, $(1\ 0\ 1)^T$, and $(-1\ 0\ -1)^T$. The former two represent irreversible modes and the latter two, reversible modes.

The reaction system shown in Fig. 2 containing a cycle can serve for illustration of condition (C3). The vector $(0\ 1\ -1\ 2\ 1\ 2)^T$ is no elementary mode since it is the sum of $(0\ 1\ -1\ 1\ 0\ 1)^T$ and $(0\ 0\ 0\ 1\ 1\ 1)^T$, both of which have a zero at the first position and have one, respectively two, zeros more than the first vector.

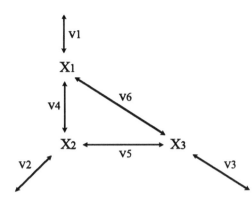

Figure 2 Reaction system containing a cycle. The X_i are intermediates. Reactions 1, 2, and 3 lead to external metabolites. This scheme can stand, for example, for the pyruvate/oxaloacetate/phos-phoenolpyruvate cycle (in that case, reactions 4, 5 and 6 involve ADP, ATP and inorganic phosphate as external species) and was also studied by Leiser and Blum (1987).

3. Detecting all elementary modes for a given system

3.1. Irreversible reactions

In case all reactions are irreversible, we can write, without loss of generality,

$$\mathbf{V} \geq 0. \tag{7}$$

Together with the steady-state equation $\mathbf{N}\ \mathbf{V} = \mathbf{0}$, this is a linear homogeneous equation/ inequality system. In convex analysis, it is shown that the region determined by such a system is a pointed convex polyhedral cone, $\mathbf{K}$ (Rockafellar, 1970; Nožička, 1974). An exemplifying cone is depicted in Fig. 3. Convex analysis also states that every point of such a cone is a non-negative combination of extreme vectors, $\mathbf{f}^{(k)}$, which are unique up to multiplication by positive scalars

$$\mathbf{K} = \{\mathbf{V}\colon \mathbf{V} = \sum_{k=1}^{p} \eta_k \mathbf{f}^{(k)} \ , \ \eta_k \geq 0 \ \forall k \} \ . \tag{8}$$

This relation has been studied also by Clarke (1980). In biochemical terms, it means that every admissible flux pattern in the system can be written as superposition of simple flux patterns, which cannot be decomposed any further.

It is worth mentioning that cone $\mathbf{K}$ can have any dimension from zero to r-rank($\mathbf{N}$), depending on how the null-space is situated relative to the positive orthant. (For the concept of rank, see Groetsch and King, 1988). Furthermore, the number of generating vectors may be greater than the dimension of the cone (cf. Fig. 3).

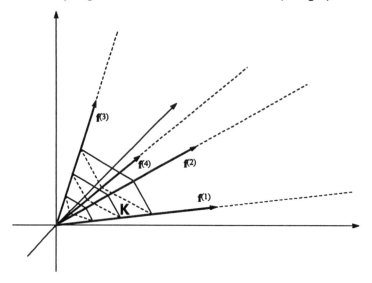

Figure 3 Pointed convex polyhedral cone, **K**, spanned by four generating vectors (edges), $f^{(1)}$ to $f^{(4)}$. Note that the cone is infinitely long. The space of reaction rates has been projected down into a three-dimensional space. Therefore, the coordinate axes represent linear combinations of reaction rates. The apex of the cone corresponds to the situation that all reaction rates in the system are zero and lies, therefore, in the origin of coordinates.

In systems with irreversible reactions only, all elementary modes are irreversible. A further important relationship is expressed in the following

THEOREM. If all reactions are irreversible, the generating vectors of the flux cone **K** determined by Eq. (8) constitute a complete set of representatives of the elementary modes under the sign restriction (7).

The basic idea of the proof is that all generating vectors satisfy, by definition, conditions (C1) and (C2). Since generating vectors of pointed convex cones are those vectors of the cone that cannot be expressed as non-negative linear combination of other vectors belonging to the cone (Rockafellar, 1970), they fulfil condition (C3). Hence, all generating vectors are representatives of elementary modes. The proof of the converse assertion is similar.

Convex analysis provides an algorithm for calculating generating vectors of convex cones (Nožička, 1974). We adapted that algorithm so as to calculate the generating vectors of the flux cone for the situation that all reactions are irreversible (Schuster and Heinrich, 1991; Schuster and Schuster, 1993). Owing to the above theorem, all of these vectors are representatives of elementary modes.

Let us illustrate this algorithm by the simple system shown in Fig. 1. The initial tableau is constructed by augmenting the transposed stoichiometry matrix with the identity matrix,

$$\mathbf{T}^{(0)} = \begin{pmatrix} 1 & 0 & 0 & \vdots & 1 \\ 0 & 1 & 0 & \vdots & -1 \\ 0 & 0 & 1 & \vdots & -1 \end{pmatrix}. \tag{9}$$

On calculating $\mathbf{T}^{(1)}$, we have to sum up the first and second rows and the first and third rows, because the respective entries have opposite sign,

$$\mathbf{T}^{(1)} = \begin{pmatrix} 1 & 1 & 0 & \vdots & 0 \\ 1 & 0 & 1 & \vdots & 0 \end{pmatrix}. \tag{10}$$

The two rows on the left-hand side are proper representatives of the elementary modes.

3.2. Reversible and irreversible reactions

Now, we consider that some of the reactions are not restricted to have a fixed orientation of net flux. Accordingly, relations (1) and (4) apply. This equation/inequality system determines a convex polyhedral cone, $\mathbf{C}$, in the V-space. Now the cone is not necessarily pointed, that is, two vectors of the cone may make an angle of 180 degree. Convex analysis states that every point of such a cone is a non-negative combination of fundamental vectors, $\mathbf{f}^{(k)}$, and basis vectors, $\mathbf{b}^{(m)}$,

$$\mathbf{C} = \{\mathbf{V}: \mathbf{V} = \sum_{k=1}^{p} \eta_k \mathbf{f}^{(k)} + \sum_{m=1}^{s} \lambda_m \mathbf{b}^{(m)} , \quad \eta_k, \lambda_m \geq 0 \ \forall k,m \}. \tag{11}$$

The basis vectors, $\mathbf{b}^{(m)}$, are those extreme rays of cone $\mathbf{C}$ for which also the negative vector, $-\mathbf{b}^{(m)}$, is contained in $\mathbf{C}$.

We can distinguish the three following cases:

a) The system has only irreversible elementary modes, although some reactions of the system are considered reversible. For the system in Fig. 1 with only one reaction treated reversible, no reversible elementary mode occurs.

b) There are irreversible as well as reversible elementary modes. Consider the reaction system of Fig. 1 with reactions 1 and 3 assumed to be reversible. It then has the reversible, respectively irreversible, elementary modes

$$\mathbf{b}^{(1)} = \begin{pmatrix} 1 \\ 0 \\ 1 \end{pmatrix}, \qquad \mathbf{b}^{(2)} = \begin{pmatrix} -1 \\ 0 \\ -1 \end{pmatrix}, \qquad (12a,b)$$

$$\mathbf{f}^{(1)} = \begin{pmatrix} 1 \\ 1 \\ 0 \end{pmatrix}, \qquad \mathbf{f}^{(2)} = \begin{pmatrix} 0 \\ 1 \\ -1 \end{pmatrix}. \qquad (13a,b)$$

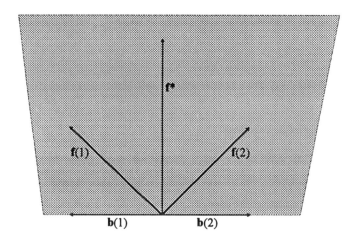

Figure 4 Cone of admissible steady-state fluxes for the system shown in Fig. 1, modified in that reactions 1 and 3 are assumed to be reversible. Notations: b(1) and b(2), basis vectors; f*, fundamental vector orthogonal to the basis vectors; f(1) and f(2), fundamental vectors representing elementary modes. The cone is in this case a half-plane, confined by a line containing the vectors b(1) and b(2). Note that this half-plane is actually situated obliquely in the space of reaction rates, crossing the positive orthant. The coordinate axes are not shown for the sake of clarity.

The cone **C** for this system is depicted in Fig. 4. One irreversible elementary mode together with the reversible elementary modes would be sufficient to span cone **C**. Thus, this example shows that for systems containing some reversible reactions, there may be more elementary modes than generating vectors. Moreover, choice of the fundamental vector is now not unique; we can choose $\mathbf{f}^{(1)}$, $\mathbf{f}^{(2)}$, or any non-negative linear combination of these and of the basis vectors, e.g., the vector $\mathbf{f}^* = (-1 \ \ 2 \ \ 1)^T$, which is orthogonal to the basis vectors, but is not related to any elementary mode.

c) There are only reversible elementary modes (and only basis vectors). Importantly, the number of reversible elementary modes need not equal the number of basis vectors of the null-space, as can be seen in the system shown in Fig. 1 with all reactions considered reversible, which has $2*3 = 6$ reversible elementary modes, while $r - \text{rank}(N) = 2$.

An algorithm for detecting the elementary modes of systems containing reversible reactions can be developed by generalizing the algorithm presented earlier (Schuster and Schuster, 1993). We start from a tableau, $\mathbf{T}^{(0)}$, containing the transposed stoichiometry matrix and the identity matrix,

$$\mathbf{T}^{(0)} = \begin{pmatrix} \mathbf{B}^{(0)} \\ \mathbf{F}^{(0)} \end{pmatrix} = \begin{pmatrix} \mathbf{I} & \mathbf{0} & \mathbf{N}_{rev}^T \\ \mathbf{0} & \mathbf{I} & \mathbf{N}_{irr}^T \end{pmatrix}, \tag{14}$$

where the decomposition of $\mathbf{N}$ into $\mathbf{N}_{rev}$ and $\mathbf{N}_{irr}$ is made according to the decomposition of $\mathbf{V}$ into $\mathbf{V}^{rev}$ and $\mathbf{V}^{irr}$. The submatrices $\mathbf{B}^{(0)}$ and $\mathbf{F}^{(0)}$ contain, as rows, preliminary reversible and irreversible elementary modes, respectively, which give the final elementary modes in the tableau $\mathbf{T}^{(n)}$. In extension to the abovementioned algorithm, now also preliminary *reversible* elementary modes have to be combined to give such modes in $\mathbf{T}^{(j+1)}$,

$$\mathbf{b}^* = t_{m,r+j+1}^{(j)} \cdot \mathbf{t}_{i\cdot}^{(j)} - t_{i,r+j+1}^{(j)} \cdot \mathbf{t}_{m\cdot}^{(j)} ,$$

$$\mathbf{t}_{i\cdot}^{(j)}, \mathbf{t}_{m\cdot}^{(j)} \in \mathbf{B}^{(j)}, i \neq m, t_{m,r+j+1}^{(j)} \neq 0, t_{i,r+j+1}^{(j)} \neq 0 . \tag{15}$$

Furthermore, some reversible elementary modes have to be combined also with irreversible elementary modes, to give fundamental vectors in $\mathbf{T}^{(j+1)}$,

$$\mathbf{f}^* = \text{sgn} t_{m,r+j+1}^{(j)} (t_{m,r+j+1}^{(j)} \cdot \mathbf{t}_{i\cdot}^{(j)} - t_{i,r+j+1}^{(j)} \cdot \mathbf{t}_{m\cdot}^{(j)}),$$

$$\mathbf{t}_{m\cdot}^{(j)} \in \mathbf{B}^{(j)}, \mathbf{t}_{i\cdot}^{(j)} \in \mathbf{F}^{(j)}, t_{m,r+j+1}^{(j)} \neq 0, t_{i,r+j+1}^{(j)} \neq 0, \tag{16}$$

in addition to the ones found by the standard algorithm. Rows for which $t_{i,r+j+1}^{(j)} = 0$ are taken over into the next tableau. The formulas (15), (16) mean that rows of the tableaux have to be linearly combined. This can be understood in that the intermediates are successively "internalized" into the system, i.e., in each step of the algorithm, preliminary elementary modes are linked pairwise so that the steady-state condition for the metabolite just considered is getting fulfilled.

Several combinations of rows of preliminary tableaux have to be ruled out by a condition guaranteeing that non-elementary modes are discarded and that each elementary mode is obtained only once. This condition states that the set of positions where two rows to be combined both have zero elements must not be a subset of the set of positions where another row has zero elements. This can be formalized as follows. For any row of a tableau, let

$$S(t_{m \cdot}^{(j)}) = \{i: \ t_{m,i}^{(j)} = 0, \ i \le r\} . \tag{17}$$

Two rows, $t_{m \cdot}^{(j)}$ and $t_{k \cdot}^{(j)}$, may only be combined if

$$S(t_{m \cdot}^{(j)}) \cap S(t_{k \cdot}^{(j)}) \not\subset S(t_{l \cdot}^{(j)}) \quad \text{for any } l \le r, \ l \ne m, k . \tag{18}$$

Since to each reversible mode $b^{(k)}$, also $-b^{(k)}$ is admissible, we have, at the end of the algorithm, to include the submatrix $-B^{(n)}$. That submatrix of $T^{(n)}$ consisting of the r left-hand side columns then contains, as rows, the elementary modes.

4. An example

We now apply the algorithm to the system shown in Fig. 2, which may serve as a model of the phosphoenolpyruvate/pyruvate/oxalacetate cycle (Leiser and Blum, 1987). Here, we assume the reactions 4, 5, and 6 to be irreversible so as to operate in clockwise direction. The starting tableau reads (zeros are omitted for clarity's sake)

$$T^{(0)} = \left(\begin{array}{ccccccccc} 1 & & & & \vdots & 1 & & & \\ & 1 & & & \vdots & & 1 & & \\ & & 1 & & \vdots & & & 1 & \\ \cdots & \cdots & \cdots & \cdots & \cdots & \cdots & \cdots & \cdots & \cdots \\ & & & 1 & & \vdots & 1 & -1 & \\ & & & & 1 & \vdots & & 1 & -1 \\ & & & & 1 & \vdots & -1 & & 1 \end{array} \right) \begin{array}{c} \\ B^{(0)} \\ \\ \\ \\ F^{(0)} \\ \\ \end{array} . \tag{19}$$

We first examine the left-hand side column of $\mathbf{N}^T$. To construct $\mathbf{F}^{(1)}$, we carry out the following operations. We combine the first and fourth rows, because the elements at positions $(1,7)$ and $(4,7)$ of $\mathbf{T}^{(0)}$ are non-zero. This linear combination is done in such a way that the element in the seventh column becomes zero. For similar reasons, we sum the first and sixth rows. We also sum the fourth and sixth rows because the elements at positions $(4,7)$ and $(6,7)$ are non-zero and of opposite sign (which is a condition for rows of a submatrix $\mathbf{F}^{(i)}$ to be combined). Finally, we keep the fifth row because the element at position $(5,7)$ is zero. $\mathbf{B}^{(1)}$ is obtained by keeping the second and third rows. This gives

$$\mathbf{T}^{(1)} = \begin{pmatrix} 1 & & & & & \vdots & 1 & \\ & 1 & & & & \vdots & & 1 \\ \cdots & \cdots & \cdots & \cdots & \cdots & \cdots & \cdots & \cdots \\ -1 & & 1 & & & \vdots & -1 & \\ & & & 1 & & \vdots & 1 & -1 \\ 1 & & & & 1 & \vdots & & 1 \\ & & 1 & & 1 & \vdots & -1 & 1 \end{pmatrix}. \tag{20}$$

In a similar way, we obtain $\mathbf{T}^{(2)}$ and

$$\mathbf{T}^{(3)} = \mathbf{F}^{(3)} = \begin{pmatrix} -1 & 1 & 0 & 1 & 0 & 0 & \vdots & 0 & 0 & 0 \\ 0 & -1 & 1 & 0 & 1 & 0 & \vdots & 0 & 0 & 0 \\ 1 & 0 & -1 & 0 & 0 & 1 & \vdots & 0 & 0 & 0 \\ 0 & 1 & -1 & 1 & 0 & 1 & \vdots & 0 & 0 & 0 \\ -1 & 0 & 1 & 1 & 1 & 0 & \vdots & 0 & 0 & 0 \\ 0 & 0 & 0 & 1 & 1 & 1 & \vdots & 0 & 0 & 0 \\ 1 & -1 & 0 & 0 & 1 & 1 & \vdots & 0 & 0 & 0 \end{pmatrix} = (\mathbf{f}^{(k)} \;\vdots\; \mathbf{0}). \tag{21}$$

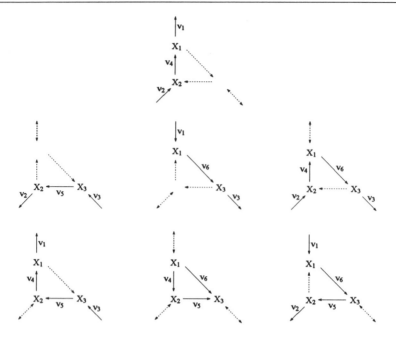

Figure 5 The seven elementary modes of the PYR/OAA/PEP cycle. The dashed arrows indicate reactions not used by the elementary mode. In contrast to Fig. 2, the reactions within the cycle are here considered irreversible.

While condition (18) has no effect on the rows of $T^{(0)}$ and $T^{(1)}$, several combinations of rows of $T^{(2)}$ have to be ruled out by that condition.

The elementary modes for the considered system are depicted in Fig. 5. $f^{(1)}$, $f^{(2)}$, and $f^{(3)}$ form a complete set of generating vectors of cone **C**. Although the remaining modes contained in $T^{(3)}$, e.g. the cyclic mode $(0\ 0\ 0\ 1\ 1\ 1)^T$, can be obtained by non-negative linear combination of generating vectors, they are elementary modes since they fulfil condition (C3) in Definition 2. Three elementary modes encompass one zero less than the remaining ones, but they are elementary in that there is no route of the same orientation that connects the same two external metabolites in a simpler way. If all reactions in the considered system are assumed to be reversible, one obtains reversible elementary modes only, e.g. $b^{(1)} = (-1\ 1\ 0\ 1\ 0\ 0)^T$, and $-b^{(1)}$.

The elementary modes of a reaction scheme describing glycolysis and gluconeogenesis have been discussed elsewhere (Schuster and Hilgetag, 1994).

5. Concluding remarks

The present analysis serves to detect essential structural features of any given biochemical network not just by inspecting the reaction scheme, but by algebraically analysing the stoichiometry matrix. This method widens the approach of calculating null-space vectors to that matrix.

We believe that the number of elementary modes is an important index characterizing biochemical systems. It indicates the richness of the system considered, by showing the variety of its physically realizable functions. Which of these functions are operative or in what proportions they operate simultaneously, is determined by the extent of inhibition and activation of enzymes, *i.e.* by the actual values of kinetic parameters, which we have not considered in our structural analysis.

An important application of the null-space is, amongst others, Metabolic Control Analysis, where this concept is frequently used to determine control coefficients, based on the generalized summation theorems (cf. Fell, 1992). Accordingly, computer programs for analysing control properties include routines for computing basis vectors of the null-space (Thomas and Fell, 1993). It could be of interest to use elementary modes instead of these basis vectors, to facilitate interpretation of the generalized summation theorems in terms of biochemical functioning. It is worth noting that contrary to the basis vectors, the elementary modes are uniquely determined.

The analysis can be used to test reaction schemes for consistency. It may occur that for a given reaction scheme, some reactions are represented by zero entries in all elementary modes calculated. If these reactions are known, from experiment, to have non-zero net fluxes, this outcome would show that the biochemical system is not properly modelled by the scheme.

Our viewpoint in defining elementary modes has been primarily theoretical; we wished to decompose the steady-state fluxes of metabolic systems into simple functional components. This approach is similar to the decomposition of quantum-mechanical states into eigenfunctions. From a biotechnological point of view, elementary modes can be interpreted as biosynthetic routes for metabolic products. One might expect that elementary modes could be conceived of as metabolic pathways. However, this is only acceptable if overlap of pathways is allowed, as can be seen in Fig. 5. In terms of *in vitro* experiments, one can interpret elementary modes as those flux patterns that remain after inhibition of a number of enzymes, such that inhibition of a further, still active, enzyme leads to cessation of any non-zero flux. In medical applications, elementary modes may be relevant for investigating enzyme deficiencies. When such a deficiency is so pronounced that the respective enzyme virtually ceases functioning, it is of interest to know what physiological functions (elementary modes) in the cell can no longer be fulfilled.

6. References

1. Clarke, B.L.: Stability of complex reaction networks. In: *Advances in Chemical Physics*, vol. 43, Prigogine, I. and Rice, S.A. (Eds.), Wiley, New York 1980, 1-216.
2. Clarke, B.L.: Stoichiometric network analysis. *Cell Biophys.* **12** (1988), 237-253.
3. Fell, D.A.: Substrate cycles: theoretical aspects of their role in metabolism. *Comments on theor. Biol.* **6** (1990), 1-14.
4. Fell, D.A.: Metabolic Control Analysis: a survey of its theoretical and experimental development, *Biochem. J.* **286** (1992), 313-330.
5. Fell, D.A., The analysis of flux in substrate cycles. In: *Modern Trends in Biothermokinetics*, Schuster, S., Rigoulet, M., Ouhabi, M., and Mazat, J.P. (Eds.), Plenum Press, New York and London 1993, 97-101.
6. Groetsch, C.W. and King, J.T.: *Matrix methods and applications.* Prentice-Hall, Englewood Cliffs 1988.
7. Leiser, J. and Blum J.J.: On the analysis of substrate cycles in large metabolic systems, *Cell Biophys.* **11** (1987), 123-138.
8. Mavrovouniotis, M.L., Stephanopoulos, G. and Stephanopoulos, G.: Computer-aided synthesis of biochemical pathways. *Biotechn. Bioengng.* **36** (1990), 1119-1132.
9. Mavrovouniotis, M.L.: Synthesis of reaction mechanisms consisting of reversible and irreversible steps. 2. Formalization and analysis of the synthesis algorithm. Ind. Eng. Chem. Res. **31** (1992) 1637-1653.
10. Nožička, F., Guddat, J., Hollatz, H., and Bank, B.: *Theorie der linearen parametrischen Optimierung.* Akademie-Verlag, Berlin 1974.
11. Rockafellar, R.: *Convex Analysis.* Princeton University Press, Princeton 1970.
12. Schuster, S. and Heinrich, R.: Minimization of intermediate concentrations as a suggested optimality principle for biochemical networks. I. Theoretical analysis. *J. Math. Biol.* **29** (1991), 425-442.
13. Schuster, S. and Hilgetag, C.: On elementary flux modes in biochemical reactions systems at steady state. *J. Biol. Syst.* (1994), in press.
14. Schuster, R. and Schuster, S.: Refined algorithm and computer program for calculating all non-negative fluxes admissible in steady states of biochemical reaction systems with or without some flux rates fixed. *Comp. Appl. Biosci.* **9** (1993), 79-85.
15. Seressiotis, A. and Bailey, J.E.: MPS: an artificially intelligent software system for the analysis and synthesis of metabolic pathways. *Biotechn. Bioengng.* **31** (1988), 587-602.
16. Thomas, S. and Fell, D.A.: A computer program for the algebraic determination of control coefficients in Metabolic Control Analysis. *Biochem. J.* **292** (1993), 351-360.

On Inferring the Kinetic Scheme of an Ion Channel from Postsynaptic Currents

M. Bier, G. Borst, K. Kits

Many drugs (e. g. valium) do their work by affecting the kinetics of the neurotransmitter gated ion channels involved in synaptic transmission. Much of the knowledge that we have about the kinetics of these channels has been obtained by patch clamp experiments on excised patches of membrane [1]. In these experiments one can control the concentration of neurotransmitter and follow the behavior of individual ion channels with a good signal to noise ratio. Mathematical methods are available [2] and continue to be developed [3] to relate open and closed times to the kinetic scheme of the involved ion channel. However, excised patches most likely contain extrasynaptic receptors (which may have different properties from synaptic ones) and a constant concentration of neurotransmitter applied to an excised patch is a very artificial situation.

We recorded postsynaptic currents (PSCs) in the whole cell configuration (Fig. 1), where instead of an excised patch we have complete cell "hanging" on our pipet [4]. The seal between the membrane and the pipet is often more stable in this setup than it is with patch recordings. In the whole cell configuration the clamped cell can be kept within the surrounding tissue and, like in the *in vivo* situation, every time a neighboring cell releases neurotransmitter a postsynaptic current (PSC) occurs (see Fig. 2). In our experiments the clamped cells where melanotropes from the intermediate lobe of the pituitary gland of *Xenopus laevis* and the observed currents derived from GABA mediated chloride channels. These neuroendocrine cells are very small (~5 μm) and we therefore had very low noise levels, so low in fact that the last 5 channels in the tail of the PSC could be seen to open and close individually (see Fig. 2). As the GABA dissociates from the receptor and diffuses out of the synaptic cleft we see the PSCs "die out". Below we only consider this so-called decay phase of the PSC. We assume that a neurotransmitter molecule doesn't rebind after it has dissociated. So the decay is a result only of the kinetics of the neurotransmitter-receptor complex. Next we show a mathematical method to "derive" a kinetic scheme from average and variance of a collection of PSCs.

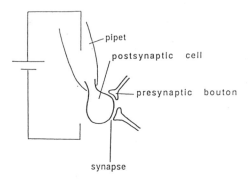

Figure 1 The setup for the experiment, every time neurotransmitter is released at a presynaptic bouton a postsynaptic current is measured.

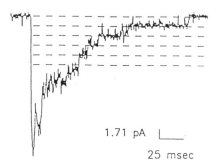

Figure 2 A typical postsynaptic current. Because of the low noise level we can see the actual discreteness of the process for the last five closing channels.

Processes on a molecular level have an innate stochasticity and this is why PSCs that start at the same amplitude decay differently. The average of a large sample of PSCs that all start at the same amplitude always appears to be well fitted with the same double exponential:

$$\overline{X}(t) = A_1 e^{-\lambda_1 t} + A_2 e^{-\lambda_2 t}, \tag{1}$$

where as a convention we take $\lambda_1 > \lambda_2$ and $t = 0$ is taken to be at the amplitude of the PSC. The quantities λ_1, λ_2 and $C = A_1/(A_1 + A_2)$ turn out to always be the same, no matter what the initial amplitude of the PSC is. The amplitude can be varied with a remaining degree of freedom $A_1 + A_2$. In terms of a kinetic scheme (i. e. a Markov process in continuous time with one absorbing state) the double exponential decay is logically equivalent with the following scheme:

$$
\begin{array}{ccc}
X_2 & \xrightarrow{\ k_{21}\ } & \\
k_{32} \uparrow \downarrow k_{23} & & X_1 \\
X_3 & \xrightarrow{\ k_{31}\ } &
\end{array}
\tag{2}
$$

where going into the absorbing state corresponds to the dissociation of the neurotransmitter. Given this scheme there are two possibilities:

1.) HIGH EFFICACY, both X_2 and X_3 are states where the channel is open.

2.) LOW EFFICACY, in one of the neurotransmitter bound states the channel is closed.

Many of the schemes that researchers have studied are special cases of either of these two models. Sequential models like $C \leftarrow O \rightleftarrows C$ and $C \leftarrow C \rightleftarrows O$ are special cases of the low efficacy model where one of the transition rates in (2) equals zero. The following scheme has been studied extensively [5] and seems consistent with a lot of experimental results on GABA mediated chloride channels.

The subscript indicates the number of bound GABA molecules. This scheme seems sensible; opening is a separate step that takes place after binding. In computer simulations to mimic real PSCs the times spent in C_1 and C_2 during the decay phase are almost negligable and the extra exponents λ that these states bring about are so large (i. e. a very fast decay) that they can never be resolved in an experiment.

So in the decay phase the behavior of this scheme effectively boils down to:

$$C_0 \overset{O_1}{\underset{O_2}{\rightleftharpoons}}$$

i. e. a special case of the high efficacy model.

We cannot discriminate between the high and low efficacy model on the basis of just the averages of samples of PSCs. We have to go to the second moment, the variance, to make the distinction. To set ourselves up mathematically we have to first relate the observed constants λ_1, λ_2 and $C = A_1/(A_1 + A_2)$ to the transition rates of the kinetic scheme.

For the kinetic scheme of Eq. 2 the time evolution of the averages in states 2 and 3 is determined by

$$\begin{pmatrix} x_2 \\ x_3 \end{pmatrix} = \begin{pmatrix} -(k_{21} + k_{23}) & k_{32} \\ k_{23} & -(k_{32} + k_{31}) \end{pmatrix} \begin{pmatrix} x_2 \\ x_3 \end{pmatrix}.$$

With the initial conditions $X_2(0) = N$ and $X_3(0) = M$ we obtain for the solution of Eq. 3

$$x_2(t) = \frac{1}{\lambda_1 - \lambda_2} \left[\{ N(\lambda_1 - k_{32} - k_{31}) - Mk_{32} \} e^{-\lambda_1 t} \right.$$
$$\left. + \{ -N(\lambda_2 - k_{32} - k_{31}) + Mk_{32} \} e^{-\lambda_2 t} \right]$$

$$x_3(t) = \frac{1}{\lambda_1 - \lambda_2} \left[\{ M(\lambda_1 - k_{21} - k_{23}) - Nk_{23} \} e^{-\lambda_1 t} \right.$$
$$\left. + \{ -M(\lambda_2 - k_{21} - k_{23}) + Nk_{23} \} e^{-\lambda_2 t} \right]$$

(3)

where λ_1 and λ_2 are the eigenvalues of the matrix.

They have the form

$$\lambda_{1,2} = \left\{ (k_{21} + k_{23} + k_{32} + k_{31}) \right.$$
$$\left. \pm \sqrt{(k_{21} + k_{23} + k_{32} + k_{31})^2 - 4(k_{21}k_{32} + k_{21}k_{31} + k_{23}k_{31})} \right\}/2$$

with

$$\lambda_1 + \lambda_2 = k_{21} + k_{23} + k_{32} + k_{31}$$
$$\lambda_1 \lambda_2 = k_{21}k_{32} + k_{21}k_{31} + k_{23}k_{31}$$

As a convention we take $\lambda_1 > \lambda_2$.

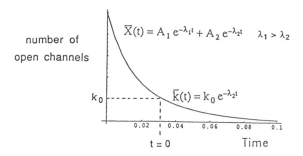

Figure 3 The idea behind our "tailvariance" analysis. Postsynaptic currents are intercepted at a fixed level of k_0 open channels, where is new $t = 0$ is set up. Each of the involved postsynaptic currents has a large enough amplitude that the fast λ_1 component is negliable when $X = k_0$. From $t = 0$ onward the variance is measured.

The quantity $C = A_1/(A_1 + A_2)$ is related to the ratio $\rho = X_2(0)/X_3(0)$, i. e. to the distribution of bound channels over the two neurotransmiter bound states at the amplitude of the PSC. This is a consequence of risephase dynamics that we don't know anything about. To rid ourselves of this unknown we shift our analysis to the tail of the PSC (see Fig. 3), i. e. to the part of the PSC where the faster decaying $A_1 \exp(-\lambda_1 t)$ component has become negligable. At the same rate as at which the $A_1 \exp(-\lambda_1 t)$ becomes negligable the ratio $X_2(t)/X_3(t)$ (i. e. the ratio of the number of channels in state 2 and the number of channels in state 3) converges to a constant ρ'.

This constant ρ' depends only on the transition rates k_{ij}. In the tail the value of ρ' is "forgotten" and both $X_2(t)$ and $X_3(t)$ decay exponentially (so variance = average) with an exponent λ_2.

Our procedure will be to take a large number of PSCs that start at a sufficiently large amplitude. All of these PSCs are intercepted in their tail at the same altitude m_0 where for all of the involved PSCs the fast component is negligable (see Fig. 3). This interception point will be the new $t = 0$. We then look at how the variance among these PSCs evolves in time. For both the high and low efficacy model we expect a variance that first increases from $\sigma^2(0) = 0$ as the PSCs diverge from one another due to their stochasticity, but then again goes to zero as at $t \rightarrow \infty$ the PSCs all die out. Next we will explain how for the low efficacy model the variance reaches a significantly higher maximum.

For the high efficacy model

the tailvariance is easily evaluated analytically. Suppose we are in the tail of the PSC and only one channel is open at $t = 0$. Then the probability that that channel is still open at time t is $p(t) = \exp(-\lambda_2 t)$. The variance is the average of the squares minus the square of the average, hence

$$\sigma_1^2(t) = 1^2 \cdot p(t) - (1 \cdot p(t))^2 = \exp(-\lambda_2 t) - \exp(-2\lambda_2 t).$$

Variances add up, so with m_0 channels at $t = 0$ instead of 1, we obtain:

$$\sigma_{k_0}^2(t) = m_0 \left(\exp(-\lambda_2 t) - \exp(-2\lambda_2 t) \right).$$

For the low efficacy model

all of the m_0 open channels are in the one state O at time $t = 0$. A number averaging $\rho' k_0$ of neurotransmitter bound channels is still in C_1 at $t = 0$ and these channels can still flow in and out of O after $t = 0$. We thus expect a higher variance for the low efficacy model. Below we show how this variance can be derived quantitatively.

We define $P_{33}(t)$ as the probability that a channel that is in X_3 at $t = 0$ is also in X_3 at time t and $P_{23}(t)$ as the probability that a channnel that is in X_2 at $t = 0$ is in X_3 at time t. $P_{33}(t)$ can be easily evaluated by substituting $N = 0$ and $M = 1$ in the equation for $X_3(t)$ under Eq. 3

$$P_{33}(t) = \frac{1}{\lambda_1 - \lambda_2}$$
$$\cdot \left\{ (\lambda_1 - k_{21} - k_{23})e^{\lambda_1 t} - (\lambda_2 - k_{21} - k_{23})e^{\lambda_2 t} \right\}.$$

Similarly, by substituting $N = 1$ and $M = 0$ in the equation $X_3(t)$ under Eq. 4 we find

$$P_{23}(t) = \frac{-k_{23}}{\lambda_1 - \lambda_2} \left(e^{-\lambda_1 t} - e^{-\lambda_2 t} \right).$$

The variance among the channels in the open state that were in X_3 at $t = 0$ is

$$\sigma_{33}^2(t) = MP_{33}(t)\{1 - P_{33}(t)\}.$$

The variance among the channels in the open state that were in X_2 at $t = 0$ is

$$\sigma_{23}^2(t) = NP_{23}(t)\{1 - P_{23}(t)\}.$$

Adding these two, we obtain the variance in the number of open channels for any M and N

$$\sigma_m^2(t) = \overline{m}(t) - MP_{33}^2(t) - NP_{23}^2(t).$$

The IPSCs will be intercepted in the tail, at a fixed amplitude. At this point only the slow component is important. From then on, the average number of open channels is $\overline{m}(t) = m_0 \exp(-\lambda_2 t)$ and the variance in the tail will be

$$\sigma_m^2(t) = m_0 e^{-\lambda_2(t)} - m_0 P_{33}^2(t) - n_0 P_{23}^2(t).$$

Although the number of open channels at the point of interception (m_0) will be the same for all IPSCs of an experiment, the number of closed channels at that point (n_0) may vary. Because the decay of both the number of closed and the number of open channels is exponential in the tail of IPSCs, the variance $(\Delta n_0)^2$ is equal to n_0. The standard deviation in $\Delta m(t)$ around $\overline{m}(t)$ caused by Δn_0 can be derived by substituting $n_0 \pm \Delta n_0$ in Eq. 3 and thus obtaining a $X_3'(t)$. Then

$$\Delta m(t) = x_3'(t) - x_3(t) = \frac{\Delta n_0 k_{23}}{\lambda_1 - \lambda_2} \left\{ \pm e^{-\lambda_1 t} \pm e^{-\lambda_2 t} \right\}$$

where $X_3(t) = \overline{m}(t)$, the average number of channels in the open state at time t, a quantity that can be measured given a sufficiently large number of PSCs. This is seen to be identical to $n_0 (P_{23}(t))^2$. Adding this, as a source of variance, we get

$$\sigma_m^2(t) = m_0 \left\{ e^{-\lambda_2(t)} - P_{33}^2(t) \right\}.$$

The variance in the number of open channels for model I is therefore

$$\sigma_m^2(t) = m_0 \left\{ e^{-\lambda_2(t)} - \left(\frac{1}{\lambda_1 - \lambda_2} \left((\lambda_1 - k_{21} - k_{23}) e^{-\lambda_2(t)} - (\lambda_2 - k_{21} - k_{23}) e^{-\lambda_2(t)} \right) \right)^2 \right\}.$$

which we can rewrite as:

$$\sigma^2_{k_0} = k_0 \left\{ e^{-\lambda_2 t} - \left(\frac{\lambda_1 - B}{\lambda_1 - \lambda_2} \right)^2 e^{-2\lambda_1 t} - \frac{2B(\lambda_1 + \lambda_2 - B) - \lambda_1 \lambda_2}{(\lambda_1 - \lambda_2)^2} e^{-(\lambda_1 + \lambda_2)t} \right.$$
$$\left. - \left(\frac{\lambda_2 - B}{\lambda_1 - \lambda_2} \right)^2 e^{-2\lambda_2 t} \right\} ,$$

where $B = k_{21} + k_{23}$. This is a one parameter (B) family of curves. But the range over which B can be varied is found by relating Eq. 3 to Eq. 1.

$$\lambda_2 < k_{21} + k_{23} < (1 - C)\lambda_1 + C\lambda_2 .$$

The area that is covered by this family of curves is colored black in Fig. 4. The figure shows that there is a significant gap between the high efficacy curve and the low efficacy family of curves. This gap gets bigger for larger values of C. For our case the gap seems big enough to not allow for any ambiguous experimental data.

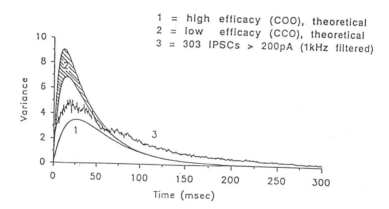

1 = high efficacy (COO), theoretical
2 = low efficacy (CCO), theoretical
3 = 303 IPSCs > 200pA (1kHz filtered)

Figure 4 The average of a sample of postsynaptic currents was fitted with $\lambda_1 = 102.4$, $\lambda_2 = 26.3$ and $C = .524$. For both high and low efficacy the theoretical tailvariance with these parameter values is plotted. The experimental tailvariance of the same sample is also plotted.

The actual experimentally observed (see also Fig. 4) variance appears to lie a little bit above the high efficacy curve. This can be explained by assuming a high efficacy model and realizing that there are additional sources of variance:

(i) Open channel noise: an open channel is not conducting at a fixed level, but is rapidly fluctuating within a noiseband of about a fifth of the single channel conductance. Open channel noise is almost completely white on the timescales that we are looking at and that is why we see the experimental variance jump up almost vertically at $t = 0$. The total amount of open channel noise is proportional to the number of open channels and we therefore see the experimental variance slowly get closer to the theoretical high efficacy variance during the first 50 msec.

(ii) a third slow exponent: a third very slow λ_3 component in (1) might give a negligable (small A3) contribution to the PSC at $t = 0$, but because of its slow decay it can become important in the tail of the tail, when the other exponents have almost completely died out, and that is why after 50 msec in Fig. 4 the measured variance lies above the theoretical curve.

We can thus conclude that there are 2 neurotransmitter-bound open states involved in the decay phase of our PSCs and that the amount of time spent in neurotransmitter bound closed states is negligable. GABA appears to be an efficient neurotransmitter.

References

[1] Hille, B. 1992. Ionic channels of excitable membranes 2nd edition, Sinauer
    Associates, Massachusetts.
[2] Colquhoun, D. and A.G. Hawkes. 1977. Relaxation and fluctuations of membrane currents that
    flow through drug operated channels. *Proc. R. Soc. Lond. B. Biol. Sci.* **199**:231-262.
[3] Silberberg, S.D. and K.L. Magleby. 1993. Preventing errors when estimating
    single channel properties from the analysis of current fluctuations.
    Biophys. J. **65**: 1570-1584.
[4] Borst, J.G.G., J.L. Lodder and K.S. Kits. 1994. Large amplitude variability of
    GABA-ergic IPSCs in melanotropes from *Xenopus laevis* : evidence that
    quantal size differs between synapses. *J. Neurophysiol.*, **71**:639-655.
[5] Busch, C. and B. Sakmann 1990, Synaptic transmission in hippocampal neurons:
    numerical reconstruction of quantal IPSCs. In *Cold Spring Harbor Symposia
    in quantitative biology*, Vol LV, pp. 69-80. Cold Spring Harbor Laboratory,
    Cold Spring Harbor, New York.
[6] Borst, J.G.G., K.S. Kits and M. Bier. 1994. Variance analysis of GABA-ergic
    IPSCs from melanotropes of *Xenopus laevis*.
    Accepted for publication in *Biophys. J.*

Mathematical Modelling of Red Blood Cell Enzyme Deficiencies as an Example for Large-Scale Parameter Changes in Biochemical Reaction Systems

R. Schuster and H.-G. Holzhütter

1. Introduction

There are numerous examples showing that the metabolism of cells can be severely impaired if the activity of only one of the participating enzymes undergoes large-scale alterations resulting, for example, from spontaneous mutations (inherited or aquired enzymopathies), administration of toxic drugs or self-inactivation of enzymes during cell aging. Beside these unavoidably occuring natural changes of enzyme-kinetic properties, there is substantial interest in medicine and biotechnology to modify the kinetic properties of enzymes in order to manipulate the metabolism and functional performance of specific cells. In all these areas one important subject of mathematically oriented theoretical research is to quantify the metabolic changes caused by changing the activity of a given enzyme in a defined manner. The theoretical approach presented in this paper is based on a comprehensive mathematical model of erythrocyte metabolism encompassing the main pathways of cellular energy and redox metabolism. The decision for the erythrocyte was made in the light of the long tradition and advanced level reached in the mathematical modelling of the metabolism of this cell type (e.g. Rapoport *et al.*, 1976; Joshi and Palsson, 1990; Schuster *et al.*, 1988). A second reason was the great number of various enzyme deficiencies which have been elucidated during the last three decades (cf. Valentine *et al.*, 1983; Tanaka and Zerez, 1990) and which represent an excellent basis for comparing computational results with 'experiments' done by nature. Hitherto deficiencies of about 20 enzymes of human erythrocytes associated with widely different degrees of severity and complexity have been identified (Valentine *et al.*, 1983; Fujii and Miwa, 1990). We define a 'homeostasis function' which takes into account the metabolic entities essential for cell integrity. This function is used for predicting the range of enzyme activities in which the metabolic alterations should be either tolerable, associated with non-chronic or chronic hemolytic diseases, or letal.

2. Presentation of the mathematical model

The mathematical model comprises the central pathways of the energy and redox metabolism of mature red blood cells: glycolysis, pentose phosphate pathway, and glutathione reduction (cf. Fig. 1). ATP and GSH consuming processes of the cell have been lumped together into the generalized reactions ATPase and GSHox, respectively. In contrast to the 'internal' reactions constituting the metabolic pathways, the 'external' reactions ATPase and GSHox represent a measure for energetic and oxidative load brought about by various processes needed to maintain cell viability (e.g. membrane pumps, reduction of lipid peroxides). The mathematical model is based on ordinary differential equations which describe the temporal behaviour of metabolite concentrations:

$$[\dot{G6P}] = v_{HK} - v_{G6PD} - v_{GPI}$$

$$[\dot{F6P}] = v_{PGI} - v_{PFK} + v_{TA} + v_{TK2}$$

$$[\dot{FDP}] = v_{PFK} - v_{ALD}$$

$$[\dot{GAP}] = v_{ALD} + v_{TPI} + v_{TK1} - v_{TA} + v_{TK2} - v_{GAPD}$$

$$[\dot{DHAP}] = v_{ALD} - v_{TPI}$$

$$[1,3\dot{D}PG] = v_{GAPD} - v_{PGK} - v_{DPGM}$$

$$[2,3\dot{D}PG] = v_{DPGM} - v_{DPGase}$$

$$[3\dot{P}G] = v_{PGK} + v_{DPGase} - v_{PGM}$$

$$[2\dot{P}G] = v_{PGM} - v_{EN}$$

$$[P\dot{E}P] = v_{EN} - v_{PK}$$

$$[A\dot{T}P] = -v_{HK} - v_{PFK} + v_{PGK} + v_{PK} - v_{AK} - v_{ATPase} - v_{PRPPS}$$

$$[A\dot{M}P] = v_{PRPPS} - v_{AK}$$

$$[6\dot{P}G] = v_{G6PD} - v_{6PGD}$$

$$[N\dot{A}DP] = v_{LDH(P)} + v_{GSSGR} - v_{G6PD} - v_{6PGD}$$

$$[G\dot{S}H] = v_{GSSGR} - v_{GSHox}$$

$$[Ru\dot{5}P] = v_{6PGD} - v_{EP} - v_{KI}$$

$$[X\dot{5}P] = v_{EP} - v_{TK1} - v_{TK2}$$

$$[R\dot{5}P] = v_{KI} - v_{TK1} - v_{PRPPS}$$

$$[S\dot{7}P] = v_{TK1} - v_{TA}$$

$$[E\dot{4}P] = v_{TA} - v_{TK2} \tag{1}$$

All enzyme reactions are considered to be reversible (for general rate equation cf. Schuster *et al.*, 1992):

$$v = V_m^* r(x)^* \left(\prod_i x_i^{c_{ij}^+} - \frac{1}{q} \prod_i x_i^{c_{ij}^-} \right) \tag{2}$$

The description of some reactions as completely irreversible made in previous models is not justified when modelling enzyme defects since high accumulation of intermediates may increase backward fluxes by several orders of magnitude. The rate equations were mostly taken from Schuster *et al.* (1988, 1989) and McIntyre *et al.* (1989). Based on literature data, we set up rate equations for those enzymes for which no rate equation was established up to now.

In the modelling of enzyme deficiencies, we restricted our considerations to alterations in the maximum activity V_{max}, which is the only parameter which has been determined in all experimental studies on enzyme deficiencies.

Our intention was to rationalize long-term alterations in the metabolic states of red cells, associated with permanent changes in the activity of a particular enzyme. Consequently, we considered stationary metabolic states, putting the time dependent variation of metabolite concentrations in equation system equal to zero. This results in a non-linear algebraic equation system for metabolite concentrations. The stability of the calculated steady states was checked by the eigenvalues of the Jacobian matrix which is obtained by the derivation of the right-hand sides of Eq. (1) with respect to metabolite concentrations.

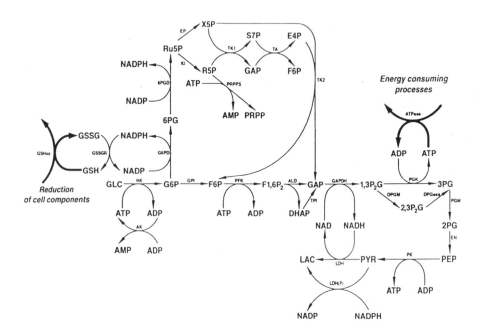

Figure 1 Reaction scheme - GLC - glucose; G6P - glucose 6-phosphate; F6P - fructose 6-phosphate; F1,6P2 - fructose 1,6-bisphosphate; GAP - glyceraldehyde 3-phosphate; DHAP - dioxyacetone phosphate; 1,3P2G - 1,3-bisphosphoglycerate; 2,3P2G - 2,3-bisphosphoglycerate; 3PG - 3-phosphoglycerate; 2PG - 2-phosphoglycerate; PEP - phosphoenolpyruvate; PYR - pyruvate; LAC - lactate; 6PG - 6-phosphogluconate; Ru5P - ribulose 5-phosphate; X5P - xylulose 5-phosphate; R5P - ribose 5-phosphate; S7P - seduheptulose 7-phosphate; E4P - erythrose 4-phosphate; GSH (GSSG) - reduced (oxidized) glutathione; HK - hexokinase; GPI - glucose 6-phosphate isomerase; PFK - phosphofructokinase; ALD - aldolase; TPI - triosephosphate isomerase; GAPD - glyceraldehyde phosphate dehydrogenase; PGK - phosphoglycerate kinase; DPGM - 2,3bis-phosphoglycerate mutase; DPGase - 2,3bis-phosphoglycerate phosphatase; PGM - 3-phosphoglycerate mutase; EN - enolase; PK - pyruvate kinase; LDH - lactate dehydrogenase; AK - adenylate kinase; G6PD - glucose 6-phosphate dehydrogenase; 6PGD - 6-phosphogluconate dehydrogenase; GSSGR - glutathione reductase; EP - epimerase; KI - isomerase; TK - transketolase; TA - transaldolase.

3. Results and Discussion

In what follows, we will predict for each enzyme a 'range of disease', encompassing those maximum activities leading to an observable impairment of the cell without cell death (Fig. 2).

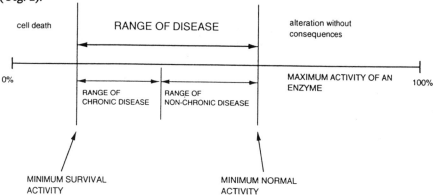

Figure 2 Definition of the 'range of disease'

Homeostasis function: The erythrocyte metabolism has to ensure survival of red cells during 120 days in blood circulation. For quantifying the proper metabolic performance of the cell, we propose a 'homeostasis function' by imposing critical values to key metabolic variables. In clinical and experimental studies, the loss of cell integrity is attributed to depletion of ATP or GSH, or to dramatic accumulation of metabolites proximal to the point of metabolic block. For GSH, experimental findings suggest that its concentration may not fall below a critical level $GSH^{crit} = 0.9 \, GSH^{total}$ in order to prevent hemolysis (Ataullakhanov *et al.*, 1981). For ATP and Ω (share of intermediates in total osmolarity of the cell) such threshold values are not available in the literature.

We have chosen the plausible limits: $ATP^{crit} = 0.5 \, ATP^{normal}$ and $\Omega^{crit} \approx 0.1$ osmolarity of the cell. For defining the homeostasis function, one has furthermore to take into account that red cells are exposed to different outer conditions during blood circulation. Therefore, the change of the key variables with respect to small changes in the load parameters k_{ATPase} and k_{GSHox} has to be included.

We combined all key variables into a single function which we named *homeostasis function:*

$$H = \prod_i \frac{\left(x_i - x_i^{\text{crit}}\right)}{x_i^{\text{normal}} - x_i^{\text{crit}}}$$

(3)

x_i: value at steady state (with changed parameters)

x_i^{crit}:   critical value

x_i^{normal}:   value at normal steady state (without enzyme deficiency)

x: [ATP], [GSH],

$\partial[\text{ATP}]/\partial \ln k_{\text{ATPase}}$, $\partial[\text{ATP}]/\partial \ln k_{\text{GSHox}}$,

$\partial[\text{GSH}]/\partial \ln k_{\text{ATPase}}$, $\partial[\text{GSH}]/\partial \ln k_{\text{GSHox}}$,

$\partial/\partial \Omega \ln k_{\text{ATPase}}$, $\partial/\partial \Omega \ln k_{\text{GSHox}}$.

This function is unity at the normal in vivo point; it always becomes zero whenever one critical value is reached.

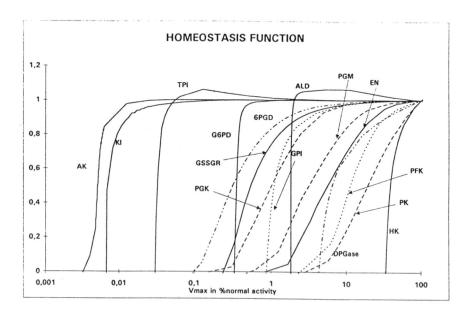

Figure 3 Homoestasis function as calculated by Eq. (3) for the enzymes shown in Fig. 1

Minimum survival activity: Upon a lowering of the activity of a given enzyme, the homeostasis function will decrease. The activity at which it becomes zero for the first time we call the *minimum survival activity*. It represents the minimum activity of an enzyme at which the metabolism of the cell is just sufficient for maintaining cell integrity (cf. Fig. 3).

Range of chronic disease: Chronic diseases are characterized by an impairment of cell functions under normal external load conditions. The range of enzyme activities giving rise to chronic diseases is defined by values of the homeostasis function smaller than one but not yet zero. We have chosen a value of 0.8 for the upper limit of this range.

Range of non-chronic disease: Non-chronic diseases are characterized by an impairment of cell functions which occur only under enhanced external load conditions. Thus, to estimate the activity range for non-chronic diseases, one has to investigate the metabolism of cells under stress conditions. Although outer stress may be very different in nature (e.g. drugs (mostly oxidants), fava beans, infections, fever, physical exercise), it mostly leads to an increase in k_{ATPase} and k_{GSHox}. For this end, we calculated upper critical boundaries for these two parameters (bifurcation points or GSH^{crit} attained) at varying the maximum activity of a given enzyme. The maximum activity at which one of these critical boundaries of k_{ATPase} and k_{GSHox} is halved in comparison with the value for normal cells defines the upper limit for the *range of non-chronic disease* (cf. Fig. 4).

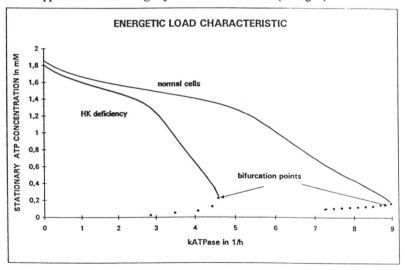

Figure 4A Energetic load characteristic: Dependence of the stationary ATP concentration on the rate constant of the ATPase for normal and hexokinase-deficient cells. Solid and dotted curves represent stable and unstable stationary states, respectively.

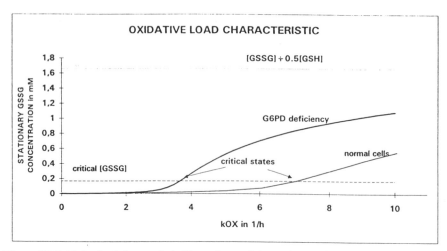

Figure 4B Oxidative load characteristic: Dependence of the stationary glutathione concentration on the rate constant of the oxidative load for normal and glucose-6-phosphate dehydrogenase-deficient cells. The lower critical glutathione concentration obtained from experimental values is represented by a dashed line.

Minimum normal activity: The minimum normal activity separates the range of disease (chronic or non-chronic) from the non-disease state. According to the above definitions, the *minimum normal activity* corresponds to the upper limit of either the range of chronic or the range of non-chronic disease depending on what value is higher.

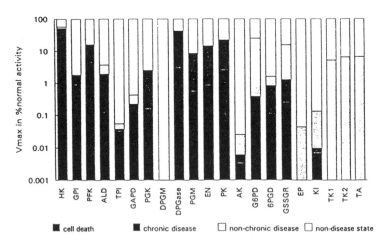

Figure 5 Calculated ranges of disease

The calculated ranges of disease are shown in Fig. 5. The minimum survival activities are very different. It can be shown by correlation analysis that these differences cannot be accounted for by the differences in the normal maximum activities of red blood cell enzymes. Furthermore, enzymes like HK, PK, PFK and G6PD have relatively large ranges of disease, whereas for quite a lot of enzymes it is rather small (TPI,GAPD,ALD,AK) or even absent (DPGM). For most of the enzymes, the range of disease practically corresponds to the range of chronic disease (PK,EN,PGM), whereas in the case of G6PD and GSSGR it is almost identical with the range of non-chronic disease. For all enzymes, having no minimum survival activity, the range of chronic disease does not exist, but in some cases deficiencies of these enzymes may cause non-chronic diseases (EP,TK1,TK2,TA). The range of disease for an individual enzyme can be compared with residual activities observed in vivo (cf. Table 1). For a group of enzymes (HK, PGK, PK, G6PD, SSGR), most of the experimental and clinical observations can be satisfactorily accounted for by the model. There is a second group of enzymes (DPGM, ALD, EN, AK, 6PGD) for which contradictory data have been reported in the literature, i. e. where the expected correlation between severity of disease and residual activity of the affected enzyme is missing.

Generally, the results of the model for these enzymes are better in line with observations found in those probationers equipped with the lowest activities reported for this enzyme. For example, Beutler et al. (1983) describes one case of AK deficiency, where two siblings had activities smaller than 0.05 %, whereby one of them was free of symptomes. The minimum normal activity of AK predicted by the model is about 0.025 % which may explain these results but not the others reported. For DPGM deficiency no range of disease is predicted by the model. This is confirmed by the clinical finding in a family with DPGM activity of about 0.28 %, but complete absence of nonspherocytic hemolytic anemia (NSHA) (Rosa et al., 1989; Tanaka and Zerez, 1990). In this light, it has to be questioned whether in all other cases of DPGM deficiencies with residual activities between 10 % and 50 %, the observed NSHA was solely due to a defect of this enzyme. For a third group of enzymes (PFK, GPI, TPI), theoretical and experimental data are conflicting. Diseases caused by PFK defects are predominantly related to the lack or instability of one of the subunit types (M, L) forming the enzyme tetramer (Valentine and Paglia, 1984).

Since the 5 isoenzymes normally found in red blood cells substantially differ in their kinetic constants for ATP- and DPG-inhibition, the lack of one subunit leads to a drastic change in PFK kinetics but a moderate activity decrease to 40 % - 60 % (Vora *et al.*, 1983), which cannot be simulated by varying V_{max} only. The discrepancies for GPI and TPI deficiencies are probably due to the remarkable instability of almost all abnormal enzyme variants (Tanaka and Zerez, 1990). Thus, it is especially doubtful whether the activities measured in the surviving cell fraction are representative for those critical activities causing cell damage.

Table 1 Calculated results and experimental data

CALCULATED RESULTS AND EXPERIMENTAL DATA		
ENZYME	CALCULATED RANGE OF DISEASE	ENZYME ACTIVITIES MEASURED IN ENZYME DEFICIENCY
	maximum activity in %normal activity	
HK	34.0 - 58.0	15.0 - 50.0
GPI	0.86 - 1.82	5.0 - 40.0
PFK	4.3 - 16.2	40.0 - 60.0
ALD	1.95 - 3.7	4.0 - 6.0,16.0
TPI	0.032 - 0.055	1.6 - 20.0
GAPD	0.22 - 0.42	30.0
PGK	0.15 - 2.5	0.0 - 10.0
DPGM	0.0	0.0, 12.0 - 50.0
DPGase	2.8 - 43.3	50.0
PGM	0.53 - 8.5	?
EN	0.82 - 14.5	6.5 ,8.5, 50.0
PK	2.4 - 22.3	5.0 - 40.0
AK	0.003 - 0.025	0.05, 0.5 - 44.0
G6PD	0.34 - 25.0	0.0 - 30.0
6PGD	0.105 - 1.6	2.4, 4.5, 30.0, 70.0
GSSGR	0.24 - 15.5	<9.0, 10.0
EP	0.0 - 0.043	?
KI	0.006 - 0.13	?
TK1	0.0 - 5.1	?
TK2	0.0 - 6.4	?
TA	0.0 - 6.6	?

Bearing the instability in mind, the low values of maximum activities, constituting the ranges of disease in GPI and TPI deficiency, are probably attained during normal cell life. Unfortunately, only experimental data from the thermostability test are available, which does not say anything about the dynamics of enzyme decay in vivo. For some enzymes, the experimental data are not enough substantiated (GAPD (only 1 case), DPGase (not directly measured), EP, KI, TK, TA). Besides the range of disease, we calculated activity ranges for chronic and non-chronic diseases.

In good agreement with experimental data (cf. Valentine *et al.*, 1983), these results show that most enzymopathies of glycolysis lead to chronic hemolytic anemia, whereas enzymes responsible for maintenance of glutathione in its reduced state (G6PD, GSSGR etc.) have a large range of activity where hemolytic crises are expected only under stress conditions.

References

Ataullakhanov, F.I., Zhabotinsky, A.M., Pichugin, A.V., and Toloknova, N.F. (1981) *Biochimija* 46, 530-541.

Beutler, E., Carson, D., Dannawi, H., Forman, L., Kuhl, W., West, C., and Westwood, B. (1983) *J. Clin. Invest.* 72, 648-655.

Fujii, H., and Miwa, S. (1990) *Am. J. Hematol.* 34, 301-310

Joshi, A., and Palsson, B.O. (1990) *J. Theor. Biol.* 141, 515-528.

McIntyre, L.M., Thorburn, D.R., Bubb, W.A., and Kuchel, R.W. (1989) *Eur. J. Biochem.* 180, 399-420.

Rapoport, T.A., Heinrich, R., and Rapoport, S.M. (1976) *Biochem. J.* 154, 449-469.

Rosa, R., Blouquit, Y., Calvin, M.C., Prome, D., Prome, J.C., and Rosa, J. (1989) *J. Biol. Chem.* 264, 7837-7843.

Schuster, R., Holzhütter, H.G., and Jacobasch, G. (1988) *BioSystems* 22, 19-36.

Schuster, R., Jacobasch, G., and Holzhütter, H.G. (1989) *Eur. J. Biochem.* 182, 605-612.

Schuster, R., Schuster, S., and Holzhütter, H.G. (1992) *Faraday Trans.* 88, 2837-2844.

Tanaka, K.R., and Zerez, C.R. (1990) *Sem. Hematol.* 27, 165-185

Valentine, W.N., and Paglia, D.E. (1984) *Blood* 64, 583-591.

Valentine, W.N., Tanaka, K.R., and Paglia, D.E. (1983) in *The Metabolic Basis of Inherited Disease*, Eds: Stanbury, J.B., Wyngaarden, J.B., Fredrickson, D.S., Goldstein, J.L., and Brown, M.S., McGraw-Hill, Inc., New York, 1606-1628

Vora, S., Davidson, M., Seaman, C., Miranda, A.F., Noble, N.A., Tanaka, K.R., Frenkel, E.P., and Dimauro, S. (1983) *J. Clin. Invest.* 72, 1995-2006.

Simulation of Dioxygen Free Radical Reactions: Their Importance in the Initiation of Lipid Peroxidation

P. Moniz-Barreto and D.A. Fell

1. Introduction

Dioxygen is an essential compound for most living organisms, the formation of reactive dioxygen intermediates appears to be commonplace in aerobically metabolizing cells and the free radicals that are produced are damaging to biological materials. The properties of dioxygen that are relevant to its role in living systems are:

1. it is a powerful oxidising agent (i. e. it has high electron affinity);
2. it has two unpaired electrons, and
3. it has low reactivity.

The kinetic inertness of dioxygen in solution can be explained by its electronic structure. Although molecular oxygen contains an even number of electrons, it has two unpaired electrons in its two HOMO (Highest Occupied Molecular Orbital), Fig 1. These give dioxygen a spin quantum number of one ($S=\sum s$) and a spin multiplicity of three ($2S+1$), that is, a triplet molecule for its lowest energy electronic configuration. Its low reactivity is due to a quantum rule that requires conservation of overall quantum spin state between reactants and products. As dioxygen has a spin state of 3, many reaction pathways for oxidation are unavailable because they would involve a forbidden change of spin state.

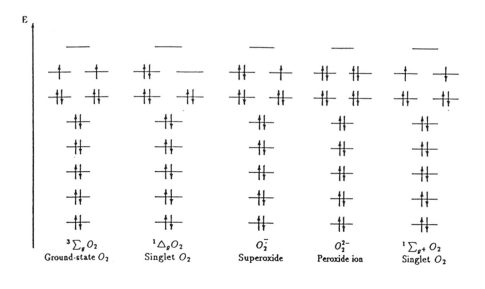

Fig 1. Bonding in the diatomic oxygen molecule, redrawn from [1].

The $\sum s$ denotes that the electrons have angular momenta in different directions (they belong to two orbitals) and Δ in the same direction (they are in the same orbital); and the superscripts of 1 and 3 mean respectively that they have opposed or parallel spins (singlet and triplet molecules) [2].

In exploiting the oxidising properties of oxygen, living systems have adopted both possible solutions to the problems arising from (3): using mechanisms that involve either the intermediate production of radicals, or the action of transition metals [3].

However, there is an associated disadvantage in that the radicals that are produced are damaging to biological structures. There are indications that this damage may play a part in several human diseases (including cancer, multiple sclerosis, Parkinson's disease) and ageing [4, 5, 6, 3, 7, 8, 9, 10]. On the other hand, the damage can be used for beneficial purposes as part of the body's mechanisms for killing foreign or abnormal cells [11].

Numerous studies have been undertaken to understand the mechanisms where free radicals are involved; however the progress has always been slow due to:

- the large number of available reactions some of which not yet well characterised
- the large range of rate constants with some still to be determined
- the very low concentration of some suspected intemediates.

Whereas techniques exist to study the involvement of free radicals in biological systems, the integration of the data with the information provided in the literature is scarce and qualitative rather than quantitative. The use of modeling and simulation is one of the best means to provide the integration of all the information into the network of processes and provide quantitative formulation of the relationships existing in the systems under study.

Some limited work has been done with computer simulations [12, 13] providing some quantitative insights into lipid peroxidation processes. However, the way the whole system is formalized in the first study does not give an insight into the particular mechanisms, and, in the second study the peroxidation initiation reaction does not seem the most appropriate.

The numerical methods by which both models were solved were not appropriate, as the systems to be solved are highly stiff. Although Babbs and Steiner [13] worked round this problem, it was at the expense of representing the mechanisms of diffusion and of amplification, whereby with small amounts of a reactive dioxygen intermediates can cause the appearance of biologically hazardous substances with concentrations many orders of magnitude greater.

The drawbacks of the previous work can then be summarized in three main points:

- choice of reactions modelled
- model formulation
- inappropriate or inadequate numerical methods.

The other problems that make difficult the simulation of free radical reactions using the known methods are:

- observation of very low concentrations for some metabolites, below 10^{-12} M, which is less than
  a molecule pereukaryote cell
- rates of change at low concentrations can be extremely low, giving severe numerical problems
  (with difficulty in distinguishing from zero reaction rate).

All these problems are largely overcome with an appropriate form of stochastic simulation. A new program was developed for numerical simulating the stochastic time evolution of coupled chemical reactions. The program is based on the method developed by Bunker [14] and separately by Gillespie [15, 16].

Models were devised to account for the importance of the Haber-Weiss and Fenton type reactions on the production of Hydroxyl radical and the role of this species in the initiation of lipid peroxidation studied.

Two sets of reactions were chosen and are described in section 3: the first includes nine reactions and represents the production and interconversion of radical species and the second the initiation, propagation and termination reactions for general lipid peroxidation.

2. Method

The algorithm involves the assumption that one has a small reaction vessel of volume V with small numbers of molecules that are not spatially assigned. A track of the total number of molecules of each species must be maintained throughout the simulation. The volume V is considered to typify the system as a whole, so that:

$$\frac{X_i}{\sum_i X_i} = \frac{c_i}{\sum_i c_i}$$

$$V = \frac{\sum_i X_i}{N \sum_i c_i}$$

where N is Avogadro's number, X_i the number of molecules of species i and c_i its concentration.

The instantaneous number of events for reaction j involving species i and k is then defined by:

$$a_j = \frac{k_j X_i X_k}{(NV)^{s_j-1}}$$

where S_j is the order of the jth reaction. The probability is defined as:

$$p_j = \frac{a_j}{\sum_i a_i}$$

the elapsed time, t, as:

$$t = \left(1/\sum_i a_i\right) * \ln(1/r_2)$$

and the reaction number m of the next event to occur is determined by the inequality:

$$\sum_{i=1}^{m-1} p_i < r_1 < \sum_{i=1}^{m} p_i$$

where r_1 and r_2 are random numbers between 0 and 1.

Illustrative applications of the algorithm showing the applicability of this approach for numerical simulation of the temporal behaviour of coupled chemical reactions will be made available on request.

3. The Models

To understand the amplification mechanism and and the importance of diffusion, we consider it of prime importance to study a set of reactions that includes interconversion of the major radical species and the link to lipid peroxidation. The reactions proposed by Koppenol (interconversion of free radicals) [17] and Kappus [18] (peroxidation) were chosen:

$$O_2^{\cdot} + HO_2^{\cdot} + H^+ \rightarrow {}^1O_2 + H_2O_2 \tag{1}$$

$$^1O_2 + {}^1O_2 \rightarrow 2O_2 \tag{2}$$

$$^1O_2 + O_2^{\cdot} \rightarrow O_2 + O_2^{\cdot} \tag{3}$$

$$H_2O_2 + H_2O_2 \rightarrow {}^1O_2 + H_2O \tag{4}$$

$$H_2O_2 + O_2^{\cdot} + H^+ \rightarrow {}^1O_2 + OH^{\cdot} + H_2O \tag{5}$$

$$OH^{\cdot} + H_2O_2 \rightarrow H_2O + O_2^{\cdot} + H^+ \tag{6}$$

$$O_2^{\cdot} + H^+ + OH^{\cdot} \rightarrow {}^1O_2 + H_2O \tag{7}$$

$$OH^{\cdot} + OH^{\cdot} \rightarrow H_2O_2 \tag{8}$$

$$HO_2^{\cdot} \rightarrow O_2^{\cdot} + H^+ \tag{9}$$

$$LipidH + OH^{\cdot} \rightarrow Lipid^{\cdot} + H_2O \tag{10}$$

$$Lipid^{\cdot} + O_2 \rightarrow LipidOO^{\cdot} \tag{11}$$

$$LipidOO^{\cdot} + LipidH \rightarrow Lipid^{\cdot} + LipidOOH \tag{12}$$

Reaction 5 is the so-called Haber-Weiss reaction, which together with reaction 6 forms a cycle with the same name.

Initial simulations were performed using the first nine reactions and the results compared with the same set where reaction 5 was replaced by:

$$O_2^{\cdot} + Fe^{3+} \rightarrow O_2 + Fe^{2+} \tag{13}$$

$$H_2O_2 + Fe^{2+} + H^+ \rightarrow OH^{\cdot} + H_2O + Fe^{3+} \tag{14}$$

to simulate the possibility of the Haber-Weiss being catalysed in the presence of ionic iron. Reactions (10) to (12) were added and the levels of peroxidation determined. The kinetic data was taken from [19, 20, 21], the initial concentrations were set according to Babbs [13] and the simulated volume was set to represent a single cell of $10^{-12} l$.

All the simulations were performed with a computer program written specifically for these models and are based on the method described in section 2.

4. Results

The initial simulations showed that the Haber-Weiss reaction is unlikely to function as a source of hydroxyl radicals even when iron is available to catalyse the reaction. For the non-catalysed model, only when the kinetic parameter for reaction 5 was raised to 10^5 (for a total time of 10^6 s) or 10^7 (for a total time of 10^5 s) does this reaction occur as a single event in the simulated cell. For the corresponding model with the catalysed reaction, similar behaviour was encountered. In the presence of high concentrations of $Fe(\text{II})$, no hydroxyl radical was produced, owing to the relatively small kinetic parameter for the reaction $(76 l mol^{-1} s^{-1}$ [21]); this reaction only appears to be an efficient source of the hydroxyl radical with values of its kinetic parameter corresponding to chelated iron reacting $(9 \times 10^5 l mol^{-1} s^{-1}$ [21]).

The inclusion of reactions 10 to 12 showed that even the production of a single hydroxyl radical would readily account for the initiation of the peroxidation process. Before any hydroxyl radical appeared, the simulations evolved with relatively large time increments, whereas as soon as it appeared and initiated the peroxidation, the time increments decreased more than 10,000 fold, showing the adaptability of the method in accommodating events on such different time scales.

5. Conclusion

- These stochastic simulations show that only one molecule of hydroxyl radical is necessary to initiate the chain reactions;

- the stochastic approach is more appropriate relative to the method where the metabolites are simulated in terms of chemical concentration, as in the latter case all the reactions will be possible even though no hydroxyl is present in the cell for most of the time.

References

[1] Halliwell, B. and Gutteridge, M. C. (1984). *Biochem. J.* **219**, 1-14.
[2] Bland, J. (1976). *J. Chem. Edu.* **53**, 274-279.
[3] Wilson, R. L. (1985). *Ciba Found. Symp.* **65**.
[4] Pryor, W. A., ed. (1976). *Free Radicals in Biology*. Academic Press, New York.
[5] Fridivich, I. (1977). in *Biochemical and Medical Aspects of Active Oxygen* (Hayaschi, O. and Asada, K., eds.), pp. 3-12. University Park Press, Tokyo.
[6] Pryor, W. A. (1978). *Photochem. Photobiol.* **28**, 787-801.
[7] Halliwell, B. and Gutteridge, J. M. C. (1986). *Arch. Biochem. Biophys.* **246**, 501-514.
[8] Cotgreave, I. A., Moldeus, P. and Orrenius, S. (1988). *Ann. Rev. Pharmacol. Toxicol.* **28**, 189-212.
[9] Halliwell, B. and Gutteridge, J. (1989). *Free radicals in Biology and Medicine*, chapter Oxygen is poisonius - an intriduction to oxygen toxicity and free radicals, pp. 1-21. Clarendon press, Oxford, 2nd edition.
[10] Sun, Y. (1990). *Free Radical Biol. & Med.* **8**, 583-599.
[11] Dormandy, T. L. (1988). Lancet, 1126-1128.
[12] Tappel, A. L., Tappel, A. A. and Fraga, C. G. (1989). *Free radical Biol. & Med.* **7**, 361-368.
[13] Babbs, C. F. and Steiner, M. G. (1990). *Free Radical Biol. &. Med.* **8**, 471-485
[14] Bunker, D., Garret, B., Kleindienst, T. and Long III, G. (1974). *Combustion and Flame* **23**, 373-379.
[15] Gillespie, D. (1976). *J. Comp. Phys.* **22**, 403-434.
[16] Gillespie, D. (1977). *J. Phys. Chem.* **81**, 2340-2361.
[17] Koppenol, W. H. and Butler, J. (1977). *FEBS Letters* **83**, 1-6.
[18] Kappus, H. (1985). *Oxidative Stress*, chapter Lipid peroxidation: Mechanisms, analysis, enzymology and biological relevance, pp. 273-309. Academic press, London, 1st edition.
[19] Dorfman, L. M. and Adams, G. E. (1973). Reactivity of the Hydroxyl Radical in Aqueous Solutions. Technical Report NRSDS-NBS-46, National Bureau of Standards, Government Printing Office, Washington.
[20] Singh, A. (1978). *Photochem. Photobiol.* **28**, 429-433.
[21] Bielsky, B., Cabelli, D., Arudi, R. and Ross, A. (1985). *J. Phys. Chem. Ref. Data* **14**, 1041-1099.

Residual Sodium and Potassium Fluxes Through Red Blood Cell Membranes

I. Bernhardt, S. Richter, K. Denner, and R. Heinrich

Humboldt-University Berlin, Institute of Biology, Experimental Biophysics, Invalidenstr. 42, 10115 Berlin and *Institute of Biochemistry, Department of Chemistry, Free University Berlin, Thielallee 63, 14195 Berlin, Germany.

Ion transport processes can be divided in four principal different mechanisms:

(1) active transport (pump mechanism, consuming ATP),
(2) channel mechanisms (regulated by the electrical potential across the membrane, specific substances, calcium ions, or mechanical pressure),
(3) carrier mechanisms, i.e. stoichiometrically coupled transport (cotransport, symport or antiport)
(4) residual or 'leak' transport.

To investigate the residual transport of an ion species it is necessary to know and to inhibit the specific transport pathways, i.e. active transport as well as transport via channels and carriers. The question arises whether the residual transport represents a 'classical' electrodiffusion described by the Nernst-Planck equation and solved for an assumed linear electrical potential profile inside the membrane by Goldman (1943) [1], or whether the residual transport is due to the existence of other, but yet unknown, specific transport pathways.

For the red blood cell membrane, a large variety of cation transport pathways have been described. Table 1 summarises all the presently known transport pathways for Na^+ and K^+ across the mammalian erythrocyte membrane. In addition, specific inhibitors for the transport pathways and also short comments including the presence of the transport pathways in red blood cells of different species are given.

It is clear from the existence of the variety of transport systems described in Table 1 that it is difficult experimentally to determine the true residual fluxes for Na^+ and K^+. At present, the best experimental measure of residual K^+ fluxes involves unidirectional K^+ uptake or loss in the presence of ouabain, bumetanide, and EGTA, with chloride replaced by nitrate or methylsulphate; to suppress the Na^+,K^+-pump, Na^+,K^+,Cl^- cotransport, Ca^{2+}-activated K^+ transport and any K^+,Cl^- cotransport, respectively. The study of the residual Na^+ fluxes is more complicated because of the greater variety of specific transport systems for Na^+ compared to K^+ (cf. Table 1, see also [2]).

The problem of how a monovalent cation passes through a biological membrane in the absence of specific transport pathways is still unsolved. However, some experimental conditions are known which lead to significant changes of the residual transport of monovalent cations through the red blood cell membrane:

(1) A reduction of the extracellular (NaCl+KCl) concentration of isotonic solutions (sucrose replacement) results in an enhancement of the unidirectional residual K^+ as well as Na^+ fluxes (both effluxes and influxes) [3, 4].

(2) The unidirectional residual Na^+ and K^+ fluxes are stimulated by high hydrostatic pressure (40 MPa) [5].

(3) Ouabain-insensitive Na^+ and K^+ fluxes show a 'paradoxical' temperature dependence in the presence of the anions salicylate or thiocyanate with a flux minimum at about 20 $^{\circ}C$ [6]. It has been shown that such a minimum also occured in chloride media in the presence of ouabain and bumetanide (in this case at 8 $^{\circ}C$) [7]. If an organic compound is used as the major cation, this paradoxical effect is more pronounced and the minimum is shifted to about 20 $^{\circ}C$ [8].

(4) Various studies have shown that the formation of disulphide bonds by different oxidative mechanisms result in enhanced fluxes for hydrophilic molecules and ions [9-11]. In addition, it is known that organic mercurials, which bind to sulfhydryl groups increase Na^+ and K^+ fluxes through the human erythrocyte membrane [12-15].

(5) A membrane expansion with a variety of amphiphiles, including charged, zwitterionic, and uncharged species can modify passive monovalent cation fluxes [16-18].

(6) The passive transport of monovalent cations through the erythrocyte membrane is influenced by the phospholipid composition of the membrane. This includes the head group as well as the fatty acid composition [19,20].

Table 1 Transport pathways for Na^+ and K^+ and specific inhibitors in red blood cell membrane (for more details see [2]).

Transport pathway	Comment	Inhibitor
Na^+,K^+-pump		Ouabain
Na^+/Li^+ (Na^+/Na^+) exchange		Phloretin
Na^+/Ca^{2+} exchange	in *carnivora* red cells	Amiloride
$NaCO^-_3/Cl^-$ exchange	via band 3 (capnophorin)	DIDS, SITS, DNDS
Na^+/H^+ exchange		Amiloride
Na^+,Cl^--dependent glycine transport		
Na^+-dependent amino acid transport (system ASC and N)	Na+-dependent glutamate transport in dog red cells	
Na^+,K^+,Cl^- cotransport		Bumetanide, Furosemide
K^+,Cl^- cotransport	Cl^- (Br)-dependent, latent in mature human red cells (can be activated by various manoeuvres), absent in HK-sheep red cells	H74, DIOA
Ca^{2+}-activated K^+ transport (Gardos channel)	absent in horse, cow, sheep and goat red cells	Quinine, Quinidine, Nitrendipine
residual Na^+ or K^+ transport		

Despite the results from these six experimental manoeuvres which alter the residual monovalent cation flux the molecular mechanisms which participate in 'leak' cation transport across the membrane are obscure.

One significant feature of some of these 'leak' fluxes is that under certain conditions they cannot be explained on the basis of the Goldman flux equation. In this respect, especially the effect of the enhancement of the unidirectional Na^+ and K^+ fluxes through the human erythrocyte membrane after lowering the extracellular (NaCl+KCl) concentration of isotonic solutions was investigated in great detail. It could be shown that the residual Na^+ and K^+ fluxes across the erythrocyte membrane cannot be explained on the basis of 'classical' electrodiffusion (see e.g. [3]). So, it is not possible to explain a drastic increase of the residual K^+ influx in human erythrocytes when reducing the extracellular (NaCl+KCl) concentration on the basis of the Goldman flux equation. Under these conditions one would predict a threefold decrease of the unidirectional cation influx. This is because of the depolarization of the membrane from the resting value of about -10 mV to about +50 mV in low (NaCl+KCl) solutions [2,3]. Additionally, in bovine erythrocytes under the same experimental conditions no enhanced K^+ flux was observed. This is despite the fact that on reducing of the extracellular (NaCl+KCl) concentration the transmembrane potential of bovine erythrocytes changes by the same amount as compared with human erythrocytes [21]. Finally, it was shown that after changing the transmembrane potential of human erythrocytes by replacing NaCl with Na-gluconate or Na-glucuronate no effect on residual K^+ efflux could be demonstrated [3].

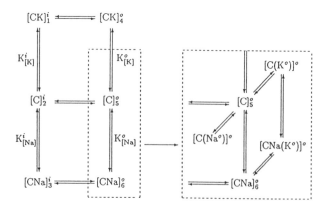

Figure 1 Model of carrier-mediated transport for two competing substrates (Na^+ and K^+) and the extension of this model by introducing one modifier site (dashed box). The rate constants of transition for the carrier of the various forms across the membrane (k_{ij} cf. eqs. (1) to (9)) are not presented in the scheme. The indices i,j characterize the numbers i and j of the carrier states. $K_{[Na]}^{i,o}$ and $K_{[K]}^{i,o}$ denote the dissociation constants for Na^+ and K^+ at both sides of the membrane.

Two attempts can be made, therefore, to explain the above described effect of enhancement of both unidirectional Na$^+$ and K$^+$ effluxes as well as influxes through the human erythrocyte membrane. Both explanations are based on the assumption that there is a still unknown specific transport pathway for monovalent cations. First, it is possible to consider a cation channel which cannot discriminate between N$^+$ and K$^+$. Such a channel which is voltage-sensitive is assumed by the Tosteson group [22, 23]. Another explanation would be the assumption of a cation carrier mechanism. However, it cannot be ruled out that both a channel and a carrier are of importance for the described effect. In addition, it has to be taken into consideration that a description of the enhancement of the residual Na$^+$ and K$^+$ fluxes on the basis of a carrier kinetic does not necesserely mean that the mechanism is a 'classical' carrier transport (also channel transport can show a carrier kinetic).

Using a carrier model presented in Fig. 1 without a modifier site, we assumed a mechanism where the substrates (Na$^+$ and K$^+$) compete for the same binding site of the transport system. It can be described by a set of equations based on the following assumptions [4] (for explanation see figure legend):

(1) The rates of association and dissociation of the substrates are much higher than the rates of translocation (mass low equilibria at the two membrane interfaces):

$$K^i_{[Na]} = \frac{[CNa]^i}{[C]^i[Na]^i}, \quad K^o_{[Na]} = \frac{[CNa]^o}{[C]^o[Na]^o}, \quad K^i_{[K]} = \frac{[CK]^i}{[C]^i[K]^i}, \quad K^o_{[K]} = \frac{[CK]^o}{[C]^o[K]^o} \tag{1}$$

(2) Under steady-state conditions the distribution of the carrier at the two membrane interfaces is time independent, and the rate constants of translocation across the membrane are independent of the direction of motion:

$$k_{14}([CK]^o - [CK]^i) + k_{25}([C]^o - [C]^i) + k_{36}([CNa]^o - [CNa]^i) = 0 \tag{2}$$

(3) The carrier is confined to the membrane and its total concentration $[C_t]$ is constant:

$$[C_t] = [C]^o + [C]^i + [KC]^o + [KC]^i + [NaC]^o + [NaC]^i \qquad (3)$$

In order to fit the parameters of the elementary steps (rate constants of translocation, dissociation constants) to the experimental results the unidirectional fluxes were calculated taking into account that:

$$J_K^{io} = k_{14}[CK]^i, \quad J_K^{oi} = k_{41}[CK]^o, \quad J_{Na}^{io} = k_{36}[CNa]^i, \quad J_{Na}^{oi} = k_{63}[CNa]^o \qquad (4,a\text{-}d)$$

However, the fitting procedure was not successful if only positive values for all parameters characterizing the carrier (parameters of elementary steps) were taken into account [4].

In addition to transfer sites which represent specific regions of transport protein molecules involved in substrate binding and translocation, so-called modifier sites which are capable of reacting with specific modifiers of transport have been discussed for carrier-mediated transport. A minimal extension of the reaction scheme given in Fig. 1 (dashed box) by introducing one modifier site (affecting mainly carrier forms loaded with Na^+) was sufficient to fit the equations derived for such a model to the experimental results (see Fig. 2, cf. also [4]). In this case the fit could be obtained with positive numerical values for the parameters of elementary steps. It has to be pointed out that the result of the fitting procedure depend on the starting point in the parameter space. There are, therefore, various combinations of parameters which lead to a similar or possibly better fit of the experimental values than the parameters obtained for the fit in Fig. 2.

A general problem in modelling ion transport with carrier kinetics is the aspect of substrate binding to the carrier molecule. To reduce the amount of unknown parameters in the model usually the binding of the substrate to the transporting molecule is assumed to be in chemical equilibrium and can be described by Eq. (1). In case some transitions of the carrier are as fast as the substrate binding or release, one obtains different dependencies of the unidirectional fluxes on the transition rate constants. This is especially true for those transitions that do not involve diffusional steps.

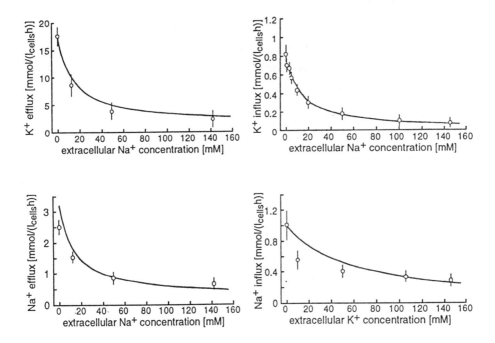

Figure 2 The effect of the extracellular (NaCl + KCl) concentration on the unidirectional residual Na$^+$ and K$^+$ fluxes (influxes as well as effluxes) through the human erythrocyte membrane. Symbols represent the mean of at least four independent experiments ± S.D. For experimental condition see [4]. The solid lines represent flux values calculated on the basis of the carrier model with modifier sites (Fig. 1).

Furthermore, the assumption of a fast equilibrium of substrate binding might not be true if there are fast steps in the transition of the carrier.

For the calculation of unidirectional fluxes usually the concentration of the *cis*-carrier-substrate complex is assumed to be proportional to the unidirectional flux of the substrate, Eqs. (4,a-d). Based on the carrier model given in Fig. 1 (without extension of modifier site) and assuming a chemical equilibrium of substrate binding to the carrier, the unidirectional K$^+$ efflux is given by:

$$J_K^{io} = \frac{K_K^i k_{14}[K]^i [C_t] \times (K_K^o k_{41}[K]^o + k_{52} + K_{Na}^o k_{63}[Na]^o)}{D} \tag{5}$$

with

$$D = (K_K^i k_{14}[K]^i + k_{25} + K_{Na}^i k_{36}[Na]^i) \times (1 + K_K^o[K]^o + K_{Na}^o[Na]^o)$$
$$+ (K_K^{ol} k_{41}[K]^o + k_{52} + K_{Na}^o k_{63}[Na]^o) \times (1 + K_K^i[K]^i + K_{Na}^i[Na]^i)$$

The Eq. (5) can be simply obtained by solving the Eqs. (1) - (3) for the inside carrier concentration with potassium loaded ([CK]^i. Finally this value is used for the flux calculation by inserting it in Eq. (4a). The unidirectional K$^+$ influx as well as the unidirectional Na$^+$ fluxes can be expressed in a similar way.

A different approach to calculate the unidirectional flux of the same model is to follow mathematically all the states of the transporting molecule that involves the transport of the subtrate. Therefore, taking into accout that the bound substrate partly flows back to its original side one obtains (for more details see [24]):

$$J_K^{io} = \frac{K_K^i k_{14}[Na]^i[C_t] \times (K_K^o k_{41}[K]^o + k_{52} + K_{Na}^o k_{63}[Na]^o)}{D} \times \frac{k_{12}k_{45}}{k_{12}k_{41} + k_{12}k_{45} + k_{14}k_{45}} \quad (6)$$

A comparison of Eqs. (5) and (6) shows that both methods result in the same dependency of the unidirectional flux on the substrate concentration at both sides of the membrane. In contrast, the dependency of the unidirectional flux on the transition rate constants is different in both cases. However, this leads only to recognizable differences if those rate constants depend differently on the transmembrane potential as shown later.

Assuming that the binding of the substrate on the transporting molecule is not in equilibrium (see above), it is not valid to set the unidirectional fluxes proportional to the concentration of the carrier-substrate complex, as this would lead to the contradicting prediction that the flux of substrate is not zero even though its *cis* concentration is zero (see Eq. (7)). In addition, one would find not only a different dependence of the unidirectional flux on the transition rate constants of the transport molecule (carrier) but also a different dependence of the flux on the external and internal substrate concentrations (see Eqs. (7) and (8)) for the flux calculated according to the method where the flux is set proportional to the carrier-substrate complex and the method assuming a back flux of the carrier-substrate complex, respectively.

$$J_K^{io} = \frac{l\ K^{i} + m\ K^{o} + n\ K^{i}\ K^{o} + o\ Na^{i}\ K^{o} + p\ K^{i}\ Na^{o}}{D_1} \tag{7}$$

with

$$D_1 = a\ K^{i} + b\ K^{o} + c\ Na^{i} + d\ Na^{o} + e\ K^{i}\ K^{o} + f\ Na^{i}\ Na^{o} + g\ K^{i}\ Na^{o} + h\ K^{o}\ Na^{i} + 1$$

$$J_K^{io} = \frac{(1 + m'\ K^{o} + o'\ Na^{o}) \times l'\ K^{i}}{D_1} \tag{8}$$

where a..h, l..o, l'..o' represent phenomenological constants, that can be expressed in terms of the transition rate constants.

In addition to these extensions and refinements, considering transport of ions, the effect of transmembrane potential has to be introduced in the model calculations. This can be done assuming that those transition rates, that involve transfers of charged transport molecules, are dependent on the transmembrane potential according to [25]:

$$k_{ij} = k_{ij}^0 \exp(zF\Delta\psi\alpha\ RT), \quad k_{ji} = k_{ji}^0 \exp(zF\Delta\psi(\alpha-1)\ RT) \text{ with } (0 \le \alpha \le 1) \tag{9}$$

In this equation the parameter α depends on the effect of the electrical transmembrane potential of the forward and backward transition of the carrier-states (symmetrical or asymmetrical reaction) [25]. Taking into account a transmembrane potential effect on carrier-mediated ion transport, one has to replace the rate constants of translocation in Eq. (5) by the corresponding expressions of Eq. (9). For the carrier model presented in Fig. 1 this analysis is under investigation.

Another consideration in modeling residual cation fluxes is a possible coupling between those fluxes and anion transport. This is of importance since the overall current (of all positive and negative charges) at steady state equals zero. In erythrocytes the permeability for anions due to band 3 protein is far greater than for cations, the flux of anions should usually not limit the flux of sodium and potassium. This might not be true in those cases, where the flux of anions is coupled to the cation flux in a cotransport process, or in case of application of a specific anion transport inhibitor.

Acknowledgment

This work was supported by the Deutsche Forschungsgemeinschaft (DFG-project Be 1655/1-1).

References

1. Goldman, D,E., *J. Gen. Physiol.* **27** (1943), 37-60
2. Bernhardt, I., Hall, A.C., Ellory, J.C., *studia biophys.* **126** (1988), 5-21
3. Bernhardt, I., Hall, A.C., Ellory, J.C., *J. Physiol.* **434** (1991), 489-506
4. Denner, K., Heinrich, R., Bernhardt, I., *J. Membrane Biol.* **132** (1993), 137-145
5. Hall, A.C., Ellory, J.C., *J. Membrane Biol.* **94** (1986), 1-17
6. Wieth, J.O., *J. Physiol.* **207** (1970), 563-580
7. Stewart, G.W., Ellory, J.C., Klein, R.A., *Nature* **286** (1980), 403-404
8. Blackstock, E.J., Stewart, G.W., *J. Physiol.* **375** (1986), 403-420
9. Deuticke, B., Poser, B., Lütkemeier, P., Haest, C.W.M., *Biochim. Biophys. Acta* **731** (1983), 196-210
10. Heller, K.B., Poser, B., Haest, C.W.M., Deuticke, B., *Biochim. Biophys. Acta* **777** (1984), 107-116
11. Deuticke, B., Lütkemeier, P., Poser, B., *Biochim. Biophys. Acta* **1109** (1992), 97-107
12. Knauf, P.A., Rothstein, A., *J. Gen. Physiol.* **558** (1971), 211-223
13. Sutherland, R.M., Rothstein, A., Weed, R.I., *J. Cell. Physiol.* **69** (1967), 185-198
14. Grinstein, S., Rothstein, A., *Biochim. Biophys. Acta* **508** (1978), 236-245
15. Haas, M., Schmidt, W.F., *Biochim. Biophys. Acta* **814** (1985), 43-49
16. Fortes, P.A.G., Ellory, J.C., *Biochim. Biophys. Acta* **413** (1975), 65-78
17. Isomaa, B., Hagerstrandt, H., Paatero, G., Engblom, A.C., *Biochim. Biophys. Acta* **860** (1986), 510-524
18. Klein, R.A., Ellory, J.C., *J. Membrane Biol.* **55** (1980), 123-131
19. Kirk, R.G., *Biochim. Biophys. Acta* **464** (1977), 157-164
20. Kuypers, F.A., Roelofsen, B., Op den Kamp, J.A.F., Van Deenen, L.L.M., *Biochim. Biophys. Acta* **769** (1984), 337-347
21. Bernhardt, I., Erdmann, A., Glaser, R., Reichmann, G., Bleiber, R., in: *Topic in lipid research. From structural elucidation to biological function.* (R.A. Klein, B. Schmitz, eds.), The Royal Society of Chemistry, London, 1986
22. Halperin, J.A., Brugnara, C., Tosteson, M.T., Van Ha, T., Tosteson, D.C., *Am. J. Physiol.* **257** (1989), C986-C996
23. Halperin, J.A., Cornelius, F., *Biochim. Biophys. Acta* **1070** (1991), 497-500
24. Stein, W.D., *J. Theor. Biol.* **62** (1976), 467-478
25. Schultz S. G., in: *Basic principles of membrane transport.* Cambridge University Press, 1980

The Role of Natural Selection and Evolution in the Game of the Pentose Phosphate Cycle

F. Montero, J.C. Nuño, M.A. Andrade, C. Pérez-Iratxeta, F. Morán, E. Meléndez-Hevia

1. Introduction

The pentose phosphate cycle has been presented as a paradigm of optimization in the design of metabolic pathways. The hypothesis of simplicity, i.e. the tendency to the least number of both steps and carbons involved in every intermediate, seems to have played a prominent role in its evolution. In fact, the number of steps and carbons of the intermediates affect the values of important characteristics of the system, such as the total flux of the pathway and its transition time.

In order to study the selection and evolution of the pentose phosphate cycle, a quasispecies algorithm has been performed. Simulations are carried out using a Monte-Carlo method. Given an initial condition, with the specificity of enzymes for different substrates randomly choosen, the system evolves towards a solution in which a cost function is optimized. In most cases the final solution reached is identical to the present pentose cycle design. However, different optimal solutions appear depending on certain thermodynamic constraints.

The biological implications of the results, as well as the generalization of the optimization method to other metabolic pathways is discussed.

2. The pentose-phosphate cycle as an optimization problem

In general, a metabolic pathway defines a transformation of an initial product A into a final product B through successive steps. This transformation is strongly enhanced by the action of very specific catalysts (enzymes) that act on the different pathway intermediates.

As has been previously discussed[1,2] many metabolic conversions could be achieved by alternative pathways. However, one is still surprised by the particular designs that well-known metabolic pathways exhibit. Thus, why do current metabolic pathways run in a specific manner among so large number of possibilities? A very plausible starting hypothesis is the following: the present-day metabolic pathways are the result of a very long evolutive process, during which particular system features have been optimized.

This optimization process might act on two different levels; firstly, on the metabolic design where different ways of converting A into B were chemically possible. Secondly, by selecting enzymes with better catalytic activity. In general, both processes are not independent of each other since the selection of a new particular metabolic design involves the selection of more efficient and specific enzymes. Searching for examples in the metabolism of cells where this optimization process has been carried out is not an easy task. The reason is that any evolutive analysis needs the knowledge of every chemically possible alternative, and then to explain why one alternative has been selected.

In this working line, during the last few years Prof. Meléndez-Hevia and coworkers have studied the non-oxidative part of the pentose phosphate cycle, trying to find any trace of an evolutive process[1-3]. Basically, the non-oxidative part of the pentose phosphate cycle describes the conversion of six sugars of five carbons each into five sugars of six carbons each. This conversion is carried out by means of the action of three specific enzymes that transfer fragments of two (*transketolase* (*TK*)) and three (*transaldolase* (*TA*) and *aldolase* (*AL*)) carbons. In these seminal papers, the problem of the pentose phosphate cycle was formulated as a mathematical game. The main conclusion obtained from this theoretical perspective was that the simplest pathway, i.e. that implying the lowest number of steps and involving intermediates formed by the lowest number of carbons, corresponds to the present-day design of the pathway. From this viewpoint, it seems that Nature has been able to find the simplest solution to the problem.

But, what does the selection of a concrete design mean? Let us assume that initially different enzymes with transketolase, transaldolase and aldolase activities were present; these enzymes were able to catalyze the transfer of two or three carbons, and the condensation of two sugars, at least one of them with three carbons, respectively. Assuming a low specificity of the enzymes for different substrates, many processes could be performed simultaneously, yielding the same stoichiometric result as the pentose phosphate cycle. A particular design of the pathway requires a very strict specificity of the enzymes for the different substrates. In particular, the present-day *TK* is able to react only with sugars of five, six and seven carbons acting as donors, and with sugars of three, four and five carbons acting as acceptors. *TA* only reacts with sugars of six and seven carbons as donors, and sugars of three and four carbons as acceptors. Finally, *AL* can only condense sugars of three carbons. Nevertheless, other specificities could have been selected and therefore display alternative designs.

Many questions arise after these considerations: why has the simplest pathway been selected? Through what selective mechanism has the system found the right solution? Why these game rules? In other words, why do *TK* and *TA* transfer fragments of two and three carbons, respectively, even though transfer of different numbers of carbons is also chemically possible? If we assume the decisive role of Darwinian selection in evolutionary processes, the first question can be answered in a straighforward way: the selection of a particular pathway is related to the optimization of some kind of function. Simplicity, the basic hypothesis introduced in the original papers by Meléndez-Hevia *et al.*, must be considered as a consequence of an optimization process. Then, the fundamental problem is to find out the function that has been optimized during evolution, from which the simplest solution can be obtained.

3. Looking for a fitness function

A large number of system macroscopic variables depend critically on the metabolic pathway design. The total flux of conversion (from 5-carbon-sugars to 6-carbon-sugars) in steady state, the transition time to reach the stationary regime and the osmolarity depend on both the pathway design and the values of the kinetic constants involved in the pathway. In previous papers[4-6], R. Heinrich and coworkers have studied these problems profoundly. A first important conclusion that can be extracted from these works is that for linear pathways, assuming a constant concentration of free enzyme, the higher the number of steps is, the smaller will be the total flux of conversion. Let us suppose that a substrate S_1 can be transformed into a product S_3 by two alternative ways, one of them through two steps, and the other one through three steps. In addition, assuming that all steps, independently of the way considered, are catalyzed by the same enzyme, this setup being schematically drawn in ref. 7. Note that under this hypothesis the enzyme has no specificity for the substrate at all. If the enzyme has the same affinity for any substrate, then it can be proven that the fluxes are differently distributed through the longer and the shorter pathway, being larger for the latter[7].

Moreover, a similar increase of affinity for the substrates of the two pathways gives rise to a larger increase of the total flux through the shorter pathway. However, these results do not explain the selection of the shorter route since the system would take advantage also of an improvement of the longer route. Nevertheless, actual systems evolve under very particular conditions: the total concentration of the enzyme is limited. Therefore, the substrates of one way act as competitive inhibitors with regard to the substrates of the other route. This inhibitory effect will be more efficient if the affinity is larger. If the concentration of total enzyme is kept constant, while increasing the affinity for the substrates of the shorter way increases the total flux, the same growth of the affinity for the substrates of the longer way decreases the total flux, therefore implying a selective disadvantage. In other words, any mutation that causes an increase in the affinity of the enzyme for substrates of the shorter route will tend to be fixed, whereas those mutations leading to an increase in the affinity for substrates of the longer way will tend to be removed.

In consequence, the total flux of conversion of a metabolic pathway could have had an important influence on its design, and in particular on the number of steps needed to get the whole conversion. Let us focus the attention to the pentose phosphate cycle. In a first moment proteins could bring about as a consequence of the translation of different parts of the genome. Some with transketolase activity, others with transaldolase activity, etc... However, these enzymes could have a very low affinity for any substrate, and would be also very unspecific. Likely, even the first pentose phosphate pathway would be a more complex scheme than the present-day cycle, involving a larger number of steps and sugars with more carbons.

Obviously, the total flux of these incipient pathways would be extremely low. By means of mutation and selection, more specific enzymes, with better activities would appear. The present-day state of this process is the optimal (by now) solution just as we know at present. However, because of the complexity of the system these ideas are difficult to be proved in an analytical way. The reactions considered are bisubstrate, there are intermediates shared by different enzymes, etc... Moreover, pathways with the same number of steps but with different design can appear. All of these considerations disable the deduction of an analytical expression that relates the total flux straighforward with the concentrations of the enzymes and the total affinity of reaction, as was the case in the linear conversions mentioned above. Anyway, the value of total flux in steady state can always be obtained by numerical integration of the corresponding differential equations that govern the temporal evolution of macroscopic variables (clearly, after the definition of a kinetic model). However, since the number of possible designs can be very high, and is *a priori* difficult to define, this simple computation can be almost impossible to be carried out. So, how to find the solution that maximizes the total flux of conversion?

4. Quasispecies as an optimization tool

In the early seventies, Manfred Eigen proposed a mathematical model to explain the selective and evolutionary features of a population formed by selfreplicative species[8]. Essentially, this seminal work focused on a theoretical model assuming a population formed by chains of monomers of length v formed by μ types of monomers. Consequently, the number of possible sequences in the system (the *sequence space*) is μ^v (an astronomically large number if we assume $\mu=4$ and $v=100$). A cost function assigns a value to each sequence: its *fitness*. The self-replication rate of a sequences depends on both its fitness and its *mutation rate*. The mutation probability (P) of a chain is given by the product of the mutation probabilities per digit (p_i). If we assume that all the digits have the same mutation rate $\left(p_i \equiv p \ \forall i \right)$, then $P = p^v$. If the population is kept constant a selection pressure is introduced in the system. Then, the basic question in Eigen's theory is: which sequences will be selected during evolution?

Usually, the fitness of a sequence is related directly to its probability of replication: the higher the fitness value, the higher the replication rate. As a result of this process, under any kind of constraint, the information of a sequence, that of the highest fitness value, is selected (the so called *master sequence*). But at the same time, because of the error-prone replication of sequences, an *error-tail* is continuously formed around the master copy. The master sequence together with its error-tail is named a *quasispecies*. As a consequence in every moment most of the population corresponds to this quasispecies. The structure of the quasispecies allows for a dynamical search in the sequence space. If, as a consequence of mutation, a new sequence with better fitness than the master sequence appears, then the current equilibrium is broken and the population is moved to a new metastable equilibrium, in which this new master sequence and its error tail dominate the population[9].

Because of this behavior, quasispecies evolution has been taken as a paradigm of optimization methods. The method, based mainly on laws of Natural Selection is currently used to seek the optimal solution in many other problems in which the right solution is what optimizes a cost function[10]. As a new approach to the understanding of metabolism, we have applied the quasispecies algorithm to study the pentose phosphate cycle. Nevertheless, to formulate the pentose phosphate cycle in terms of an optimization problem is not straightforward. Firstly, the pathway parameters must be codified in an adequate manner. Secondly, the cost function should be found that drives the evolution of the pathway.

Table 1 Kinetic Model. In this representation C_n means sugars with n carbons. TK and TKC_n stand for the free enzyme and the complexes formed from the enzyme and the C_n fragment, respectively. A similar notation has been used for TA and for AL.

$$\text{TRANSKETOLASE } (TK)$$

$$C_n + TK \underset{tk_{-n}}{\overset{tk_n}{\rightleftharpoons}} TKC_n \underset{tk'_{-n}}{\overset{tk'_n}{\rightleftharpoons}} TKC_2 + C_{n-2}$$

$$\text{TRANSALDOLASE } (TA)$$

$$C_n + TA \underset{ta_{-n}}{\overset{ta_n}{\rightleftharpoons}} TAC_n \underset{ta'_{-n}}{\overset{ta'_n}{\rightleftharpoons}} TAC_3 + C_{n-3}$$

$$\text{ALDOLASE } (AL)$$

$$C_n + AL \underset{al_{-n}}{\overset{al_n}{\rightleftharpoons}} ALC_n \underset{al'_{-n}}{\overset{al'_n}{\rightleftharpoons}} ALC_3 + C_{n-3}$$

$$C_3 + AL \underset{ll_{-1}}{\overset{ll_1}{\rightleftharpoons}} ALC_3$$

The latter problem was already discussed in the previous section. We argued that the main function of the pentose phosphate cycle is the formation of hexoses from pentoses through different intermediate steps. Every step is controlled by an enzyme with its correponding kinetic constant values. Multiple ways of performing this function can be imagined without violating the laws of biochemistry. All of them are different combinations of intermediates which interact with their corresponding reaction velocities. The kinetic model shown in Table 1 takes into account the above-mentioned ideas, and it has been used in the present work.

Through this kinetic scheme, the temporal evolution for every intermediate, as well as the net production of hexoses at the steady state, can be evaluated by numerical integration of the corresponding differential equations.

According to these considerations, we suggested that the cost function could be given by the output flux of hexoses at the steady state Φ_i, assuming the total affinity of the global reaction to be constant. Therefore, to each possible pathway P_i, is assigned a value of this function, Φ_i.

Moreover, by restricting the possible values of the kinetic parameters to 0 or 1, we get a codification of the possible pathways. In a first approach, we assume that $tk_n = tk_{-n} = tk'_n = tk'_{-n}$, and similar assumptions have been made for ta_n, and for al_n. In a general sense, each species (*pathway design*) is defined by a chain of digits (*genome*), each of them standing for the enzymatic action of the three present enzymes (*transketolase*, *transaldolase*, and *aldolase*) on sugars with different numbers of carbons. For example, if the existence of sugars up to 8 carbons is assumed, the genome standing for the present PPC would be:

	C_5	C_6	C_7	C_8
TK	1	1	1	0
TA		1	1	0
AL		1	0	0

which means that sugars of 5, 6, and 7 carbons can act as donors of a fragment of 2 carbons by action of the *TK*, sugars of 6 and 7 carbons will be donors of fragments of 3 carbons by action of the *TA*, and sugars of 6 carbons will be substrate of *AL*. Schematically,

$$P_i = \left(tk_5, tk_6, tk_7, tk_8, ta_6, ta_7, ta_8, al_6, al_7, al_8\right).$$

every element of the above sequence being 1 or 0, depending on whether or not the enzyme is active on the correponding sugar, respectively. In Figure 2 more examples of genomes for other different designs are shown.

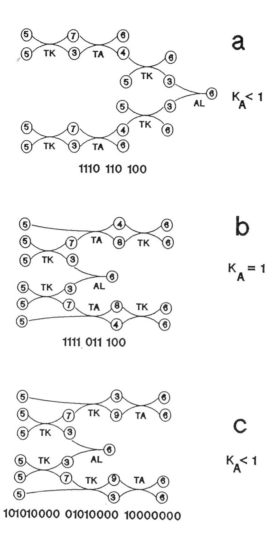

Figure 2 Results of several quasispecies experiment curried out for different K_A values. Genome size was 10 in **a** and **b**; i. e. sugars from three to eight carbons, while in case **c** a size genome of 25 was used, i. e., sugars from three to thirteen. The sequences representing the selected designes are shown below the schemes.

In every step, sequences can reproduce themselves and, as a consequence, can mutate (that is, one of the digits can be modified) giving rise to other different species, i.e. another pathway. The net production of hexoses is calculated for this new chain. After the action of natural selection the final result of this game is likely to be the pathway with the larger flux of hexose production.

5. Preliminary results

In a first approach only sugars with less than eight carbons are considered. However, this simplification is useful not only from a practical point of view, but indeed, it is consistent with the formose reaction where only sugars with 3, 4, 5, 6, 7 and 8 carbons are formed in valuable quantities[11]. Likely, when the metabolic pathway started to be designed it could only use those sugars synthesized in prebiotic conditions.

In these experiments the total affinity was kept constant, i.e. the concentrations of both sugars of five and six carbons were fixed. The affinity is considered to be positive when there is a net production of sugars of six carbons each, which in the steady state must be equal to the net amount of sugars of five carbons each consumed.

Because the critical dependence on the pentose cycle with the condensation and cleavage of sugars in the presence of the aldolase, different equilibrium constant for aldolase, K_A, were taken into account in sucessive experiments. This has been made considering different values for the kinetic constant ll and ll_1 in the model shown in Table I. These experiments might analyse the effect of this step in the design of the pathway.

A significant result of these experiments is shown in Figure 1. The zero generation shows some of the genomes (pathway design codification) initially present in the population randomly generated.

Initial population

k:1011 a:111 l:001 f: 0.132824
k:1100 a:110 l:101 f: 0.000000
k:0000 a:010 l:010 f: 0.000000
k:1010 a:000 l:000 f: 0.000000
k:1000 a:010 l:010 f: 0.000000
k:0110 a:000 l:011 f: 0.000000
k:0111 a:100 l:000 f: 0.000000
k:1111 a:101 l:010 f: 0.004396
k:0001 a:000 l:010 f: 0.000000
k:0010 a:100 l:100 f: 0.000000

Generation 40

k:1111 a:110 l:101 f: 0.270777
k:1111 a:110 l:101 f: 0.270777
k:1101 a:110 l:101 f: 0.000000
k:1111 a:110 l:101 f: 0.270777
k:1101 a:110 l:101 f: 0.000000
k:1111 a:110 l:101 f: 0.270777
k:1111 a:111 l:111 f: 0.250790
k:1111 a:111 l:111 f: 0.250790
k:1110 a:010 l:101 f: 0.000000
k:1111 a:110 l:101 f: 0.270777

Generation 20

k:1011 a:111 l:001 f: 0.132824
k:1100 a:110 l:101 f: 0.000000
k:1111 a:111 l:001 f: 0.246018
k:1011 a:111 l:001 f: 0.132824
k:1111 a:111 l:001 f: 0.246018
k:1111 a:111 l:001 f: 0.246018
k:0111 a:100 l:000 f: 0.000000
k:1111 a:111 l:001 f: 0.246018
k:1111 a:101 l:010 f: 0.004396
k:0001 a:000 l:010 f: 0.000000

Generation 60

k:1110 a:110 l:100 f: 0.279679
k:1110 a:110 l:100 f: 0.279679
k:1110 a:111 l:100 f: 0.189591
k:1110 a:110 l:100 f: 0.279679
k:1110 a:110 l:100 f: 0.279679
k:1110 a:110 l:000 f: 0.000000
k:1110 a:110 l:100 f: 0.279679
k:1100 a:110 l:100 f: 0.000000
k:1110 a:110 l:100 f: 0.279679
k:1110 a:110 l:100 f: 0.279679

Figure 1 Summarized results of a quasispecies simulation of a genome population with the characteristics described under the text. The value of the digits (0 or 1) represents either the activity or no activity of enzymes (i. e., *TK*, *TA* and *AL*, respectively) for sugars with different number of carbons. In this experiment a K value for the *AL* reaction of 0.001 was taken.

As can be seen, most of them give rise to null fluxes. Only few sequences have a non-zero flux, though very low. It is worth to notice how the quasispecies distribution varies (changing the master copy) in the sucessive generations until the system finds a master copy that might be the fittest (at least, no long-time simulation found another, better one). Then, it seems that the algorithm works efficiently for this particular setup.

Table 2 Sequences selected by quasispecies experiments for different values of the equilibrium constant in the AL reaction.

K_A	selected sequence
10	1111 011 100
1	1111 011 100
0.1	1111 011 100
0.01	1110 110 100
0.001	1110 110 100

In Table 2 the selected sequence for different values of the aldolase equilibrium constant is presented. As can be seen, for $K_A < 0.1$ the selected copy corresponds to pathway shown in Figure 2a, i.e. the present-day design of the pentose phosphate cycle. On the contrary, if $K_A > 0.1$, the selected sequence stands for the metabolic pathway shown in Figure 2b. Notice that the latter design, although involving the same number of steps as the pentose cycle, needs the presence as intermediates of sugars with a larger number of carbons (8). In any case, the selection of one of the pathways depends critically on the actual value of the aldolase equilibrium constant. In experiments carried out with sugars up to 13 carbons (data not shown) the metabolic pathway schematically drawn in Figure 2c is selected for particular values of the aldolase equilibrium constant.

6. Concluding remarks

We would like to make several remarks at two different levels: on the one hand, about the optimization algorithm used in the simulations; on the other hand, about the preliminary results above discussed.

It seems to be beyond any doubt that this kind of algorithm is really useful when the sequence space is so large that is not possible to compute the cost function for all the sequences in a reasonable period of time. In the first example analysed, i.e. the model with sugar up to eight carbons, that is not the case. The dimension of the sequence space is 2^{10} =1024, and any personal computer is able to compute the flux of every sequence in not too much time. Even so, these experiments are very illustrative in the sense that it allows to check the reliability of the method. When the model considers sugars up to 13 carbons, the size of the sequence space increases significantly ($2^{35} \approx 10^7$), but the algorithm is able to find the fittest sequence in an attainable number of generations.

One may think that the results obtained from this model would not correspond to reality because of the simplification carried out in the values of the kinetic constants (only two values are allowed, 0 and 1). In this sense, it would be very interesting to improve the model allowing the kinetic constants to take more than two values between 0 and 1. For instance, if 11 values are considered (0, 0.1, ..., 0.9, 1), the dimension of the sequence space would be 11^{10} for the simplest model. This number is too large to be tractable even with a powerful computer. Therefore, these kind of algorithms, likely combined with others as genetic algorithms[12], can be very efficient to find the optimal solution in these problems.

A second remark refers to the particular results derived in this contribution. In previous papers[1-3], the pentose phosphate pathway was presented as the simplest solution obtained from a mathematical game. Later on, it was proved that for linear pathways, the shorter a pathway is, the greater will be the flux[7]. But, when the pathways are not linear and moreover, some substrates are shared in bisubstrate reactions, the solution is not so simple. In fact, in the experiments discussed here, alternative designs are selected depending on particular thermodynamic constraints. It seems that, additionally to the simplicity of the pathway, its thermodynamic characteristics should also be considered. In other words, the answer to the question why the present-day pentose phosphate cycle has been selected not only depends on simplicity but on the values of equilibrium constants (real or apparent) of all the steps involved in the system.

As a final point, we would like to notice that other important aspects of the pentose phosphate pathway have been neglected. That is the case of enzymes isomerase and epimerase, and the existence of different isomers interchangeable under their action. Obviously, these details must be taken into account in a deeper study, and probably they could shed light to new features of the pentose phosphate cycle.

Acknowledgements

This work is partially supported by DGICYT (Spain), projects PB92-0908 and PB90-0846.

References

1. Meléndez-Hevia, E. and Isidoro, A. (1985) *J. Theor. Biol.* **117**, 251-263
2. Meléndez-Hevia, E. and Torres, N.V. (1988) *J. Theor. Biol.* **132**, 97-111
3. Meléndez-Hevia, E. (1990) *Biomed. Biochim. Acta* **49**, 903-916
4. Heinrich, R. and Hoffmann, E. (1991) *J. Theor. Biol* **151**, 249-283
5. Schuster, S. and Heinrich, R. (1987) *J. Theor. Biol.* **129**, 189-209
6. Heinrich, R. and Sonntag, I. (1982) *Biosystems* **15**, 301-316
7. Meléndez-Hevia, E., Waddell, T.G. and Montero, F. (1994)
   J. Theor. Biol. **166**, 201-220.
8. Eigen, M. (1971). *Naturwissenschaften* **58**, 465-523
9. Eigen, M., McCaskill, J. and Schuster, P. (1988). *J. Phys. Chem.* **92**, 6881-6891
10. Fontana, W., Schnabl, W. and Schuster, P. (1989). *Phys. Rev. A* **40**, 3301-3321
11. Reid, C. and Orgel, L. (1967) *Nature* **216**, 455
12. *Handbook of genetic algorithms.* (1991). Ed. Davis, L. Van Nostrand Reinhold.

The Forward and Inverse Problems of Electrocardiography

B.M. Horáček

The cardiac electric sources arising during the orderly process of propagated excitation (which precedes the mechanical contraction of the heart muscle) also produce a concomitant flow of electric current in the surrounding conductive tissues of the thorax and, thus, time-varying potentials on the body surface; these potentials are routinely recorded as clinical electrocardiograms. The diagnostic interpretation electrocardiograms are given depends upon how the relationship between cardiac electric sources and body-surface potentials is understood. The classical interpretation of electrocardiograms—described in every textbook of electrocardiography—is based on a model that regards the body as part of an infinite homogeneous conductor and assumes that the cardiac electric sources can be represented by a single, time-varying dipole at a fixed location (see, e.g., Macfarlane [1]). While it is true that this simplified model has been—and still is—widely and successfully applied in clinical electrocardiography, advances made during the last two decades have provided much more detailed insight into the relationship between cardiac electric sources and their body-surface manifestations.

The possibility that the measurements of cardiac electric activity obtained by means of new techniques of modern electrocardiography (for a review, see De Ambroggi et al. [2]) might be interpreted to better clinical effect has been created by the sound mathematical formulation—and, in particular, the numerical representation—of this relationship. The authors of the seven chapters that follow address fundamental problems associated with numerical calculations performed in an effort to determine the relationship between cardiac electric sources and body-surface potentials, both in the causal (forward) sense [3, 4, 5] and in the inverse sense [6, 7, 8, 9]. Like yin and yang, the forward and inverse problems of electrocardiography are complementary opposites: the former deals with how bioelectric phenomena in the heart engender electrocardiographic potentials, the latter with efforts to estimate equivalent cardiac electric sources that would account for the electrocardiographic measurements (for reviews, see Gulrajani et al. [10, 11]).

1 Forward problem

Current mathematical studies dealing with the genesis of electrocardiograms focus on modelling and simulation of the bioelectric activity of the heart—both at the cellular and at the macroscopic level (for a review, see Colli Franzone and Guerri [12]). The anisotropic conductivity of cardiac tissues and features of the anatomical architecture of the heart, such as the transmural rotation of fibers from the epicardium to the endocardium, influence the cardiac propagated excitation and the generation of extracardiac electric potentials. It is therefore desirable to study the propagation phenomena and the associated extracardiac fields in models that represent the cardiac anatomical structure as realistically as possible. The crucial practical problem is, then, how to mimic complex electrophysiological and anatomical characteristics of cardiac tissue accurately enough, while still making the simulations computationally tractable.

The approach that Colli Franzone et al. [3] propose uses an eikonal equation to simplify calculations of propagated excitation in a model of ventricular tissue that takes into account the anisotropic conductivity induced by the fiber structure of the ventricular myocardium. They compare two mathematical models of the depolarization process in the heart: one described by a reaction-diffusion system, the other by an approximate eikonal equation, which they have derived from the reaction-diffusion system by applying to it a singular-perturbation technique. They show that the eikonal model requires much less computer time and memory allocation than the reaction-diffusion model, and that it yields results that agree closely with those of the latter.

Horáček et al. [4] describe another approach to simplifying calculations of propagated excitation in the ventricular tissue with transmural rotation of fibers. Their hybrid model of cardiac excitation—which simulates electrotonic interactions on the basis of the anisotropic bidomain theory, but merely represents the action potential descriptively—combines the advantages of propagation models based on differential equations and of those based on cellular automata. Under the assumption of an equal anisotropy ratio, this model reduces the propagation problem for subthreshold potentials to one of solving a single nonlinear parabolic equation in the transmembrane potential. It performs as well as the reaction-diffusion model, at a small fraction of the computational cost.

Nenonen and Horáček [5] show how the extracardiac electric and magnetic fields are calculated, based on the anisotropic bidomain theory, for sources arising in the anisotropic ventricular myocardium; they compare the influence of the anisotropic electrical conductivity on the electric field and on the magnetic field.

2 Inverse problem

Unfortunately, the uniqueness of the inverse solution is not guaranteed. In trying to solve the inverse problem, one must first postulate a cardiac electrical generator and a volume-conductor model of the torso. One can then solve the forward problem for the particular sources in a particular volume conductor, and the restrictions implied by the specification of sources and their load enable one to solve the inverse problem, i.e., to estimate parameters of the postulated cardiac electrical generator. The inverse solution considered here [6, 7, 8, 9] is the one that is formulated as a reconstruction of maps of epicardial potentials and/or activation times from observed body-surface potential distributions, by means of boundary-element [6, 7, 9] or finite-element [8] methods.

Even with the restrictions described above (implicit constraints), any noise in measured potentials or any inaccuracy in the characterization of the volume conductor would be amplified by the inverse solution. To overcome this instability, additional, explicit constraints have to be incorporated in the inverse procedure by including a function that represents a penalty on any departure of the solution from these added constraints. This method of restricting the solution space of inverse problems is commonly called 'regularization'. In choosing the regularization parameter, one must strike a balance between the requirements that the solution be accurate and that it be stable.

Van Oosterom and Huiskamp [6] argue eloquently that any constraints used when computing the cardiac electric activity from potentials measured on the body surface must be based on sound biophysical and physiological principles. They illustrate their point through examples based on their inverse procedure for calculating activation times, in which implicit and explicit spatio-temporal constraints are incorporated, and they compare their inverse procedure with that aimed at computing epicardial potentials.

Greensite and Qian [7] show that, provided some practical requirements are satisfied, one can obtain a nominally well-posed formulation of the inverse problem of electrocardiography by using the time constraint that potential at the leading edge of a propagated excitation wave is a step discontinuity.

Johnson and MacLeod [8] present new methods for achieving better estimates of epicardial potentials from body-surface potentials through an inverse procedure in the finite-element model. They outline an automatic adaptive-refinement algorithm that minimizes the spatial discretization error in the transfer matrix and increases the accuracy of the inverse solution, and they introduce a local-regularization procedure that partitions the global transfer matrix into sub-matrices, allowing for varying amounts of smoothing.

Clements et al. [9] compare two methods for selecting an appropriate value of regularization parameters for the inverse computation of epicardial potentials in the boundary-element model. They use the generalized singular-value decomposition to obtain the explicit formula for the regularized solution of the least-squares problem produced by Tikhonov's regularization method.

Collectively, the following seven papers provide a fairly representative cross section of topics pertaining to the forward and inverse problems in contemporary electrocardiographic research.

References

[1] P.W. Macfarlane. Lead systems. In P.W. Macfarlane and T.D.V. Lawrie, editors, *Comprehensive Electrocardiology*, pp. 315–352, Pergamon Press, Oxford, 1989.

[2] L. De Ambroggi, E. Musso, and B. Taccardi. Body-surface mapping. In P.W. Macfarlane and T.D.V. Lawrie, editors, *Comprehensive Electrocardiology*, pp. 1015–1049, Pergamon Press, Oxford, 1989.

[3] P. Colli Franzone, L. Guerri, and M. Pennacchio. Simulation of the spread of excitation in the myocardial tissue. In this volume.

[4] B.M. Horáček, J. Nenonen, J. Edens, and L.J. Leon. A hybrid model of propagated excitation in the ventricular myocardium. In this volume.

[5] J. Nenonen and B.M. Horáček. Extracardiac field due to propagated excitation in the ventricular myocardium. In this volume.

[6] A. van Oosterom and G. Huiskamp. Spatio-temporal constraints in inverse electrocardiography. In this volume.

[7] F. Greensite and J.-J. Qian. When is the inverse problem of electrocardiography well-posed. In this volume.

[8] C.R. Johnson and R.S. MacLeod. Local regularization and adaptive methods for the inverse problem. In this volume.

[9] J.C. Clements, R. Carroll, and B.M. Horáček. On regularization parameters for inverse problems in electrocardiography. In this volume.

[10] R.M. Gulrajani, F.A. Roberge, and G.E. Mailloux. The forward problem of electrocardiography. In P.W. Macfarlane and T.D.V. Lawrie, editors, *Comprehensive Electrocardiology*, pp. 197–236, Pergamon Press, Oxford, 1989.

[11] R.M. Gulrajani, F.A. Roberge, and P. Savard. The inverse problem of electrocardiography. In P.W. Macfarlane and T.D.V. Lawrie, editors, *Comprehensive Electrocardiology*, pp. 237–288, Pergamon Press, Oxford, 1989.

[12] P. Colli Franzone and L. Guerri. Models of the spreading of excitation in myocardial tissue. *CRC Crit. Rev. Biomed. Eng.*, 20:211–253, 1992.

Simulation of the Spread of Excitation in the Myocardial Tissue

P. Colli Franzone, L. Guerri, and M. Pennacchio

1 Introduction

The mathematical modeling and simulation of the bioelectric activity of the heart, at the cellular or at the macroscopic level, is the subject of much past and present research. These models exhibit a wide range of complexity; for a sampling of the vast literature available, see the references in [1].

Here we are concerned with two macroscopic models of the depolarization process, i.e. the initial phase of the cardiac bioelectric cycle. The myocardial tissue is represented by an anisotropic bidomain in order to take into account the anisotropic conductivity induced by the fiber structure of the myocardium.

The first model is described by a reaction-diffusion system, and the dependent variables are the transmembrane and the extracellular (interstitial) potentials v, u and, if a conducting medium is adjacent to the myocardium, the extracardiac potential. Because a sharp front-like variation of v moves across the myocardium during depolarization, accurate numerical simulations require small space- and time-steps, which makes the reaction-diffusion model rather costly in terms of computer time and memory allocation. Consequently, this type of model has been used only for simulations in small myocardial volumes.

The occurrence of a steep moving front suggests the application of a singular perturbation technique to the reaction-diffusion system, which yields an approximate model called *eikonal*. For this model, we introduce the activation time $\psi(\mathbf{x})$, defined as the time instant such that $v(\mathbf{x}, \psi(\mathbf{x})) = (v_r + v_p)/2$, with v_r, v_p being the resting and plateau values of v. The eikonal model is characterized by a nonlinear elliptic equation in $\psi(\mathbf{x})$, the transmembrane potential v depends on $t - \psi(\mathbf{x})$, and the extracellular potential u satisfies a linear elliptic equation with a source term related to v.

The eikonal model requires much less computer time and memory allocation than the reaction-diffusion model. Since the two models, applied to small myocardial volumes, yield results in close agreement, only the eikonal model has been used for large-scale simulations described in this chapter.

2 Scope

Two mathematical models of the depolarization process are considered. The first model is described by a reaction–diffusion system and the dependent variables are the transmembrane and the extracellular potentials v, u. A second approximate model, called eikonal, is derived from the reaction-diffusion system by applying to it a singular perturbation technique. This model is characterized by a nonlinear elliptic equation in the activation time $\psi(\mathbf{x})$. The transmembrane potential $v(\mathbf{x}, t)$ depends on $t - \psi(\mathbf{x})$ and the extracellular potential $u(\mathbf{x}, t)$ is the solution of a linear elliptic equation with a source term related to $v(\mathbf{x}, t)$.

3 The reaction-diffusion model

The myocardial tissue is represented by an anisotropic bidomain, i.e. by the superposition of the anisotropic intra- and extracellular media, connected everywhere by the distributed cellular membrane.

The indices i, e will refer to the intra- and extracellular medium; $u_i, u_e, v \ (= u_i - u_e)$ are the intracellular, extracellular and transmembrane potentials; $\mathbf{a} = \mathbf{a}(\mathbf{x})$ is the unit vector parallel to the fiber direction in $\mathbf{x}$; $\sigma_l^{i,e}, \sigma_t^{i,e}$ are the conductivity coefficients in the direction parallel or perpendicular to $\mathbf{a}$; under the assumption of axial symmetry, $\sigma_t^{i,e}$ is the same for all directions perpendicular to $\mathbf{a}$.

$$M_{i,e} = \sigma_t^{i,e} I + \left(\sigma_l^{i,e} - \sigma_t^{i,e} \right) \mathbf{a}\,\mathbf{a}^T$$

are the conductivity tensors.

Setting $u = u_e$ and $M = M_i + M_e$, the reaction-diffusion model is described by the system of partial differential equations:

$$c_m v_t + I(v) = -\nabla \cdot M_i \nabla v + \nabla \cdot M_i \nabla u$$
$$\nabla \cdot (M_i \nabla v + M \nabla u) = 0 \tag{1}$$

in the myocardial volume H; $c_m = \chi C_m$, where χ is membrane surface per unit volume, and C_m is membrane capacitance per unit surface. Also, $\nabla \cdot M_0 \nabla u_0 = 0$ in the conducting extracardiac volume adjacent to H. $I(v)$ is related to the ionic current and should be described by a set of ordinary differential equations (Hodgkin–Huxley or FitzHugh–Nagumo type).

For the depolarization phase, one can consider a simpler form for $I(v)$: $I(v) = gf(v)$, where $g = \chi G$, with G being the membrane-conductivity coefficient and f a cubic-like function of v with three real zeros $v_r < v_{th} < v_p$ (resting, threshold, plateau values). In the particular case of f being a cubic polynomial, and with v shifted so that $v_r = 0$, f takes the form:

$$f = v(1 - v/v_{th})(1 - v/v_p). \tag{2}$$

Boundary and interface conditions for v, u, u_0 must be considered, as well as the initial condition $v(\mathbf{x}, 0) = v_0(\mathbf{x})$ [2]. In the depolarization phase, v exhibits a steep front traveling across the myocardium. The front is characterized by a thin layer (about 1 mm) in which v changes from v_r to v_p in about 1–2 ms.

For good accuracy in the numerical simulations, small space- and time-steps are required (of the order of 0.2 mm and 0.2 ms). Thus a very long sequence of discretized nonlinear problems would have to be solved is order to describe just the depolarization phase, which lasts about 50–60 ms in the dog heart and about 80–100 ms in the human heart. To overcome these computational difficulties, cellular-automata type models have been investigated by some [3, 4].

4   The eikonal model

Our approach to overcoming these computational difficulties is based on the singular perturbation analysis, which can be applied because of the steep front of v. We introduce the activation time $\psi(\mathbf{x})$, defined as the time instant such that $v(\mathbf{x}, \psi(\mathbf{x})) = (v_r + v_p)/2$. In the depolarization phase, $\psi(\mathbf{x})$ is a well-defined function and the level surface $t = \psi(\mathbf{x})$ represents the midsurface of the corresponding excitation layer.

The change of scales and the developments related to the perturbation analysis of the reaction-diffusion system lead to the first-order eikonal equation:

$$\nabla \cdot \mathbf{q} + (Kg^{1/2})(\mathbf{p} \cdot \mathbf{q})^{1/2} = c_m \qquad (3)$$

where $\mathbf{p} = \nabla\psi$, $\mathbf{q} = \alpha^{-2}(\alpha_i^2 M_e + \alpha_e^2 M_i)\,\mathbf{p}$, $\alpha_{i,e} = \mathbf{p}^T M_{i,e}\mathbf{p}$, $\alpha = \alpha_i + \alpha_e$.
This equation is elliptic and nonlinear. The boundary condition is $\mathbf{n}^T M_i \mathbf{p} = 0$ on ∂H, with the exception of those parts of ∂H and H where the value of ψ, related to an initial stimulation, is assigned.

In the eikonal model, the transmembrane potential v is approximated by:

$$v(\mathbf{x}, t) = V(\rho(t - \psi(\mathbf{x}))) \qquad \rho = g/(\mathbf{p} \cdot \mathbf{q})^{1/2}. \qquad (4)$$

The constant K and the function $V = V(\tau)$ constitute the solution of the eigenvalue problem:

$$KV_\tau + f(V) = V_{\tau\tau} \qquad -\infty < \tau < \infty, \qquad (5)$$

with $V(-\infty) = v_r = 0$, $V(0) = v_p/2$, and $V(+\infty) = v_p$.

When $f(V)$ is a cubic polynomial, K and V can be given in explicit form. Finally, $\nabla \cdot M\nabla u = -\nabla \cdot M_i \nabla v$ in H, and $\nabla \cdot M_0 \nabla u_0 = 0$ in the conducting medium adjacent to H, and, in addition, boundary conditions have to be satisfied (see [2]).

Since $\psi(\mathbf{x})$ is smoother than $v(\mathbf{x}, t)$, a greater space step can be used to solve the eikonal equation (about 3–4 times that needed for the reaction-diffusion system), leading to a reduction in the number of grid nodes by a factor of $3^3 - 4^3$. The time sequence can be selected so that the evolution of v and u can be followed.

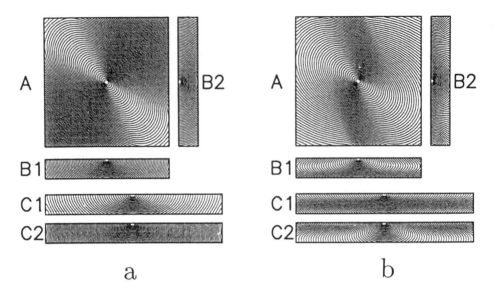

Figure 1. Propagation isochrones for a central epicardial stimulation in tissue slabs with parallel fibers (a) and with fibers that rotate 120° counterclockwise from epicardium to endocardium (b). Isochrones are displayed in 2-ms intervals on the epicardial plane (A), and on vertical sections through the center; sections B1 and B2 are perpendicular to the sides; C1 and C2 intersect, respectively, the lower left and lower right vertices of A.

5 Numerical simulations

We have applied previously the reaction-diffusion and eikonal models to a small myocardial volume ($1.5 \times 1.5 \times 0.5$ cm) and found that the results are in excellent agreement [2]. Therefore, the eikonal model can be applied with confidence to greater volumes, as will be done for the cases to be described.

Two myocardial volumes were considered:

a) a rectangular parallelepiped $6.5 \times 6.5 \times 1.0$ cm with the short distance taken as vertical (along the z − axis) and the lower and upper horizontal planes corresponding to the endocardial and epicardial surfaces.

Concerning the fiber direction, two cases were considered:
i) parallel fibers making an angle of 45° clockwise with the x − axis, and
ii) linear rotation of fibers, dependent on z, so that, for z fixed, the fibers were parallel. The fiber rotation was of 120° counterclockwise from epicardium to endocardium, starting at 45° clockwise from the x − axis on the epicardium.

b) a curved parallelepiped cut out from an ellipsoidal shell with axial symmetry relative to the z − axis. This curved parallelepiped was defined by:

$$x = a \cos \theta \cos \varphi, \; y = a \cos \theta \sin \varphi, \; z = c \sin \theta, \; \text{with } \theta_1 \leq \theta \leq \theta_2 \text{ and } \varphi_1 \leq \varphi \leq \varphi_2$$

$$a = a_1 + r(a_2 - a_1), \; c = c_1 + r(c_2 - c_1), \; \text{with } 0 \leq r \leq 1.$$

For r fixed, we have a quadrilateral portion of an ellipsoidal surface. On this surface, **a** is tangent and makes an angle $\gamma = \gamma(r)$ with the vector tangent to the line of the surface defined by θ constant and φ variable; $\gamma(r)$ is linear, with $\gamma(0) = -45°$ and $\gamma(1) = 75°$ (i.e. a counterclockwise rotation of 120°). The inner and outer surfaces (with $a = a_1$ and $a = a_2$) correspond to the endocardium and epicardium, respectively. We chose $a_1 = 1.5$ cm, $a_2 = 2.7$ cm, $c_1 = 4.4$ cm, $c_2 = 5.5$ cm, $\theta_1 = -3\pi/8$, $\theta_2 = 0$, $\varphi_1 = -\pi/3$, $\varphi_2 = \pi/3$.

The parameters chosen for the numerical simulations were: $\chi = 2 \cdot 10^3$ cm^{-1}, $C_m = 0.8 \; \mu\text{F cm}^{-2}$, $v_{th} = 33$ mV, $v_p = 100$ mV, $f(v)$ was a cubic function, $\sigma = 10^{-3} \; \Omega^{-1}$ cm^{-1}, $\sigma_l^e = 2\sigma$, $\sigma_t^e = 1.35\sigma$, $\sigma_l^i = 3\sigma$, $\sigma_t^i = 0.315\sigma$, $G = 3.448 \cdot 10^{-4} \; \Omega^{-1}$ cm^{-2}.

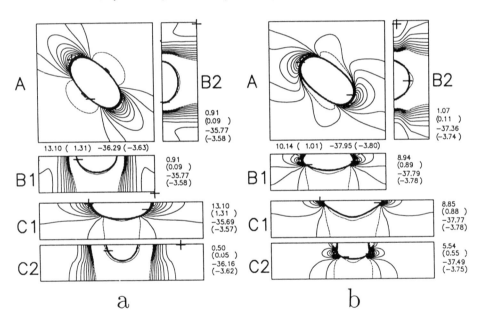

Figure 2. Potential maps 30 ms after a central epicardial stimulation in tissue with parallel (a) and rotating (b) fibers. Layout as in Fig. 1. The zero isopotential line is dashed; the maximum and minimum (mV), and the intervals between consecutive positive and negative lines (in parenthesis) are indicated for each panel.

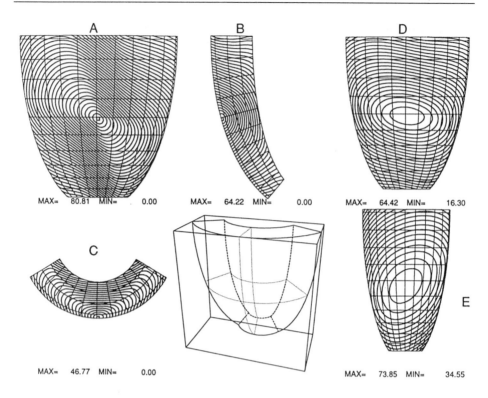

Figure 3. Propagation isochrones for a central epicardial stimulation in a curved slab. Panels A, D and E refer to r-sections ($r = 1.0, 0.5, 0$); panels B and C refer to φ-sections and θ-sections, indicated by dotted lines in the perspective view of the curved slab. Intervals between isochrone lines are of 2 ms.

The numerical approximation of the eikonal model in the case of a curved slab has been described in [5]; it has been obtained by a suitable extension of the finite-element technique used for a flat slab.

Comparison of the isochrones of Figs. 1a and 1b reveals the influence of the fiber rotation on the shape and propagation of the wavefront; the wavefronts exhibit a marked change of curvature due to the variable fiber angle that the front encounters during its intramural propagation. After endocardial breakthrough, the velocity of the front near the endocardium increases steadily, so that the endocardial foot of the front races ahead of the epicardial foot (see panel C2 of Fig. 1b, where the lower edge is parallel to the fiber direction). Thus, after endocardial breakthrough, the spread of excitation tends to be directed toward the epicardial face, from which the excitation started, as shown in the vertical sections displayed in panels B1 and C2 of Fig. 1b.

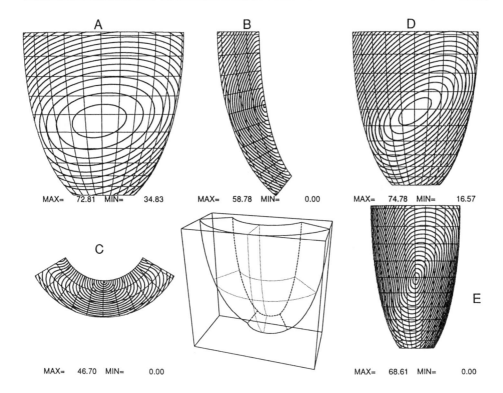

Figure 4. Propagation isochrones for a central endocardial stimulation in a curved slab. Same layout as in Fig. 3.

This does not occur in panel B1 or C2 of Fig. 1a, which depicts the slab with no fiber rotation. The depolarization of the slab in Fig. 1b is completed in 104 ms, compared to the 152 ms required in the case of parallel fibers.

Figs. 2a and 2b display the epicardial extracellular potential 30 ms after epicardial stimulation. Fiber direction is constant in Fig. 2a and rotates counterclockwise in Fig. 2b; in both cases, the endocardium is in contact with a slab of isotropic medium (blood) having the same dimension of the myocardial slab ($6.5 \times 6.5 \times 1.0$ cm) and $\sigma_b = 6 \cdot 10^{-3} \Omega^{-1}$ cm^{-1}.

The two epicardial potential patterns show two maxima lying ahead of the narrow ends of the front propagating along a fiber. These maxima are surrounded by a family of positive equipotential lines.

In Fig. 2a (no fiber rotation), the maxima are aligned along the major axis of the ellipsoidal wavefront, coinciding with the fiber direction. In Fig. 2b (counterclockwise fiber rotation), the underlying fiber rotation manifests itself as a progressive counterclockwise expansion and rotation of the positive area surrounding the maxima. Intramural sections (particularly Fig. 2b, panels B1, C2) show that this rotation and expansion was an epicardial reflection of the fiber rotation.

Examination of increasingly deeper sections (not shown) reveals a rotation of the wavefront shape and of the segment joining the maxima, related to the fiber rotation. However, the wavefront rotation lagged behind that of the fibers, while no such lag was observed for the segment joining the maxima.

Fig. 3 displays the isochrone maps on r-sections (panels A, D and E, $r = 1.0, 0.5, 0$), from the epicardium to the endocardium, following an epicardial stimulation. Intramural φ- and θ-sections are shown in panels B and C.

The rotation of the front shape on the successive sections is related to, but lags behind, the fiber rotation through the curved slab. The epicardial isochrone map shows a faster spreading along fibers. Across fibers, the front's spreading is slower and the front is rather flat near the stimulation site. Subsequently, as the front expands, its motion across fibers is accelerated, a sign of the influence of the excitation spreading in deeper layers of rotating fibers.

Fig. 4 displays the isochrone maps elicited by an endocardial stimulation. Panels D and E refer respectively to r-sections ($r = 1.5, 0.5, 0$), and panels B and C refer to intramural φ-sections. In the θ-section displayed in panel C of Figs. 3 and 4, the curved inner and outer edges lie, respectively, on the endocardial and epicardial surfaces. Comparing these panels, we see that in the epicardial stimulation the isochrones exhibit a greater change of curvature and that the front propagation after breakthrough is markedly slower than in the endocardial stimulation.

References

[1] P. Colli Franzone and L. Guerri. Models of the spreading of excitation in myocardial tissue. *CRC Crit. Rev. Biomed. Eng.*, 20:211–253, 1992.

[2] P. Colli Franzone and L. Guerri. Spreading of excitation in 3-D models of the anisotropic cardiac tissue. I. Validation of the eikonal model. *Math. Biosci.*, 113:145–209, 1993.

[3] L.J. Leon and B.M. Horáček. Computer model of excitation and recovery in the anisotropic myocardium. *J. Electrocardiol.*, 24:1–41, 1991.

[4] B.M. Horáček, J. Nenonen, J. Edens, and L.J. Leon. A hybrid computer model of propagated excitation in the anisotropic human ventricular myocardium. In this volume.

[5] A. Meuli. Simulazione numerica di modelli matematici della propagazione del fronte di eccitazione e del potenziale in elettrocardiologia. Doctoral dissertation, Pavia, 1992.

A Hybrid Model of Propagated Excitation in the Ventricular Myocardium

B.M. Horáček, J. Nenonen, J.A. Edens, and L.J. Leon

1 Introduction

The anisotropic conductivity of cardiac tissue and features of the anatomical architecture of the heart, such as the transmural rotation of fibers from the epicardium to the endocardium or their spiral rotation near the apex [1, 2], have a profound influence on the heart's propagated excitation and the generation of extracardiac electric potential and magnetic field—as has been substantiated by many experimental findings (e.g. [3, 4]). Therefore it is of great interest to study the propagation phenomena and the associated electromagnetic field in mathematical models that represent *realistically* the anisotropic heart. We have addressed this problem, and the result of our efforts is a model [5, 6, 7] whose salient features related to the propagation algorithm are highlighted in this paper; a companion paper [8] deals with features related to extracardiac electric and magnetic fields.

Our ultimate goal is to simulate accurately how cardiac anisotropy affects the electrophysiological events in cardiac tissue that are reflected in electrocardiograms and magnetocardiograms. To achieve this goal, our computer model of cardiac excitation and recovery has to combine the advantages, and avoid the drawbacks, of the two classes (see [9]) of propagation models: those based on differential equations and those based on cellular automata.

The theoretical framework for our model is provided by the general anisotropic bidomain theory [10, 11]; however, we have to adopt the restrictive assumption of an equal anisotropy ratio to make the propagation problem tractable, and another restrictive assumption of isotropic interstitial medium has to be made to make calculation of extracardiac fields (see [8]) tractable. The assumption of an equal anisotropy ratio helps to reduce the propagation problem to one of solving a single nonlinear parabolic equation in the transmembrane potential. By using this equation when an element has its transmembrane potential under a threshold value, we avoid the drawbacks (demonstrated in [12]) of the approach based on Huygens' principle; during recovery phase the element's behavior is ruled by a cellular automaton. The latter feature substantially reduces the model's computational requirements.

2 Scope

In this chapter, we present results that validate our propagation algorithm. The validation is based on simulations in which cardiac excitation propagates through a slab that mimics the transmural fiber rotation in the ventricular wall. Our algorithm performs as well as the reaction-diffusion model of Colli Franzone and Guerri [13], at a small fraction of the computational cost. On the other hand, it is probably not significantly more computationally efficient than the eikonal-equation models of Colli Franzone and Guerri [13], and Keener [14].

3 Theory

Our model is defined in the anisotropic bidomain region H, which represents myocardium as a system of cylindrical fibers with anisotropic conductivity; a *longitudinal* fiber direction is determined by a unit vector $\mathbf{a}_\ell(\mathbf{x})$. The local basis consisting of an orthogonal set $\{\mathbf{a}_1(\mathbf{x}), \mathbf{a}_2(\mathbf{x}), \mathbf{a}_3(\mathbf{x})\}$ is chosen at $\mathbf{x}$ so that $\mathbf{a}_3(\mathbf{x})$ is parallel with $\mathbf{a}_\ell(\mathbf{x})$; $\mathbf{a}_1(\mathbf{x})$ and $\mathbf{a}_2(\mathbf{x})$ are said to be in the *transverse* direction at $\mathbf{x}$. In the local basis, the intracellular conductivities along the axes are $\sigma_1^i, \sigma_2^i, \sigma_3^i$, and the corresponding interstitial conductivities are $\sigma_1^e, \sigma_2^e, \sigma_3^e$; due to fibers' axial symmetry, $\sigma_1^{i,e} = \sigma_2^{i,e} = \sigma_t^{i,e}$ are conductivities in the transverse direction, and $\sigma_3^{i,e} = \sigma_\ell^{i,e}$ are conductivities in the longitudinal direction ($\sigma_t^{i,e}$ and $\sigma_\ell^{i,e}$ are independent of $\mathbf{x}$).

The tensors D_i^* and D_e^* describe respectively the intracellular and interstitial anisotropic conductivities [12]; in the local basis, these tensors are diagonal. To allow for variable fiber direction, we set the global coordinate system in which the local basis is defined as $A = (\mathbf{a}_1(\mathbf{x}), \mathbf{a}_2(\mathbf{x}), \mathbf{a}_3(\mathbf{x}))$. To represent the tensors D_i^* and D_e^* in that global coordinate system, we have to rotate the local basis by multiplying it on each side by a rotation matrix

$$D_{i,e} = A D_{i,e}^* A^T, \tag{1}$$

where the superscript T denotes matrix transpose. In our model, $\mathbf{a}_\ell(\mathbf{x})$ is stored as azimuth angle Φ and colatitude angle Θ of the global spherical coordinate system and (1) becomes

$$D_{i,e} = \begin{pmatrix} \cos\Theta\cos\Phi & -\sin\Phi & \sin\Theta\cos\Phi \\ \cos\Theta\sin\Phi & \cos\Phi & \sin\Theta\sin\Phi \\ -\sin\Theta & 0 & \cos\Theta \end{pmatrix} \begin{pmatrix} \sigma_t^{i,e} & 0 & 0 \\ 0 & \sigma_t^{i,e} & 0 \\ 0 & 0 & \sigma_\ell^{i,e} \end{pmatrix} \begin{pmatrix} \cos\Theta\cos\Phi & \cos\Theta\sin\Phi & -\sin\Theta \\ -\sin\Phi & \cos\Phi & 0 \\ \sin\Theta\cos\Phi & \sin\Theta\sin\Phi & \cos\Theta \end{pmatrix}.$$

Electrical potential and current density are defined in H as macroscopic quantities, which may be considered to be averages over small volumes of cardiac tissue encompassing several cells. The current densities are given by Ohm's law:

$$\mathbf{J}_i = -D_i \nabla \phi_i, \quad \mathbf{J}_e = -D_e \nabla \phi_e, \tag{2}$$

where ϕ_i and ϕ_e are, respectively, the electrical potentials in the intracellular and interstitial domains. The total macroscopic current density $\mathbf{J}$ is then

$$\mathbf{J} = \mathbf{J}_i + \mathbf{J}_e = -D_i\nabla\phi_i - D_e\nabla\phi_e. \tag{3}$$

The conservation law requires that the divergence of total current density vanish

$$\nabla \cdot \mathbf{J} = \nabla \cdot (\mathbf{J}_i + \mathbf{J}_e) = 0, \tag{4}$$

or, equivalently,

$$\nabla \cdot (D_i\nabla\phi_i + D_e\nabla\phi_e) = 0, \tag{5}$$

i.e., current can flow from one domain to the other, but there are no net sources or sinks of current in H.

The intracellular and interstitial domains are coupled through a distributed cellular membrane; the outward transmembrane current per unit volume is, in accordance with the conservation law (equation (4)),

$$i_m = -\nabla \cdot \mathbf{J}_i = \nabla \cdot \mathbf{J}_e, \tag{6}$$

or, equivalently,

$$i_m = \nabla \cdot D_i\nabla\phi_i = -\nabla \cdot D_e\nabla\phi_e. \tag{7}$$

The transmembrane potential is defined in H as

$$v_m = \phi_i - \phi_e. \tag{8}$$

The terms $D_i\nabla\phi_e$ and $-D_i\nabla\phi_e$ can be added to (3) and, by using (8) and denoting $D = D_i + D_e$, we get for the total current density

$$\mathbf{J} = -D_i\nabla v_m - D\nabla\phi_e = \mathbf{J}^i - D\nabla\phi_e, \tag{9}$$

where $\mathbf{J}^i$ (called an *impressed current density* [15, 16], whose dimension is a current dipole moment per unit volume) was substituted for $-D_i\nabla v_m$. The impressed current density is driven by the electrochemical generators in cardiac cells, while the term $-D\nabla\phi_e$ represents passive return currents in the tissue. From the condition that the divergence of $\mathbf{J}$ must vanish (equation (4)) or, alternatively, by substitutions from (8) into (7), we can obtain a fundamental partial differential equation in ϕ_e, with a source term that involves the gradient of the transmembrane potential v_m (cf. equations 4 and 7 in [10] and equation 7 in [17])

$$\nabla \cdot D\nabla\phi_e = -\nabla \cdot D_i\nabla v_m$$
$$\nabla \cdot D\nabla\phi_e = \nabla \cdot \mathbf{J}^i = -I_v. \tag{10}$$

Thus, (10) relates the distribution of the interstitial potential ϕ_e to the current sources. Equation (10) also indicates that H can be regarded as a composite medium that is characterized by the bulk conductivity tensor D and in which there is a distributed impressed current density $\mathbf{J}^i$ that equals $-D_i \nabla v_m$ and whose divergence is the impressed scalar current per unit volume $I_v = -\nabla \cdot \mathbf{J}^i$ [15].

From cable theory [18], we have

$$i_m = \chi[C_m \frac{\partial v_m}{\partial t} + I_{ion} - I_{app}], \tag{11}$$

where χ is the membrane surface area per unit volume of the tissue, C_m is the membrane capacitance per unit area of the membrane surface, I_{ion} is the ionic current per unit area of the membrane surface, and I_{app} represents an applied current stimulus to start the excitation. By combining (7) and (11) and by defining

$$c_m = \chi C_m, \quad i_{ion} = \chi I_{ion}, \quad i_{app} = \chi I_{app},$$

we get an equation that relates the spatial distribution of intracellular potential and the membrane dynamics:

$$c_m \frac{\partial v_m}{\partial t} + i_{ion} - i_{app} = \nabla \cdot D_i \nabla \phi_i. \tag{12}$$

By substituting $\phi_i = v_m + \phi_e$, we can rewrite (12) in terms of v_m and ϕ_e:

$$c_m \frac{\partial v_m}{\partial t} - \nabla \cdot D_i \nabla v_m + i_{ion}(v_m) = \nabla \cdot D_i \nabla \phi_e + i_{app}. \tag{13}$$

In this form, we have in H a system composed of a nonlinear parabolic equation (13) in v_m, coupled with an elliptic equation (10) in ϕ_e.

Under the condition of an equal anisotropy ratio [17]—which requires that $\sigma_1^e/\sigma_1^i = \sigma_2^e/\sigma_2^i = \sigma_3^e/\sigma_3^i = k$, or $D_e = kD_i$, where k is a scalar constant—we get from (7) and (10) (cf. equation 21 in [17])

$$i_m = \frac{k}{1+k} \nabla \cdot D_i \nabla v_m = \frac{1}{1+k} \nabla \cdot D_e \nabla v_m = \nabla \cdot D_k \nabla v_m, \tag{14}$$

where $D_k = kD_i/(1+k) = D_e/(1+k)$. It follows from (7) and (14) that

$$\phi_i = \frac{k}{1+k} v_m. \tag{15}$$

Substituting $D = D_i + kD_i$ (for the bulk conductivity tensor under the condition of equal anisotropy ratio) in (10) yields

$$\nabla \cdot D_i \nabla \phi_e = -\frac{1}{1+k} \nabla \cdot D_i \nabla v_m; \tag{16}$$

therefore,

$$\phi_e = -\frac{1}{1+k} v_m. \tag{17}$$

As pointed out by Plonsey and Barr [17], (15) and (17) are generalizations of the relationships noted by Hodgkin and Rushton [19] for a one-dimensional cable (where $k = \sigma_e/\sigma_i$); in particular, it follows from (2), (15) and (17) that the intracellular current density $\mathbf{J}_i$ and interstitial current density $\mathbf{J}_e$ satisfy the relationship

$$\mathbf{J} = \mathbf{J}_i + \mathbf{J}_e = 0, \tag{18}$$

just as in cable theory, and substitution from (14) into (12) yields a single nonlinear parabolic equation in the transmembrane potential v_m

$$c_m \frac{\partial v_m}{\partial t} = \nabla \cdot D_k \nabla v_m - i_{ion}(v_m) + i_{app}, \tag{19}$$

on which we base our propagation algorithm (described in [5, 6]).

To conclude this brief account of the theoretical basis of our model, we will examine the nature of the term $\nabla \cdot D_k \nabla v_m$ in (19) and make a remark about the net contribution to the membrane current that is solely due to the effect of anisotropy. Utilizing the property of axial symmetry, we can write the conductivity tensors $D_{i,e}$ as

$$D_{i,e} = (\sigma_\ell^{i,e} - \sigma_t^{i,e})\mathbf{a}_\ell\mathbf{a}_\ell^T + \sigma_t^{i,e}I, \tag{20}$$

where I is the identity matrix. Thus, an anisotropic medium can be thought of as having isotropic conductivities $\sigma_t^{i,e}$ throughout, plus the "boost" $(\sigma_\ell^{i,e} - \sigma_t^{i,e})$ along the fiber direction. According to (14) and (20), the transmembrane current i_m is then

$$i_m = \nabla \cdot D_k \nabla v_m = \frac{k}{1+k}\{\nabla \cdot (\sigma_\ell^i - \sigma_t^i)\mathbf{a}_\ell\mathbf{a}_\ell^T \nabla v_m + \nabla \cdot \sigma_t^i \nabla v_m\}. \tag{21}$$

From this, using the identity $\mathbf{a}_\ell\mathbf{a}_\ell^T \nabla v_m = (\nabla v_m \cdot \mathbf{a}_\ell)\mathbf{a}_\ell$, we can identify the isotropic and anisotropic components of transmembrane current i_m as follows: isotropic component of i_m equals $(k/(1+k))\sigma_t^i \nabla^2 v_m$, where ∇^2 is a laplacian operator; and anisotropic component of i_m equals $(k/(1+k))\nabla \cdot (\sigma_\ell^i - \sigma_t^i)(\nabla v_m \cdot \mathbf{a}_\ell)\mathbf{a}_\ell$. These last two expressions, in their discretized form, governed the electrotonic interaction of excitable elements in our model.

4 Implemention

The geometry of the model is that of a realistically shaped human ventricular myocardium, with section-planes parallel to those chosen by Durrer et al. [20]. This geometry was reconstructed from 1-mm sections of the human heart [21]. The assignment of the principal fiber direction was performed separately for left-ventricular, right-ventricular and papillary-muscle elements, with the fiber direction rotating counterclockwise from epicardium to endocardium [1].

The software has been written in Fortran 77. The model comprises 1,000,000 elements, each of which has been allotted 64 bits. For each element, there is a 42-bit *dynamic section* (including two bits in masks) and a 24-bit *static section*. The bits of the dynamic section store the state-transition number, clocks, potentials v_i and v_{i+1}, where i is a time-step, and the excitation time. The static section stores the fiber orientation (as the colatitude Θ and the azimuth Φ) and the element's type.

Figure 1. Isochrones of propagated excitation started at the upper layer in block I of Colli Franzone and Guerri. Isochrones are plotted at 1-ms increments, with every 10th line thicker, on horizontal sections which are 1 mm apart from the top (upper panel) to the bottom (lower panel) of a 1 cm × 1 cm × 0.3 cm block that is subdivided into 51 × 51 × 16 cubic elements with 0.2-mm sides. The fibers in the panels cross the horizontal side of the figure at angles of (from top to bottom) $0°, 42°, 47°$ and $90°$, according to the rotation function defined in [13]. The left, middle and right columns show the propagated excitation started at the right vertex, midpoint of edge, and left vertex, respectively. Compare with Fig.2 in [13].

5   Simulations

Our earlier papers describe simulations of propagated excitation and recovery in two-dimensional and three-dimensional arrays of excitable elements—including simulations in the idealized ellipsoidal left ventricle [5] and in the entire human ventricular my-ocardium [7]. In this paper we add results that validate our propagation algorithm. Recently, Colli Franzone and Guerri [13] validated their eikonal-equation model of propagated excitation by using as the "gold standard" a reaction-diffusion system im-plemented on a parallelepipedal lattice of limited size. To validate our propagation algorithm, we performed simulations of propagated excitation in a volume H with ro-tational anisotropy, which closely resembled their Block Type I, consisting of rotating layers of parallel fibers representing a portion of the ventricular wall (see [13] page 172).

Our "Block I" was a slab consisting of $51 \times 51 \times 16$ cubic elements with 0.2-mm sides; i.e., the centre-to-centre dimensions of this block were 1 cm $\times$ 1 cm $\times$ 0.3 cm. (In our companion paper [8], we describe simulations with a larger, 1-cm-thick block, whose dimensions are more in accordance with the thickness of the wall of the left ventricle in the human heart.) The fiber rotation from the top to the bottom of the block was modelled by the same rotation function as that used by Colli Franzone and Guerri (see equation 4.17 in [13]); this function describes the experimental data of Streeter [1].

Three different excitation sequences were started by stimulation at the upper layer of the block; they were initiated at the right vertex, the midpoint of the edge, and the left vertex. The excitation sequences were initiated by a $10\mu\text{A/cm}^2$ stimulus I_{app}, delivered at $3 \times 3 \times 2$ cubic elements for vertex stimulation and at $4 \times 3 \times 2$ cubic elements for edge stimulation. The propagation parameters used in all simulations are listed in Table 1 (cf. parameters in [17, 22]). The simulations were run on the Stellar GS1000 computer system (Stardent Computer Inc., Concord, MA, U.S.A.).

Table 1. Propagation parameters

Equal anisotropy ratio k	1
Spatial step h	0.2 mm
Time step δt	0.025 ms
Resting potential v_r	-84 mV
Threshold potential v_{th}	-60 mV
Peak potential v_p	$+20$ mV
Membrane capacitance C_m	$1\mu\text{F/cm}^2$
Surface-to-volume ratio χ	500 cm^{-1}
Transverse intracellular conductivity σ_t^i	0.05 S/m
Longitudinal intracellular conductivity σ_ℓ^i	0.20 S/m

Figure 1 shows the evolution of three excitation sequences as isochrones of activation; the activation time is defined as the instant at which the transmembrane potential v_m equals $(v_p + v_r)/2$. Isochrones are constructed, at 1-ms increments, as straight lines connecting neighbouring nodes with the same activation time. Figure 2 in [13], which has an identical layout as our Fig. 1, shows results obtained by Colli Franzone and Guerri by means of simulations in the reaction-diffusion model. A detailed comparison of our results with those obtained by the reaction-diffusion model reveals striking resemblance. Although we have not yet performed a quantitative comparison by calculating the relative error and correlation coefficient, it can be easily ascertained by *examining* the two figures that the activation-time errors produced by our model in the corners distal from the stimulation site do not exceed 1 millisecond. The run time of our simulations was typically under one hour, and, though we have no direct comparison of run times yet, we feel safe in saying that our algorithm is at least 10 times faster than the reaction-diffusion system used in [13]. It will be interesting to compare the merits of the eikonal-equation approach [13, 14] and our approach based on (19).

6   Conclusions

Our model allows simulations involving large and complex anatomical structures, such as human ventricles. This was achieved at the cost of bypassing simulation of membrane ionic processes during the cardiac action potential. However, at subthreshold levels of transmembrane potential, the model's propagation algorithm physiologically simulates electrotonic interactions.

Acknowledgements

The work reported herein was supported in part by research grants from the Heart and Stroke Foundation of Nova Scotia and from the Medical Research Council of Canada.

References

[1] D.D. Streeter Jr. Gross morphology and fiber geometry of the heart. In R.M. Berne, N. Sperelakis, and S.R. Geiger, editors, *Handbook of physiology - Section 2: The cardiovascular system, Volume I: The heart*, pages 61–112, American Physiological Society, Bethesda, MD, 1979.

[2] P.M.F. Nielson, I.L. LeGrice, B.M. Smaill, and P.J. Hunter. A mathematical model of the geometry and fibrous structure of the heart. *Am J Physiol*, 260:H1365–H1378, 1991.

[3] S. Watabe, B. Taccardi, R.L. Lux, and P.D. Ershler. Effect of non-transmural necrosis on epicardial potential fields: correlation with fiber direction. *Circulation*, 82:2115–2127, 1990.

[4] D.J. Staton, R.N. Friedman, and J.P. Wikswo Jr. High-resolution SQUID magnetocardiographic mapping of action currents in canine cardiac slices. *Circulation*, 84:II–667, 1991.

[5] L.J. Leon and B.M. Horáček. Computer model of excitation and recovery in the anisotropic myocardium. *J Electrocardiol*, 24:1–41, 1991.

[6] J. Nenonen, J.A. Edens, L.J. Leon, and B.M. Horáček. Computer model of propagated excitation in the anisotropic human heart: I. Implementation and algorithms. In *Computers in cardiology*, pages 545–548, IEEE Computer Society Press, Los Alamitos, CA, 1992.

[7] J. Nenonen, J.A. Edens, L.J. Leon, and B.M. Horáček. Computer model of propagated excitation in the anisotropic human heart: II. Simulation of extracardiac fields. In *Computers in cardiology*, pages 217–220, IEEE Computer Society Press, Los Alamitos, CA, 1992.

[8] J. Nenonen and B.M. Horáček. Simulation of the extracardiac electromagnetic field due to propagated excitation in the anisotropic ventricular myocardium. In this volume.

[9] R. Plonsey and R.C. Barr. Mathematical modeling of electrical activity of the heart. *J Electrocardiol*, 20:219–226, 1987.

[10] D.B. Geselowitz and W.T. Miller III. A bidomain model for anisotropic cardiac muscle. *Ann Biomed Eng*, 11:191–206, 1983.

[11] P. Colli Franzone, L. Guerri, and C. Viganotti. Oblique dipole layer potentials applied to electrocardiology. *J Math Biol*, 17:93–124, 1983.

[12] P. Colli Franzone, L. Guerri, and S. Tentoni. Mathematical modelling of the excitation process in myocardial tissue: influence of fiber rotation on wavefront propagation and potential field. *Math Biosci*, 101:155–235, 1990.

[13] P. Colli Franzone and L. Guerri. Spreading of excitation in 3-D models of the anisotropic cardiac tissue. I. Validation of the eikonal model. *Math Biosci*, 113:145–209, 1993.

[14] J.P. Keener. An eikonal-curvature equation for action potential propagation in myocardium. *J Math Biol*, 29:629–651, 1991.

[15] R. Plonsey. *Bioelectric phenomena*. McGraw-Hill, New York, 1969.

[16] D.B. Geselowitz. On bioelectric potentials in an inhomogeneous volume conductor. *Biophys J*, 7:1–11, 1967.

[17] R. Plonsey and R.C. Barr. Current flow patterns in two-dimensional anisotropic bisyncytia with normal and extreme conductivities. *Biophys J*, 45:557–571, 1984.

[18] J.J.B. Jack, D. Noble, and R.W. Tsien. *Electric current flow in excitable cells*. Clarendon Press, Oxford, 1983. (2nd ed).

[19] A.L. Hodgkin and W.A. Rushton. The electrical constants of crustacean nerve fiber. *Proc R Soc Biol Sci*, 133:444–479, 1946.

[20] D. Durrer, R.Th. van Dam, G.E. Freud, M.J. Janse, F. L. Meijler, and R. C. Arzbaecher. Total excitation of the isolated human heart. *Circulation*, 41:899–912, 1970.

[21] W.J. Eifler, E. Macchi, H.J. Ritsema van Eck, B.M. Horáček, and P.M. Rautaharju. Mechanism of generation of body surface electrocardiographic P-waves in normal, middle, and lower sinus rhythms. *Circ Res*, 48:168–182, 1981.

[22] F. A. Roberge, A. Vinet, and B. Victorri. Reconstruction of propagated electrical activity with a two dimensional model of anisotropic heart muscle. *Circ Res*, 58:461–475, 1986.

Extracardiac Field due to Propagated Excitation in the Ventricular Myocardium

J. Nenonen and B.M. Horáček

1 Introduction

The bidomain theory was originally developed to relate macroscopically the transmembrane potential and the extracellular potential (see, e.g., [1, 2, 3]). For pragmatic reasons, most three-dimensional propagation models based on the bidomain concept had to be reduced to oversimplified cellular automata. In such models, the excitation wavefront travels through the tissue at a preassigned conduction velocity, obeying Huygens' principle of wave propagation (see, e.g., [3, 4, 5]). In addition, excitation isochrones are usually fixed on the basis of experimental data measured from isolated human hearts [6]. Extracardiac potentials and magnetic fields are calculated from a set of lumped current dipoles, which are obtained by summing the contributions of the transmembrane potential gradients within the spatial region represented by each dipole.

Anisotropic conductivity was incorporated into later formulations of the bidomain theory [7, 8]. Leon and Horáček [9] reported a hybrid propagation model that combines the anisotropic bidomain theory with the cellular automata concept. The model, which was developed as a means of simulating propagation and extracardiac potentials in complex three-dimensional arrays of cardiac cells, simulates electrotonic interactions on the basis of the anisotropic bidomain theory, but merely represents the action potential descriptively. Algorithms of this model were described in detail [10], and its capabilities were tested by simulations of the ventricular pre-excitation in a realistic model of human ventricles [11].

2 Scope

The anisotropic bidomain theory and the salient features of the hybrid computer model of propagated excitation are described in a previous chapter [1]. In this chapter, we report simulations of the extracardiac electric potential and magnetic field. We present results computed for small propagated wavefronts (diameter < 15 mm) in three-dimensional rectangular arrays of excitable cells. The focus is on comparing the relative influence of the anisotropic electrical conductivity on external electric and magnetic fields. Extracardiac potentials and magnetic fields are evaluated in an unbounded, isotropic composite medium. We also show the influence of a macroscopic conductivity boundary by evaluating the fields in a semi-infinite volume conductor.

3 Methods

The bidomain theory assumes two interpenetrating resistive domains, intra- and extracellular, which occupy the same volume H (myocardium) and which are separated by a distributed cellular membrane [2, 3]. The bidomain volume H can be immersed either in a homogeneous or in an inhomogeneous isotropic volume conductor. We denote the intracellular potential by ϕ_i, the extracellular potential by ϕ_e, and the transmembrane potential by $v_m = \phi_i - \phi_e$. The principal (longitudinal) fiber direction of each excitable (cylindrical) cell in H is defined by the unit vector $\mathbf{a}_\ell$. In addition, the electrical conductivities in the longitudinal and transverse directions are $\sigma_\ell^{i,e}$ and $\sigma_t^{i,e}$, where i and e refer to the intracellular and interstitial media, respectively. Then the global intracellular and interstitial conductivities can be expressed as the tensors D_i and D_e, respectively (see [1]): $D_{i,e} = \sigma_t^{i,e} I + (\sigma_\ell^{i,e} - \sigma_t^{i,e})\mathbf{a}_\ell \mathbf{a}_\ell^T$, where I is the identity tensor.

The total (quasistatic) current density is $\mathbf{J} = \mathbf{J}_i + \mathbf{J}_e$, where $\mathbf{J}_i = -D_i \nabla \phi_i$, and $\mathbf{J}_e = -D_e \nabla \phi_e$. It is to be noted that here these current densities are macroscopic, i.e. they represent averages over a small volume (~ 1 mm^3) of myocardial tissue. Under the assumption of the (constant) equal anisotropy ratio (i.e., $D_e = kD_i$) [12], the transmembrane potential v_m can be solved from a single parabolic partial differential equation (diffusion equation 19 in the companion paper [1]).

To evaluate the extracardiac electric potential ϕ_o, we have to assume that the composite medium H is homogeneous with the scalar electrical conductivity σ_e. Then we substitute $\mathbf{J}^i = -D_i \nabla v_m$, where $\mathbf{J}^i$ is the *impressed current density*, which is driven by the electrochemical generators in cardiac cells [13]. A more detailed discussion of the source term $\mathbf{J}^i$ is presented in the companion paper [1].

The extracardiac potential ϕ_o is obtained from the divergence of $\mathbf{J}^i$, as shown in [13, pp. 211–215] (cf. the isotropic source term $-\sigma_i \nabla v_m$ used in [3]). It has been shown [9, 11] that the result for ϕ_o is written as

$$\phi_o(\mathbf{r}) = -\frac{1}{4\pi\sigma_e}\left\{ \sigma_t^i \int_H \nabla v_m \cdot \nabla r^{-1} dV + (\sigma_\ell^i - \sigma_t^i)\int_H (\nabla v_m \cdot \mathbf{a}_\ell)\mathbf{a}_\ell \cdot \nabla r^{-1} dV \right\}, \quad (1)$$

where $\mathbf{r}$ is the vector from the source point in H to the field point. We can identify the result as a combination of two dipole sources: the first term $\mathbf{p}_n = -\sigma_t^i \nabla v_m dV$ is refered as the *normal dipole*, and the second term $\mathbf{p}_\ell = -(\sigma_\ell^i - \sigma_t^i)(\nabla v_m \cdot \mathbf{a}_\ell)\mathbf{a}_\ell dV$ as the *longitudinal dipole* [11]. The latter term accounts for the anisotropic properties of myocardial tissue. Correspondingly, ϕ_o in (1) can be separated into two components: the normal-component potential distribution (the one given, under the isotropic assumption, by the solid-angle formula of the uniform-double-layer theory, e.g. [5]), and the longitudinal-component distribution due to anisotropic properties.

The external magnetic field $\mathbf{B}(\mathbf{r})$ due to the total current density $\mathbf{J}$ can be evaluated from the Biot-Savart law. If we, in an analogous manner with the previous considerations of the extracardiac potential, regard the term $\mathbf{J}^i = -D_i \nabla v_m$ as the impressed source current density embedded in an unbounded homogeneous composite medium [11] , the result can be written as

$$\mathbf{B}(\mathbf{r}) = -\frac{\mu_o}{4\pi} \left\{ \sigma_t^i \int_H \nabla v_m \times \nabla r^{-1} dV + (\sigma_\ell^i - \sigma_t^i) \int_H (\nabla v_m \cdot \mathbf{a}_\ell) \mathbf{a}_\ell \times \nabla r^{-1} dV \right\} . \quad (2)$$

Again, the first term on the right evaluates the contribution of the normal component, and the second integral evaluates the contribution of the longitudinal component.

Although (1) and (2) are valid for an unbounded conducting medium, additional terms can be added to them that describe the contributions of the boundaries where the extracardiac conductivity σ_e changes. The resulting equations can be discretized and solved by applying the collocation method [4]. If the composite medium is modeled as a homogeneous semi-infinite space, the solution becomes simpler. It is well known that the electric potential by a current dipole in this case becomes twice the infinite-medium potential. The magnetic-field component perpendicular to the boundary plane remains unaffected, while the tangential field components can be evaluated by analytical formulas [14]. Moreover, the dipole component perpendicular to the boundary plane gives a zero magnetic field.

Briefly, the propagation model that is introduced in the companion paper [1] combines features from both the anisotropic bidomain theory and the cellular automata concept [9, 10]. The cellular automaton is run for desired number of time steps. The time instants where the cells move to the absolute refractory state are stored and displayed at selected time intervals. The computer algorithms and implementation were presented in detail in [10]. The extracardiac potentials and magnetic fields are then computed in an unbounded, homogeneous composite medium, using (1) and (2). In addition, the influence of a plane boundary between the air and a semi-infinite volume conductor is taken into account as described above.

4   Results

First simulations were performed in an array of $100 \times 100 \times 100$ excitable cells, where the intercellular distance was $h = 0.2$ mm and where all fibers were oriented in the same direction $\mathbf{a}_\ell$, defined by global spherical polar angles Θ (a positive angle from the z-axis) and Φ (a positive angle from the x-axis on the xy-plane). Propagation parameters were chosen on the basis of literature reports, e.g. [12, 16]; they are listed in Table 1 of the companion paper [1]. The excitation was propagated for 10 ms with

Table 1. Propagation velocities

σ_ℓ^i [S/m]	σ_t^i [S/m]	ϑ_ℓ [m/s]	ϑ_t [m/s]	$\vartheta_\ell/\vartheta_t$
0.05	0.05	0.33	0.33	1.00
0.10	0.05	0.46	0.33	1.40
0.20	0.05	0.65	0.33	1.97
0.45	0.05	1.00	0.33	3.03

different values of σ_ℓ^i. Resulting average propagation velocities (longitudinal ϑ_ℓ and transversal ϑ_t) estimated from the isochrones are shown in Table 1.

As the propagation velocities yielded by the conductivity values $\sigma_\ell^i = 0.2$ S/m and $\sigma_t^i = 0.05$ S/m are reasonably close to the values determined from experimental data reported by Durrer et al. [6], the associated set of parameter values was used in later simulations. The resulting activation isochrones for an intramural stimulus are displayed in Fig. 1 at one cross section.

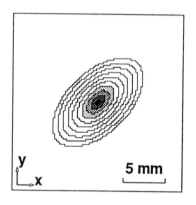

Figure 1. Propagation isochrones at one (xy) cross section in the case of uniform anisotropy. The dark region indicates the initial excitation. The time separation between adjacent isochrones is 1 ms. Here $\Theta = 90°$, $\Phi = 45°$.

Staton et al. [17] have reported magnetic fields measured from 1-mm-thick in vitro preparations of uniform anisotropy. To compare their results with our predictions, we propagated the excitation for 10 ms in an array of $100 \times 100 \times 5$ elements where $a_\ell = e_x$ in all cells. The external magnetic field was evaluated in a semi-infinite volume conductor; the resulting distribution is displayed in Fig. 2 together with the measured one.

Next we propagated the excitation for 10 ms in an array of $100 \times 100 \times 50$ elements, where the external current stimulus was applied subendocardially (2 mm from the endocardium), epicardially, and intramurally. Three different cases of anisotropy were studied: (a) uniform anisotropy case as above, (b) rotational anisotropy, which

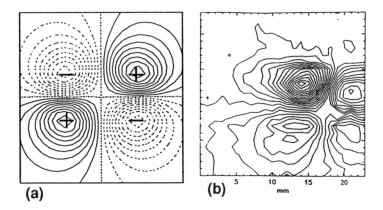

(a)   (b)

Figure 2. (a) The magnetic field normal (z) component due to excitation in a 1-mm-thick slab of tissue with uniform anisotropy ($\Theta = 90°$, $\Phi = 0°$) in a semi-infinite volume conductor, computed at a distance of 2 mm from the surface of the slab. (b) The corresponding distribution of the magnetic field normal component measured from a 1-mm-thick in vitro preparation [17]. Both distributions have the same polarity and an isocontour separation of 25 pT.

mimicks the typical fiber arrangement in the free wall of the left ventricle (see [18]), and (c) spiral anisotropy, where $\mathbf{a}_\ell$ of each cell is at a constant angle α with the radius vector drawn from the origin (see [19]). In all cases, the global fiber angle Θ was fixed at 90°, while in (a), the angle Φ was fixed at 0°, in (b), Φ was varied as a linear function of the z-coordinate, so that $\Phi = -45°$ for $z = 0$ and $\Phi = 45°$ for $z = 10$ mm, and in (c), the values of 25°, 45° and 65° were used for the spiral angle α. Fiber orientations on the "endocardial" surface are depicted in Fig. 3.

We computed the resulting electric potentials and magnetic fields at various distances (1–40 mm) from the "epicardial" surface of the slab. At each distance, we computed the contributions of the normal component and the longitudinal component in an unbounded volume conductor, using (1) and (2), in an 11 × 11 point grid in which the separation between the gridpoints increased as a function of the distance from the "epicardial" surface. The fields were also evaluated in a semi-infinite volume conductor as described above. Figs. 4–7 display the distributions of the normal component, the longitudinal component and the combined potentials and magnetic fields calculated at the distance of 5 mm from the "epicardial" surface for the different anisotropy cases.

In Figs. 4–7, (a)–(c) display, respectively, the isotropic, axial and combined component of the electric potential and the magnetic field in an unbounded medium, and (d) shows the combined distributions in a semi-infinite medium. The electric potentials are displayed as isocontour lines with linear curve separation. In the magnetic field plots, the length and head size of each arrow are related to the total field magnitude at the grid point. Open arrows: negative normal (z) component, dark arrows: positive normal component. The maximum field magnitudes (in 10^{-12} tesla) are show below each vector plot.

We evaluated the magnitudes of the longitudinal- and normal-component distributions of the potential and magnetic field in the unbounded medium as $||f|| = \sqrt{\sum_i f_i^2}$. The ratios of the magnitudes at different distances from the "epicardial" surface are presented in Fig. 8.

Although both normal- and longitudinal-components contributed to the extracardiac magnetic field in the unbounded case, it was notable that the contribution of the normal component vanished in all anisotropy cases in the semi-infinite volume conductor. Thus, only the anisotropic terms contributed to the external magnetic field, while both components still contributed to the external potential. If the semi-infinite volume conductor is replaced by a thin infinite slab, the same magnetic fields are obtained as in the semi-infinite case (e.g. [14]).

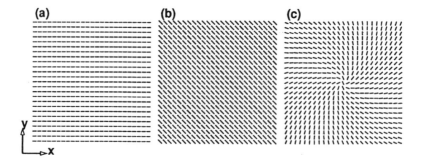

Figure 3. Fiber orientation on the "endocardial" surface. (a) Uniform anisotropy, with $\Theta = 90°$ and $\Phi = 0°$ on all xy-planes; (b) rotational anisotropy, with $\Theta = 90°$ and Φ varying linearly from $-45°$ on the endocardium to $+45°$ on the epicardium; (c) spiral anisotropy, with $\alpha = 25°$ and Φ varying as a function of the xy-coordinates.

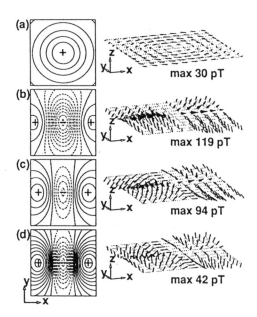

Figure 4. Extracardiac potential and magnetic field distributions in the case of uniform anisotropy at $z = 5$ mm above the "epicardial" surface. A subendocardial stimulus was applied; see text for further details. Isocontour separation is 50 μV.

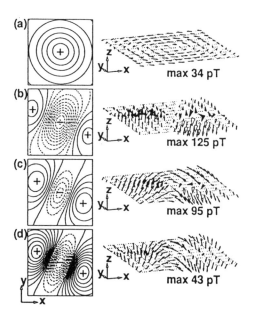

Figure 5. Extracardiac potential and magnetic field distributions in the case of rotational anisotropy at $z = 5$ mm. A subendocardial stimulus was applied; see text for further details. Isocontour separation is 50 μV.

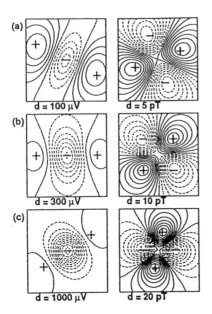

Figure 6. The external potential and normal component of magnetic field at z = 5 mm above the "epicardial" surface in the case of rotational anisotropy. (a) Subendocardial, (b) intramural and (c) epicardial stimulus. Isocontour separation is indicated below each picture.

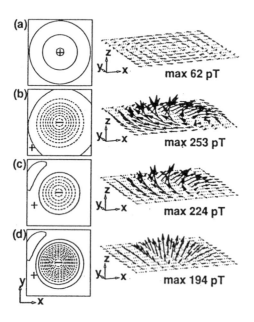

Figure 7. Extracardiac potential and magnetic field distributions in the case of spiral anisotropy ($\alpha = 25°$) at z = 5 mm above the "epicardial" surface. Isocontour separation is 100 μV. See the caption of Figs. 4 and 5.

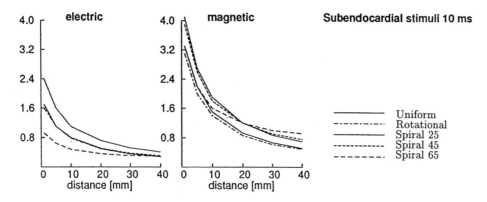

Figure 8. Ratios of the magnitudes of anisotropic and isotropic field component as a function of the distance from the epicardial surface.

5   Discussion

In this study, the viewpoint is macroscopic, and each excitable element actually represents a large number ($\sim$1000) of myocardial cells. We used a fixed shape for the action potential above the threshold value, but the shape and duration can easily be altered [10] to match experimentally measured action potentials or action potentials simulated with one- or two-dimensional cellular-scale models based on Hodgkin-Huxley-type differential equations [15].

The isochrones in simulations in which anisotropy is uniform have elliptical shapes, and the estimated average propagation velocities in Table 1 are in fairly good agreement with the square-root relation $\vartheta_\ell/\vartheta_t = \sqrt{\sigma_\ell^i/\sigma_t^i}$ [5, 20]. Moreover, the velocities are in agreement with the velocities estimated from in vitro experimental data [6].

Some results to validate the propagation algorithm were presented in the companion paper [1]. The electric potential distribution in rotational anisotropy (Fig. 5) has a striking similarity with the distributions measured from canine heart [21]: a minimum flanked by two maxima. In addition, the magnetic-field normal component in a 1-mm-thick slab with uniform anisotropy is in good agreement with the results in corresponding experimental measurements [17] (Fig. 2). Further validation was obtained in our earlier simulation study of ventricular pre-excitation and in comparisons of simulated body-surface potential and magnetic-field maps to measured ones [11].

According to our simulations, the anisotropic contributions have a significant influence on the extracardiac electric potential and magnetic field, particularly near the "epicardial" surface. As the distance of the field point increases, the anisotropic components diminish in both the extracardiac potentials and the magnetic fields faster than the corresponding isotropic components. In a semi-infinite volume conductor only the anisotropic terms contribute to the magnetic field.

In this study, relatively small excitation wavefronts were involved. Since the anisotropic terms may have less influence when the wavefronts are larger, it is not surprising that simulation studies using anisotropic isochrone shapes but isotropic field equations have shown good agreement between simulated and measured fields (see the review by Gulrajani [5]). However, in the case of intramural stimulus the extracardiac fields are solely due to anisotropic terms, while the isotropic field equations would predict zero fields.

Roth et al. [19] used analytical expressions to predict that there should be current distributions detectable only by magnetic measurements, particularly in cases with spiral anisotropy. Their predictions were based on different anisotropy ratios, which cannot be repeated with our model. Although our basic assumptions do not implicate electrically silent magnetic fields, the results indicate that the extracardiac magnetic field is more sensitive to the underlying fiber orientation than is the electric potential. Besides, in all geometries studied, the influence of the anisotropic (longitudinal) component was always much larger in the magnetic fields than in the electric potentials, particularly in cases with spiral anisotropy (Fig. 8).

Although our fiber geometries are simplified, the results indicate that measurements of extracardiac magnetic fields may reveal new complementary information to electric potential recordings. It still remains to be seen how this information could be extracted from practical measurements. A more realistic geometry model of the whole human heart allows more specific predictions, particularly concerning the extracardiac fields measured on the body surface [1, 11].

Acknowledgements

This work was supported in part by the Academy of Finland and by research grants from the Heart and Stroke Foundation of Nova Scotia and from the Medical Research Council of Canada.

References

[1] B.M. Horáček, J. Nenonen, J.A. Edens, and L.J. Leon. A hybrid computer model of propagated excitation in the anisotropic human ventricular myocardium. In this volume.

[2] L. Tung. *A bidomain model for describing ischemic myocardial dc potentials*. PhD thesis. Massachusetts Institute of Technology, 1978.

[3] W.T. Miller III, and D.B. Geselowitz. Simulation studies of the electrocardiogram: I. The normal heart. *Circ. Res.* 43:301–315, 1978.

[4] B.M. Horáček. Digital model for studies in magnetocardiography. *IEEE Trans Magn* 9:440–444, 1973.

[5] R.M. Gulrajani. Models of the electrical activity of the heart and computer simulation of the electrocardiogram. *CRC Crit Rev Biomed Eng* 16:1–61, 1988.

[6] D. Durrer, R.Th. van Dam, G.E. Freud, M.J. Janse, F.J. Mejler, and R.C. Artzbaecher. Total excitation of the isolated human heart. *Circulation* 41:899–912, 1970.

[7] P. Colli-Franzone, L. Guerri, and C. Viganotti. Oblique dipole layer potentials applied to electrocardiology. *J Math Biology* 17:93–124, 1983.

[8] D.B. Geselowitz, and W.T. Miller III. A bidomain model for anisotropic cardiac muscle. *Ann Biomed Eng* 11:191–206, 1983.

[9] L.J. Leon, and B.M. Horáček. Computer model of excitation and recovery in the anisotropic myocardium. *J Electrocardiol* 24:1–41, 1991.

[10] J. Nenonen, J.A. Edens, L.J. Leon, and B.M. Horáček. Computer model of propagated excitation in the anisotropic human heart: I. Implementation and algorithms. In K.L. Ripley and A. Murray, editors, *Computers in Cardiology*, pp. 545–548, IEEE Computer Society Press, Los Alamitos, CA, 1991.

[11] J. Nenonen, J.A. Edens, L.J. Leon, and B.M. Horáček. Computer model of propagated excitation in the anisotropic human heart: II. Simulations of extracardiac fields. In K.L. Ripley and A. Murray, editors, *Computers in Cardiology*, pp. 217–220, IEEE Computer Society Press, Los Alamitos, CA, 1991.

[12] R. Plonsey, and R.C. Barr. Current flow patterns in two-dimensional anisotropic bisyncytia with normal and extreme conductivities. *Biophys J* 45:557–571, 1984.

[13] R. Plonsey. *Bioelectric Phenomena.* McGraw–Hill, New York, 1969.

[14] B.N. Cuffin, and D. Cohen. Magnetic fields of a dipole in special volume conductor shapes. *IEEE Trans Biomed Eng* 24:372–381, 1977.

[15] G.W. Beeler, and H. Reuter. Reconstruction of the action potential of ventricular myocardial fibers. *J Physiol* 268:177–210, 1977.

[16] F.A. Roberge, A. Vinet, and B. Victorri. Reconstruction of propagated electrical activity with a two-dimensional model of anisotropic heart muscle. *Circ Res* 58:461–475, 1986.

[17] D.J. Staton, R.N. Friedman, and J.P. Wikswo Jr. High-resolution SQUID imaging of octupolar currents in anisotropic cardiac tissue. *IEEE Trans Appl Supercond*, In press.

[18] D.D. Streeter Jr. Gross morphology and fiber geometry of the heart. In R.M. Berne, N. Sperelakis, and S.R. Geiger, editors. *Handbook of Physiology – Section 2: The cardiovascular system, Volume I: The Heart*, pp. 61–112. American Physiological Society, Bethesada, MD, 1979.

[19] B.J. Roth, W.Q. Guo, and J.P. Wikswo Jr. The effects of spiral anisotropy on the electric potential and the magnetic field at the apex of the heart. *Mathem Biosc* 88:191–221, 1988.

[20] A. Muler, and V. Markin. Electrical properties of anisotropic nerve-muscle bisyncytia – III. Steady form of the excitation front. *Biofizika* 22:671–675, 1977.

[21] P. Colli-Franzone, L. Guerri, C. Viganotti, E. Macchi, S. Baruffi, S. Spaggiari, and B. Taccardi. Potential fields generated by oblique dipole layers modeling excitation wavefronts in the anisotropic myocardium. Comparison with potential fields elicted by paced dog hearts in a volume conductor. *Circ Res* 51:330–346, 1982.

Spatio-Temporal Constraints in Inverse Electrocardiography

A. van Oosterom and G. Huiskamp

1 Introduction

One cannot uniquely determine the bioelectric sources in the heart solely from the observed potentials on the body surface [1] without first postulating models of cardiac electric sources and of a volume conductor that characterizes the torso's passive electric properties. With both models in place, one can solve the forward problem by computing the potential distribution on the body surface from the given sources [2], and the inverse problem [3] by estimating source-model parameters from the observed body-surface potentials (given the restrictions implied by the source model). The inverse problem can, therefore, be viewed as a parameter-estimation problem.

The uniqueness of any given inverse solution cannot be guaranteed, because the inverse problem is one of the so-called "ill-posed" problems. If *a priori* information on the sources is available, one may mitigate the "ill-posedness" by restricting the nature of the sources to be within a certain class (implicit constraint). In addition, one can incorporate explicit constraints in the estimation procedure by including an appropriate penalty function, which represents a penalty for the solution's departing from an explicit constraint.

The method of restricting the solution space of inverse problems is commonly called "regularization", and the penalty function is usually some function of the solution parameters, which must be close to zero. The solution itself depends on the value of the regularization parameter λ ($\lambda \geq 0$), which sets the weight of the constraint in the final solution: for large values of λ, solutions permitted by the constraint dominate; for small values of λ, the unconstrained solutions dominate. There are several methods for selecting an appropriate value of λ [4]. A useful perpective on this problem can be gained by formulating and implementing constraints on the solution in terms of classical optimization theory.

2 Scope

We discuss the effectiveness of using spatio-temporal constraints in the inverse problem of computing the heart's electric activity from observed body-surface potentials. We treat the handling of constraints within the context of general optimization theory, as well as within the context of regularization. In particular, we deal with the problem of choosing the value of the regularization parameter λ, balancing the constrained and the unconstrained part of the solution. We illustrate these general principles by applying them to an inverse procedure aimed at computing the activation sequence at the heart surface, and we contrast it with an application to the inverse computation of epicardial potentials. Both inverse procedures are formulated in a way that facilitates the interpretation of these inverse problems as parameter-estimation problems. The associated forward models are first briefly summarized.

3 General formulation of the forward problem

The forward problem deals with formulating models of cardiac electric sources and of the volume-conduction effects of body tissues, and with computing the resulting body-surface potentials. A general formulation of the forward problem is

$$\Phi(\mathbf{y}, t; \mathbf{p}(\mathbf{x}, t)) = \int_V M(\mathbf{y}, \mathbf{x}; \mathbf{p}(\mathbf{x}, t)) s(\mathbf{x}; \mathbf{p}(\mathbf{x}, t)) dv(\mathbf{x}), \tag{1}$$

where $\Phi(\mathbf{y}, t; \mathbf{p}(\mathbf{x}, t))$ is the potential at $\mathbf{y}$ and time t; $s(\mathbf{x}; \mathbf{p}(\mathbf{x}, t))$ is the source strength in position $\mathbf{x}$ at time t; $\mathbf{p}(\mathbf{x}, t)$ is the parameter (vector) specifying the source at position $\mathbf{x}$ and time t; and $M(\mathbf{y}, \mathbf{x}; \mathbf{p}(\mathbf{x}, t))$ is the transfer function of the medium as related to the source and its character. It is usually assumed that the latter function is determined completely by the spatial distribution of the electric conductivity of the medium, the geometrical relationship between the observation point $\mathbf{y}$ and source position $\mathbf{x}$, and the nature of the source (*e.g.* dipole or dipole layer).

We will deal with two of several possible source models: one involves distribution of electric potential over the closed surface surrounding the heart; the other involves a uniform double layer over that surface.

The potentials on the surface bounding a volume conductor are—at least in the absence of measurement and modelling noise—uniquely related to those at some interior closed surface enclosing all primary sources [5]. In electrocardiography, the interior surface commonly chosen is one that closely encompasses the heart and is usually referred to as the "epicardial surface", S_e. (We prefer the term "pericardium" for S_e.) For this case, the specific variant of the general formulation (1) of the forward problem is

$$\Phi(\mathbf{y}, t; \Psi(\mathbf{x}, t)) = \int_{S_e} M(\mathbf{y}, \mathbf{x}) \Psi(\mathbf{x}, t) dS(\mathbf{x}). \tag{2}$$

Here a *linear* relationship exists between potentials on the body surface and the epi-cardial potentials $\Psi(\mathbf{x}, t)$, which act as the source-parameter function. The number of parameters is equal to the number of surface elements that are considered to be acting independently.

The second choice for the source model is a uniform double layer over the closed surface S_v bounding the ventricles. If cardiac sources during ventricular depolarization are assumed to be of uniform strength over the wave front, (1) can be cast as [6, 7]

$$\Phi(\mathbf{y}, t; \tau(\mathbf{x})) = \int_{S_v} M(\mathbf{y}, \mathbf{x}) H(t - \tau(\mathbf{x})) dS(\mathbf{x}), \tag{3}$$

where $M(\mathbf{y}, \mathbf{x})$ represents the transfer function from elementary double-layer sources $dS(\mathbf{x})$ at the ventricular surface S_v to torso points $\mathbf{y}$; $\tau(\mathbf{x})$ is the local depolarization time at $\mathbf{x}$; H is the Heaviside function which switches on the elementary sources $dS(\mathbf{x})$ from $\tau(\mathbf{x})$ onward. The parameter function $\tau(\mathbf{x})$ (activation time) is a function of space. The implied relationship between the observed potentials and the source-parameter function $\tau(\mathbf{x})$ is *non-linear*.

By defining QRS-integrals $w(\mathbf{y}) = \int \Phi(\mathbf{y}, t) dt$ and applying the same integration to the right side of (3), we get

$$w(\mathbf{y}; \tau(\mathbf{x})) = -\int_{S_v} M(\mathbf{y}, \mathbf{x}) \tau(\mathbf{x}) \ dS(\mathbf{x}). \tag{4}$$

Here a *linear* relationship between the time integral of the observed potentials, $w(\mathbf{y}; \tau(\mathbf{x}))$, and the source parameters $\tau(\mathbf{x})$ exists.

4   The inverse problem of electrocardiography

The inverse problem is the estimation of cardiac electric sources from measured po-tentials $V(\mathbf{y}, t)$, a computed transfer function $M(\mathbf{y}, \mathbf{x})$, and a given source model. In treating this problem as a parameter-estimation problem, one finds the source param-eters $\mathbf{p}$ by minimizing, with respect to $\mathbf{p}$, the residual function

$$Res(\mathbf{p}) = \|V(\mathbf{y}, t) - \Phi(\mathbf{y}, t; \mathbf{p})\|^2, \tag{5}$$

thus yielding the least-squares residual difference between measured potentials V and model predictions Φ. Throughout this chapter $\|\quad\|^2$ denotes the square of the Eu-clidean norm. In practical applications, the potentials are treated in matrix form; recorded data taken at T time instants (*temporal* sampling) and L electrode positions (*spatial* sampling) are represented by a $(L \times T)$ matrix Φ, in which row vector ϕ_{ℓ}. represents the time signal at electrode position ℓ and column vector $\phi_{.t}$ represents the potential distribution at discrete time t.

Estimating epicardial potentials

If the parameter vector **p** solely includes source strengths, Φ depends linearly on this parameter vector of dimension P. In other words, the source strengths constitute the elements of the parameter vector. The formulation of (2), the distributed voltage source, as discussed above, is such a case. This leads to a linear least-squares (LLS) problem. For discretized versions of Φ and V, one can represent the associated transfer function $M(\mathbf{y}, \mathbf{x})$ as a $L \times P$ dimensional matrix M, and the source Ψ as a $P \times T$ dimensional matrix representing the parameters (the epicardial potentials) at all T time instants. The formulation of the inverse problem then is: solve, in the least-squares sense, Ψ from

$$V \approx M\Psi. \tag{6}$$

For overdetermined, full-rank matrices M, the well-known unique least-squares solution is

$$\Psi = (M^t M)^{-1} M^t V, \tag{7}$$

with A^t and A^{-1} denoting the transpose and inverse of matrix A. Alternatively, for underdetermined or rank-deficient matrices, multiple solutions exist. Here, the least-squares solution having a minimum norm is often taken. This is, in fact, a particular constrained solution. This solution can be obtained, for instance, by means of subroutine HFTI [8].

Estimating the depolarization sequence at the ventricular surface

For the distributed surface source expressed by (3), the minimization problem is: find τ which minimizes the object function

$$Res(\tau) = \|V(\mathbf{y}, t) - \Phi(\mathbf{y}, t; \tau)\|^2. \tag{8}$$

This is a non-linear least-squares (NLLS) problem whose solution can be found by means of one of several iterative minimization procedures, *e.g.* those of Marquardt [9] or Box-Kanemasu [10]. These methods require "sensible" initial estimates of the solution, and, in this problem, the linear relationship between the time integral of the observed potentials, $w(\mathbf{y}; \tau(\mathbf{x}))$, and the source parameters $\tau(\mathbf{x})$ (see (3)) provides an effective initial estimate [7, 11].

5   Constraints

The ill-posed nature of the inverse problem requires additional measures in order to guarantee a stable and/or physiologically meaningful solution. The fact that both source models considered here are specified on a surface (S_e or S_v) that is assumed to be known *a priori*, represents an implied spatial constraint. In the formulation of (3), an implied *temporal* constraint is imposed by the fact that, within the depolarization phase, the equivalent double-layer elements remain active after having been "switched on" by the activation process [11].

Additional, explicit constraints are often required, particularly if—as tends to be true for the source models consisting of distributed epicardial potentials incorporated in (2) and distributed surface-source model (3)—a relatively large number of parameters must be estimated. These may be either spatial, temporal, or spatio-temporal constraints. Explicit constraints may be of a statistical nature, be related to the source model used, and be based on the physics or on the electrophysiology involved [3]. Their implementation may be viewed from regularization perspective or an optimization-theory perspective.

Explicit constraints: regularization perspective

Explicit constraints can be incorporated in the least-squares procedure by minimizing with respect to **p** the object function

$$O(\mathbf{p}; \lambda) = Res(\mathbf{p}) + \lambda Reg(\mathbf{p}), \tag{9}$$

with Reg being the penalty function restricting the solution, and λ setting the weight of the constraint.

In some cases, such as (6), the transfer function M can be expressed by a matrix, operating on another matrix S representing the sources (Ψ in (6)). If, moreover, the constraint can be formulated by a matrix R that forces the solution to be such that RS is close, in a least-squares sense, to an *a priori* assumed matrix C, (9) works out as

$$\begin{pmatrix} M \\ \lambda R \end{pmatrix} S \approx \begin{pmatrix} V \\ \lambda C \end{pmatrix}. \tag{10}$$

The least-squares solution to this system of linear equations is

$$S = (M^t M + \lambda^2 R^t R)^{-1}(M^t V + \lambda^2 R^t C). \tag{11}$$

Note that, for $\lambda = 0$, the solution is, indeed, identical to the unconstrainded solution (7), whereas, for $\lambda \rightsquigarrow \infty$, it tends to be dominated more and more by the constraint $S = (R^t R)^{-1} R^t C$.

As in the unconstrained situation, if the system is underdetermined, the minimum norm solution can be found by means of some appropriate computer code, *e.g.*, applying subroutine HFTI [8] to the system of (10). Alternatively, this system can be solved by means of truncated singular-value decomposition [12].

Regularization involves the minimization of $O(\mathbf{p}; \lambda)$ of (9) with respect to **p**. Let $\bar{\mathbf{p}}$ denote the solution to this problem. Because this solution depends on the value of λ, $\bar{\mathbf{p}} = \bar{\mathbf{p}}(\lambda)$ and $O(\lambda) = O(\bar{\mathbf{p}}(\lambda))$, and hence

$$O(\lambda) = Res(\lambda) + \lambda Reg(\lambda), \tag{12}$$

with $Res(\lambda) = Res(\bar{\mathbf{p}}(\lambda))$ and $Reg(\lambda) = Reg(\bar{\mathbf{p}}(\lambda))$.

Different types of constraints, leading to specific functions Reg, and different strategies towards finding appropriate values for λ, have been used [4]. One often accepts values of λ for which the solution does not completely satisfy the constraint. This is the case when inspection of the series (resulting for different values of λ) of solutions demonstrates increasingly more undesirable characteristics. Such cases suggest, however, that more a priori knowledge about the solution is available, and attempts should be made towards incorporating it directly in the inverse procedure.

Explicit constraints: optimization-theory perspective

A different perspective on the problem is gained by formulating the problem at hand strictly in terms of classical optimization theory (see, e.g., [13]), i.e., by minimizing the function $Res(\mathbf{p})$ with respect to $\mathbf{p}$, subject to the constraint

$$Reg(\mathbf{p}) = c. \tag{13}$$

In the inverse problem, the functions Res and Reg, which in classical optimization theory are arbitrary functions, take on the specific meaning indicated above. Note that in this formulation the value of c is a predetermined constant.

Optimization theory [13] shows that the solution $\mathbf{p}$ follows from the equating to zero of all partial derivatives with respect to all of the extended set of variables $(\mathbf{p}, \lambda)$ of the object function

$$O(\mathbf{p}, \lambda) = Res(\mathbf{p}) + \lambda(Reg(\mathbf{p}) - c). \tag{14}$$

In this formulation, the equating to zero of the derivative of O with respect to the variable λ in fact leads to (13), the original constraint. Solving the resulting set of equations in $(\mathbf{p}, \lambda)$ yields the optimum solution, which comprises both $\bar{\mathbf{p}}$—which exactly satisfies the constraint—and λ.

Having cast the problem in terms of classical optimization theory, we want to highlight two of the relevant results, which we will use when discussing the selection of λ. Firstly, for the derivative of $O(\lambda)$ with respect to λ, one has (see, e.g., [13])

$$\frac{dO}{d\lambda} = Reg - c. \tag{15}$$

Secondly, it can be shown that (see [13], result 1.46)

$$\frac{\partial Res(\bar{\mathbf{p}})}{\partial c} = -\lambda. \tag{16}$$

Since, for the optimal solution $\mathbf{p} = \bar{\mathbf{p}}$, the value of Reg is equal to c, one has

$$\frac{\partial Res}{\partial Reg} = -\lambda. \tag{17}$$

Incorporating the constraint (13) is often difficult in practice, but one can best overcome this difficulty by iteratively choosing different values for λ and minimizing the object function of (14) with respect to $(\mathbf{p}; \lambda)$ until (13) is satisfied. The practical implementation of a constraint therefore appears to be unaffected by whether one views it as an optimization problem or as a regularization problem.

Explicit spatial constraints

For the inverse solution of epicardial potentials, the need to implement explicit spatial constraints has become obvious from several experimental studies [14, 15, 5]. The types of constraints studied [4] range from those involving the norm of the computed epicardial potentials, their spatial gradients or higher-order derivatives (Laplacian, Hessian) or combinations of these.

For the solution in terms of the instantaneous epicardial potentials at time instant t_i, $\mathbf{p} \equiv \psi_i$, and a much-used type of explicit constraint in (9) results from choosing $Reg(\Psi_i) = ||\Psi_i||^2$. The associated values of λ are allowed to differ for each time instant t_i. This amounts to restricting the norms of the instantaneous solution vectors, also refered to as zero-order Tikhonov regularization. Alternatively, this can be effected by means of a truncated-SVD-based inverse of the matrix M [3].

Explicit temporal constraints

Explicit temporal constraints can also be incorporated through the formulation in (9). In some models, the time dependency can be incorporated directly into the source model and can thus be considered as an implicit constraint. Barr *et al.* [16] used this approach in their on-off model. Recently, Oster and Rudy [17] have studied the temporal aspect of the sequence of computed epicardial potential distributions. In their work, it is implemented as an explicit constraint, as in (10). In one of their attempts, they used measured epicardial potentials $\psi_m(i)$ at discrete time instants t_i as column vectors c_i constituting C (equation (10)), and with $R = I$, the identity matrix. For large values of λ, the instantaneous solutions $\psi(i)$ could, predictably, be forced to resemble closely the measured values $\psi_m(i)$. In an attempt which proved to be more interesting, weighted differences between subsequent estimates $\psi(i-1)$, $\psi(i)$, $\psi(i+1)$ were used to constrain the solutions. The preliminary results indicate that valuable constraints may be derived from this procedure.

Explicit spatio-temporal constraints

The source model implied in (3) has been studied extensively in our group [11]. Besides a major implied spatial and temporal constraint, additional measures were found necessary in order to ensure a solution that could be related to electrophysiology. We found that, of the several penalty functions (norm of the gradient, norm of the Hessian, norm of the surface Laplacian), the norm of the surface Laplacian of the computed activation function $\tau(\mathbf{x})$ integrated over S_v to be most effective. By its nature, this is a spatio-temporal constraint. The magnitude of its constrained value, c, can be related to the overall speed of depolarization.

6   On choosing regularization parameter and/or function

Several *ad hoc* strategies for choosing the proper value for λ have been described in the literature [4]. One example is the CRESO measure introduced by Colli Franzone *et al.* [18, 15]. CRESO stands for "composite residual and smoothing operator" and an associated operator $F(\lambda)$ was defined as [18]

$$F(\lambda) = Reg(\lambda) + 2\lambda \frac{d}{d\lambda} Reg(\lambda), \qquad (18)$$

or by an equivalent expression [4]

$$F(\lambda) = \frac{d}{d\lambda}(\lambda Reg(\lambda)) - \frac{d}{d\lambda} Res(\lambda). \qquad (19)$$

In the CRESO method, the optimum value of λ is taken to be the smallest positive value for which F(λ) exhibits a local maximum. Just why this might work has not been indicated, but a specific set of test data for which it was found to work has been documented [15]. In that work, little difference was found between the results based on the CRESO λ values and those obtained by optimizing the correspondence between observed and measured epicardial potentials. Rudy *et al.* [5] drew the same conclusion after analyzing further a subset of the material studied by Colli Franzone *et al* [15].

Several other methods for finding λ are discussed by Hansen [19]. Of these, the L-curve is directly related to the CRESO method. It entails plotting $Res(\lambda)$ against $Reg(\lambda)$ for different values of λ. For well-defined problems, this procedure leads to L-shaped curves, in which the λ value at the "knee" of the L-shape is taken to be the most appropriate one.

The importance of using a proper value of λ is, indeed, paramount in any regularization procedure. One can circumvent this problem by following the classical approach to constrained minimization, but to do so shifts the emphasis towards formulating an expression based on sound, *a priori* knowledge of the type of solution, *i.e.* towards finding the appropriate constraining function *Reg*, as well as the value of c at which it should be fixed. This problem having been settled, the value of λ follows from the iterative procedure outlined above for treating the problem formulated in (14).

The studies reported in the literature on computing the distributions of epicardial potentials have yielded surprisingly large relative errors, notwithstanding the fact that a relatively simple volume conductor was involved and that due attention was paid to the problem of setting an appropriate value for λ. In fact, time-varying values of λ were used. We speculate that this problem is caused by the fact that none of the regularization measures used could be related *a priori* to the nature of the solution to be found.

In Colli Franzone *et al.* [15], the regularization function which was reported to be optimal was the norm of the gradient of the epicardial potentials; the differences between this regularization function (first-order Tikhonov regularization) and other orders of Tikhonov regularization were only marginal. This also emerged from the analysis of Rudy *et al.*, who subsequently used zero-order Tikhonov regularization.

By contrast, the inverse procedure aimed at estimating the activation function $\tau(\mathbf{x})$ seems to be behaving favorably: although a single value of λ is used for the entire QRS complex, the relative residual differences at the torso surface remain well below 0.2 over the major part of the QRS complex. The reason for this, we believe, lies in the fact that in this inverse procedure the regularization function used can be related to properties based on electrophysiology and can be fixed *a priori*. To justify this claim, we will summarize in the next section some aspects of our inverse procedure (for details, see [20]).

7   Constraints in estimating the activation function

The activation function $\tau(\mathbf{x})$ at the ventricular surface S_v can be found by using (3). This involves a least-squares solution for $\tau(\mathbf{x})$, using a non-linear minimization procedure applied to the function $Res(\tau(\mathbf{x}))$. The least-squares residual difference resulting in this way has a relative RMS value of less than 0.1 over about 90 % of the QRS interval. The resulting function $\tau(\mathbf{x})$ over S_v exhibited numerous minima (indicating local breakthroughs) and maxima (indicating local extinctions of the activation wave front). We have looked for a regularization function that would restrict the number of local extrema, while leaving large gradients unaffected, as has been observed over the epicardial surface during normal excitation [21]. The surface Laplacian of the activation function $\tau(\mathbf{x})$ is such a function. The actual regularization used is the square of the Euclidean norm of the local surface Laplacian of $\tau(\mathbf{x})$ integrated over S_v. An algorithm was developed [22] for the numerical implementation of the procedure. This resulted in a matrix L acting on the vector τ representing the activation function $\tau(\mathbf{x})$, and $Reg(\tau) = \|L\tau\|^2 = \tau^t L^t L\tau$.

This function Reg is the integral over the ventricular surface S_v of the square of the surface Laplacian of $\tau(\mathbf{x})$. Since the dimension of the surface Laplacian of $\tau(\mathbf{x})$ is s.m^{-2}, the dimension of Reg is s^2.m^{-2}. For normal activation, its value is related to the overall squared reciprocal velocity of activation over S_v. We studied the behavior of this regularization function by observing the behavior of the object function O as well as the functions Res, λReg and Reg as a function of λ. We made a scan at discrete values of λ and, for each value of λ, found the function vector τ minimizing $O(\lambda)$, by means of a standard non-linear optimization procedure [23]. Since the approximate level of λ was unknown, a wide range was selected. The width of this range was determined as follows. The function $O(\lambda)$ is monotone non-decreasing, starting from $Res(\lambda = 0)$ up to a constant level $Res(\lambda = \infty)$, for which the solution $\tau(\mathbf{x})$ is a constant. After exhibiting some local extrema as well as a wide global maximum, the function $\lambda Reg(\lambda)$ was observed to tend towards zero for large values of λ. Because of (17), $Reg(\lambda)$ tends towards zero even faster for large values of λ. By taking crude steps in λ values, one can establish this limit, and a λ value for which this limit is reached up to, say, 90 % may be taken as the upper value for the stepping process. In our inverse procedure, the λ values were initially increased by equal factors. Without any further clue, some appropriate value for λ remained obscure. CRESO did not work, even when a far more refined linear scan over an interval close to $\lambda = 0$ was made. In fact, it could not work for the problem at hand, because the iterative method used for solving the non-linear system produces slight irregularities in the behavior of the functions studied. The minimization process may get stuck in a local minimum for some values of λ, producing insignificant local extrema in $F(\lambda)$. We found the way out here by inspecting one healthy subject's solution vector (by plotting it over S_v) and comparing its nature to known invasive physiology [21]. We selected that range of λ values and corresponding Reg values for which we found the nature of the solution to be in good agreement with published invasive data. Next, we carried out a fine tuning by inspecting a small range of λ values around the value identified. Here the number of local extrema over the heart surface was monitored as well. The final value for $c = Reg$ was selected from the middle of a λ interval demonstrating a constant number of local extrema as a function of λ (about 13 local extrema were observed in [21]). This number corresponded to the number observed in invasive electrophysiology; it was reached at a value of $c = 36$ $s^2.m^{-2}$.

A final step in the procedure was testing it in an application to data from different individuals. Here the value of c was set *a priori* at the level determined in the above procedure. The encouraging result obtained was that the solutions for all of the four independent cases analyzed so far have about the same quality.

8   Discussion

It may seem that both regularization and constrained solution lead in practice to essentially the same procedure, since both approaches involve choosing a λ value. However, the CRESO type of approach suggests that the optimal λ value is necessarily contained in the behavior of Reg as a function of λ. In the approach based on constrained solutions, which we believe to be the better ones, the level of the constraint is set *a priori*. It requires that both the type of the function Reg and the level at which it is constrained actually reflect true *a priori* knowledge about the solution. For the inverse solution in terms of epicardial potentials, zero-order Tikhonov regularization has not been found to be very effective; it is reported to result in "smooth" solutions [4]. However, it merely reduces the norm of the solution, and, while this *may* also lead to smooth solutions, whether it does depends on the nature of the dominant singular values of the transfer matrix involved. The norm of the gradient, again applied to the epicardial potentials [15] does not seem to be significantly more effective. The reason may, again, be that its magnitude cannot be fixed *a priori*. By contrast, the surface Laplacian applied to the activation function $\tau(\mathbf{x})$ does seem to be far more effective. Phillips [24] has previously demonstrated the effectiveness of this operator in a one-dimensional application. However, before choosing it, one should be confident that it is related to the problem at hand, bringing out features of the solution that are known *a priori*. For the inverse determination of $\tau(\mathbf{x})$, this is clearly the case. The value of the surface Laplacian of $\tau(\mathbf{x})$ is related to the number of extrema in the activation function as well as to the general nature of the activation within the myocardium. After setting the desired level c for regularization function in a design study, subsequent use of the same value of c in different subjects has been found to produce activation sequences with the same degree of physiological realism.

In combination with the spatial and temporal constraint as implied in the forward model used, the explicit spatio-temporal constraint as effected by means of the surface Laplacian has proved to yield a high-quality inverse solution which can be directly related to electrophysiology [20].

Acknowledgement

The authors are indebted to Prof. Dr. J. Strackee for his constructive criticism on the manuscript of this chapter.

References

[1] H. von Helmholtz   Über einige Gesetze der Vertheilung elektrischer Ströme in körperlichen Leitern mit Anwendung auf die thierisch-elektrischen Versuche. *Pogg. Ann. Physik und Chemie*, 89:211–233, 1853.

[2] R.M. Gulrajani, F.A. Roberge, and G.E. Mailloux. The forward problem of electrocardiography. In P.W. Macfarlane and T.D.V. Lawrie, editors, *Comprehensive Electrocardiology*, chapter 8. Pergamon Press, Oxford, 1989. volume I.

[3] R.M. Gulrajani, F.A. Roberge, and P. Savard. The inverse problem of electrocardiography. In P.W. Macfarlane and T.D.V. Lawrie, editors, *Comprehensive Electrocardiology*, chapter 9. Pergamon Press, Oxford, 1989. volume I.

[4] B.J. Messinger-Rapport and Y. Rudy. Computational issues of importance to the inverse recovery of epicardial potentials in a realistic heart-torso geometry. *Math. Biosc.*, 96:1–35, 1989.

[5] Y. Rudy and B.J. Messinger-Rapport. The inverse problem in electrocardiology: Solutions in terms of epicardial potentials. *CRC Crit. Rev. Biomed. Eng.*, 16:215–268, 1988.

[6] A. van Oosterom and J.J.M. Cuppen. Computing the depolarization sequence at the ventricular surface from body surface potentials. In Antalóczy Z and Préda I, eds., *Electrocardiology 1981*, pages 101–106, Akadémiai Kiadó, Budapest, 1982.

[7] J.J.M. Cuppen and A. van Oosterom. Model studies with the inversely calculated isochrones of ventricular depolarization. *IEEE Trans. Biomed. Eng.*, BME-31:652–659, 1984.

[8] C.L. Lawson and R.J. Hanson. *Solving Least Squares Problems*. Prentice-Hall, Englewood Cliffs, N.J., 1974.

[9] D.W. Marquardt. An algorithm for least-squares estimation of non-linear parameters. *J. Soc. Indust. Appl. Math.*, 2:431–441, 1963.

[10] J.V. Beck and K.J. Arnold. *Parameter Estimation in Engineering and Science*. Wiley, New York, 1977.

[11] G.J.M. Huiskamp and A. van Oosterom. The depolarization sequence of the human heart surface computed from measured body surface potentials. *IEEE Trans. Biomed. Eng.*, BME-35:1047–1058, 1988.

[12] G.E. Forsythe, M.A. Malcolm, and C.B. Moler. *Computer methods for mathematical computations*. Prentice-Hall, Englewood Cliffs, N.J., 1977.

[13] G. Walsh. *Methods of Optimization*. John Wiley & Sons, London, 1975.

[14] R.C. Barr and M.S. Spach. Inverse calculation of QRS-T epicardial potentials from body surface potential distributions for normal and ectopic beats in the intact dog. *Circ. Res.*, 42:661–675, 1978.

[15] P. Colli Franzone, L. Guerri, S. Tentonia, C. Viganotti, S. Baruffi, S. Spaggiari, and B. Taccardi. A mathematical procedure for solving the inverse potential problem of electrocardiography. Analysis of the time-space accuracy from in vitro experimental data. *Math. Biosc.*, 77:353–396, 1985.

[16] R.C. Barr, T.C. Pilkington, J.P. Boineau, and C.L. Rogers. An inverse electrocardiographic solution with an on-off model. *IEEE Trans. Biomed. Eng.*, BME-17:49–57, 1970.

[17] H.S. Oster and Y. Rudy. The use of temporal information in the regularization of the inverse problem of electrocardiography. *IEEE Trans. Biomed. Eng.*, BME-39:65–75, 1992.

[18] P. Colli Franzone, L. Guerri, B. Taccardi, and C. Viganotti. Inverse epicardial mapping in the human case. In Schubert E, editor, *Proc Symp Electrophysiology of the Heart*, pages 19–33. Plenum, New York, 1980.

[19] P.C. Hansen. Numerical tools for the analysis and solution of Fredholm integral equations of the first kind. *Inverse Problems*, 8:849–872, 1992.

[20] G.J.M. Huiskamp and A. van Oosterom. Finding a solution to the inverse problem in electrocardiography. In Ripley KL, editor, *Computers in Cardiology*, pages 391–394, Washington, IEEE Computer Society Press, 1988.

[21] D. Durrer, R.T. van Dam, G.E. Freud, M.J. Janse, F.L. Meijler, and R.C. Arzbaecher. Total excitation of the isolated human heart. *Circulation*, 41:899–912, 1970.

[22] G.J.M. Huiskamp. Difference formulas for the surface Laplacian on a triangulated surface. *Journal of Computational Physics*, 95:477–496, 1991.

[23] The Numerical Algorithms Group Ltd., Oxford. *subroutine E04KBF*, NAG Fortran mark II edition, 1985.

[24] D.L. Phillips. A technique for the solution of certain integral equations of the first kind. *J Ass Comp Mach*, 9:84–96, 1962.

When is the Inverse Problem of Electrocardiography Well-Posed

F. Greensite and J.-J. Qian

1 Introduction

The inverse problem of electrocardiography [1] can be formulated as a reconstruction of maps of epicardial potentials and/or activation times from observed body-surface potential distributions. The problem is formulated as a Fredholm equation of the first kind, where the kernel has a finite L^2-norm. Therefore, the body-surface potential data—being the "convolution" of the source with a well-behaved transfer function—is a smoothed version of the desired source and the least-squares solution is unstable in the presence of measurement and modelling noise. Thus, when the inverse problem of electrocardiography is formulated in this manner it is ill-posed, i.e. the solution does not depend continuously on the data, and regularization methods produce severely filtered solution candidates [2, 3, 4, 5]. Almost since the inverse problem of electrocardiography was first stated, mechanisms for incorporating inherent time constraints have been proposed, but none has led to a well-posed formulation. We will demonstrate in this chapter that the ill-posedness of the inverse problem of electrocardiography resolves when one supplements the usual bidomain assumptions with the assumption that phase 0 of the ventricular action potential is a step discontinuity. This extends work found in [6, 7, 8].

2 Scope

We incorporate into the inverse problem of electrocardiography the time constraint that phase 0 of the action potential be a step discontinuity. This is shown to lead, if one makes the usual anisotropic bidomain assumptions, to a nominally well-posed formulation of the inverse problem. The practical requirements for applying this result are that only a small number of activation extrema on the ventricular surface occur simultaneously, and that predicted discontinuities of body-surface potential time derivatives be recognizable in noisy electrocardiographic signals.

3   Demonstration of well-posedness from discontinuity of action potential and bidomain assumptions

We shall restrict our attention to the interval of ventricular depolarization (the QRS interval). Yamashita and Geselowitz [9] have shown that, under anisotropic bidomain conditions,

$$\phi(y,t) = -\int_V G_i \nabla \phi_m(x,t) \cdot \nabla \psi(x,y) dV_x, \tag{1}$$

and, after integrating by parts,

$$\phi(y,t) = -\int_S \phi_m(x,t) G_i(x) \nabla \psi(x,y) \cdot d\mathbf{S}_x + \int_V \phi_m(x,t) \nabla \cdot (G_i(x) \nabla \psi(x,y)) dV_x, \tag{2}$$

where $\phi(y,t)$ is body-surface potential; $\phi_m(x,t)$ is transmembrane potential (action potential); $\nabla \psi(x,y)$ is the electric field at x when a unit current is injected at y with the heart-conductivity tensor taken as the sum of the intracellular- and extracellular-conductivity tensors; $G_i(x)$ is the intracellular conductivity tensor; S is the surface surrounding the ventricular muscle V; and $d\mathbf{S}_x$ is the differential surface element multiplied by the outward unit normal of S at x. Yamashita and Geselowitz were interested in conditions under which the volume integral on the right side of (1) is zero, and they showed that this happens under the assumption of the equal anisotropy of the extracellular- and intracellular-conductivity tensors (the assumption that the intracellular- and extracellular-conductivity tensors are proportional). We shall make no such assumption here.

If we assume that phase 0 is a step discontinuity and that phase 0 and phase 1 cannot be distinguished at the time resolution applicable to this approach (see below), then the ventricular action potential may be written as

$$\phi_m(x,t) = a(x,t) + b(x) \cdot H(t - \tau(x)), \tag{3}$$

where $H(\cdot)$ is the Heaviside function (zero for arguments less than zero, unity for arguments greater than zero) and $\tau(x)$ is the ventricular-activation function (time phase 0 occurs at point x). For convenience, we will take $a(x,t)$ to be a constant a, this being approximately the case for normal ventricular muscle ($a(x,t) \approx -90$ mV). However, it will be seen that the following argument also remains valid whenever $a(x,t)$ varies more slowly than the steepness of phase 0 of the action potential.

Substituting (3) into (2) gives

$$\phi(y,t) = \int_S A(x,y) H(t - \tau(x)) dS_x + \int_V B(x,y) H(t - \tau(x)) dV_x \tag{4}$$

$$\equiv \phi_1(y,t) + \phi_2(y,t), \tag{5}$$

where $A(x,y)$ and $B(x,y)$ are transfer functions derived from $G_i \nabla \psi$ (which, for convenience, are assumed to have absorbed the phase 0 amplitude $b(x)$). The constant a can have no effect because it is the gradient of (3) which ultimately is the source via equation (1). (Were a taken to be a non-constant smooth function, one would have to add a third integral to the right side of (4), but it would be a smooth function of time and would therefore not affect the following derivation.) We may also note that, if one assumes equal anisotropy, (4) reduces to the equation which is the basis of the approach of Cuppen, van Oosterom, and Huiskamp [5, 10] (who have previously emphasized the desirability of expressing the inverse electrocardiography problem in terms of the activation function $\tau(x)$).

In what follows, we will take $\tau(x)$ to be restricted to whichever domain is germain to the discussion; e.g. when we refer to $\tau(x)$ in the context of the first integral on the right side of (4), it is understood that its extrema are those which pertain to this function defined on the two-dimensional manifold S, and not to a function defined on the three-dimensional manifold V.

Each of the integrals on the right side of (4) can be looked at as a linear transformation of $H(t - \tau(x))$. The latter function takes only the values zero or one and is thus a characteristic function for the depolarized point set. Consequently, the sources of ϕ_1 and ϕ_2 are simply point sets in S and V such that $H(t - \tau(x)) = 1$. Hence, ϕ_1 reflects the evolution of a set of curves (the surface isochrones) on a finite surface, and ϕ_2 reflects the evolution of a set of surfaces (the depolarization wave fronts) in a finite volume. The two integrals on the right side of (4) can be thought of as operators over each of these two point sets, respectively. Note that the first of these point sets undergoes topological changes at the extrema of the activation function $\tau(x)$ on the ventricular surface; at the time when an epicardial breakthrough or sink occurs, the number of distinct pieces or holes of the surface defined as the depolarized portion of the ventricular surface will increase or decrease. The importance of this topological change is that we can compute in more stable fashion when and where it occurs, as demonstrated below.

From (5), we have

$$\phi_1(y, t) = \int_S A(x, y) H(t - \tau(x)) dS_x.$$

Function ϕ_1 is smooth at t if t is not a critical time of $\tau(x)$ (because then the boundary of the non-zero portion of the integrand propagates smoothly, given that the bidomain and conductivity tensors are smooth). Differentiating,

$$\frac{\partial \phi_1(y, t)}{\partial t} = \int_S A(x, y) \delta(t - \tau(x)) dS_x.$$

Because of the delta function, at each time point, the surface integral can be collapsed to a line integral over the set of curves which together bound the depolarized portion of S at that time. After an epicardial breakthrough (a minimum of $\tau(x)$), the number of curves comprising the domain of the line integral increases by one (the extra curve being the new isochrone local to the breakthrough point); after an activation sink (a maximum of $\tau(x)$) the number of curves decreases by one. If x_c is an epicardial breakthrough, and $t_c = \tau(x_c)$, then it is of interest to consider the difference in the value of the derivative just before and just after t_c. Assuming that there is only a single breakthrough x_c at time t_c, there will be n connected curves comprising an isochrone just before t_c, and $n + 1$ connected curves just after t_c. Hence,

$$\frac{\partial \phi_1(y, t)}{\partial t}\Big|_{t_c^-}^{t_c^+} = \lim_{t \to t_c^+} \int_{S_{x_c}} A(x, y)\delta(t - \tau(x))dS_x, \tag{6}$$

where S_{x_c} is a small neighborhood of x_c. We evaluate the above integral as follows: Let $t > t_c$ be in a small neighborhood of t_c. To first order

$$\int_{S_{x_c}} A(x, y)H(t - \tau(x))dS_x = A(x_c, y)\int_{S_{x_c,t}} dS_x, \tag{7}$$

where $S_{x_c,t}$ is the subset of S local to x_c depolarized by time t. Let

$$C_{x_c,t} = \{x \in S_{x_c} : \tau(x) = t\}$$

(i.e., $C_{x_c,t}$ is the boundary of $S_{x_c,t}$, or the local component of the isochrone at t). Then the integral on the right side of (6) is to first order the area enclosed by the projection of $C_{x_c,t}$ on the plane tangent to $S_{x_c,t}$ at x_c. Let (ξ_1, ξ_2) be a Cartesian coordinate system on this plane with origin at x_c. For $t > t_c$ in a small neighborhood of t_c, the projection of $C_{x_c,t}$ to this plane is given to first order by

$$\{(\xi_1, \xi_2) : t - t_c = (h_{11}/2)\xi_1^2 + h_{12}\xi_1\xi_2 + (h_{22}/2)\xi_2^2\}$$

where h_{11}, h_{12}, and h_{22} are components of the Hessian of $\tau(x)$ at x_c (the above is simply taken from the Taylor expansion of $\tau(x)$ around x_c). The area enclosed by this ellipse is

$$\frac{2\pi(t - t_c)}{\sqrt{h_{11}h_{22} - h_{12}^2}} = \frac{2\pi(t - t_c)}{\sqrt{\kappa}}. \tag{8}$$

Note that, in the coordinate system we have chosen, the term under the radical (the determinant of the Hessian matrix) is the Gaussian curvature of $\tau(x)$ at x_c [11]. To calculate (6), we take the time derivative of (7), which, in view of (8), gives

$$\frac{\partial \phi_1(y, t)}{\partial t}\Big|_{t_c^-}^{t_c^+} = \frac{2\pi}{\sqrt{\kappa}}A(x_c, y).$$

Function ϕ_2 is given by the volume integral in (4). We have

$$\frac{\partial \phi_2}{\partial t}(y, t) = \int_V B(x, y)\delta(t - \tau(x))dV_x. \tag{9}$$

In view of the delta function, the right side of the above equation can be collapsed to a surface integral whose domain is the depolarization wave front. This wave front will contain a growing hole starting at a breakthrough time; thus, we may take the wave front to be a smoothly propagating surface minus a "hole" whose area grows in size after the breakthrough time according to (8). Since the size of the hole is continuous across the breakthrough time, the right side of (9) will be continuous around the breakthrough time (so, be it noted, is the domain of the surface integral in (7); however, it was the *derivative* of that expression that we desired to compute). Therefore, the discontinuity in the time derivative of ϕ is entirely due to ϕ_1, and

$$J_{t_c}(y) \equiv \left. \frac{\partial \phi(y,t)}{\partial t} \right|_{t_c^-}^{t_c^+} = \frac{2\pi}{\sqrt{\kappa}} A(x_c, y). \tag{10}$$

In principle, one calculates $J_{t_c}(y)$ (the "Jump Map") by taking the time derivative of the body-surface potential, noting the simultaneous step discontinuities, and, for each such time, making the map of the step-discontinuity sizes at all torso electrode sites y (of course, the derivation of the step discontinuities depends on the assumption that phase 0 is a step discontinuity). $A(x,y)$ is presumed to be estimated by a forward-problem solution; x_c is then calculated as the point which maximizes the correlation of $J_{t_c}(y)$ and $A(x,y)$ (note that, if $b(x)$ is constant, its value will not affect the selected point that maximizes this correlation).

Equation (10) can be contrasted with equation 13 of [7], which expresses the Jump Map relation in terms of a transfer function relating epicardial potentials instead of $A(x,y)$. The latter equation explicitly assumed that one can ignore discontinuities in the derivative of epicardial potential at a point x on the ventricular surface remote from the breakthrough point. The equation (10) is therefore more accurate than that equation; in fact, it is also consistent with (though more general than) an equation derived through the use of the oblique-dipole-layer hypothesis (equation 12 in [6]).

A Morse function is a twice-differentiable function that has nonzero curvature at its critical points. Because Morse functions are dense in the set of twice-differentiable functions, it is no limitation to restrict the admissible solutions for $\tau(x)$ to the set of Morse functions that take values in the QRS interval (this set is also uniformly bounded). It is easily shown that a uniformly bounded set of Morse functions having fewer than n extrema has uniformly bounded Total Variation, where

$$\text{Total Variation of } \tau(x) = \int_S |\nabla \tau(x)| dS_x. \tag{11}$$

The uniform boundedness of the Total Variation follows from the fact that the number of ridges in a Morse function is limited by the number of extrema it has. This then limits how frequently the function can "go up and down" (if the domain of $\tau(x)$ was one-dimensional, the assumption concerning Morse functions would be unnecessary). A set of uniformly bounded functions having uniformly bounded Total Variation is compact [12]. Thus, by computing the extrema of $\tau(x)$, we have limited the admissible candidates for $\tau(x)$ to a compact set. A continuous operator whose domain is restricted to a compact set has a continuous inverse. Hence, a "properly constructed" solution to the approximate equation of Cuppen and van Oosterom,

$$\phi(y,t) = \int_S A(x,y) H(t - \tau(x)) dS_x,$$

will be stable because we have limited sufficiently the set of possible solutions. Furthermore, the extrema of such a solution happen to be the extrema of the solution to the exact equation (3), since all members of the admissible set have the correct extrema.

It should be emphasized that the above result depends on the identification of the set of "derivative discontinuities" in noisy analog signals of body-surface potentials. To be recognized, the "discontinuities" must therefore be sufficiently large with respect to the signal-to-noise ratio. Given this set of discontinuity times, the activation map on the ventricular surface depends continuously on the body-surface potential distribution.

4 Discussion

Since phase 0 of the action potential is not infinitely steep, (2) should be replaced by

$$\phi_m(x, t) = a(x, t) + p(x, t) * [b(x) \cdot H(t - \tau(x))], \tag{12}$$

where $p(x, t)$ is a unit pulse that is at least one millisecond long, and "$*$" indicates convolution. In fact, in view of the bidomain setting, the pulse must be more than one millisecond long. This is true because the "action potential" at a bidomain point must be averaged over the spatial extent of that "point" and is related to the thickness of the depolarization wave front. If the depolarization wave front were infinitesimally thin, a ventricular electrogram would have a step discontinuity (since potential is discontinuous across a double-layer wave front). Instead, its "intrinsic deflection" has a duration of three to four milliseconds. This is consistent with a wave-front thickness on the order of one millimeter, and propagation velocity of approximately 0.5–1 millimeter per millisecond [13]. Since ϕ_m derives from the difference between the intracellular and extracellular potentials (the latter being the ventricular electrogram), it follows that phase 0 should be about 3.5 milliseconds long. Thus, the Heaviside function in (4) should be convolved with a pulse of this duration. One finds, after incorporating the convolution integral into (4), and then reversing the resulting order of integration and repeating the derivation of (10), that the previously derived step discontinuity in $\partial \phi / \partial t$ is convolved with a 3.5–millisecond pulse. Note that the spatial averaging due to the dispersed nature of activation implies that action potential phase 0 and phase 1 will also be averaged together, which justifies not explicitly featuring phase 1 in (3).

Thus, the time-resolution increment in this problem seems to be roughly 3.5 milliseconds. If only a single extremal point is associated with an interval of this duration, then the convolved version of (10) remains generally valid (although, strictly speaking, the contribution to the "Jump Map" defined by (9) is no longer zero and should also be considered, this contribution vanishes under conditions of equal anisotropy). The extremal point can still be found by taking it to be the point giving the greatest correlation between the transfer function and a "Jump Map" arising from the convolved step.

On the other hand, a non-infinitesimal time-resolution interval implies that several discrete breakthroughs could occur during this same interval. Under such circumstances, we would have to replace (10) with

$$J_{t_c}(y) = 2\pi \sum_i \frac{A(x_{c_i}, y)}{\sqrt{\kappa_i}} = \int_S A(x, y)\mu(x)dS_x,$$ (13)

where it is understood that, in the second equality, $\mu(x)$ is the sum of delta functions at the surface breakthrough and sink sites which occur during the given 3.5-millisecond interval. One could consider the source in the above integral equation to be vastly less complex than the usual formulation of the inverse problem (the source is only a few isolated points). Okamoto $et\ al.$ [2] has estimated that it may be possible to compute in a stable fashion up to fifteen source parameters if one has body-surface potential measurements that are 99% accurate. Each extremal point requires three parameters; two define its location on S, and one defines its magnitude. Thus, presuming a stable differentiation scheme, it might be possible to localize up to five "simultaneous" extrema during each time increment, since the transfer function involved in the Jump Map equation (13) is of essentially the same type as those considered by Okamoto $et\ al..$

The proper role of time constraints inherent in the inverse problem of electrocardiography has been obscure. A natural way to incorporate them would be in the manner suggested by our derivation of the result that the use of time constraints potentially renders the problem well-posed.

Acknowledgements

The work reported herein was supported in part by a grant from the RSNA Research and Education Fund.

References

[1] R.M. Gulrajani, F.A. Roberge, and G.E. Mailloux. The inverse problem of electrocardiography. In: P. Macfarlane and T.D.V. Lawrie, editors, $Comprehensive\ Electrocardiology$. Pergamon, Oxford, 1989, pages 237–288.

[2] Y. Okamoto, Y. Teramachi, and T. Musha. Limitation of the inverse problem in body surface potential mapping. $IEEE\ Trans.\ Biomed.\ Eng.$ BME-30:749–754, 1983.

[3] R.C. Barr, M.S. Spach. Inverse calculation of QRS-T epicardial potentials from body surface potential distributions for normal and ectopic beats in the intact dog. $Circ.\ Res.$ 42:661–675, 1978.

[4] P. Colli Franzone, L. Guerri, S. Tentoni, C. Viganotti, S. Baruffi, S. Spaggiari, and B. Taccardi. A mathematical procedure for solving the inverse potential problem of electrocardiography. $Math.\ Biosci.$ 77:353–396, 1985.

[5] G.J.M. Huiskamp and A. van Oosterom. The depolarization sequence of the human heart surface computed from measured body surface potentials. *IEEE Trans. Biomed. Eng.* BME-35:1047–1058, 1988.

[6] F. Greensite. Some imaging parameters of the oblique dipole layer cardiac generator derivable from body surface electrical potentials. *IEEE Trans. Biomed. Eng.* BME-39:159–164, 1992.

[7] F. Greensite. A new method for regularization of the inverse problem of electrocardiography. *Math. Biosci.* 111:131–154, 1992

[8] F. Greensite. Demonstration of "discontinuties" in the time derivatives of body surface potentials, and their prospective role in noninvasive imaging of the ventricular surface activation map. *IEEE Trans. Biomed. Eng. in press*

[9] Y. Yamashita and D. Geselowitz. Source-field relationships for cardiac generators on the heart surface based on their transfer coefficients. *IEEE Trans. Biomed. Eng.* 32:964–970, 1985.

[10] J. Cuppen and A. van Oosterom. Model studies with the inversely calculated isochrones of ventricular depolarization. *IEEE Trans. Biomed. Eng.* BME-31:652–659, 1984.

[11] M. de Carmo. *Differential geometry of curves and surfaces.* Prentice Hall, Englewood Cliffs, NJ, 1976, pages 162–163.

[12] E. Giusti. *Minimal surfaces and functions of bounded variation.* Birkhauser, Boston, MA, 1984, page 17.

[13] A. van Oosterom. Cell models – macroscopic source descriptions. In: P. Macfarlane and T.D.V. Lawrie, editors, *Comprehensive Electrocardiology.* Pergamon, Oxford, 1989, pages 161–169.

Local Regularization and Adaptive Methods for the Inverse Problem

C.R. Johnson and R.S. MacLeod

1 Introduction

The intrinsic electrical activity of the heart gives rise to an electric field within the volume conductor of the thorax. Thus, the potentials on the thorax surface are related to potentials upon the heart's surface via the resistive properties of the intermediary tissues. The general inverse problem in electrocardiography can be stated as follows: given a set of body-surface potentials and the geometry and conductivity properties of the body, calculate the potentials on, and the source currents within, the heart. Mathematically, this can be posed as an inverse source problem in terms of the primary-current sources within the heart and described by Poisson's equation for electrical conduction:

$$\nabla \cdot \sigma \nabla \Phi = -I_v \quad \text{in } \Omega \tag{1}$$

with the boundary condition

$$\sigma \nabla \Phi \cdot \mathbf{n} = 0 \quad \text{on } \Gamma_T \tag{2}$$

where Φ are the potentials, σ is the conductivity tensor, I_v are the cardiac current sources per unit volume, and Γ_T and Ω represent the surface and the volume of the thorax, respectively. The goal is to recover the magnitude and location of the cardiac sources. In general, (1) does not have a unique solution. To solve the general source problem, one usually divides the volume of the heart into subregions and makes simplifying assumptions regarding the form of the source term (such as dipoles or other equivalent sources) within those subregions. One then tries to recover information regarding the magnitude and direction of the simplified model sources.

Alternatively, one can solve for the potentials on a surface bounding the heart. Thus, instead of solving Poisson's equation, one solves a generalized Laplace equation with Cauchy boundary conditions:

$$\nabla \cdot \sigma \nabla \Phi = 0 \tag{3}$$

with boundary conditions:

$$\Phi = \Phi_0 \quad \text{on } \Sigma \subseteq \Gamma_T \quad \text{and} \quad \sigma\nabla\Phi \cdot \mathbf{n} = 0 \quad \text{on } \Gamma_T. \tag{4}$$

While this version of the inverse problem has a unique solution [1], the problem is still mathematically ill-posed; i.e., because the solution does not depend continuously on the data, small errors in the measurement of body-surface potentials can yield unbounded errors in the solution.

An accurate solution to the inverse problem in electrocardiography would provide a noninvasive procedure for evaluating myocardial ischemia [2, 3], localizing ventricular arrhythmias and the site of accessory pathways in Wolff-Parkinson-White (WPW) syndrome, and, more generally, determining patterns of excitation and recovery of excitability.

2   Scope

One of the fundamental problems in theoretical electrocardiography can be characterized as an inverse problem. We present new methods for achieving better estimates of heart-surface potential distributions from body-surface potentials through an inverse procedure. First, we outline an automatic adaptive-refinement algorithm that minimizes the spatial-discretization error in the transfer matrix and increases the accuracy of the inverse solution. Second, we introduce a new local regularization procedure, which works by partitioning the global transfer matrix into sub-matrices, allowing for varying amounts of smoothing. This allows regularization parameters for each sub-matrix to be specifically "tuned" by means of an *a priori* scheme based on the L−curve method. This *local* regularization method provides substantially greater accuracy than do global-regularization schemes. We conclude with specific examples of these techniques, using thorax models derived from MRI data.

3   Mathematical theory

Finite element approximation

To solve the boundary-value problem in (1) or (3) in terms of the epicardial potentials, we need to pose the problem in a computationally tractable way. This involves approximating the boundary-value problem on a finite dimensional subspace and re-formulating it in terms of a linear matrix equation. For this study, we used the finite-element method (FEM) to approximate the field equation and construct the set of matrix equations. Briefly, one starts with the Galerkin formulation of (1). Note that this includes the Dirichlet and Neumann boundary conditions,

$$(\sigma\nabla\Phi, \nabla\overline{\Phi}) = -(I_v, \overline{\Phi}), \tag{5}$$

where $\overline{\Phi}$ is an arbitrary test function, which can be thought of physically as a virtual potential field, and the notation $(\phi_1, \phi_2) \equiv \int_\Omega \phi_1 \phi_2 \, d\Omega$ denotes the inner product in $L_2(\Omega)$. We now use the finite-element method to turn the continuous problem into a discrete formulation. First, we discretize the solution domain, $\Omega = \bigcup_{e=1}^{N} \Omega_e$, and define a finite dimensional subspace,

$$V_h \subset V = \{\overline{\Phi} : \overline{\Phi} \text{ is continuous on } \Omega, \nabla \overline{\Phi} \text{ is piecewise continuous on } \Omega\}$$

. We define parameters of the function $\overline{\Phi} \in V_h$ at node points, $\alpha_i = \overline{\Phi}(x_i), i = 1, 2, \ldots, N$ and define the basis functions $\Psi_i \in V_h$ as linear piecewise continuous functions that take the value 1 at node points and 0 elsewhere. We can then represent the function $\overline{\Phi} \in V_h$ as

$$\overline{\Phi} = \sum_{i=1}^{N} \alpha_i \Psi(x_i) \tag{6}$$

such that $\overline{\Phi} \in V_h$ can be written in a unique way as a linear combination of the basis functions $\Psi_i \in V_h$. The finite-element approximation of the original boundary-value problem (1) can be stated as

$$\text{Find } \Phi_h \in V_h \text{ such that } (\sigma \nabla \Phi_h, \nabla \overline{\Phi}) = -(I_v, \overline{\Phi}). \tag{7}$$

Furthermore, since $\Phi_h \in V_h$ satisfies (7), then we have $(\sigma \nabla \Phi_h, \nabla \Psi) = -(I_v, \Psi_i)$. Finally, since Φ_h itself can be expressed as the linear combination

$$\Phi_h = \sum_{i=1}^{N} \xi_i \Psi_i(x) \quad \xi_i = \Phi_h(x_i), \tag{8}$$

we can then write (7) as

$$\sum_{i=1}^{N} \xi_i (\sigma_{ij} \nabla \Psi_i, \nabla \Psi_j) = -(I_v, \Psi_j) \quad j = 1, \ldots, N. \tag{9}$$

The finite-element approximation of (1) can equivalently be expressed as a system of N equations with N unknowns, $\xi_1, \ldots, \xi_N$ (the potentials). In matrix form, the above system is expressed as $A\xi = b$, where $A = (a_{ij})$ is the global stiffness matrix and has elements $(a_{ij} = (\sigma_{ij} \nabla \Psi_i, \nabla \Psi_j))$, while $b_i = -(I_v, \Psi_i)$ is the vector of source contributions.

Adaptive methods

Discrete approaches to bioelectric inverse and imaging problems have centered around classical numerical techniques for solving partial differential equations: finite difference, finite element, and boundary-element methods. To date, instead of developing and utilizing specific, quantitative methods, most investigators have assigned spatial-discretization levels by relying largely on previous experience and their common sense, utilizing what has apparently worked before and perhaps reducing the discretization level in regions that are known *a priori* to contain high gradients. This situation has arisen, at least in part, because of the computational load represented by the large, complex, inhomogeneous, even-anisotropic geometries that characterize many problems in bioelectric field imaging.

We have found, however, that by using *a posteriori* estimates from the finite-element approximation of the governing equations, along with a minimax theorem to determine the stopping criterion, we can locally refine the discretization and reduce the errors in the forward solution. It has always been assumed—and our findings support this notion—that improving the accuracy of the forward solution also improves the subsequent inverse solution. The novel aspect of our approach is that it uses local approximations of the error in the numerical solutions to drive an automatic adaptive-mesh refinement [4, 5].

Mathematically, we can obtain an error estimator by starting from the *weak formulation* of the finite-element approximation (7). We can prove that $\Phi_h \in V_h$ (where Φ_h is the finite-element solution and related to ξ by (8)) is the best approximation of the exact solution Φ in the sense that

$$\|\nabla\Phi - \nabla\Phi_h\| \leq \|\nabla\Phi - \nabla\tilde{\Phi}\| \quad \forall\, \tilde{\Phi} \in V_h \tag{10}$$

where

$$\|\nabla\tilde{\Phi}\| = (\int_\Omega \nabla\tilde{\Phi} \cdot \nabla\tilde{\Phi})^{\frac{1}{2}}. \tag{11}$$

In particular, we can choose $\tilde{\Phi}$ to be the interpolant of Φ

$$\pi_h\Phi(N_i) = \Phi(N_i) \qquad i = 1,\ldots,M \quad \pi_h\Phi \in V_h. \tag{12}$$

This says that the error in the finite-element approximation is bounded from above by the interpolation error. With certain constraints on our elements, we can prove the following, well-know relationships:

$$\|\nabla\Phi - \nabla\pi_h\Phi\| \leq Ch \tag{13}$$

and

$$\|\Phi - \pi_h\Phi\| \leq Ch^2 \tag{14}$$

where C is a positive constant that depends on the size of the second partial derivative of Φ and the smallest angle formed by the sides of the elements in V_h. These estimates show that the error in both Φ and the gradient of Φ tend to zero as the discretization parameter, h, tends to zero. We can use these relationships to provide an adaptive algorithm for decreasing discretization error.

To make this a little more precise, we define the semi-norm

$$|\tilde{\Phi}|_{H^r(\Omega)} = (\sum_{|\alpha|=r} \int_\Omega |D^\alpha\tilde{\Phi}|^2 \, dx)^{\frac{1}{2}} \tag{15}$$

where

$$D^\alpha \tilde{\Phi} = \frac{\partial^{|\alpha|} \tilde{\Phi}}{\partial x_1^{\alpha_1} \partial x_2^{\alpha_2}} \qquad |\alpha| = \alpha_1 + \alpha_2 \tag{16}$$

are α-order partial derivatives and

$$H^K(\Omega) = \{\tilde{\Phi} \in L_2(\Omega) : D^\alpha \tilde{\Phi} \in L_2(\Omega), \ |\alpha| \leq K\} \tag{17}$$

defines the Sobolev spaces. Given these definitions, we can then obtain the error estimates analogous to those in (13) and (14) for the finite-element approximation [6, 7]:

$$|\Phi - \Phi_h|_{H^1(\Omega)} \leq |\Phi - \pi_h \Phi|_{H^1(\Omega)} \leq C[\sum_{K \in T_n} (h_k |\Phi|_{H^2(K)})^2]^{\frac{1}{2}} \tag{18}$$

or, in the $L_2(\Omega)$ norm,

$$\|\Phi - \Phi_h\|_{L_2(\Omega)} \leq \|\Phi - \pi_h \Phi\|_{L_2(\Omega)} \leq Ch^2 |\Phi|_{H^2(\Omega)}. \tag{19}$$

For the potential gradients, we have [8]

$$\|\nabla\Phi - \nabla\Phi_h\|_{L_\infty(\Omega)} \leq C\|\nabla\Phi - \nabla\pi_h\tilde{\Phi}\|_{L_\infty(\Omega)} \leq C \max[h_K \max_{\alpha=2} \|D^\alpha\Phi\|_{L_\infty(\Omega)}]. \tag{20}$$

Given these previous error estimates for the finite-element approximation, we can now use the estimates to decide where in our original finite-element mesh the approximation is not accurate and create a recursive, adaptive algorithm to re-discretize in the appropriate areas. Suppose we want the accuracy of our finite-element approximation to be within some given tolerance, δ, of the true solution, i.e.

$$|\Phi - \Phi_h|_{H^1(\Omega)} \leq \delta. \tag{21}$$

Then, we redefine our mesh until

$$\sum_{K \in T_h} (h_k |\Phi_h|_{H^2(K)})^2 \leq \frac{\delta^2}{C^2}. \tag{22}$$

Here, the sum is over all the K elements in the tessellated geometry, T_h, and we test to see if the error is less than the *normalized* tolerance. If the value of the error estimate exceeds the tolerance, the element is "flagged" for further refinement. The refinement can occur either by further subdividing the egregious element (so-called h-adaption) or by increasing the basis function of the element (p-adaption), or both (hp-adaption). One should note that the $H^2(\Omega)$ norm in (22) requires the second partial derivative of the finite-element approximation. As stated in (8), we have assumed a linear basis function for our fundamental element. Thus, there is no continuity of the second derivative, and we must approximate it using a centered difference (or other) formula based on first-derivative information.

Of significant practical importance is the choice of a reasonable tolerance, δ. Since the general location and bounds of the potential field are dictated by known physiological considerations, we can generate an initial mesh based upon simple field-strength–distance arguments. To estimate an upper bound of the potential (or electric) field, we can find the field strength on a cylindrical surface which encloses the actual geometry. The result is an estimate of the potential field and potential gradients which are used to produce the initial graded mesh. The goal of this calculation is to assure that the errors far from the sources are minimal. We can then refine our finite-element mesh by choosing the tolerance, δ, to be some fraction of the estimated error corresponding to the elements furthest from the sources. Thus, we have used a minimax principle, minimizing the maximum error based upon initial (conservative) estimates of the potential field and potential gradients [4].

Regularization

Traditional schemes for solving the inverse problem, (3), have involved reformulating the linear equation $A\xi = b$ into $\xi_T = K\xi_E$, where ξ_T and ξ_E are the potentials on the body surface (torso) and the heart's surface (epicardium), respectively, and K is the $T \times E$ transfer matrix of coefficients relating the measured body-surface potentials to the potentials measured on the heart. From this statement of the forward problem, the inverse problem can then be expressed as $\xi_E = K^{-1}\xi_T$

Unfortunately, K is ill-conditioned, and regularization techniques must be used to restore continuity of the solution back onto the data. To date, all regularization schemes have focused on treating K globally and using Tikhonov, singular-value decomposition (SVD), or Twomey algorithms with a single regularization parameter. Briefly, one wishes to find an approximate vector, $\xi_{E\varepsilon}$, such that $\xi_{E\varepsilon} \to \xi_E$ as $\varepsilon \to 0$, where ε is the error between the approximate and true values. It is possible to reformulate the K matrix into an expression involving sub-matrices of K. We have found that it is then possible, and advantageous, to apply regularization not globally, but at the *local* level of the sub-matrices.

Investigators have long noticed that the discontinuities in the inverse solution appear irregularly distributed throughout the data. Tikhonov and other regularization schemes are, in effect, filters, which restore continuity by attenuating the high (spatial) frequency components of the solution. Since regularization is usually applied globally, the results can leave some regions overly damped or smoothed while others remain poorly constrained. Our idea, then, is to apply regularization only to sub-matrices that require it and to apply different amounts of regularization to different sub-matrices.

We begin by expressing $A\xi = b$ for the Cauchy problem in (3) as

$$
\begin{pmatrix} A_{TT} & A_{TV} & A_{TE} \\ A_{VT} & A_{VV} & A_{VE} \\ A_{ET} & A_{EV} & A_{EE} \end{pmatrix} \begin{pmatrix} \xi_T \\ \xi_V \\ \xi_E \end{pmatrix} = \begin{pmatrix} 0 \\ 0 \\ 0 \end{pmatrix}. \tag{23}
$$

We can then rearrange the matrix to solve directly for the epicardial potentials in terms of the measured body-surface potentials,

$$\xi_E = (A_{TV} A_{VV}^{-1} A_{VE} - A_{TE})^{-1} (A_{TT} - A_{TV} A_{VV}^{-1} A_{VT}) \xi_T, \qquad (24)$$

where the subscripts T, V, and E stand for the nodes in the regions of the torso, the internal volume, and epicardium, respectively. Since torso and epicardial nodes are always separated by more than one element, the A_{TE} submatrix is zero, and we can then rewrite (24) as

$$\xi_E = (A_{VE}^{-1} A_{VV} A_{TV}^{-1})(A_{TT} - A_{TV} A_{VV}^{-1} A_{VT}) \xi_T. \qquad (25)$$

Note that we now have inverses of three sub-matrices, A_{VE}, A_{TV}, and A_{VV}, as well as several other matrix operations to perform. If one estimates the condition number, κ, of the three sub-matrices, using the ratio of the maximum to minimum singular values from a singular-value decomposition (SVD), one finds that the condition number varies significantly (from $\kappa \approx 200$ for A_{VV} to $\kappa \approx 1 \times 10^{16}$ for A_{VE} in the two-dimensional model described below). Thus, one can regularize each of the sub-matrices differently, even leaving some of the other sub-matrices untouched. The overall effect of this *local regularization* is more control over the regularization process.

Since the goal of the inverse problem in electrocardiography is to estimate the potentials on the epicardial surface accurately and noninvasively (i.e., non-surgically), we need a method for choosing an optimal regularization parameter *a priori*. Traditional schemes have been based on discrepancy principles [9], quasi-optimality criterion [9, 10], and generalized cross-validation [11]. Recently, Hansen [12] has extended the observations of Lawson and Hanson [13] and proposed a new method for choosing the regularization parameter, based on an algorithm that locates the "corner" of a plot of the norm (or semi-norm) of the regularized solution, $\|Lx_\alpha\|$, versus the norm of the corresponding residual vector, $\|Ax_\alpha - b\|$. In this way, one can evaluate the compromise between the minimization of the two norms. We have used the L−curve algorithm along with our local Tikhonov-regularization scheme to improve the accuracy of solutions to the inverse problem.

For the global Tikhonov-regularization scheme, we can find an estimate for $\xi_{E\epsilon}$ by minimizing a generalized form of the Tikhonov functional

$$M^\alpha[\xi_E, \tilde\xi_T, \tilde K] = \|K\xi_E - \tilde\xi\|^2_{\phi_T} + \alpha \|C(\xi_E - \xi'_E)\|^2_{\phi_E} \quad \alpha > 0 \qquad (26)$$

in terms of the epicardial potentials,

$$\xi^\alpha_{E\epsilon} = [\tilde K^T \tilde K + \alpha C^T C]^{-1} [\tilde K^T \tilde\xi_T + \alpha C^T C \xi'_E], \qquad (27)$$

where $\tilde K$ is the approximation of the true transformation matrix, K, $\tilde\xi_T$ are the measured body-surface potential values, ξ'_E are *a priori* constraints placed on the epicardial potentials based on physiological considerations, C is a constraint matrix (either the identity matrix, a gradient operator, or a Laplacian operator), and α is the regularization parameter. For our *local* regularization scheme, we replace the global matrix, K, by the two sub-matrices A_{VE} and A_{TV} from (25). (A_{VV} in (25) has a stable inverse and, thus, does not need regularization.)

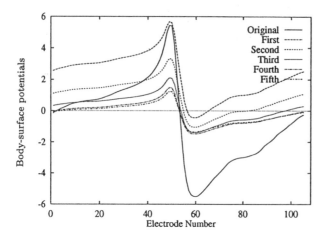

Figure 1. Effects of automatic mesh refinement.

4   Results and discussion

To study the forward and inverse problems in electrocardiography, we developed a series of two- and three-dimensional boundary-element and finite-element models based upon magnetic resonance images from a human subject. We segmented each of 116 MRI scans into contours that defined outer boundary, fat, muscle, lung, and heart regions. We then added additional node points so that we could digitize each layer and pairs of layers tessellated into tetrahedra, using a Delaunay triangulation algorithm [14, 15]. The resulting model of the human thorax contained approximately 675,000 volume elements, each with a corresponding conductivity tensor. For the initial studies on the effects of adaptive control of errors and local regularization, we used a two-dimensional finite-element model extracted from our three-dimensional model of the human thorax. The two-dimensional model was constructed from a single transaxial MRI located approximately 4 cm above the apex of the heart.

As a test of the adaptive algorithm, we placed a simulated source distribution on the surface of the heart model, then, using the procedure described previously, computed forward solutions for different levels of mesh refinement and compared them at the outer torso boundary. Figure 1 shows the potential at the outer boundary versus distance around the two-dimensional contour. The original mesh contained approximately 1500 elements; the final mesh, after five iterations of the adaptive algorithm, contained approximately 7000. The maximum estimated error in the calculated potential was over 30% greater in the original mesh than in the final mesh, and the maximum estimated error in the potential gradient was over 13% larger in the original mesh than in the refined mesh.

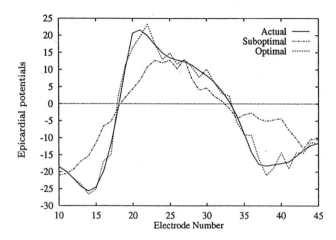

Figure 2. Local-regularization technique.

Increased accuracy does not come without a price; the h-adaption increases the number of degrees of freedom, and thus the computational costs. One must balance accuracy against CPU time. However, since the relative error falls off rapidly with increasing degrees of freedom and then levels out after only a few iterations of the adaptive algorithm, it is not difficult to choose a reasonable balance.

As a test of the local regularization, we applied epicardial potentials recorded during open-chest surgery (from a patient diagnosed with WPW syndrome) as the Dirichlet boundary conditions of the two-dimensional model described above. The tissue conductivities were assigned as follows: fat = 0.045 S/m, fatpad = 0.045 S/m, lungs = 0.096 S/m, skeletal muscle (along the fibers) = 0.3 S/m, skeletal muscle (across the fibers) = 0.1 S/m, and an average thorax value = 0.24 S/m. Forward solutions calculated by means of the adapted mesh served as torso boundary conditions, $\Phi = \Phi_0$ on $\Sigma \subseteq \Gamma_T$ for the inverse solution.

For the sub-matrices A_{VE} and A_{TV}, the $L-$curve algorithm determined the optimal *a priori* regularization parameter. We then applied the local Tikhonov-regularization technique to the two sub-matrices and computed the inverse-solution matrix. Because the size of the two-dimensional finite-element model was relatively small, the calculations stayed in dynamic memory on an IBM RISC-6000 model 580 (for the three-dimensional models, we will use a distributed iterative algorithm to achieve realistic computation times). We then compared the resulting inverse solutions to those we calculated using standard global Tikhonov techniques and to the original measured epicardial potentials.

Applying the local regularization technique recovered the potentials to within 12.6% RMS error. Previous studies have reported the recovery of epicardial potentials with errors in the range of 20–40% [16, 17, 18, 19]. Figure 2 shows the inverse solution calculated by means of the local regularization technique compared with the recorded heart potentials as a function of position on the epicardium. The global solution tended to be smoother, not able to follow the extrema as well as the local solution could. The local solution also showed areas of local error, which suggests that a different partitioning of the sub-matrices may provide even better accuracy.

The limitations of our study are that we have used the same finite-element mesh to calculate solutions for both the forward and inverse problem, and we have not, at this point, added noise to the input torso boundary values. Currently, we are investigating the effects of added noise on a variety of experimental data, as well as developing more general and computationally efficient ways to decompose the transfer matrix. Of particular interest is a method which employs a parallel-block LU decomposition algorithm. Other investigations include methods with which to automatically determine which partitioning of the matrix yields the most accurate solutions and how the computations may be best distributed over multiple processors.

Acknowledgments

This research was supported in part by awards from the Whitaker Foundation, NIH Grant HL47505, the Nora Eccles Treadwell Foundation, and the Richard A. and Nora Eccles Harrison Fund for Cardiovascular Research. The authors would like to thank K. Coles and C. Gitlin for their comments and suggestions.

References

[1] Y. Yamashita. Theoretical studies on the inverse problem in electrocardiography and the uniqueness of the solution. *IEEE Trans Biomed Eng*, BME-29:719–725, 1982.

[2] R.S. MacLeod, C.R. Johnson, M.J. Gardner, and B.M. Horáček. Localization of ischemia during coronary angioplasty using body surface potential mapping and an electrocardiographic inverse solution. In *Computers in Cardiology*, pages 251–254. IEEE Press, 1992.

[3] R.S. MacLeod, M.J. Gardner, R.M. Miller, and B.M. Horáček. Application of an electrocardiographic inverse solution to localize ischemia during coronary angioplasty. *J Cardiovasc Electrophysiol*, 6:2–18, 1995.

[4] C.R. Johnson and R.S. MacLeod. Nonuniform spatial mesh adaption using a posteriori error estimates: applications to forward and inverse problems. *Applied Numerical Mathematics*, 14:311–326, 1994.

[5] J.A. Schmidt, C.R. Johnson, J.C. Eason, and R.S. MacLeod. Applications of automatic mesh generation and adaptive methods in computational medicine. In J.E. Flaherty and I. Babuska, editors, *Modeling, Mesh Generation, and Adaptive Methods for Partial Differential Equations*. Springer-Verlag, 1994 (to appear).

[6] P.G. Ciarlet and J.L Lions. *Handbook of Numerical Analysis: Finite Element Methods*, volume 1. North-Holland, Amsterdam, 1991.

[7] C. Johnson. *Numerical solution of Partial Differential Equations by the Finite Element Method*. Cambridge University Press, Cambridge, 1990.

[8] R. Rannacher and R. Scott. Some optimal error estimates for piecewise linear finite element approximations. *Math. Comp.*, 38:437–445, 1982.

[9] V.A. Morozov. *Methods for Solving Incorrectly Posed Problems*. Springer-Verlag, New York, 1984.

[10] R. Kress. *Linear Integral Equations*. Springer-Verlag, New York, 1989.

[11] G.H. Golub, M.T. Heath, and G. Wahba. Generalized cross-validation as a method for choosing a good ridge parameter. *Technometrics*, 21:215–223, 1979.

[12] P.C. Hansen. Analysis of discrete ill-posed problems by means of the L-curve. *SIAM Review*, 34(4):561–580, 1992.

[13] C.L. Lawson and R.J. Hanson. *Solving Least Squares Problems*. Prentice-Hall, Englewood Cliffs, NJ, 1974.

[14] C.R. Johnson, R.S. MacLeod, and P.R. Ershler. A computer model for the study of electrical current flow in the human thorax. *Computers in Biology and Medicine*, 22(3):305–323, 1992.

[15] C.R. Johnson, R.S. MacLeod, and M.A. Matheson. Computer simultions reveal complexity of electrical activity in the human thorax. *Comp. in Physics*, 6(3):230–237, May/June 1992.

[16] P. Colli Franzone, G. Gassaniga, L. Guerri, B. Taccardi, and C. Viganotti. Accuracy evaluation in direct and inverse electrocardiology. In P.W. Macfarlane, editor, *Progress in Electrocardiography*, pages 83–87. Pitman Medical, 1979.

[17] P. Colli Franzone, L. Guerri, S. Tentonia, C. Viganotti, S. Spaggiari, and B. Taccardi. A numerical procedure for solving the inverse problem of electrocardiography. Analysis of the time-space accuracy from *in vitro* experimental data. *Math Biosci*, 77:353, 1985.

[18] B.J. Messinger-Rapport and Y. Rudy. Regularization of the inverse problem in electrocardiography: A model study. *Math Biosci*, 89:79–118, 1988.

[19] P.C. Stanley, T.C. Pilkington, and M.N. Morrow. The effects of thoracic inhomogeneities on the relationship between epicardial and torso potentials. *IEEE Trans Biomed Eng*, BME-33:273–284, 1986.

On Regularization Parameters for Inverse Problems in Electrocardiography

J.C. Clements, R. Carroll, and B.M. Horáček

1   Introduction

Regularization methods for applied inverse and ill-posed problems have received considerable attention (see e.g. [1, 2]). We shall be concerned here with determining the optimal regularization parameter required for the solution of the discrete ill-posed problem $A\phi_H = \phi_B$, which arises in electrocardiography when one seeks to recover epicardial potentials ϕ_H from body-surface potentials ϕ_B via a transfer-coefficient matrix A [3, 4]. In saying that the problem is *ill-posed*, we mean that the matrix A is ill-conditioned, with singular values decaying to zero in such a way that no practical separation point can be identified in the singular value spectrum of A.

In the case of the discrete ill-posed problem $A\phi_H = \phi_B$, regularization methods replace the original problem with a "slightly perturbed" well-posed problem which incorporates a smoothing constraint to ensure that the computed regularized solutions are not overly sensitive to small perturbations of A and ϕ_B. We adopted the regularization method of Tikhonov [5], which replaces $A\phi_H = \phi_B$ by the least squares problem $\min\{\|A\phi_H - \phi_B\|^2 + t\|R\phi_H\|^2, \ \phi_H \in E^{N_1}\}$, where $\|..\|$ denotes the N_1-dimensional Euclidean norm, t is the regularization (smoothing) parameter, and R is the regularizing (smoothing) operator. We use both zero-order and second-order Tikhonov regularization; that is, the regularizing operators are $R = I$, the identity operator, and $R = H$, the discretized Laplace operator, respectively. In choosing the regularization parameter t, one must strike a balance between the requirements that the solution be accurate and that it be stable with respect to perturbations of A and ϕ_B. We use the generalized singular value decomposition (GSVD) of A and R to obtain the explicit formula for the regularized solution of the least squares problem that Tikhonov's method produces, and we use this formula to reduce the conditions for estimating the optimal t by the L-curve method [6] and the composite residual and smoothing operator (CRESO) method [7] to computing the maximum value of the simple functions. This provides a fast, reliable procedure for computing optimal regularization parameters.

2 Scope

We derive formulas for computing the regularization parameters that satisfy the L-curve and CRESO criteria when Tikhonov regularization is applied to the solution of a discrete ill-posed problem of the form $Ax = b$. The formulas are based on the GSVD of the matrix $A = UDZ^{-1}$ and the regularizing operator $R = VMZ^{-1}$, and they reduce the problem of determining the parameter values to computing the maximum of a simple function which depends only on the regularization parameter t, the generalized singular values μ_j, and the scalar quantities $u_j \cdot b$, where the u_j are the columns of U. Consequently, solving the problem for different choices of b and the same transfer-coefficient matrix A then only requires the additional calculation of the scalar products. In a sample application, we use this new approach to solve a specific ill-posed inverse problem in electrocardiography for which the solution is known *a priori*, and we examine the relative errors obtained using the $L-$curve and CRESO methods, with both zero-order and second-order Tikhonov regularization.

3 Tikhonov regularization: the L-curve and CRESO methods

Let $x = (x_1, \ldots, x_n)$ be a (column) vector in the real Euclidean space E^n. For $1 \leq p \leq \infty$, we define the norm $\|x\|_p$ by

$$\|x\|_p \equiv \left(\sum_{i=1}^n |x_i|^p \right)^{\frac{1}{p}}, \qquad 1 \leq p < \infty, \qquad \|x\|_\infty \equiv \max\{|x_i|, \ i = 1, \ldots, n\},$$

where $\|x\|_2 = (x \cdot x)^{\frac{1}{2}} = (x^T x)^{\frac{1}{2}} \equiv \|x\|$ and $x \cdot x$ is the usual scalar product. Let A be an $m \times n$ $(m \geq n)$ matrix of rank $r \leq n$, $A \in E^{m \times n}$.

Suppose that the discretization of a specific problem leads to the least squares problem $\min\{\|Ax - b\|^2, x \in E^n\}$ for a fixed $b \in E^m$. One regularization method for solving such problems when the singular value spectrum of A has a well-ordered structure is singular value decomposition itself, truncated to remove the zero singular values in the case of the pseudoinverse matrix A^+ or with the smallest singular values weighted only slightly in other cases [8, 9]. However, if the singular values of A decay to zero in such a way that no practical separation point can be identified, then rank-deficient methods are no longer useful. An approach which has proved successful in solving such problems is Tikhonov regularization. Here a bound is imposed on $\|Rx\|$ for some suitably chosen regularizing matrix $R \in E^{p \times n}(p \leq n)$ and leads to the quadratically constrained least squares problem

$$\min_{x \in E^n} \|Ax - b\|^2, \quad s.t. \ \|Rx\|^2 \leq \gamma, \ \gamma > 0. \tag{1}$$

We shall assume that R has full row rank p and is such that the problem $\min \|Rx\|^2 \leq \gamma$ has a solution $x \in E^n$. We shall also assume that the constraint in (1) is binding, that is that $\gamma < \|Rx_0\|^2$ for the usual least squares solution x_0. Then for any $A \in E^{m \times n}$ there exists a solution x_t of the problem stated as (1) that satisfies the generalized normal equations

$$(A^T A + tR^T R)x_t = A^T b, \tag{2}$$

with the Lagrange multiplier t being determined by the secular equation $\|Rx_t\| = \gamma$ [8]. Moreover, if rank $\begin{pmatrix} A \\ R \end{pmatrix} = n$, where $\begin{pmatrix} A \\ R \end{pmatrix} \in E^{(m+p) \times n}$, then x_t is unique. If $R^T R$ is positive definite, for example, then $A^T A + tR^T R$ is a symmetric, positive definite and invertible matrix for every $t > 0$, and $x_t = (A^T A + tR^T R)^{-1} A^T b$ is the unique solution of (2) for each t. With γ specified, the regularization parameter t can be determined by iterating (2) until the secular equation is satisfied. The difficulty with this method is that the most appropriate value for the bound γ on $\|Rx\|$ is usually not known *a priori*. However, since the equations in (2) are just the normal equations for the least squares problem

$$\min_{x \in E^n} \{\|Ax - b\|^2 + t\|Rx\|^2\} = \min_{x \in E^n} \left\{ \left\| \begin{pmatrix} A \\ \sqrt{t}R \end{pmatrix} x - \begin{pmatrix} b \\ 0 \end{pmatrix} \right\|^2 \right\}, \tag{3}$$

the approach employed when γ is not known *a priori* is to compute that Tikhonov regularized solution x_t of (3) which corresponds to some "optimal" value of the regularization parameter $t > 0$. Two well-known methods for estimating this optimal t are the L-curve and CRESO methods.

The L-curve associated with (3) depicts the relationship between the residual and the solution seminorm as the continuous curve $C_0 : \{(\|Ax_t - b\|, \|Rx_t\|)|t \geq 0\}$ [6, 10, 11, 12]. A suitable choice for t in (3) must be a good compromise between minimizing the residual and regularizing the solution. The criterion proposed by Hansen and O'Leary [11] is that value of t which corresponds to the "corner" point on the L-curve possessing maximum curvature. This condition makes good sense because a smaller t will produce a larger solution seminorm with only a marginally smaller residual, while a larger t will produce a larger residual and only a marginally smaller solution seminorm. For computational and display purposes, however, we found that it is actualy more effective to work with the log-log version of the L-curve defined by $C_1 : \{(\ln(\|Ax_t - b\|), \ln(\|Rx_t\|))|t \geq 0\}$. We show in the next section that $\ln(\|Rx_t\|)$ is a smooth, monotonically decreasing function of $\ln(\|Ax_t - b\|)$ and that C_1 always possesses a point of maximum curvature $\kappa(t^*)$.

Using the CRESO method, one chooses that t which corresponds to the first inflection point on the curve defined by $C_2 : \{(t, t\|Rx_t\|^2 - \|Ax_t - b\|^2)|t > 0\}$ [7]. Thus, the most suitable t here is deemed to be that at which $\dfrac{d}{dt}(t\|Rx_t\|^2)$ and $\dfrac{d}{dt}\|Ax_t - b\|^2$ are changing at precisely the same rate, which leads to the problem of determining the first local maximum $\rho(t^*)$ of the function $\rho(t) = \dfrac{d}{dt}(t\|Rx_t\|^2) - \dfrac{d}{dt}\|Ax_t - b\|^2 = \|Rx_t\|^2 + 2t\dfrac{d}{dt}\|Rx_t\|^2$.

4 Generalized singular value decomposition

Since R has full row rank p, it follows that A and R can be written in the form

$$A = UDZ^{-1}, R = VMZ^{-1} \tag{4}$$

with

$$D = \begin{pmatrix} D_p & O_{p \times (n-p)} \\ O_{(n-p) \times p} & I_{n-p} \end{pmatrix} \in E^{n \times n}, \qquad M = \begin{pmatrix} M_p & O_{p \times (n-p)} \end{pmatrix} \in E^{p \times n}$$

where $Z \in E^{n \times n}$ is a nonsingular matrix, $D_p = \mathrm{diag}(\alpha_1, \ldots, \alpha_p) \in E^{p \times p}$, $M_p = \mathrm{diag}(\beta_1, \ldots, \beta_p) \in E^{p \times p}$, and $U \in E^{m \times n}$ and $V \in E^{p \times p}$ have orthonormal columns, $U^T U = I_n$, $V^T V = I_p$, and $O_{p \times (n-p)}$ is the $p \times (n - p)$ additive identity [8]. In addition, the generalized singular values $\mu_j = \dfrac{\alpha_j}{\beta_j}, j = 1, \ldots, p$ satisfy

$$0 \le \alpha_1 \le \cdots \le \alpha_p \le 1, \quad 1 \ge \beta_1 \ge \ldots \ge \beta_p > 0 \quad \text{and} \quad (\alpha_j)^2 + (\beta_j)^2 = 1, j = 1, ..., p.$$

Denoting the columns of Z by z_j and substituting (4) into (2) gives, after some matrix algebra, $x_t = Z(D^T D + tM^T M)^{-1} D^T U^T b$, or equivalently

$$x_t = \sum_{j=1}^{p} \frac{\mu_j}{\mu_j^2 + t} \left(\frac{1}{\beta_j} \right) (u_j \cdot b) z_j + \sum_{j=p+1}^{n} (u_j \cdot b) z_j, \tag{5}$$

where u_j and z_j are the columns of U and Z, respectively.

Since each regularized solution of (3) is given explicitly by (5) for all $t > 0$, $\|Ax_t - b\|$ and $\|Rx_t\|$ can now be expressed simply in terms of t, the singular values $\mu_j, j = 1, \ldots, p$ and the scalar quantities $u_j \cdot b, j = 1, \ldots, n$. Let $x(t)$ and $y(t)$ be the positive, real-valued functions defined for all $t > 0$ by $x(t) \equiv \ln(\|Ax_t - b\|)$ and $y(t) \equiv \ln(\|Rx_t\|)$. Straightforward matrix calculation gives

$$x(t) = \frac{1}{2}\ln \left(\sum_{j=1}^{p} \left[\frac{t}{\mu_j^2 + t}(u_j \cdot b) \right]^2 + \|(I_m - UU^T)b\|^2 \right) \tag{6}$$

$$y(t) = \frac{1}{2}\ln \left(\sum_{j=1}^{p} \left[\frac{\mu_j}{\mu_j^2 + t}(u_j \cdot b) \right]^2 \right). \tag{7}$$

Then $x(t)$ and $y(t)$ are continuously differentiable functions with $x'(t) > 0$ and $y'(t) < 0$ for all $t > 0$, and the higher-order derivatives can all be efficiently computed as explicit functions of t, $\mu_j, j = 1, \ldots, p$ and $u_j \cdot b, j = 1, ..., n$. It follows that $\ln\|Rx_t\|$ is a strictly monotone decreasing function of $\ln(\|Ax_t - b\|)$ for all $t > 0$, that the residual norms corresponding to zero and infinite regularization are given by

$$\|Ax_0 - b\| = \|(I_m - UU^T)b\|, \quad \|Ax_\infty - b\| = \left\{ \sum_{j=1}^{p} [u_j \cdot b]^2 + \|(I_m - UU^T)b\|^2 \right\}^{1/2} \tag{8}$$

and that the regularization-operator norms corresponding to zero and infinite regularization are given by

$$\|Rx_0\| = \left\{ \sum_{j=1}^{p} \left[\frac{1}{\mu_j} (u_j \cdot b) \right]^2 \right\}^{1/2} , \quad \|Rx_\infty\| = 0. \tag{9}$$

Thus, the curvature function associated with the log-log L-curve

$$\kappa(t) = \frac{x'(t)y''(t) - y'(t)x''(t)}{((x'(t))^2 + (y'(t))^2)^{3/2}} \tag{10}$$

is explicitly defined for all $t > 0$, and the endpoints of the curve given by Eqs. 8 – 9 are finite provided $\mu_1 > 0$. For the L-curve method, the condition of Hansen and O'Leary [11] corresponds to determining the maximum value $\kappa(t^*)$ of (10). Similarly, it follows from Eqs. 5 – 7 that the CRESO method reduces to determining the first local maximum $\rho(t^*)$ of the function

$$\begin{aligned} \rho(t) &= \frac{d}{dt} \left(t \|Rx_t\|^2 \right) - \frac{d}{dt} \|Ax_t - b\|^2 = \|Rx_t\|^2 + 2t \frac{d}{dt} \|Rx_t\|^2 \\ &= \sum_{j=1}^{p} \mu_j^2 \frac{\mu_j^2 - 3t}{(\mu_j^2 + t)^3} (u_j \cdot b)^2 \end{aligned} \tag{11}$$

An interesting observation concerning the CRESO method can be made at this point. Since (11) has continuous derivatives of any order for $t > 0$, at the first local maximum t^* it is necessary that

$$\rho'(t^*) = \sum_{j=1}^{p} 6\mu_j^2 \frac{(t^* - \mu_j^2)}{(\mu_j^2 + t^*)^4} (u_j \cdot b)^2 = 0 \tag{12}$$

and

$$\rho''(t^*) = - \sum_{j=1}^{p} 6\mu_j^2 \frac{3t^* - 5\mu_j^2}{(\mu_j^2 + t^*)^5} (u_j \cdot b)^2 < 0. \tag{13}$$

If we denote the minimum and maximum values of $(\mu_j)^2$, $j = 1, ..., p$ by $(\mu_m)^2$ and $(\mu_M)^2$ respectively, then it is clear from (12) that t^* must lie in the interval $[(\mu_m)^2, (\mu_M)^2]$ and from (13) that $\rho''(t^*) > 0$ if $t^* < (5/3)(\mu_m)^2$. Thus, if $\mu_m > 0$ and $(\mu_M)^2 < (5/3)(\mu_m)^2$, no maximum can exist and there will be no t^* for which the CRESO condition is satisfied. A simple class of such examples is given by the matrix $A = U \begin{pmatrix} 1 & 0 \\ 0 & \sqrt{5/3} \end{pmatrix} V^T$ with $R = I$. Even more interesting is the fact that for the same A, the CRESO condition may be satisfied for some b values and not for others. For example, if A is the Hilbert matrix $A = \begin{pmatrix} 1 & 1/2 \\ 1/2 & 1/3 \end{pmatrix}$ with singular value decomposition $A = UDV^T$ and $R = I$, then a CRESO solution exists for, say, $b = u_1 + 10u_2$ but not for $b = 10u_1 + u_2$. This suggests that some care should be exercised when employing the CRESO method to compute regularization parameters.

5   Formulation of the forward and inverse problems

The following formulation is based on the boundary-element method [4, 13]. Let Ω be a doubly connected, homogeneous, isotropic volume conductor with boundary $\partial\Omega = \Gamma_1 \cup \Gamma_2$ defined by the smooth nonintersecting closed surfaces Γ_1 and Γ_2 with unit outer normals n_1 and n_2 as in Figure 1. In the specific application to electrocardiography, Γ_1 is the epicardial surface and Γ_2 the torso surface. $C^k(\Omega)$ is the linear space of k-times continuously differentiable functions on Ω, and the three-dimensional Laplace and gradient differential operators are denoted by Δ and ∇, respectively. Let $r = |p - q|$ be the scalar function defined by the Euclidean distance between any two points $p = (x, y, z)$ and $q = (\xi, \eta, \zeta)$ in $\Omega \cup \partial\Omega$ and let $\phi(p)$ be a $C^2(\Omega) \cap C^1(\Omega \cup \partial\Omega)$ electrostatic-potential function defined on $\Omega \cup \partial\Omega$ and satisfying the nonconducting boundary condition $\nabla\phi \cdot n_2 = 0$ on Γ_2. Let $dq = d\xi\, d\eta\, d\zeta$ and $d\sigma$ denote the volume and surface-area differentials, respectively. It is assumed here that there are no current sources within the volume conductor, that is, that $\Delta\phi = 0$ in Ω.

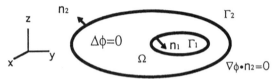

Figure 1. The volume conductor Ω bounded by the closed surfaces Γ_1 and Γ_2.

The forward problem consists of solving for $\phi(p)$ on Γ_2 given continuous boundary data $\phi = \phi_1$ on Γ_1. This problem is well-posed in the sense that, if Ω is sufficiently smooth, then there always exists a unique solution which depends continuously on the bounday data ϕ_1. The inverse problem consists of solving for $\phi(p)$ on Γ_1 given $\phi = \phi_2$ on Γ_2 and is mathematically ill-posed in the sense that a unique solution exists but may not depend continuously on ϕ_1 [1].

Green's second identity

$$\int_\Omega \left[\phi\Delta\frac{1}{r} - \frac{1}{r}\Delta\phi \right] dq = \int_{\partial\Omega} \left[\phi\nabla\frac{1}{r} - \frac{1}{r}\nabla\phi \right] \cdot n\, d\sigma \tag{14}$$

gives [14, 15] for any point p on the surface $\partial\Omega$ of Ω

$$2\pi\phi(p) = \int_{\Gamma_1} \phi\, d\omega - \int_{\Gamma_2} \phi\, d\omega - \int_{\Gamma_1} \frac{1}{r}\nabla\phi \cdot n\, d\sigma, \tag{15}$$

where n_1 has been redirected into the Green's volume Ω and $d\omega$ denotes the incremental solid-angle term $d\omega = \nabla\frac{1}{r} \cdot n\, d\sigma$. The surface integrals here are evaluated using the Cauchy principal value and hence, when being computed numerically, must exclude the singularity at $r = 0$.

One approach to the numerical solution of the boundary integral equation (15) is the collocation method [13, 16]. Let ϕ_1 and ϕ_2 be the potentials on Γ_1 and Γ_2, respectively. Let Γ_1 and Γ_2 each be subdivided into triangular surface elements defined by N_1 node points $\boldsymbol{P}_n$, $n = 1, \ldots, N_1$ on Γ_1 and N_2 node points $(\boldsymbol{Q}_m)$, $m = 1, \ldots, N_2$ on Γ_2. Let $\phi_{1n} = \phi_1(\boldsymbol{P}_n)$, $n = 1, \ldots, N_1$ and $\phi_{2m} = \phi_2(\boldsymbol{Q}_m)$, $m = 1, \ldots, N_2$ denote the potentials at these node points. Then for each $n = 1, \ldots, N_1$ and $m = 1, \ldots, N_2$, (15) gives

$$\phi_{1n} - \frac{1}{2\pi} \int_{\Gamma_1} \phi_1 dw_{11}^n + \frac{1}{2\pi} \int_{\Gamma_2} \phi_2 dw_{12}^n + \frac{1}{2\pi} \int_{\Gamma_1} \frac{1}{r_n} \nabla \phi_1 \cdot \boldsymbol{n} d\sigma = 0 \qquad (16)$$

$$\phi_{2m} - \frac{1}{2\pi} \int_{\Gamma_1} \phi_1 dw_{21}^m + \frac{1}{2\pi} \int_{\Gamma_2} \phi_2 dw_{22}^m + \frac{1}{2\pi} \int_{\Gamma_1} \frac{1}{r_m} \nabla \phi_1 \cdot \boldsymbol{n} d\sigma = 0 \qquad (17)$$

where $r_n = |\boldsymbol{P}_n - q|$ and dw_{rs}^k denotes the differential solid angle subtended by an elemental area of the integration surface Γ_s at the kth node point on Γ_r. The above integrals can be discretized in terms of coefficients α_{rs}^{kj} and β_{rs}^{kj}, which depend only on the geometry of the triangular surface elements of Γ_1 and Γ_2, giving

$$\phi_{1n} - \sum_{j=1}^{N_1} \alpha_{11}^{nj} \phi_{1j} + \sum_{j=1}^{N_2} \alpha_{12}^{nj} \phi_{2j} + \sum_{j=1}^{N_1} \beta_{11}^{nj} \psi_{1j} = 0 \qquad (18)$$

$$\phi_{2m} - \sum_{j=1}^{N_1} \alpha_{21}^{mj} \phi_{1j} + \sum_{j=1}^{N_2} \alpha_{22}^{mj} \phi_{2j} + \sum_{j=1}^{N_1} \beta_{21}^{mj} \psi_{1j} = 0 \qquad (19)$$

for each $n = 1, \ldots, N_1$ and $m = 1, \ldots, N_2$ where $\psi_{1j} = \nabla \phi_1(\boldsymbol{P}_j)$, $j = 1, \ldots, N_1$ [13, 17, 18]. The systems of equations defined by (18) for $n = 1, \ldots, N_1$ and by (19) for $m = 1, \ldots, N_2$ can be written in the form

$$(I_n - A_{11})\boldsymbol{\phi}_1 + A_{12}\boldsymbol{\phi}_2 + B_{11}\boldsymbol{\Psi}_1 = O \qquad (20)$$

$$(I_m + A_{22})\boldsymbol{\phi}_2 - A_{21}\boldsymbol{\phi}_1 + B_{21}\boldsymbol{\Psi}_1 = O \qquad (21)$$

for matrices A_{rs} and B_{rs}, where $\boldsymbol{\phi}_1 = (\phi_{11}, \ldots, \phi_{1N_1})$, $\boldsymbol{\phi}_2 = (\phi_{21}, \ldots, \phi_{2N_2})$ and $\boldsymbol{\Psi}_1 = (\psi_{11}, \ldots, \psi_{1N_1})$. From (20), $\boldsymbol{\Psi}_1 = -B_{11}^{-1}((I_n - A_{11})\boldsymbol{\phi}_1 + A_{12}\boldsymbol{\phi}_2)$, and substituting this expression into (21) yields

$$\boldsymbol{\phi}_2 = A\boldsymbol{\phi}_1 = ((I_m + A_{22}) - B_{21}B_{11}^{-1}A_{12})^{-1}(B_{21}B_{11}^{-1}(I_n - A_{11}) + A_{21})\boldsymbol{\phi}_1, \qquad (22)$$

which is the composition of the $N_2 \times N_1$ forward transfer matrix A for this problem.

6   A sample application

In the example considered here, the volume conductor Ω is a geometrically realistic homogeneous human torso, with the epicardial surface Γ_1 and torso surface Γ_2 subdivided into triangular elements defined by $N_1 = 202$ and $N_2 = 352$ nodes, respectively; details regarding the calculation of the geometrical-coefficient matrices are given elsewhere [17, 18].

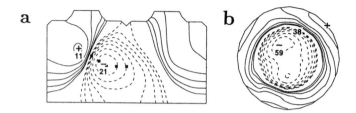

Figure 2. The potential distributions obtained by the forward simulation using a single-dipole source. (a) body-surface potential map, with the left side representing the anterior and the right side the posterior torso surface; (b) epicardial-surface potential map in a polar display, with centre corresponding to the apex and the circular border to the atrio-ventricular ring. Each map shows the location of the maximum (+) and minimum (−) and their magnitude in units of μV.

To determine an *a priori* solution Φ_H, epicardial and body-surface potentials were calculated for a single-dipole source located in the region bounded by Γ_1. Considering all of the region bounded by Γ_2, including the region interior to Γ_1, as a homogeneous volume conductor with conductivity σ, (14) gives

$$c(\boldsymbol{p})\phi(\boldsymbol{p}) = \frac{\boldsymbol{m} \cdot \boldsymbol{r}}{\sigma r^3} - \int_{\Gamma_2} \phi d\omega, \tag{23}$$

where $\boldsymbol{q}$ is the source point, $\boldsymbol{m}$ is the dipole moment, $r = |\boldsymbol{r}| = |\boldsymbol{p} - \boldsymbol{q}|$ and $c(\boldsymbol{p}) = 0, 2\pi, 4\pi$ depending on whether $\boldsymbol{p}$ is exterior to Γ_2, lies on Γ_2, or is interior to Γ_2, respectively [14, 15]. The Fredhom integral equation (23) is singular in two respects: the kernel $K(\boldsymbol{p}, \boldsymbol{q}) = \nabla(1/r) \cdot \boldsymbol{n}$ is singular and the resolvent is singular because the corresponding homogeneous equation (i.e. (23) with $\boldsymbol{m} \equiv 0$) has nontrivial constant solutions [15]. Discretizing (23) as above and substituting in each of the N_2 nodes on Γ_2 yields a singular linear system

$$\phi_{2m} = \frac{1}{2\pi\sigma}\frac{\boldsymbol{m} \cdot \boldsymbol{r}_m}{r_m^3} - \sum_{j=1}^{N_2} \alpha_{22}^{mj}\phi_{2j} + \epsilon_m, \quad m = 1,\dots,N_2, \tag{24}$$

or in matrix notation $(I_m + A_{22})\boldsymbol{\Phi}_B = \boldsymbol{u} + \boldsymbol{\epsilon}$, where the ϵ_m terms are associated with the errors in the integral approximations. Since $I_m + A_{22}$ is singular for this system, Wielandt deflation technique is used to determine a stable nonsingular system whose solution approximates a solution $\boldsymbol{\Phi}_B = (\phi_2(\boldsymbol{Q}_1),\dots,\phi_2(\boldsymbol{Q}_{N_2}))$ of (24) corresponding to a particular choice of reference potential [14, 16]. Using $\boldsymbol{\Phi}_B$, the corresponding epicardial potentials $\boldsymbol{\Phi}_H = (\phi_1(\boldsymbol{P}_1),\dots,\phi_1(\boldsymbol{P}_{N_1}))$ at the N_1 node points on Γ_1 can then be computed from Eqs. 23 – 24 to give the required "known" solution (see Fig. 2).

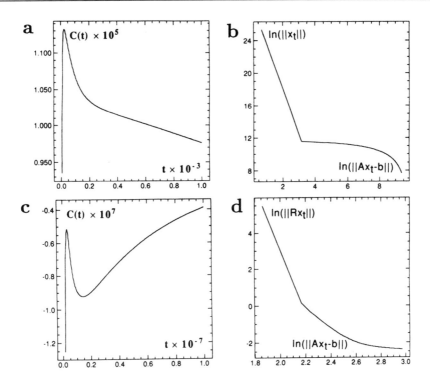

Figure 3. Plots of the CRESO functions and L–curves for Tikhonov regularization. (a) CRESO function for $R = I$; (b) L–curve for $R = I$; (c) CRESO function for $R = H$; (d) L–curve for $R = H$.

The L-curve and CRESO methods were then applied to the solution of $A\phi_H = \Phi_B$ for the Φ_B obtained above to determine the regularized epicardial potentials ϕ_H for the specific forward transfer coefficient matrix A associated with the realistic torso. Both zero-order and second-order Tikhonov regularization was employed corresponding to the smoothing operators $R = I$ and $R = H$. The discretized Laplace surface differential operator H was constructed by a method described by Oostendorp et al. [19] applied to the tesselated heart surface Γ_1. For simplicity, the measures used to compare the effectiveness of the two methods were chosen to be the relative error $RE = \|\phi_H - \Phi_H\| / \|\Phi_H\|$ and the correlation coefficient CC. Actually, these terms incorporate both the error encountered in the forward simulation as well as that resulting from the inverse solution procedure. In each case, trial and error (i.e. the method of exhaustion) was used to determine that value of the regularization parameter t which produced the best possible RE.

Table 1. The CRESO, L–Curve and Exhaustion Results Obtained Using
Zero-Order and Second-Order Tikhonov Regularization

Regularizing operator	Parameter-selection method	T	Relative error (%)	Correlation coefficient (%)
I	Exhaustion	2.695×10^{-4}	40.65	91.37
	L–curve	5.691×10^{-5}	46.45	89.01
	CRESO	2.026×10^{-5}	57.09	84.59
H	Exhaustion	5.882	15.19	98.85
	L–curve	10.00	15.75	98.79
	CRESO	1.696×10^{-9}	90452.	-1.36

I, the identity operator; H, the Laplace differential operator;
CRESO, the composite residual and smoothing operator;
$T = t^*$, the regularization parameter obtained for each specific regularizing
operator and parameter-selection method.

The GSVD of A and R was carried out using the LAPACK driver routine DG-GSVD [20] while the NAG routine E04BBF [21] was used to maximize the functions given by Eqs. 10 and 11. The CRESO, L-Curve and exhaustion results obtained are given in Table 1. For both regularizing operators, the CRESO function defined by (11) exhibited a clear maximum as indicated in Figure 3a,c while the corners of the L-curves are distinctly evident in Figure 3b,d. The computed solutions obtained using each of the methods can be gauged qualitatively by comparing their epicardial maps in Figure 4 to the epicardial map of the known solution Φ_H in Figure 2b.

7 Discussion

The results in Table 1 indicate that the choice of regularizing operator can be important for inverse problems in electrocardiography. Certainly $R = H$ does much better than $R = I$ for the L-curve method applied to the specific example considered here. An explanation of the role to be played by the regularizing operator follows from the GSVD expansion, given by (5), of the solution itself. When $u_j \cdot b$ is large, the filter function $\mu_j/(\mu_j^2 + t)$ should not damp this component of b because it may contain a substantial amount of information. On the other hand, when β_j is very small, the filter function should be close to zero (highly damped) since errors in this component will be greatly magnified. Thus, the effectiveness of Tikhonov regularization depends on the relative rates at which $u_j \cdot b$ and the singular values β_j of R decay to zero.

The analysis presented in this chapter and the results in Table 1 suggest that the L-curve method is more accurate and reliable than the CRESO method. One additional advantage associated with the L-curve method is that it has been shown to be more robust than generalized cross-validation method in the presence of correlated

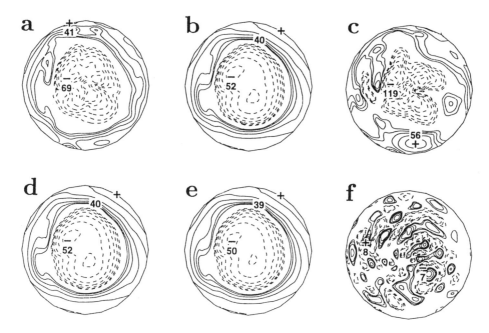

Figure 4. Epicardial maps obtained be the Exhaustion, L−curve and CRESO methods using zero-order and second-order Tikhonov regularization. (a) best Tikhonov solution with operator I; (b) L−curve solution with operator I; (c) CRESO solution with operator I; (d) best Tikhonov solution with operator H; (e) L−curve solution with operator H; (f) CRESO solution with operator H.

errors [11]. In any event, Eqs. 5– 10 provide a simple, fast and accurate procedure for computing the regularization parameters for both methods.

Acknowledgements

The work reported herein was supported in part by research grants from the Natural Sciences and Engineering Council of Canada, the Medical Research Council of Canada and from the Heart and Stroke Foundation of Nova Scotia.

References

[1] V.A. Morozov. *Regularization Methods for Ill-Posed Problems.* CRC Press, Boca Raton, 1993.

[2] A.N. Tikhonov and A.V. Goncharsky, editors. *Ill-Posed Problems in the Natural Sciences.* MIR, Moscow, 1987.

[3] Y. Rudy and B.J. Messinger-Rapport. The inverse solution in electrocardiography. *CRC Crit Rev Biomed Eng*, 16:215–268, 1988.

[4] Y. Rudy and H.S. Oster. The electrocardiographic inverse problem. In T.C. Pilkington, B. Loftis, J.F. Thompson, S.L-Y. Woo, T.C. Palmer, and T.F. Budinger, editors, *High-Performance Computing in Biomedical Research*, pages 135–155, CRC Press, Boca Raton, 1993.

[5] A.N. Tikhonov and V.Y. Arsenin. *Solution of Ill-posed Problems.* Wiley, New York, 1977.

[6] P.C. Hansen. Analysis of discrete ill-posed problems by means of the L-curve. *SIAM Review*, 34:561–580, 1992.

[7] P. Colli Franzone, L. Guerri, B. Taccardi, and C. Viganotti. Finite element approximation of regularized solution of the the inverse potential problem of electrocardiography and application to experimental data. *Calcolo*, 22(I):91–186, 1985.

[8] Å. Björck. Constrained least squares problems. In P.G. Ciarlet and J.L. Lions, editors, *Handbook of numerical analysis, Vol. I: Finite difference methods—solution of equations in R^n*, pages 589–615, Elsevier, New York, 1990.

[9] G.H. Golub and C.F. Van Loan. *Matrix Computations.* Johns Hopkins Univ. Press, Baltimore, 1983.

[10] P.C. Hansen. Regularization, GSVD and truncated GSVD. *BIT*, 29:491–504, 1989.

[11] P.C. Hansen and D.P. O'Leary. The use of the L-curve in the regularization of discrete ill-posed problems. *SIAM J Sci Statist Comput*, 14:1487–1503, 1993.

[12] B. Hofmann. *Regularization for Applied Inverse and Ill-Posed Problems.* Teubner-Texte Mathe., 85, Teubner, Leipzig, 1986.

[13] R.C. Barr, M. Ramsey III, and M.S. Spach. Relating epicardial to body surface potential distributions by means of transfer coefficients based on geometry measurements. *IEEE Trans Biomed Eng*, BME-24:1–11, 1977.

[14] A.C.L. Barnard, I.M. Duck, and M.S. Lynn. The application of electromagnetic theory to electrocardiology: I. Derivation of the integral equations. *Biophys J*, 7:443–462, 1967.

[15] R. Courant and D. Hilbert. *Methods of Mathematical Physics.* Wiley, New York, 1962.

[16] A.C.L. Barnard, I.M. Duck, M.S. Lynn, and W.P. Timlake. The application of electromagnetic theory to electrocardiology. II. Numerical solution of the integral equations. *Biophys J*, 7:463–491, 1967.

[17] R.S. MacLeod. *Percutaneous Transluminal Coronary Angioplasty as a Model of Cardiac Ischemia.* PhD thesis, Dalhousie Univ., Halifax, N.S. Canada, 1990.

[18] R.S. MacLeod, M.J. Gardner, R.M. Miller, and B.M. Horáček. Application of an electrocardiographic inverse solution to localize ischemia during coronary angioplasty. *J Cardiovasc Electrophysiol*, 6:2–18, 1995.

[19] T.F. Oostendorp, A. van Oosterom, and G. Huiskamp. Interpolation on a triangulated 3D surface. *J Comp Physics*, 80:331–343, 1989.

[20] E. Anderson, Z. Bai, C. Bischof, J.W. Demmel, J.J. Dongarra, J. Du Croz, A. Greenbaum, S Hammarling, A. McKenney, S. Ostrouchov, and D. Sorensen. *LAPACK User's Guide.* Society for Industrial and Applied Mathematics, Philadelphia, 1992.

[21] *The NAG Fortran Library Manual – Mark 13.* The Numerical Algorithms Group Ltd., Oxford UK, 1988.

The Forces Exerted on the Duodenum and Sphincter of Oddi During ERCP, and Sphincterotomy

R. Ratani and P. Swain

1 Introduction

Little is known about the forces transmitted to the tip of the endoscope or about forces exerted by manipulating accessories in the biopsy channel of a flexible endoscope. Every endoscopist is concerned about the possibility of perforation of viscera while manoeuvring the endoscope. Perforation is an occasional complication of endoscopic retrograde cholangiopancreatography (ERCP) and sphincterotomy [1-11]. Pushing the endoscope, angulating the tip of the endoscope, rotatory movements (torque) of the endoscope, pushing accessories through the channel of duodenoscope, flexing the tip of the cannula or sphincterotome by movements of the bridge, and bowing the sphincterotome are some commonly performed manoeuvres to cannulate the sphincter of Oddi and perform sphincterotomy. The aim of this study was to measure the forces exerted on the duodenum and sphincter of Oddi during endoscopic sphincterotomy and to evaluate the force required to actually perforate the stomach and duodenum in bench experiments.

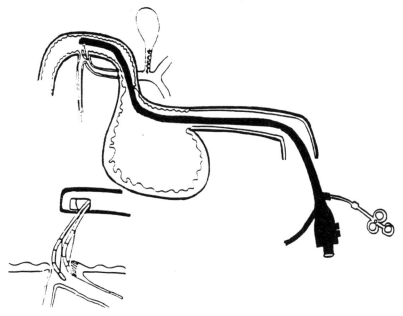

Figure 1 Sphincterotomy is performed after inserting sphincterotome in the ampulla of Vater by combination of bowing, pulling on sphincterotome and bridge movements while radiofrequency current is passed through sphincterotome wire.

2 Materials and Methods

The following equipment was used:
- Olympus duodenoscope (JIFT10)
- Olympus bow sphincterotome-Erlangen type (Olympus No: KD 20Q with 3cm exposed wire length)
- balances (counterbalance and electronic)
- retort stand, flexible Goretex tubing 20 cm in length and 12.5mm in internal diameter
- postmortem specimens of human liver. stomach and duodenum.

The duodenoscope was passed through a length of flexible Goretex plastic tubing of 12.5 mm internal diameter and steadied over a balance with the help of a retort stand as shown in Figure 2. The tubing allowed free distal transmission of the forces applied to the proximal end of the endoscope during various manoeuvre. The maximum compressive forces exerted during various movements of the tip of the endoscope (upward, downward, right and left angulation movements with as well as without application of brake to fix the angulation; torque applied after the tip of the endoscope in flexion) were measured.

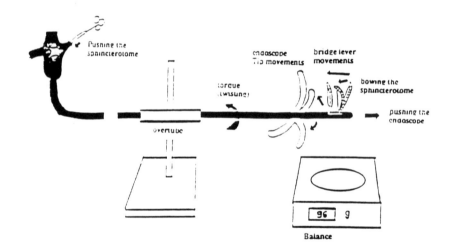

Figure 2 Measurement of forces exerted by endoscopic manoeuvres

The end on force exerted at the tip by pushing the endoscope was measured by holding it up to 25 cm from the tip, perpendicular to the balance. The maximum force exerted by the sphincterotome passed through the channel of the endoscope with 'up' movements of the bridge was noted with up to 4 cm of the sphincterotome brought out of the distal end of the endoscope. The forces exerted by bowing the sphincterotome was noted at the end of the maximum force exerted with 'up' movement of the bridge when 4 cm of sphincterotome was brought out of the tip. All measurements were repeated 5 times and are presented as a range and a mean.

For measurement of the minimum forces required to perforate the duodenum and stomach, fresh postmortem human material was loosely tied over a cylinder like the skin of a drum and the cylinder was placed on a counterbalance and the tip of the duodenoscope was pushed down with increasing force until perforation occurred.

3 Results

Pushing sphincterotome through the biopsy channel of the endoscope exerted only 10-15 g [mn 12.5 g] when measured on the balance. Sphincterotome handle movements added unto 8-12 g [mn 10 g] of force and varied with the length and tension of exposed sphincterotome wire. Bridge lever movements produced forces between 11-38 g [mn 26 g]. The force exerted by the bridge lever movement varied with the length of sphinc-terotome protruding; increasing as the length of sphincterotome out of the endoscope increased up to three centimetres (11 to 15 g [mean 12.5 g] for 1 cm; 30 to 33 g [mean 32 g] for 2 cm; 31 to 38 g [mn 34 g] for 3 cm) and then decreasing with a further increase in the length (17 to 20 g [mn 19 g] for 4 cm). Torque movements (twisting the shaft of the endoscope) produced force ranging between 35 to 47 g [41 mn]:- 37 to 47 g [44 mn] with brake, 35-42 g [39 mn] without brake. The force exerted by endoscope tip angulation varied from 76 to 105 g [92 mean]:- up with brake 98-104 g [98 mn], without break 90-94 g [92 mn]; down with brake 101-105 g [102 mn], without brake 85-101 g [92 mn] 87-96 g [90 mn] without brake 93-101 g [97 mn]. Considerable forces (300 g and above) could be exerted by pushing the endoscope held upto 25 cm from the tip. The forces exerted by these endoscopic manoeuvres are indicated in Table 1.

Table 1.

Forces exerted during endoscopic sphincterotomy

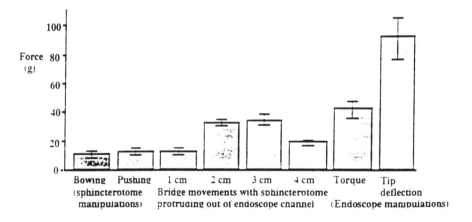

The results show that endoscopic manoeuvres can be arranged in order of the force that they exert on tissue. Bowing the sphincterotome (bs), pushing the sphincterotome (pusp), bridge movements (bm), torque (to) and tip deflection (td) produce increasing force on the duodenal wall. Analyzed statistically, significant differences were shown comparing (td>to,bm,pusp,bs), (to>bm,pusp,bs), (bm>pusp,bs) [p<0.05].

We measured the effect of force on the rate of sphincterotomy in bench experiments on post-mortem tissue. The sphincterotome was passed through a side-viewing endoscope and into exposed bile ducts on the cut surface of post-mortem liver tissue resting on a balance. The force exerted on wire was significantly associated with increase rate of cutting.

Experiments using weak force (10 g) at setting cut 5, time 0.5 sec, wire in contact 0.5 cm - length of the cut was 4.6 mm and it was 9.8 mm with strong force (50 g) (n = 10 experiments, p < 0.05).

Force required to perforate human postmortem duodenum ranged from 300 g to 360 g [332 mn], and to perforate stomach ranged from 300 g to 400 g [362 mn]. Force exceeding 300 g could only be exerted by forward movements of the tip when the endoscope was held in a fixed position up to 25 cm from the tip which may occur when the endoscope is held fixed in the duodenum. Considerably larger forces could be exerted when the endoscope was held closer to its tip.

4 Discussion

These measurements show that the forces transmitted to tissue during flexible endoscopy by the endoscope or by an instrument such as a sphincterotome are relatively small with considerable losses because of the flexibility of the endoscope, the flexibility of the cannula and the compliance of the gastrointestinal tube through which the endoscope passes. Damping of forces transmitted at flexible endoscopy is one reason why flexible endoscopy has proved so safe.

The forces exerted during ERCP and sphincterotomy are generally well below those required to produce perforation. Compressive force exerted on the control end of a flexible endoscope results in poor transmission of that force to the distal end unless the endoscope is confined. Force exerted on flexible instruments confined within the biopsy channel of the flexible endoscope is very inefficiently transmitted.

Perforations encountered in endoscopic practice may be pharyngeal- when posterior pharyngeal wall may get crushed by the tip of the endoscope proximal to cricopharynx or a thin walled diverticulum may get perforated by perforated by endoscope. As cricopharynx is 15 cm from the incisor, the end on pushing force applied to the endoscope being held closer to the tip may transmit considerable forces when the closed upper oesophageal sphincter or stricture are providing firm resistance.

The oesophagus made friable and non distensible by strictures with scar tissue, malignancy or the presence of diverticula is also susceptible to endoscopic perforation for the same reason. Most oesophageal perforations occur with dilatation using bougies or balloons where considerably more force is applied than at conventional endoscopy.

The stomach being muscular and mobile is relatively resistant to perforation by forces exerted during bench experiments.

The duodenum is a relatively more vulnerable organ, being thin walled and fixed retroperitoneally in a 'C' shaped curvature. Retroperitoneal perforation may occur while performing diagnostic and therapeutic ERCP procedures at the papilla due to repeated injury by pointed accessories. Most perforations occur as a consequence of sphincterotomy which if too long can cut into the retroperitoneal space without the exertion of undue force. Tissue damage produced by diathermy markedly reduce the force required to perforate the duodenum. Other factors increasing the risk of perforation are peripapillary diverticula [7,8], use of needle knife [9,10] and whiplash injury by the endoscope especially when forcefully retrieving stones [11]. Some videoduodenoscopes have a longer rigid tip than fibreoptic duodenoscopes and this has been blamed as a possible cause of duodenal perforation in a difficult procedure of biliary stone extraction while pushing down and away from the ampulla to pull out a reluctant gallstone thus causing a whiplash injury [12].

The compressive forces exerted by pushing on accessories such as a sphincterotome were surprisingly weak (10-15 g). The poor transmission of force at ERCP is occasionally a source of therapeutic failure when tight strictures cannot be dilated or stented and may also explain difficulties that occur in penetrating tissue during attempts at needle biopsy through flexible endoscopes. If it were desirable to increase the force exerted at flexible endoscopy on accessories such as stents or dilators, hydraulic or rotational (screw) forces might allow more force to be exerted than can be achieved with transmitted compressive"push" forces.

5 Conclusion

These measurements illustrate the poor transmission of force through flexible endoscopic system. Although individual forces produced during sphincterotomy are well below the threshold to produce perforation, rapid combination of these forces may be responsible for perforation of a viscus.

Measurement of the difference in force maximally exerted at endoscopy and that required to perforate a viscus allows us to assess the margin of safety. It is difficult to exert sufficient force to perforate without splinting the endoscope close to its tip.

It is occasionally necessary to exert more force at endoscopy for therapeutic purposes e.g. to dilate and stent a biliary stricture. This indicates the need for instrument that allows such forces to be exerted at distal end without having to push proximally, to increase the efficacy and safety of the procedure.

6 Summary

Movements such as pushing the endoscope, endoscope tip angulation, torque, movements of the bridge, sphincterotome tip bowing and pushing the sphincterotome are used during endoscopic sphincterotomy. In bench experiments we measured the force exerted on the tissue by such manoeuvre using Olympus Erlangen type bow sphincterotome (3 cm exposed wire length) following its passage through a 2.8 mm diameter channel of a duodenoscope.

Results: Pushing sphincterotome through channel exerted only 10-15 g. The force exerted by endoscope tip angulation varied from 76 to 105 g and was significantly higher than that produced by bridge lever movements: 11-38 g (p<0.05). Torque movements produced force ranging between 35 to 47 g. Force exerted by the bridge lever movement varied with the length of sphincterotome tip out of the endoscope, increasing as the length of sphincterotome out of the endoscope increased up to three centimetres (11 to 15 g for 1 cm; 30 to 33 g for 2 cm; 31 to 38 g for 3 cm) and then decreased with a further increase in the length (17 to 20 g for 4 cm). Sphincterotome handle movements add up to 10 g of force and vary with the length and tension of exposed sphincterotome wire. Force required to perforate human post-mortem duodenum ranged from 300 g to 360 g. Force exceeding 300 g could only be exerted by forward movements of the tip when the endoscope was held in a fixed position up to 25 cm from the tip which may occur when the endoscope is held fixed in the duodenum.

During sphincterotomy using radiofrequency current, bridge lever movements are mainly used to control direction and length of cut. When light force (10 g) was compared with strong force (50 g) during experimental 'sphincterotomy' cuts on isolated liver at optimal electrosurgical setting the rate of cut was significantly increased ($p < 0.05$).

Conclusions: pushing the sphincterotome, bridge and bow movements exert very little force during cannulation or sphincterotomy; endoscope angulation exerts much greater force; it is difficult to exert sufficient force to perforate the duodenum without splinting the endoscope close to its tip. These observations may be helpful to endoscopists trying to use controlled force during endoscopic sphincterotomy.

References

[1] P.B. Cotton, G. Lehman, J. Vennes, J.E. Geenan, R.C.G. Russell, W.C. Meyers, C. Liguory, N. Nickl. Endoscopic Sphincterotomy complications and their management: an attempt at consensus. *Gastrointest Endosc.* 37: 383-393, 1991.

[2] F. Dunham, N. Bourgeois, M. Gelin, J. Jeanmart, J. Toussaint, M. Cremer. Retroperitoneal perforations following endoscopic sphincterotomy; clinical course and management. *Endoscopy.* 14 :92-96, 1982.

[3] P. Byrne, J.W.C. Leung, P.B. Cotton. Retroperitoneal perforation during duodenoscopic sphincterotomy. *Radiol.* 15 :383-4, 1984.

[4] F. Tam, T. Prindeville, B. Wolfe. Subcutaneous emphysema as a complication of endoscopic sphincterotomy of the ampulla of Vater. *Gastrointest. Endosc.* 35: 447-449, 1989.

[5] D.F. Martin, D.E.F. Tweedle. Retroperitoneal perforation during therapeutic ERCP. *Gut.* 31: A608, 1990.

[6] S. Sherman, T.A. Ruffolo, R.H. Hawes, G.A. Lehman. Complications of endoscopic sphincterotomy. A prospective series with emphasis on the increased risk associated with sphincter of Oddi dysfunction and non dilated bile ducts. *Gastroent.* 101: 1068-1075, 1991.

[7] D. Vaira, J.F. Dowsett, A.R.W. Hatfield, S.R. Cairns, A.A. Polydorou, P.B. Cotton, P.R. Salmon, R.C.G. Russell. Is duodenal diverticulum a risk factor for sphincterotomy? *Gut.* 30: 939-942, 1989.

[8] E. Shemesh, E. Klein, A. Czerniak, A. Coret, L. Bat. Endoscopic sphincterotomy in patients with gallbladder in situ: the influence of periampullary diverticula. *Surgery.* 107: 163-166, 1990.

[9] D.E.F. Tweedle, D.F. Martin. Needle knife papillotomy for endoscopic sphinctertomy and cholangiography. *Gastrointest. Endosc.* 37: 518-421, 1991.

[10] J.H. Siegel, J.S. Ben-Zvi, W. Pullano. The needle knife: a valuable tool in diagnostic and therapeutic ERCP. *Gastrointest. Endosc.* 35: 499-503, 1989.

[11] R.N.M. Van Someren, M.J. Benson, M.J. Glynn, C.C. Ainley, C.P. Swain. Incidence and outcome of duodenal perforation following therapeutic ERCP. *Gastroenter.* 104: A381, 1993.

[12] P.B. Cotton. Take care with the tip of your videoduodenoscope. *Gastrointest. Endosc.* 35: 582, 1989.

Small Bowel Propulsion: Transport of a Solid Bolus

R. Miftakhov and G. Abdusheva

1 Introduction

Peristalsis is the main mode of propulsion which enables the passage of solids and liquids in most visceral organs. It is a complex electromechanical reaction, the internal mechanisms of which are still not well understood. Analysis of these phenomena is an extremely difficult problem from both theoretical and an experimental points of view. Publications on the modelling of peristaltic motility are sparse, and include separate analyses of the organ's (e.g. small bowel) shape change under conditions of symmetric deformation, e.g., [2], [7]. To our knowledge, only two theoretical analyses of peristaltic transport of a rigid body enclosed in a membrane exist. Kydoniefs [6] considered the problem of determining the deformed configuration of an elastic, homogenous, isotropic cylindrical tube enclosing a rigid ellipsoidal body. Bertuzzi et al [3] simulated the propulsion of a solid bolus in a membrane whose mechanical properties were similar to the small intestine. However, these models do not describe the real processes of electromechanical conjugation, peristaltic wave generation and propagation, and stress-strain states in the organ during the mechanical reaction.

In our study, we sought to simulate the peristaltic propulsion of a rigid sphere (bolus) in a biological tube - an analogue of the small intestine. Some results of strain-states analysis and peristaltic transport of the bolus obtained from numerical simulation are discussed.

2 Biomechanical model

Theoretical investigations of biological shells usually have employed either small-strain shell theory, large-strain membrane theory, or three dimensional elasticity theory. The first two of these approaches are inadequate and the third lacks the advantages of a two-dimensional formulation. Partly because of its complexity, large-strain shell theory has been avoided in these studies.

The following basic assumptions are made in the model construction (Figure 1):

(i). The small bowel is modelled as a soft orthotropic cylindrical tube. The wall is composed of two muscle layers, covered by a connective tissue network. Muscle fibres in the outer layer are orientated in the longitudinal direction of the organ and muscle fibres of the inner layer are arranged in an orthogonal, circular, direction. Mechanical properties are different for the two layers but are assumed to be uniform along the wall.

(ii). The tubular segment of length, l, and radius, r_0 contains a rigid sphere of the radius R, and undergoes isometric contractions throughout. The intraluminal pressure, p, changes during the mechanical reaction according to the adiabatic law.

(iii). The contact forces between the wall and rigid sphere are assumed to be orthogonal to the sphere surface. The motion of the bolus is subjected to dry and viscous friction.

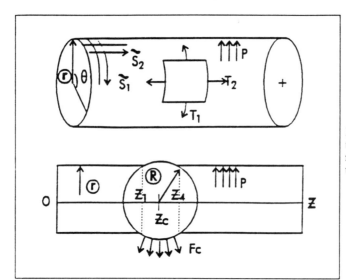

Figure 1. A segment of the small bowel with a rigid bolus enclosed under consideration

(iv). The muscle layers contract independently but in a coordinated way with the generation of active forces (T^a). Reciprocal relationships exist in their work. The first contractions start in the longitudinal muscle layer. When the contractile force reaches a maximum the activation of the circular muscle layer starts [5].

(v). Both muscle layers are assumed to be a syncytium with cable electrical properties. The longitudinal layer has anisotropic and the circular layer has isotropic electrical properties.

(vi). The mechanical activity of the tube is under the control of a pacemaker cell located at the left boundary which generates an excitatory stimulus of given intensity.

The governing system of equations consists of:

(i). The small intestinal segment dynamics [8,9]:

$$\gamma_0 \, V_t = (T_1 e_{ij}\sqrt{g_{22}})s_1 + (T_2 e_{ij}\sqrt{g_{11}})s_2 + p\sqrt{g} \, n_j \qquad (2.1)$$

where:

$$T_{1(2)} = k \frac{\partial \varepsilon_{c(l)}}{\partial t} + \varphi_{c(l)} \, T^a \, (\lambda_{c(l)}) \, T^p \, (\lambda_c, \lambda_l)$$

The following notations are used: γ_0 - the linear density of a biomaterial in an undeformed sate; λ_c, λ_l - the rate of elongation (hereafter the subscripts l and c refer to the longitudinal and circular muscle layers, respectively; $\varepsilon_{c,l} = \lambda_{c,l} + 1$; e_{ij} - the direction cosines of the outward normal n_j, to the surface with respect to the cylindrical j - axis (i = 1,2; j = r, s, z); g_{ij}, g - the components and determinant of the fundamental tensor; V (V_r, V_s, V_z) - the velocity vector and its radial circumferential and longitudinal components; $T_{1,2}$ - the components of the tensor of membrane forces; T^p, T^a - the passive and active components, respectively, of the total force ($T_{c,l}$); k - the rheological parameter; p - intraluminal pressure; s_1, s_2 - the langragian coordinates of the bioshell.

The passive $T^p{}_{c,l}$ components are calculated from:

$$T^p{}_{(c,l)} = \frac{\partial \gamma_0 W}{\partial(\lambda_{(c,l)}\text{-}1)} \qquad (2.2)$$

where: W - the strain energy density function has the form [4]:

$$\gamma_0 W = \frac{1}{2}\left[c_1(\lambda_1 - 1)^2 + 2c_3(\lambda_1 - 1)(\lambda_c - 1) + c_2(\lambda_c - 1)^2\right] + $$
$$+ c_{10}\exp\left(c_4(\lambda_1 - 1)^2 + c_5(\lambda_c - 1)^2 + 2c_6(\lambda_1 - 1)(\lambda_c - 1)\right)$$

For the active force components $(T^a{}_{c,l})$ components we have [10]:

$$T^a{}_{(c,l)} = c_{7(c,l)}\,\lambda^2{}_{(c,l)} + c_{8(c,l)}\,\lambda_{(c,l)} + c_{9(c,l)} \tag{2.3}$$

Here c_{1-10} are the mechanical constants of a biocomposite.
(ii). The dynamics of propagation of the electrical waves of depolarization along the anisotropic longitudinal muscle layer (φ_l) is defined as:

$$C_m(\varphi_l)_t = I_{m1}(s_1, s_2) + I_{m2}(s_1 - s_1', s_2 - s_2') - I_{ion} \tag{2.4}$$

where I_{m1}, I_{m2} are the transmembrane ion currents per unit volume:

$$I_{m1}(s_1, s_2) = M_{vs}\left\{\frac{-2(\mu_{s1}-\mu_{s2})}{(1+\mu_{s1})(1+\mu_{s2})}\mathrm{arctg}\left(\frac{ds_2}{ds_1}\sqrt{\frac{G_{s2}}{G_{s1}}}\right) + \frac{g_{0s2}}{G_{s2}}\right\}$$

$$\times \left(\left(\frac{g_{0s1}}{\varphi_c}(\varphi_l)_{s1}\right)s_1 + \left(\frac{g_{0s1}}{\varphi_l}(\varphi_l)_{s2}\right)s_2\right)$$

$$I_{m2}(s_1 - s_1', s_2 - s_2') = M_{vs}\iint_s \frac{\mu_{s1}-\mu_{s2}}{2\pi\,(1+\mu_{s1})(1+\mu_{s2})G}\times$$

$$\times\frac{(s_1-s_1')^2/G\mu_{s1} - (s_2-s_2')^2/G\mu_{s2}}{\left[(s_1-s_1')^2/G_{s1} + (s_2-s_2')^2/G_{s2}\right]^2}\ \times$$

$$\times \left(\left(\frac{g_{os1}}{\lambda_c}(\varphi_l)_{s1}\right)s_1 + \left(\frac{g_{os2}}{\lambda_c}(\varphi_l)_{s2}\right)s_2\right)ds_1'\,ds_2'$$

here:

$$\mu_{s1} = g_{os1}/g_{is1}, \quad \mu_{s2} = g_{os2}/g_{is2}$$

$$G_{s1} = \frac{g_{os1}+g_{is1}}{\lambda_c}, \quad G_{s2} = \frac{g_{os2}+g_{is2}}{\lambda_1}, \quad G = \sqrt{G_{s1}\,G_{s2}}$$

and the following notations are accepted: C_m - the capacitance of smooth muscle; g_{is1}, g_{is2}, g_{os1}, g_{os2} - the maximal intracellular (the subscript (i)) and interstitial space (the subscript (o)) conductivity of the longitudinal and circular muscle layers in the longitudinal and circumferential directions, respectively; M_{vs} - the membrane volume to surface ratio; I_{ion} - the total ionic current is defined as in [8,9,11,12].

In the case of propagation along the isotropic circular muscle layer (φ_c):

$$C_m(\varphi_c)_t = \frac{M_{vs}}{(1+\mu_{s1})}\left(\frac{g_{os1}}{\lambda_c}(\varphi_l)s_1\right) s_1 + \left(\frac{g_{os2}}{\lambda_c}(\varphi_l)s_2\right) s_2 - I_{ion} \qquad (2.5)$$

where the above-mentioned abbreviations are used.

During all stage of dynamic reaction the points of the wall lie on the surface of thesphere:

$$(Z_c - u_z)^2 + (r_0 + u_r)^2 + (r_0 + u_s)^2 - R^2 = 0, \qquad z\epsilon[z_1, z_2] \qquad (2.6)$$

where: $u(u_z, u_r, u_s)$ - the displacement vector; Z_c - the position of the centre of the sphere at time t; z_1, z_2 - the boundary points of contact of the sphere and the wall. The kinematic equation of the motion of the bolus along the tube is governed by:

$$\eta\frac{dZ_c}{dt} + F_d = 2\pi r_0 \int_{z_1}^{z_2}\int_{r_0}^{r} F_c \, d\xi \, d\xi \qquad (2.7)$$

where F_c, F_d are the contact force and the force of dry friction, respectively, η is the coefficient of viscous friction.

At the initial moment of time the whole system is in the resting state. The left and right boundaries of the tube are supposed to be rigidly fixed. The discharge of the pacemaker cell causes the development of the wave of depolarization in the longitudinal smooth muscle. When the maximum of the total force in the longitudinal muscle layer is achieved activation of the circular muscle layer started. The right boundary of the tubular segment remains in the resting state throughout: $\varphi_{c,1} = 0$.

3 Numerical procedure

The following sequence of events defining the dynamics of peristaltic propulsion of a solid sphere was concerned: the discharge of a pacemaker cell → the action potential generation and propagation along the longitudinal smooth muscle layer → the development of mechanical reaction of the longitudinal smooth muscle → the depolarization wave generation and propagation along the smooth muscle layer → the development of a mechanical reaction of the circular smooth muscle → the peristaltic propulsion of a sphere along the tube.

The set of mechanical and electrical parameters and constants used in the calculations are: $l = 1.0$ (cm), $r_0 = 1.0$ (cm); $\gamma_0 = 26.14$ (g/cm^2); $c_{71} = -40$, $c_{81} = 120$, $c_{91} = -50$, $c_{7c} = -25$, $c_{8c} = 75$, $c_{9c} = -31.2$ (mN/cm·mV); $k = 33.3$ (mN·sec/cm); $p_0 = 1.0$ mPa; $C_m = 1.0$ (μF/cm^2); $M_{vs} = 0.05$. Some of the mechanical and electrical parameters used in the model we have been unable to estimate for the small bowel tissue. With these uncertancies in mind, the above missing data have been adjusted during the numerical simulation: $c_1 = c_2 = 10.4$, $c_3 = 2.59$, $c_4 = 3.79$, $c_5 = 12.7$, $c_6 = 0.587$; in a case of electrical anisotropy the conductivity constants were: $g_{\widetilde{is}_1} = 2$, $g_{\widetilde{is}_2} = g_{0\widetilde{s}_1} = 20$, $g_{0\widetilde{s}_2} = 80$ (mSm/cm^2); in a case of isotropy the conductivity constants were: $g_{\widetilde{is}_1} = 2$, $g_{\widetilde{is}_2} = g_{0\widetilde{s}_1} = g_{0\widetilde{s}_2} = 20$ (mSm/cm^2).

A flow chart of the numerical algorithm is shown in Figure 2. The main program defines a set of input parameters and constants, the initial configuration and the stress strain state of the tube. It calls the following subroutines:

EXCLONG(EXCCIRC) - These subroutines compute the dynamics of the depolarization wave φ_1 long the longitudinal and circular smooth muscle bisyncytia. An explicit finite - difference scheme of the second order approximation over the spinal and time variables is used.

BIOSHELL - It calculates the stress-strain states of the intestinal segment. An explicit hybrid finite - difference scheme of the second order approximation over the space and time is used.

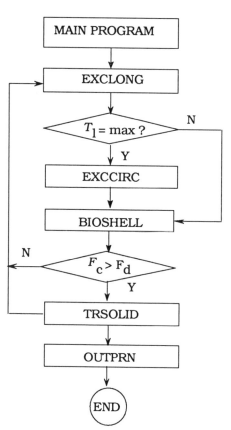

Figure 2. Flow chart of the algorithm of numerical simulation of the peristaltic reflex of the small bowel.

TRSOLID - It calculates the stress distribution over the enclosed sphere and its propulsion along the tube.

OUTPRN - Outputs and prints the results.

4 Results

The discharge of a pacemaker cell initiated an excitatory wave of depolarization φ_1 of an amplitude 69 mV. It had the constant length of 0.5 - 0.55 cm and propagated along the longitudinal smooth muscle syncytium at a velocity of ≈ 2.5 cm/s. In response to the excitation, a short-term phase of relaxation emerged in the longitudinal smooth muscle layer. The rapidly progressing phase was due to the dynamic reaction of the pseudoelastic connective tissue components of the intestinal wall. The wave T^P_1 of a maximum amplitude of 1.14 mN/cm and length of 0.4 - 0.5 cm was registered at t = 0.1 s. As it was propagated its amplitude increased and at t = 0.4s the right part of the biological tube underwent a biaxial deformation where the max T^P_1 = 7.09 mN/cm was observed.

On the activation of the contractile components of the smooth muscle fibres, a phasic contraction developed. At t = 0.4 s a wave of an amplitude of 29.2 mN/cm was registered near the left boundary. It was propagated along the surface of the tube at the velocity of the electrical wave of depolarization. From t > 0.32 s the tonic phase of contraction began to

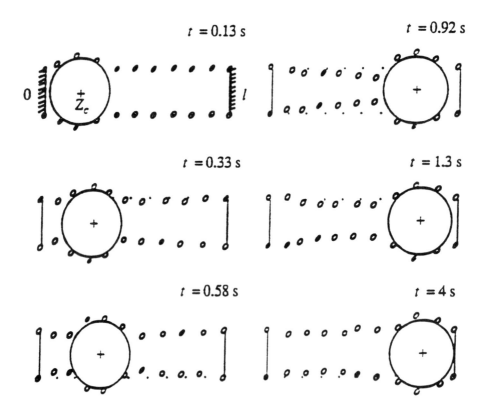

Figure 3. Sequence of peristaltic tube configurations with the enclosed rigid in it.

predominate in the right- hand part with $T^a_1 = 26.9$ mN/cm. For t = 0.4 s more than a half of the tube proved to be uniformly stressed. The maximum 216.9 mN/cm of the total force in the vicinity of the pacemaker area was registered at $t_m = 0.1$ s.

The excitatory wave of depolarization φ_c of an amplitude 69-72 mV was initiated at t = 0.1 s. It was propagated along the electrically isotropic circular smooth muscle syncytium at a velocity of 1.9 - 2.0 cm/s and initiated the mechanical reaction of contraction in the circular muscle components. At the time t = 0.14 s the wave T^a_c had an amplitude of 11.6 mN/cm. With the development of tonic contractions in the circular muscle (t > 0.55 s) more than a half of the intestinal segment went into a uniform stress-strained state.

The dynamics of sphere transport along the tube are shown in Figure 3. The sphere started moving as the contact force exceeded the force of dry friction and underwent propulsive propagation at an average velocity of 0.6 cm/s which was less than the average velocity of propagation of the excitatory waves of depolarization φ_l and φ_c. As mentioned above, the tube underwent nonsymmetric deformation in the early stages of mechanical activity.

With the proposed model we sought to validate the general physiological principles of peristaltic propulsion. The qualitative similarity in the behaviour of theoretical and *in vivo* and *in vitro* observations [1], shows that the model can adequately explain the dynamics of peristaltic transport of 'solid content' in the small intestine. However, in this study we were not able to analyze the motions of the sphere *per se*, which would correspond to the modelling of the processes of grinding and mixing. It is also important to note that the propulsive activity largely depends on the enteric nervous system function - a planar neural network, that controls the intensity of the mechanical reaction of contraction-relaxation in the smooth muscle layers. Combined analysis of these mechanisms is the aim of our further investigations.

References

[1]   N.K. Ahluwalia, D.G. Thompson, J. Bancewicz, and L. Heggie. Relations between propulsion and pressure activity in the human small intestine. *Gut.* 231: 39, 1991.

[2]   A. Bertuzzi, R. Mancinelli, G. Ronzoni, and S. Salinari. Peristaltic transport of a solid bolus. *J. Biomech.* 16: 459-464, 1983.

[3]   A. Bertuzzi, R. Mancinelli, G. Ronzoni, and S. Salinari. A mathematical model of intestinal motor activity. *J Biomech.* 11: 41-47, 1978.

[4]   Y.C. Fung. Perspectives of soft tissue mechanics. *Biomech. Principles and Applic.* 94-114, 1982.

[5]   S.R. Kottegoda. An analysis of possible nervous mechanisms involved in the peristaltic reflex. *J. Physiol.* 200: 687- 712, 1969.

[6]   A.D.Kydoniefs. Finite axisymmetric deformations of an initially cylindrical elastic membrane enclosing a rigid body. *Quart. J. of Mech. and Appl. Math.* 22: 319-331, 1969.

[7]   E.O. Macagno, J. Christensen, and C.L. Lee. Modelling the effects of wall movement of absorbt on in the Intestine. *American J. Physiology,* Vol. 243, G541-G550, 1982.

[8]   R.N. Miftakhov. Nonlinear electromechanical waves cylindrical shell. *Biomech.of Soft Tissues.* 40-72, 1986 (in Russian) .

[9]   R.N. Miftakhov. Numerical modelling of the small intestine motility. In: *Proceedings of the Fist Int. Conf. Biomechanics in Medicine and Surgery,* Riga, USSR, 147-183, 1989 (in Russian).

[10] R.N. Miftakhov, G.R. Abdusheva, and D.L. Wingate. A biomechanical model of small bowel motility. In: *Proc. of the 14th Ann. Int. Conf. of the IEEE/BMS ,* Paris, 1637-1639, 1992.

[11] R. Plonsey, and R.G. Barr. Current flow patterns in two-dimensional anisotropic bisyncytia with normal and extreme conductivities. *Biophys. J.* 45: 557-571, 1984.

[12] F. Ramon, N.C. Anderson, R.W. Joyner, and J.W. Moore. A model for propagation of action potentials in smooth muscle. *J. Theor. Biol.* 59: 381-408, 1976.

Regulatory Role of the Adrenergic Synapse Within a Neural Circuit

R. Miftakhov, G. Abdusheva, A. Mougalli

1 Introduction

Coordination of function in the small bowel is managed by the autonomic nervous system which provides the neurogenic regulation by the directed propagation of the electric nerve-pulse within structural elements (neurones). According to the morphofunctional principles of organization it belongs to the class of neuronal networks in which the interaction among neurones is maintained not only by electrical but also by the chemical mechanisms of transsynaptic propagation. The physiological significance of excitatory - inhibitory interactions among neurones is to quantify the transmitter output so that a precise amount of information coded as an electrochemical signal can be passed or gaited in the synaptic zone.

The mathematical modelling of the neuronal networks is related the construction of self-organizing multilevel neuronal structures, analogous to the brain elements [1], [4], [6]. The dynamics of such networks are provided by the *a priori* given memory matrix that totally defines the logical capabilities of the system, but excluding from consideration the chemical mechanisms of the coding and multiplication of the transferred information. This significantly limits the applications of these models in physiological and clinical applications for the analysis of the influence of pharmacological agents. Suffice it to note that mathematical models of the autonomic nervous system based on actual neuromorphology have not been constructed before.

The aim of this study was the mathematical modelling and numerical simulation of nerve-pulse transmission in an excitatory - inhibitory neural circuit composed of cholinergic and adrenergic neurones.

2 Model of a Neural Circuit

The basic assumptions of the model are:

(i). The neural circuit is composed of excitatory (cholinergic) and inhibitory (adrenergic) neurones that are arranged as shown in Figure 1; each neurone is modelled with a geometrically non-uniform cable core of a given length and radius; its diameter increases by a factor of two in the terminal area;

(ii). The electrical impulse propagation along the unmyelinated axons, and in the vicinity of presynaptic nerve terminals is described by the modified Hodgkin - Huxley equations;

(iii). The electrochemical coupling at the inhibitory axo-axonal adrenergic synapse includes: the activation of Ca^{2+} ion influx into the presynaptic terminal; the extrusion of the fraction of noradrenaline (NA) from the free releasable store into the synaptic cleft; NA binding with the α_1 - adrenoceptors located on the postsynaptic membrane; NA binding with the α_2 - adrenoceptors located on the presynaptic membrane; removal of NA by uptake

- 1 and uptake - 2 mechanisms where it is degradated by catechol-O-methyltransferase enzyme; the generation of the inhibitory postsynaptic potential, (IPSP).
(iv). The chemical reactions of noradrenaline transformation are adequately described by first order Michaelis - Menten kinetics;
(v).   The resulting potential at the axo-axonal junction is defined as a linear sum of the action potential propagating along the cholinergic axon and the inhibitory postsynaptic potential generated by the adrenergic neurone.

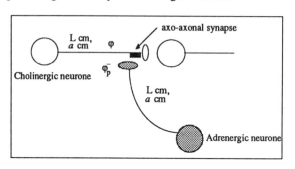

Figure 1. A hypothetical neuronal circuit under consideration.

For the proposed model, the above considerations lead to:
(i). Equations of the action potential, (φ), propagation along the axons of cholinergic and adrenergic neurones [3].

$$C_m^a \frac{\partial \varphi}{\partial t} = \frac{1}{2R_a} \frac{\partial}{\partial s} \left(a^2(s) \frac{\partial \varphi}{\partial s} \right) - I_{ionic} \qquad (2.1)$$

$$a(s) = \begin{cases} a, & 0 < s \le L - L_0 \\ 2a, & s > L - L_0, \end{cases}$$

Here: C_m^a - the axonal membrane capacitance; R_m - the resistance; $a(s)$ the radius of axon; L, L_0 - the length of the axon and synapse, respectively, I_{ionic} - the total ionic current; s - the lagrangian coordinate; t - time;
(ii). Equations of chemical kinetics for the adrenergic synapse function [8] (Fig.2):

$$d[Ca^{2+}] / dt = [Ca^{2+}]_{out} \varphi - k^A_{+9}[Ca^{2+}]$$

$$d[N_s] / dt = - k_c [Ca^{2+}] [NA_s]$$

$$d[NA_c] / dt = k_c [Ca^{2+}][NA_c] - (k^A_d + k^A_{+5} + k_{+6} + k^A_{+1}[AR_{\alpha 2}^0]) [NA_c] +$$
$$+ (k^A_{+1} [NA_c] + k^A_{-1}) [NA_c - AR_{\alpha 2}]$$

$$d[NA_p] / dt = k^A_d [NA_c] + k^A_{-3} + [NA_p - AR_{\alpha 1}] + k^A_{-7} [NAp - COMT] -$$
$$- (k^A_{+3}[AR_{\alpha 1}^0] + k^A_{+7} [COMT^0]) [NA_p]$$

$$d[NA_c - AR_{\alpha 2}] / dt = k^A_{+1} [NA_c][AR_{\alpha 2}^0] - (k^A_{-1} + k^A_{+2}) [NA_c - AR_{\alpha 2}] -$$
$$- k^A_{+1}[NA_c] [NA_c - AR_{\alpha 2}] \qquad (2.2)$$

$$d[NA_p - AR_{\alpha 1}] / dt = k^A_{+3} [NAp][AR_{\alpha 1}^0] - (k^A_{-3} + k^A_{+4}) [NAp - AR_{\alpha 1}] -$$

$$- k^A_{+3} [NA_p] [NA_p - AR_{\alpha 1}]$$

$$d[NA_p\text{-}COMT]/dt = k^A_{+7}[NA_p][COMT^0]-(k^A_{-7} + k^A_{+8})[NA_p\text{-} COMT] -$$

$$- k^A_{+7}[NA_p] [NA_p - COMT]$$

$$d[S_A]/dt = k^A_{+2}[NA_c\text{-}AR_{\alpha 2}]+k^A_{+4}[NA_p\text{-}AR_{\alpha 1}]+k^A_{+8}[NA_p][NA_p\text{-} COMT]$$

$$[AR_{\alpha 2}] = [AR_{\alpha 2}{}^0] - [NA_c - AR_{\alpha 2}]$$

$$[AR_{\alpha 1}] = [AR_{\alpha 1}{}^0] - [NA_p - AR_{\alpha 1}]$$

$$[COMT] = [COMT^0] - [NA_p - COMT]$$

where: $k^A_{-, +(i)}$- the velocity constants of backward and forward chemical reactions (i = 1 -4; 7, 8); k^A_{+9}- the affinity constant; k^A_d- the diffusion constant; $k^A_{+(j)}$- the velocity of re-uptake of the fraction of NA_c from the synaptic cleft by uptake - 1, 2 mechanisms (j = 5, 6); NA_s - vesicular noradrenaline, and its content in the cleft, (NA_c), and on the postsynaptic, (NA_p), membrane; $AR_{\alpha 1}$, $AR_{\alpha 2}$ - adrenoceptors, COMT - catechol - O-methyltransferase enzyme; $[NA_p\text{-} AR_{\alpha 1}]$, $[NA_p - AR_{\alpha 2}]$ - noradrenaline - adrenoceptor complexes; S_A- products of chemical reactions.

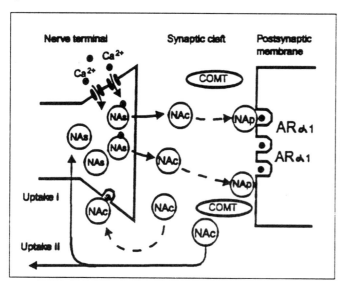

Figure 2. Conversions of noradrenaline at the axo-axonal synapse

(iii). Equation of the inhibitory postsynaptic potential, φ^-_p, development:

$$C_p\frac{d\varphi^-_p}{dt} + \varphi^-_p \left(-\Omega [NA_p\text{-} AR_{\alpha 1}] + R^{-1}_v\right) = \varphi^-_R/R_v \qquad (2.3)$$

Here: C_p- the capacitance of the subsynaptic membrane; Ω - the empirical constant; R_v - the general resistance of the extrasynaptic structures; φ^-_R - the resting potentials.
(iv). The resulting potential at the axo-axonal synaptic zone (s = s*) is calculated

$$\varphi_{sum} = \varphi\big|_{s=s*} + \varphi_{\bar{p}} \tag{2.4}$$

(v). Initial conditions assume that:
- the neural circuit is in the resting state: $\varphi = 0$; $\varphi_{\bar{p}} = \varphi_R^-$;
- the initial concentrations of the reacting components are known

$$[Ca^{2+}]_{out} = 1.0; [Ca^{2+}] = 1.10^{-4}; [NA_s] = 80; [NA_c] = 3.16 \ 10^{-5};$$

$$[AR_{\alpha1}^0] = 0.1; [AR_{\alpha2}^0] = 5.10^{-2}; [COMT^0] = 0.13 \ (mM); \tag{2.5}$$

$$[NA_c - AR_{\alpha1}] = [NA_p - AR_{\alpha2}] = NA_p] = [NA_p - COMT] = [S_A] = 0.$$

(vi). Boundary conditions assume that cholinergic and adrenergic neurones are excited by the external electric impulse

$$\varphi(0,t) = \begin{cases} \varphi_0, & 0 < t \le 0.002 \ s, \quad \varphi(L,t) = 0 \\ 0, & t > 0.002s, \end{cases} \tag{2.6}$$

Equations (1) - (3) have been solved numerically. A hybrid second order accuracy in both time and space finite - difference scheme [7], and the second order Runge-Kutta scheme have been used. The results of calculations are obtained at the following values of electrical, morphological and chemical parameters of the neurones: $C_p = 5.10$; $C_m^a = 1.0 \ (\mu F/cm^2)$; R_a = 34.5; $R_v = 10^3$ (ohm.cm); $\varphi_0 = 100$; $\varphi_R^- = 0$ (mV); $\Omega = 8.10^3$ (mS/M); $a = 1.2 \ \mu m$; $L = 2.5$; $L_0 = 0.1$ (cm); $k_c = 1$; $k_d^A = 40$; $k_{-1}^A = 2.6$; $+k_{+2}^A = 4$; $k_{+3}^A = 50$; $k_{+4}^A = k_{+5}^A = 5$; $k_{+6}^A = k_{+8}^A = 8$; $k_{-7}^A = 200$; $k_{+9}^A = 2 \ (ms^{-1})$; $k_{+1}^A = 2.5.10^4$; $k_{+3}^A = 1.5.10^5$; $k_{+7}^A = 2.10^6 (mM.ms)^{-1}$.

3 Solution Procedure

The following sequience of events defining the dynamics of the neural circuit are considered: discharge of the soma of the cholinergic/adrenergic neurone; the action potential generation and propagation along the unmyelinated cholinergic/adrenergic nerve axon to the synaptic terminal; electrochemical processes of nerve-pulse transmission via the axo-axonal synapse and inhibitory postsynaptic potential generation; interaction of the propagating action potential along the cholinergic axon with the inhibitory postsynaptic potential at the axo-axonal synapse.

A flow chart of the numerical algorithm is shown in Figure 3. The main program defines a set of input parameters and constants, the initial state. It calls the following subroutines:

NERVE: This is a subroutine for solving system of equations (1), intial and boundary conditions (5)-(6) describing the dynamics of the propagation of the nerve-pulse along the cholinergic/adrenergic axons.

SYNAPSE: It is for solving the electrochemical coupling problem at the axo-axonal synaptic zone and computation of the IPSP on the postsynaptic membrane. The equations (2), (3) and (5) are solved.

INTERAC: This subroutine computes the resulting potential at the axo-axonal synapse, following (4).

OUTPRN - Outputs and prints the results.

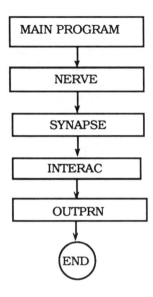

Figure 3. Flow chart of the numerical algorithm

4 Results

During the numerical simulation of interaction between the action potential propagated along the axon of cholinergic neurone and the inhibitory postsynaptic potential at the axo-axonal synapse the following time cases were considered: (a) a simultaneous approach of the impulses at the axo-axonal synapse, (b) the excitatory action potential arrives at the synaptic zone before the development of the IPSP, and vice versa.

(a) Normal physiological conditions

Changes in the resulting potential at the synaptic zone are shown in Figure 4. The critical value of the time lag, Δt, at which there was no influence of the IPSP on the dynamics of the excitatory potential propagation, equalled 1.68 ms. The amplitude and velocity of the spike φ remained unchanged. It reached the cholinergic synapse where it initiated further mechanisms of cholinergic nerve-pulse transmission. As the Δt interval was shortened to 0.6 ms the inhibitory influence began to take effect. The intensity of depolarization phase decreased with a decrease in the maximum amplitude of φ_{sum} to 44.2 mV. The period of repolarizaton went more quickly with the development of stable hyper-polarization. At $\Delta t = 0.39$ ms the amplitude and the duration of φ reduced significantly. A spike of duration of 0.,85 ms and a maximum amplitude of 6.5 mV was generated. This level of depolarization was insufficient to initiate further propagation of the excitation along the axon of cholinergic neurone and, as a result, the blockade of propagation developed. In the case of the simultaneous ($\Delta t = 0$) or the later generation of the wave φ at the axo-axonal synapse than the IPSP generation, a complete suppression of the propagation of electrical signal was observed.

(b) Influence of drugs

The process of noradrenaline release from the vesicular store is controlled by Ca^{2+}-dependent cytosolic mechanisms. A decrease in the initial extracellular concentration of calcium ions to 0.5 mM, and 0.1 mM reduced the concentration of free Ca^{2+} ions in the nerve terminal. In both cases NA overflow was completely abolished and no response in the postsynaptic membrane was observed.

The effect of action of TTX was modelled by the change in maximal conductance of the axonal membrane for sodium ions. According to the numerically obtained results the application of TTX in the vicinity of the presynaptic terminal significantly diminished the Na^+ - influx and, as a result, blocked the nerve-pulse propagation in/through the affected zone. The nerve-terminal of the adrenergic neurone remained in an unexcited state throughout: $\varphi_p^- = 0$ and no influence on the propagation of nerve-pulse was observed.

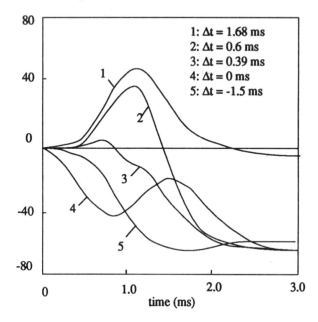

1: $\Delta t = 1.68$ ms
2: $\Delta t = 0.6$ ms
3: $\Delta t = 0.39$ ms
4: $\Delta t = 0$ ms
5: $\Delta t = -1.5$ ms

Figure 4. Changes of the resulting potential(φ_{sum}) at the axo-axonal synapse of the neural circuit

The addition of the inhibitor of COMT caused a significant increase in the concentration of free NA fraction on the postsynaptic membrane. The response was nonlinear and dependent on the concentration of the added compound. Thus in the case of treatment of the adrenergic neurone with 0.15 mM of inhibitor $[NA_p - AR_{\alpha 1}] = 15.5\ \mu M$ was formed and the IPSP of the amplitude of - 83.3 mV was generated. After the addition of 0.2 mM of inhibitor maximum concentration of $[NA_p - AR_{\alpha 1}] = 19.8\ \mu M$ was observed and $\varphi_p^- = -84.6$ mV was obtained. These changes had no influence on the time characteristics of the interaction process at the axo-axonal synapse but significantly reduced the amplitude of the resulting potential to 8% from the 'norm'.

The addition of α_1 - adrenoceptor blockers in concentration of 0.15 mM and 0.2 mM caused a block of 55.4% and 67.5%, respectively, of the total amount of receptors on the postsynaptic membrane. The velocity and the maximum concentration of active $[NA_p - AR_{\alpha 1}]$ - complex development also decreased. The maximum content of the noradrenaline - adrenoceptor complex equalled 4.81 μM and 3.49 μM was recorded. This influenced the level and intensity of IPSP generation. The amplitude decreased to $\varphi_p^- = -71.42$ mV and -66.24 mV, respectively. These changes again had no effect on the dynamical characteristics of excitation - inhibition interaction but caused an increase in the amplitude of the resulting potential to 10 - 12% from the 'norm'.

5 Conclusion

The mathematical model presented, may be considered as a first step in the modelling of the enteric nervous system and the numerical investigation of its function in normal and pathological conditions. It reproduces with sufficient accuracy the effect of the temporary interaction between the IPSP and the excitatory impulse. The minimum time lag in the generation of the IPSP which has no effect on the dynamics of propagating electrical signal along the cholinergic neurone is: $\Delta t \geq 1.68$ ms. Experimental recordings from the neuromuscular synapse obtained by a method of intracellular registration give $\Delta t = 1.5$ ms, see Edwards *et al.* [5]. The decrease in Δt diminishes the amplitude of the spike φ and significantly shortens it. The results of numerical simulation (curves φ_{sum}) are similar to the experimental results of interaction between excitatory and inhibitory postsynaptic potentials in the some of motor neurones, e.g. Akari *et al.* [2]. The model should prove to be valuable for predicting the effects of new pharmacological agents, and thus diminishing the requirement for time - consuming (and expensive) *in vitro* and *in vivo* experimentation.

References

[1] L.F. Abbott, E. Farhi, and S. Gutmann. The pathway integral for dendritic tree. *Biol. Cybern.* 66: 49-60, 1991.

[2] T. Akari, J.C. Eccels, and M. Ito. Correlation of the inhibitory postsynaptic potential of motor neurons with the latency and time course of inhibition of monosynaptic reflexes. *J. Physiol.* 154: 354-377, 1960.

[3] M.B. Bekenblit, and T.M. Shura-Bura. The velocity of impulse propagation in the expanding and narrowing fibres. *Biophysica* 85: 871-875, 1981 (in Russian).

[4] R. Budelli, J. Torres, E. Catsigeras, and H. Enrich. Two-neurons network. *Biol. Cybern.* 66: 95-101, 1991.

[5] F.R. Edwards, G.D.S. Hirst, and E.M. Silinsky. Interaction between inhibitory and excitatory synaptic potentials at a peripheral neurone. *J. Physiol.* 259: 647-663, 1976.

[6] O. Ekeberg, P. Wallen, A. Lansner, H. Traven, L. Brodin, and S. Griller. A computer based model for realistic simulations of neural networks. *Biol. Cybern.* 65: 81-90, 1991.

[7] D.J. Evans, and A.R.B. Abdullah. Group explicit method for parabolic equations. *Inter. J. Comp. Math.* 14: 73-105, 1983.

[8] R.N. Miftakhov. Numerical modelling of adrenergic nerve-pulse transmission. *Investigations on Theory of Shells.* XXI: 100-111, 1988 (in Russian).

Duodenal Bioelectrical Waxing and Waning Activity

S. Salinari and R. Mancinelli

1 Introduction

Recordings of the electrical activity of gut muscles show spontaneous periodic oscillations called slow waves or Electrical Control Activity (ECA). The slow waves are related to oscillations of the membrane electrical potentials; thus, they are myogenic in nature and persist after anatomic and pharmacological denervation. In intact segment the frequency of slow waves decreases aborally in the stepwise fashion, with variable lengths of intestine having the same frequency (plateau areas). When the intestine is cut into small segments the slow wave frequency of each consecutive segment decreases aborally in linear fashion. These phenomena were largerly investigated in literature and were explained on the assumption that intestinal smooth muscle can be regarded as a series of loosely coupled oscillators having successively decreasing intrinsic frequencies. In a frequency plateau, individual oscillators are driven by the oscillator with the highest intrinsic frequency which assumes the role of pacemaker for the next frequency plateau. It follows that there are multiple pacemaker areas, of decreasing frequency, situated at various levels of the intestine. This areas can be variable in time and position since each cell or group of cells of the intestinal tissue can act as a pacemaker. Superimposed on the slow waves and in relationships with them, burst of spikes, associated with contraction, also appear in the recordings. Spiking activity is due to a fast membrane depolarization of the smooth muscle cells and occurs generally at the peak of the slow waves since the excitability of the tissue is maximal at this time (1). Thus, the slow waves have a pacing role with respect to the mechanical activity, but, in general, they do not give rise to a contraction. The phase locking of the slow waves at the plateau areas of an intestinal segment seems to the basis for the occurence of propagated spike burst.

Fig.1.1 illustrates the electrical and mechanical activity recorded on a segment of rat colon and illustrates the correlation between the slow waves, the propagation of the fast activity and the mechanical response. The control activity and spikes have been recorded from two sites of the intestinal segment to evidence the propagation of the fast component from oral to aboral end.

In addition to these rhythms of seconds in duration, changes in amplitude and frequency over period of minutes and hours can also occur, but the mechanisms responsible of this type of activity are still partially unknown. A typical pattern of variation is "spindling" or waxing and waning of slow wave amplitude that can be recorded both "in vivo" and "in vitro".

Diamant and Bortoff (2) found that waxing and waning areas in general separate two zones of frequency plateau and attributed spindling to the interaction of two competing pacemakers, of slightly different frequencies, from two adjacent plateau areas. They also related this pattern to propagation: in a frequency plateau the propagation tends to be uniform, while it is locally disrupted in a zone of waxing and waning. As a consequence, they suggested that spindling activity can serve as a "stopcocks" to the flow of chyme during digestion.

Recently Suzuki et al. (3) confirmed the presence of two interacting pacemakers during spindling by Fourier power analysis. They also investigated on the mechanisms which can be responsible of favoring waxing and waning activity. They found that some depolarizing agents as Potassium and Ouabain can induce spindles and suggested that the depolarization of the membrane potential may change frequency in a pacemaker region and asincrony among pacemakers and spindles can result. They also found that spindles can be abolished by a treatment which reduces electrical coupling between the smooth muscle cells such as bathing in hypertonic sucrose solutions or in solutions of reduced pH. Nevertheless changes in Calcium concentration seem to have no effect on spindles. In the same paper Suzuky et al. also correlated the waxing and waning of slow waves to the contractile activity, in particular for rhythms of 1-2 min period. They hypothesized that, during spindles, slow waves *per se* can induce contractions due to a membrane depolarization of sufficient amplitude. Smith et al (4) also noted, in canine colon, that slow potential oscillations may be the primary electrical event in the excitation-contraction coupling and attributed that to the interaction of slow waves with a faster myoenteric potential oscillations.

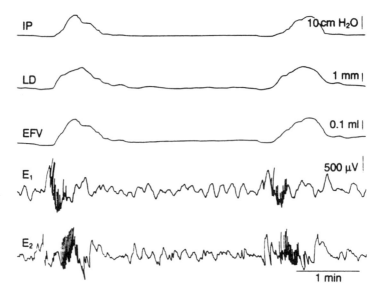

Figure 1.1
Mechanical and electrical intestinal activity, recorded during peristalsis, in isolated distal colon of the rat. IP intraluminal pressure, LD longitudinal displacement, EFV ejected fluid volume. E_1, E_2 extracellular electrical activity (τ = 0.55). Bath temperature 37°C, ph 7.4

Postorino et al. observed, in isolated rat duodenum (5), a spontaneous waxing and waning activity both in electrical and mechanical records. This spontaneous spindling activity appeared to be affected by the presence of tetrodotoxin in the bath, while no changes were induced in the Electrical Control Activity. This observation suggested that nervous elements might be responsible for maintaining variable pacemaker activities. This hypothesis agrees with the one proposed by Suzuky et al. (3) that hypothesized, as a possible mechanism responsible of differences between "in vivo" and "in vitro" recordings, a greater frequency gradient "in vivo" due to the augmentation of the intrinsic frequency gradient by neural or hormonal input.

In a previous paper (6) the authors observed that, in isolated rat duodenum without spontaneous spindling, a typical waxing and waning of mechanical waves occasionally occurred by adding phospholipids vescicles to the bath. To verify the hypothesis that fatty acids, components of phospholipids, might be responsible for this pattern, several fatty acids, with different hydrocarbon chain length, were tested. In particular Capronic acid (saturated C6), at 10^{-5} M, was capable of evoking a clear waxing and waning activity. Typical spindles also appeared when ouabain was added to the incubation bath, or when potassium concentration of the medium was increased. This pattern was observed in experimental recordings performed on isolated segments and strips of muscular tissue. The authors hypothesized that fatty acids might evoke spindling activity by directly interacting with the mechanisms underlaying the slow wave generation. However Yajima (7) noted, in isolated colon of the rat, that mechanical contractions, elicited by the infusion of short chain fatty acids in the bath, were completely suppressed in presence of tetrodotoxin, suggesting a possible involvement of some neural paths.

The study of so complex mechanisms underlaying the spindling activity can be considerably aided by a modelling approach; first, this requires an analytical description of the pacemaker cells and of the electrical coupling between them. As waxing and waning affects electrical and mechanical patterns, both slow and fast electrical activity (bursting cells) has to be taken into account in modeling the single pacemaker. Since 1985, Honerkamp et al.(8) proposed and described a general mechanism underlying bursting. This mechanism consisted of two coupled non linear oscillators with different frequencies where the slower oscillator alternatively switches the faster one on and off. In particular they tested the model on an extended Bonhofer-van der Pol oscillator and on a modified version of the Hodgkin-Huxley equations.

Later Kopell and Ermentrout, (9) and (10), considered a general class of models for bursting depending on the interaction of a slowly changing subcellular oscillation and a spiking mechanism. In particular, in (10), Kopell and Ermentrout considered in detail bursting cells in smooth muscle and suggested that the oscillations underlying the bursting behaviour of active smooth muscle can be divorced from the spiking mechanism. They observed that bursting appears to come from an interaction between the slow waves and the spike generation, where the slow waves appear to be metabolic in origin, mediated by the intracellular Calcium concentration.

In this paper a van der Pol oscillator was adopted to model the slow wave activity, while the spiking activity was represented by a simplified form of Hodgkin-Huxley equation. Concerning the interaction between pacemakers, it has been well established that smooth muscle cells are electrotonically coupled, probably by low resistance nexuses. In this paper coupling between two pacemaker cells is taken into account by a junction resistance, that models the influence of a membrane potential variation in a cell on the membrane potential of a near cell. The leakage current can be also considered by an high value resistance. Parameter values of spiking model are taken from Ramon et al. (11) that modified the Hodgkin-Huxley equation to simulate the occurence of action potential in smooth muscular cells. Parameters of slow oscillator are taken from literature and from experimental data relative to electrical recordings on intestinal muscle segments. Experimental findings compared with the behaviours predicted by the model give preliminary indications about the validity of the hypotheses to the basis of the model.

2 Experimental procedures and data

Male albino rats, weighing 250-300 g, were killed by general anaesthesia. A segment of about 2 cm of the duodenum, near the pylorus, was quickly removed and placed in a 100 ml thermostatically controlled organ bath continuously perfused with Krebs solution areated with 5% CO_2 and 95% O_2, pH 7.4 and maintained at 37 °C. The aboral end of the duodenal segment was connected via a catheter to a pressure transducer. The oral end was secured by a surgical silk to an isometric force displacement transducer. The intestinal segment was distended by injecting small volumes of physiological solution (0.3-0.5 ml) and kept under a resting load of 1.5 g; the preparation was allowed to equilibrate for a period of at least 30 min. Extracellular electrical activity was detected by means of a glass pressure electrode (tip diameter 150-200 μm) filled with 3M KCl solution and suspended on Ag-Ag Cl wire coil. Electrical signals were processed in parallel by two different preamplifiers: one with a lower cut-off frequency of 0.31 Hz for recording both low and fast components; the other one with a cut-off frequency of 3.97 Hz for recording only the fast component.

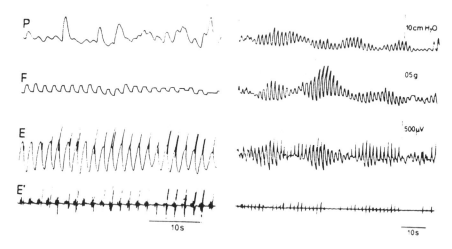

Figure 2.1
Left: spontaneous mechanical and electrical activity in an isolated duodenal segment.
P pressure, F force, E extracellular electrical activity, E' extracellular fast electrical activity.
Right: spontaneous waxing and waning activity of mechanical changes (top) and electrical waves (bottom)

Capronic acid (C6) was filled in the bath at a concentration ranging from 10^{-5} M - 10^{-3} M. For each dose, different preparations were used and at least five trials were recorded. The experiments were performed after an adaptation period of about two hours and the acid was added to the bathing solution after a control period not less than therty minutes.

Fig.2.1 shows, on the left, the spontaneous mechanical and electrical activity in an isolated duodenal segment; in the same figure, on the right, spontaneous tracings with waxing and waning of both mechanical changes and electrical slow waves are also presented.

The effects of capronic acid at 10^{-4} M, in a preparation without spontaneous spindling, are shown in Fig.2.2. This acid induces a clear and persisting spindling activity.

Fourier analysis of the extracellular electrical recordings showed a single frequency peak of 0.55±0.07 Hz, when the pattern of slow waves was regular, but two frequency peaks whenever spindles were recorded. Frequency differences between two interfering rhythms ranged from 0.004 to 0.05 Hz.

3 Mathematical model

Model of the pacemaker
The model of the single pacemaker is based on the approach proposed by Kopell and

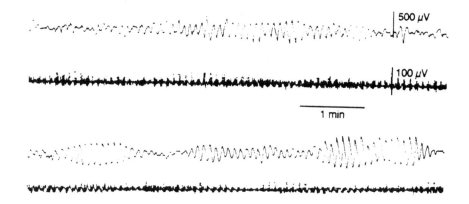

Figure 2.2
Extracellular electrical slow and fast activity before (top) and after (bottom) the injection in
the bath of Capronic acid at 10^{-4} M.

Ermentrout in (9) and (10) for bursting cells. In these papers they distinguished two types
of slow cellular oscillations: the membrane oscillations due to periodical changes of ionic
conductances under the control of transmembrane potential and the *subcellular oscillations*
that remain also in the absence of changes in this potential. In this context they analyzed a
general class of models postulating both a subcellular oscillator and a mechanism which
links it to the membrane potential. The model allows the subcellular oscillator to be
modifiable by changes of membrane potential to take into account many biological
applications. The equations are:

$$\dot{x} = f(x) + \varepsilon^2 \, g(x,y,\varepsilon)$$
$$\dot{y} = \varepsilon \, h(x,y,\varepsilon) \tag{3.1}$$

where $x \in R^p$, $y \in R^q$, $\varepsilon \ll 1$ and f, g and h are smooth functions. The two main
hypotheses are: i) $x = f(x)$ has an attracting invariant circle with a single critical point in $x=0$
(a sink-saddle); ii) $y = h(0,y,0)$ has a stable limit-cycle solution. In (3.1) the variable x can
be interpreted as the vector of transmembrane potential and ionic components that describe
the dynamics of electrically excitable tissue (spiking mechanism). The variable y can
represents the slow oscillations involving processes not in cell membrane. The function g
represents the coupling of the slow waves to the spiking mechanism.

The electrical behaviour of the smooth muscle membrane may be described by the
general model, deduced by the Hodgkin-Huxley equation:

$$I_m = C \frac{dV_m}{dt} + I_k + I_{Na} + I_L + I_p = \tag{3.2}$$

$$= C \frac{dV_m}{dt} + g_k (V_m - E_k) + g_{Na} (V_m - E_{Na}) + I_L + I_p$$

where Im and Vm are the transmembrane current and the transmembrane potential respectively, C is the membrane capacitance, I_k is the potassium current, I_{Na} is the sodium current, I_L is a leakage current and Ip is a polarizing current which can represent the slow oscillator. g_k is the K^+ conductance, g_{Na} the Na^+ conductance and E_k and E_{Na} the equilibrium potentials for K^+ and Na^+ respectively. All the currents are referred to a unit area membrane ($\mu A\ cm^{-2}$).

In this paper this equation was specialized on the basis of some simplifying hypotheses. In particular a simplified form of the Hodgkin-Huxley model analyzed by Colding-Jørgesen in (12) was adopted. In this model it was assumed that the charging of the membrane takes place so fast that the time, and the necessary current, for the charging, could be ignored, so the membrane capacity was taken to be zero. The leakage current I_L was also taken equal to zero. The sodium inactivation was omitted and, finally, the change in sodium conductance was assumed to be instantaneous with respect to the change in potassium conductance. Thus the time dependency, in the spike equation, is exclusively introduced through the potassium conductance as a first order relation.

The current Ip was assumed, as a first approximation, to be a forcing term and a Van der Pol oscillator, which was widely used in literature to represent the slow oscillations, was adopted to model it. Then, in this first approach, the model merely describes a slowly forced excitable system.

Under these simplifying hypotheses the equations (3.2) become (Fig.3.1 a):

$$I_m = g_k\ (V_m\text{-}E_k) + G_{Na}\ (V_m\text{-}E_{Na}) + I_p$$
$$\dot{g_k} = (G_k\text{ -}g_k)/\tau_k \tag{3.3}$$

where the steady state values of sodium and potassium conductances are depending on V_m and are taken as sigmoid functions. Then $G_j\ (V_m)$, where j stands for k or Na, is given by $G_j = aa_j + A_j/[1+exp(V_m\text{-}V_j)]$.

I_p is the solution of a Van der Pol equation:

$$\dot{I_p} = dI_p - \lambda\ (\text{-}I_p + I_p^3/3)$$
$$\dot{dI_p} = \text{-}\omega^2\ I_p \tag{3.4}$$

Note that the missing capacity current limits the freedom of the membrane potential. At any time V_m must fulfil the eq. (3.3), even if it has to change istantaneously to do it. Parameters of eq.(3.3) are derived by the paper of Ramon et al. (11) that adapted the Hodgkin and Huxley equation to model the action potential in smooth muscle. Resting membrane potential was assumed in the range -70 + -50mV and the spike maximum equal to 20mV for I_m=0. Equations (3.4) represent the forcing input which elicits the slow periodic oscillations of the membrane potential in the smooth muscle and are in the form of a Van der Pol equation. The parameters λ and ω, are chosen so that the slow oscillations of the membrane potential fit frequency and shape of the experimentally recorded slow wave. In order to take into account some experimental observations (see also Concluding Remarks), equations (3.4) were successively modified to allow the slow oscillation to be significantly affected by the excitable system. In particular, considering a feedback from the membrane potential, we obtain:

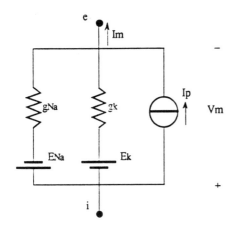

e

I_m

Figure 3.1 Equivalent circuit of the smooth muscular cell membrane: I_m and V_m are the transmembrane current and the transmembrane potential , I_p is a polarizing current which can represent the slow oscillator, g_k is the K^+ conductance, g_{Na} the Na^+ conductance and E_k and E_{Na} the equilibrium potentials for K^+ and Na^+ respectively.

$$\dot{I}_p = dI_p - \lambda \,(-\,(I_p - I_{p0}) + (I_p - I_{p0})^3/3) - gr\,V_m$$

$$\dot{dI}_p = -\omega^2\,(I_p - I_{p0}) \qquad\qquad\qquad\qquad (3.5)$$

Model of interaction between two pacemaker cells

The interaction of two pacemaker cells is modeled with a junction resistance R and a leakage resistance r according to the scheme adopted by Torre in (13) (Fig.3.2). As a consequence the coupling between two pacemaker cells can be described in the following way:

$$(V_{m2} - V_{m1})/R - V_{m1}/r = g_k\,(V_{m1} - E_k) + G_{Na}\,(V_{m1} - E_{Na}) + I_{p1}$$

$$\dot{g}_k = (G_k - g_k)/\tau_k$$

$$(V_{m1} - V_{m2})/R - V_{m2}/r = g_k\,(V_{m2} - E_k) + G_{Na}\,(V_{m2} - E_{Na}) + I_{p2} \qquad (3.6)$$

$$\dot{g}_k = (G_k - g_k)/\tau_k$$

where R is the junction resistance, r is the leakage resistance and $G_j = aa_j + A_j/[1+\exp(V_m - V_j)]$

$$\dot{I}_{p1} = dI_{p1} - \lambda_1\,(-\,(I_{p1} - I_{p01}) + (I_{p1} - I_{p01})^3/3) \qquad\qquad\text{or}$$

$$\dot{I}_{p1} = dI_{p1} - \lambda_1\,(-\,(I_{p1} - I_{p01}) + (I_{p1} - I_{p01})^3/3) - gr\,V_{m1}$$

$$\dot{dI}_{p1} = -\omega_1^2\,(I_{p1} - I_{p01}) \qquad\qquad\qquad\qquad (3.7)$$

$$\dot{I}_{p2} = dI_{p2} - \lambda_2\,(-\,(I_{p2} - I_{p02}) + (I_{p2} - I_{p02})^3/3) \qquad\qquad\text{or}$$

$\overset{\bullet}{Ip2} = dIp2 - \lambda 2 \left(- (Ip2 - Ip02) + (Ip2 - Ip02)^3/3\right) - gr\ Vm2$

$$(3.8)$$

$\overset{\bullet}{dIp2} = -\omega 2^2 (Ip2 - Ip02)$

where $Ip1$, $Ip2$ are the slow polarizing currents and $Vm1$, $Vm2$ are the membrane potentials of the first and the second pacemaker respectively.

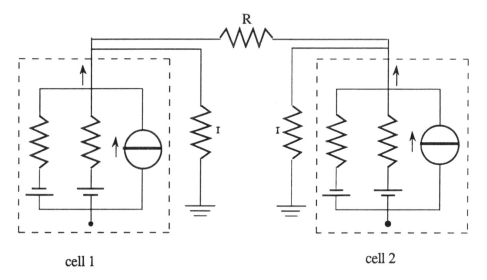

cell 1   cell 2

Figure 3.2
The interaction of two pacemaker cells is modeled with a junction resistance R and a leakage resistance r.

4 Results

The model of the two interacting pacemaker cells, described by the eqs. (3.6) - (3.8) of the above section, was implemented on a PC 486 DX2 using a MATLAB software package. At each time, the values of the potassium conductance g_k and of the polarizing currents $Ip1$ and $Ip2$ are obtained integrating the differential equations in (3.6) and (3.7) by a Runge-Kutta routine with variable time step. The corresponding values of the membrane potentials $Vm1$ and $Vm2$ were obtained by solution of the instantaneous equations in (3.6) using a quasi-Newton minimization routine.

The set of parameters used for the simulation of the model is reported in Table 4.1. Parameters of the spike equation were assumed as in Ramon et al.[11].

In particular, the values of Ek, ENa are the equilibrium potential of potassium and sodium ions in smooth muscular cells. Note that troughout this paper the transmembrane voltage was defined as intracellular minus extracellular voltages ($Vm = Vi - Ve$) and all membrane voltages in HH-model were changed accordingly to this sign convention. To simplify the often complex mathematical nature of the functions Gj (Vm) sigmoid functions

Table 4.1

Parameters of fast oscillator	Parameters of slow oscillator
$E_k = -82$ mV	$\lambda_1 = 1$
$E_{Na} = 40$ mV	$\lambda_2 = 1$
$a_K = 0.009$ mmΩ/cm^2	$\omega_1 = 3.454 \pm 0.44$ rad s^{-1}
$A_K = 0.014$ mmΩ/cm^2	$|\omega_1 - \omega_2| = 0.025 + 0.35$ rad s^{-1}
$a_{Na} = 0.005$ mmΩ/cm^2	$g_r = 0 + 0.1$ mmΩ cm^{-2}
$A_{Na} = 0.008$ mmΩ/cm^2	
$\gamma = 0.25$ mV^{-1}	
$V_j = -35$ mV	
$a = 0.998$	
$\tau_k = 20$ ms	
$V_r = -70 + -50$ mV	
$R = 0 + 10000$ KΩ/cm^2	
$\tau = \infty$	

($G_j = aa_j + A_j/[1+\exp(V_m - V_j)]$) were used. aa_K and aa_{Na} are the values of the resting sodium and potassium conductances. A_{Na} was computed considering that, for a quick depolarization of the membrane to 20mV, the sodium conductance changes instantaneously, while the potassium conductance maintains its resting value. The value A_K was determined assuming the ratio A_K /A_{Na} equal to the ratio a_K/a_{Na}. Parameters γ and V_j were empirically chosen to achieve the more suitable form for the sigmoid function between aa_j and A_j. The potassium time constant τ_k was set in order to obtain a realistic spike duration.

Concerning the slow oscillations, the parameters of the Van der Pol equation, λ, ω, I_{p0} were chosen so that the frequency and shape of the slow wave were in accordance with the experimental observations.

Fig.4.1 shows an example of waxing and waning activity, obtained with a junction resistance $R=50$ KΩ/cm^2, a frequency values for the two pacemakers of 0.57 and 0.62Hz and a feedback coefficient $g_r = 0.0\ 09$ mmΩ cm^{-2}. For comparison, in the same figure, an experimental record of spontaneous electrical activity in isolated rat duodenum with spindles is also presented. Note that the experimental data are relative to extracellular recordings while the model was implemented using intracellular parameters. However, the effects of the extracellular electrodes can be taken into account, as a first approximation, considering that the configuration of the slow waves, recorded with bipolar electrodes approximates the first time derivative of the transmembrane potential change attenuated by the volume conductor action.

The influence of the junction resistance and of the frequency difference between pacemakers has been tested. In particular it was noted that spindles occur only in a limited range of values of R. For example, in the case of Fig.4.1, the membrane potential presents a waxing and waning activity for values of R ranging approximately from 50 KΩ/cm^2 to 5000 KΩ/cm^2. For higher values of the junction resistance, the two pacemakers oscillate independently, while lower values of R determine a persistent oscillation of the two pacemakers at the same frequency.

If the frequencies of the two pacemakers are closer, the waxing and waning activity is present in a more restrict range of resistance values. For example, being the frequency difference between the two pacemakers equal to 0.02 Hz and gr equal to 0.009, spindles are

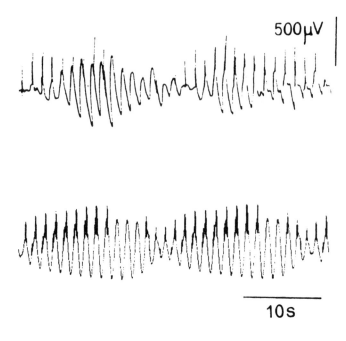

Figure 4.1
Top: spontaneous waxing and waning of the electrical activity in rat duodenum.
Bottom: model simulation using parameters of Table 4.1

present in a range of junction resistance R from 300 $K\Omega/cm^2$ to 5000 $K\Omega/cm^2$

The effect of a feedback from the membrane potential to slow oscillator was also roughly tested. In particular it was noted that the occurrence of a persistent oscillation is strictly dependent on the presence of this feedback: if the feedback is absent, gr=0, a waxing and waning is observed also for a junction resistance equal to zero while values of gr large enough cause quiescence.

5 Concluding remarks

In (3) Suzuki et al. hypothesized that three conditions are mainly responsible of favoring waxing and waning activity, that is a variation in slow wave frequency at a point, electrotonic coupling between muscle fibers and a slight membrane depolarization. The simple model presented in this paper allowed to test the influence of the coupling and of the frequency difference between the pacemaker cells. In particular, since the waxing and waning activity, in simulation tests, occurs only for limited ranges of frequency and junction resistance, it could confirm the observations that the occurrence of spindles is variable in time and position during an experiment. In fact, also small changes of frequency and/or coupling between cells, due to metabolic or neural agents could cause the disappearence or the occurence of a waxing and waning activity.

The condition regarding the influence of a membrane depolarization on the waxing and waning occurrence requires an accurate investigation about the mechanisms responsible for the slow waves in the smooth muscle.

A first possible mechanism, hypothesized in literature, is a metabolic oscillator, which could be connected to membrane activity through the Na-K pump. Some experimental evidences, as reported in (14, 15), support the hypothesis that sodium pump is involved in slow waves and that agents which affect metabolism affect also oscillators. However a metabolic mechanism can not take into account some experimental behaviours as the little effect of metabolic inhibitors on the frequency of slow waves and the problem of synchronization, which can be difficulty obtained by purely metabolic mechanisms. Finally, a metabolic oscillator could not explain the sensitivity of slow waves to membrane perturbations and the possibility of entrainement by external pulse stimuli.

Some of these questions, in particular the effect of membrane perturbation on the subcellular oscillations, have been explained, in literature, taking into account the role of calcium in the generation of slow waves. Calcium oscillations can affect the membrane activity through the Na-K pump, moreover the presence of a calcium channel, the conductance of which increases with depolarization, allows to simply understand the dependence of slow wave frequency on the membrane potential (16).

The simple model proposed in this paper is not suitable for taking into account in detail all the complex aspects described above. In fact, in this model, a black-box approach was adopted for the slow wave modelization, using a Van der Pol oscillator, while the influence of the membrane potential was taken into account by a feedback term. This case is similar to that theoretically analyzed by Kopell and Ermentrout (10) in *Proposition 4.2* and, for a single pacemaker, confirms their theoretical results. In particular if the feedback term is equal to zero, depolarization and hyperpolarization have no effect on the subcellular oscillation and the membrane depolarization influences only the fraction of the slow wave period occupied by bursting. When the feedback term is present in the slow equation, the depolarization of the membrane decreases the amplitude and the period of the slow waves.

In the case of a coupling between two pacemakers of different frequencies, for gr=0, a waxing and waning verifies at a frequency equal to the difference of frequencies of the two pacemakers up to high values of the junction resistance. If a feedback from the membrane to slow oscillator is hypothesized, the behaviour of the system is more complex. It seems that there exists a limit value of gr coefficient, depending on the junction resistance and on the frequency difference, for which the two slow oscillators are entrained. Therefore a number of different patterns can be obtained varying the values of gr and R : entrainement, beating of different time period and also quiescence. This suggested that a model which takes into account the possibility of feedback from the membrane to the subcellular oscillator, could better fit the variety of behaviours observed in the experiments. However conclusive results about the effects of membrane depolarization on slow wave frequency and amplitude, and the study of the influences of different substances and drugs on the electrical activity of smooth muscle cells require a more accurate description of the mechanisms to the basis of slow oscillation.

References

[1]   C.Mendel, A.Pousse and J.F.Grenier: Temporal relationship between the spike burst and the slow wave during the fasted and fed states. In: *"Gastrointestinal Motility"* (G.Labo and M.Bertolotti Eds.), 215-218, Cortina Int., Verona, Italy, 1983.

[2]   N.E.Diamant and A.Bortoff: Nature of intestinal slow-wave frequency gradient. *Am. J. Phisiol.* 216,2, 301-307, 1969.

[3]   N.Suzuki, C.L.Prosser and W.Devos: Waxing and waning of slow waves in intestinal musculature. *Am. J. Physiol.* 250 (Gastrointest. Liver Physiol. 13), G28-G34, 1986.

[4]   T.K.Smith, J.B.Reed and K.M.Sanders: Interaction of two electrical pacemakers in muscularis of canine proximal colon. *Am. J. Physiol.* 252 (Cell Physiol. 12), C290-C299, 1987.

[5]   A.Postorino, R.Mancinelli, C.Racanicchi, E.B.Adamo and R.Marini: Spontaneous electromechanical activity in rat duodenum in vitro. *Archiv. Int. Physiol.* Bioch. 98, 35-40, 1990.

[6]   R.Mancinelli, G.P.Littarru, G.B.Azzena, G.L.Gessa: Azione degli acidi grassi sulla motilità del duodeno di ratto. *Atti della Società Italiana di Biologia Sperimentale.* 8-10 September 1993, Pavia (Italy).

[7]   T.Yajima: Contractile effect of short chain fatty acid on the isolated colon of the rat. *J.Physiol.*, 386,667-678, 1985.

[8]   J.Honerkamp, G.Mutschler, R.Seitz: Coupling of a slow and a fast oscillator can generate bursting. *Bull. Math. Biol.* 47,1-21, 1985.

[9]   G.B.Ermentrout and N.Kopell: Parabolic bursting in an excitable system coupled with a slow oscillator. *SIAM J. Appl. Math.* 46, 2, 233-253, 1986.

[10] N.Kopell and G.B.Ermentrout: Subcellular oscillations and bursting. *Math. Biosci.* 78, 265-291, 1986.

[11] F.Ramon, N.C.Anderson, R.W.Joyner and J.W.Moore: A model for propagation of action potentials in smooth muscle. *J. Theor. Biol.* 59, 381-408, 1976.

[12] M.Colding-Jorgensen: Fundamental properties of the action potential and repetitive activity in excitable membranes illustrated by a simple model. *J.Theor.Biol.* 144, 37-67, 1990.

[13] V.Torre: A theory of synchronization of heart pacemaker cells. *J.Theor. Biol.* 61, 55-71, 1976.

[14] J.A.Connor: On exploring the basis of slow potential oscillations in the mammalian stomach and intestine. *J. Exp. Biol.* 81, 153-173, 1979.

[15] D.D.Job: Effects of antibiotics and selective inhibitors of ATP on intestinal slow waves. *Am. J. Physiol.* 220, 299-306, 1971.

[16] C.L.Prosser and A.W.Mangel: Mechanisms of spikes and slow wave pacemaker activity in smooth muscle cells. In: *"Cellular pacemakers I "* (D.O.Carpenter Ed.), 273-302, Wiley-Interscience, New York, USA, 1982.

Drug Delivery Systems: Technology and Clinical Applications

A. Hoekstra

General introduction

The development of novel drug delivery systems has become an important activity, not only for researchers at Universities, but also for pharmaceutical and some medical device industries. The reasons for this are that the costs for the development of new drugs are becoming extremely high, and that many drugs are available, which given in a better dosage form, should lead to considerably improved therapeutic results.

The next two chapters shall provide further insight into the area of drug delivery technology. The first chapter begins with the concepts of drug delivery and targeting, ranging from very simple infusion pumps to highly complex microprocessor devices, as well as their potential clinical applications. The second chapter focuses on access port design, advantages and applications, and prevention of procedure-related complications based on preclinical and clinical results.

An important goal in the area of improved drug delivery systems is to localize the drug in a specific body-compartment or to target the drug to specific organs or cells. Different systems to reach this goal are for example: the use of liposomes as drug vehicles, water-soluble drug carrier systems, implantable drug delivery systems such as micropheres (biodegradable polymeric implants), and implantable pumps (non-biodegradable implants), which can be reloaded (insulin pump). Another exiting method, although very difficult, for the delivery of biologically active molecules is the use of microencapsulated viable mammalian cells. A typical example is the encapsulation of insulin-producing pancreatic islet cells for the treatment of diabetics (hybrid artificial pancreas). Blood glucose concentration control in diabetic subjects can theoretically be obtained by an insulin delivery device, which is directly controlled by a continuous measurement of the present glucose concentration in the body (blood or tissue). Such a closed-loop system, often referred to as artificial beta cell or artificial pancreas, principally consists of three components: a glucose sensor, an insulin delivery pump and a microprocessor control unit regulating the insulin administration based on the measured amount of glucose. However, the main difficulty for the realization of such a system is the development of a reliable long-term continuously measuring glucose sensor.

The concept of "total implantation" of vascular access devices, substituting for the "percutaneous exit" catheters, has added appreciably to the range of applications and the safe duration of catheterization. New, high biocompatible materials have extended the catheter indwelling time from a few days to a year or longer. The parallel development of small, lightweight, compact, and reliable infusion pumps, carried or worn by the patients, have provided mobility and relatively unrestricted patient activity. These important benefits have reduced significantly the frequency, duration, and cost of hospitalization associated by extended drug therapy in the treatment of selected chronic and transient diseases. Although incomplete, I hope that the topics covered in the next two chapters will provide the reader with up-to-date information on the state of the art in the selected fields of drug delivery technology.

Recent Developments in Drug Delivery Systems

A. Hoekstra

1 Introduction

The method of administering medicines has become a central consideration in drug development. As products are developed to specifically target diseases, the ability to administer drugs with greater efficiency and efficacy becomes more and more important. The properties of an ideal drug dosage form include a precise, specific attack on the diseased tissue without excessive or toxic effects, either locally or systemically. Most fall short of this ideal and, consequently, there is scope for new (improved) drug delivery systems.

Improved drug delivery is a growth area in pharmaceutical medicine, not only for the pharmaceutical industry, but also for researchers at universities and some medical device industries. The development of new drug delivery technologies is driven by the desire to maximize the safety and efficacy of therapeutic agents. Efforts are focused on the identification of the optimal route of administration and temporal pattern.

With the appearance of therapeutically active peptides and proteins developed through recombinant DNA technology, there is an increasing need to find acceptable novel, non-parenteral routes of delivery. Most peptides and proteins are poorly absorbed via the oral route due to rapid degradation in the gastrointestinal tract and/or poor transport across the epithelial barriers. Hence alternative delivery methods to circumvent the necessity of frequent injections are being explored.

Technologies such as oral and transdermal drug delivery continue to mature and offer additional capabilities, while research continues toward the establishment of new capabilities using such technologies as bioerodible polymers, electrotransport, liposomes, monoclonal antibodies, and microspheres. Furthermore, in the past, various systems from single access ports to micropumps are now available for administration of repeated bolus doses, or continuous or programmable infusions. Some drugs will benefit from being administrated in a pulsatile fashion in order to avoid effects associated with tolerance[1] or to achieve a delivery pattern that takes into account the chronobiological variation of the disease condition[2]. This contribution summarizes a selection of major drug delivery challenges.

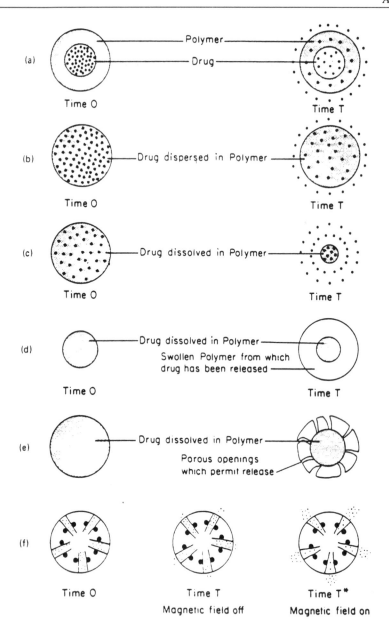

Figure 1   Mechanisms of drug release from subdermal implants of nonbioerodable and bioerodable polymers: a) diffusion reservoir; b) diffusion-matrix (nonbioerodable polymers or fused drug implants); c) chemical; d) solution, solvent-activated systems (bioerodable polymers); e) osmotically controlled systems (bioerodable polymers); and f) magnetic systems. (Adapted from: Zatuchni GI, Goldsmith A, Shelton JD, Sciarra JJ, eds., Long-Acting Contraceptive Delivery System, Harper & Row, NY, 1984)

2 Polymeric Drug Delivery Systems

Innovative research into mebrane technology has been directed into three major categories of controlled drug release: diffusion through a polymeric membrane or a matrix, osmotic diffusion of water through a membrane, and controlled erosion of a polymeric matrix wherein the drug is entrapped. Diffusion of a drug through the polymer is the rate-limiting step. It remains constant (zero-order release) as long as the concentration outside the system is negligibly small compared with the inside (saturated solution) and the membrane characteristics do not change. Thus zero-order release of a drug from a diffusional system is only possible by using a rate-controlling membrane or a polymer matrix of special geometry. Implantable bioerodible systems are based on biodegradation of polymer containing drug or drug-polymer complex. Polymers used are, for instance, copolymers of lactic acid and glycolic acid and polypeptides of glutamic acid and leucine. Degradation of these polymers is a continuous process, thereby releasing the drug proportional to the square root of time. Stable synthetic polymers which biologically degrade *in vivo* eliminates the need for surgical removal of any implanted system, which increases patient comfort and/or improves compliance. Several types of nonbioerodable and bioerodable implants have been described (Fig.1). The controlled-release devices, which are already available commercially, can maintain the drug in the desired therapeutic range and localize delivery of the drug to a particular body compartment, which lowers the systemic drug level and consequently reduces the need for follow-up care. Nevertheless, there are a number of clinical situations where this approach of constant rate drug delivery may not be sufficient. These include the delivery of insulin for patients with diabetes mellitus, antiarrhythmics for patients with heart rhythm disorders, gastric acid inhibitors for ulcer control, nitrates for patients with angina pectoris, as well as selective ß-blockade, birth control, general hormone replacement, immunization, and cancer chemotherapy[3].

Recent studies in the field of chronopharmacology indicate that the onset of certain diseases exhibit strong circadian temporal dependency. Thus, drug delivery patterns can be further optimized by pulsed or self-regulated delivery, adjusted to the staging of biological rhythms[3].

Several research groups have been developing responsive systems which would more closely resemble the normal physiological process in which the amount of drug released can be affected according to physiological needs. The responsive polymeric drug delivery systems can be classified as open- and closed-loop systems. The open-loop systems are known as pulsed or externally regulated and the closed-loop systems as self-regulated controlled drug delivery devices. The externally controlled devices apply external triggers for pulsed delivery such as: magnetic, ultrasonic, thermal, and electric; while in the self-regulated devices the release rate is controlled by feedback information from the environment, without any external intervention, e.g. delivery of insulin in response to glucose levels in blood.

The self-regulated systems utilize several approaches as rate-control mechanisms such as enzyme-substrate reactions, pH-sensitive drug solubility, pH-sensitive polymers, competitive binding and metal concentration-dependent hydrolysis (Fig. 2). Current developments and ideas have been extensively reviewed by Kost and Langer[3]. It should be emphasized that these drug delivery systems are still in the early developmental stage and much research will have to be conducted for such systems to become practical clinical alternatives.

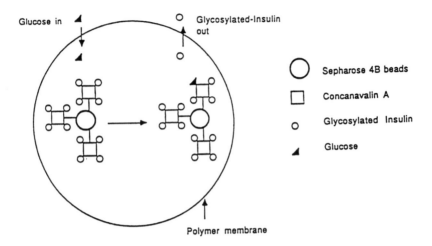

Figure 2   Self-regulating insulin delivery system based on the concept of the competitive and complimentary
binding behaviour of a plant lectin, Con A with glucose and glycosylated insulin.

3   Electrotransport-Mediated Drug Delivery

Electrically assisted transport (electrotransport) is one means of enhancing and controlling the
transdermal delivery of drugs for either local or systemic therapy. The principles of electrotransport,
often referred to as ionthophoresis, have been well known for over a century. Transdermal drug
delivery via electrotransport involves three physical processes: passive diffusion, electromigration,
and electro-osmosis. Passive infusion involves the movement of a drug in response to a
concentration gradient, that is, from an area of high concentration to an area of low concentration.
Electromigration, on the other hand, is the movement of ions in response to an electric field (Fig. 3).

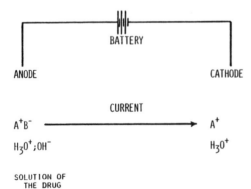

Figure 3
Movement of active drug ions A^+ from
anode to cathode under the influence of
current.

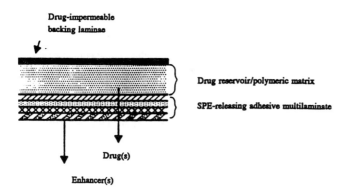

Drug-impermeable
backing laminae

Drug reservoir/polymeric matrix

SPE-releasing adhesive multilaminate

Drug(s)

Enhancer(s)

Figure 4 Cross-sectional view of a skin permeability-enhancing transdermal drug delivery (SPE-TDD) system, showing various major structural components. The skin permeation enhancer is released from the surface adhesive polymer as it comes into contact with the skin, while drug molecules are released at controlled rate from polymer matrix.

4 Microencapsulation As Drug Delivery Method

Encapsulation is a general term used for a number of different technologies in drug delivery. These include microcarriers and microspheres, liposomes, cyclodextrins, and red blood cells. In this technique, drugs, including proteins, are enclosed in another molecule or membrane which protects the drug until it is released. Depending on the characteristics of the encapsulating drug, targeting might also be achieved. A formidable task in the treatment of insulin-dependant diabetes mellitus is to transplant pancreatic islets only. There are several ways of isolating the insulin-producing islets of Langerhaus from the pancreas with preservation of their secretory function. Different approaches have been considered, each based on the principle of shielding the transplanted islets from the host's immune system by encapsulating them in a semi-permeable membrane (alginate-polylysine-alginate) allowing insulin out and nutrients in, but excluding large protein molecules like antibodies[41]. Future application of these so-called "hibrid" artificial ß cells is much dependent on isolation procedures and immuno response prevention (Fig. 5).

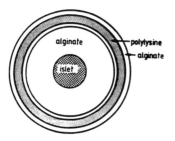

alginate — polylysine
— alginate
islet

Figure 5
Schematic composition of an alginate-polylysine-alginate microencapsulated islet.

5 Liposomes As Drug Vehicles

Since their originally description, liposomes have been discussed as vehicles that could be used as carriers of pharmaceutically active agents[13]. Liposomes are microscopical vesicles composed of one or more lipid bilayers arranged in concentric fashion enclosing an equal number of aqueous compartments[14]. Various amphipathic molecules have been used to form the liposomes, and the method of preparation can be tailored to control their size and morphology. Drug molecules can either be encapsulated in the aqueous space or intercalated into the lipid bilayer (Fig. 6); the exact location of a drug in the liposome will depend upon its physiochemical characteristics and the composition of the lipids[15,16].

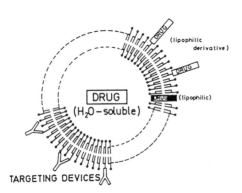

Figure 6

Schematic representation of a unilamellar liposome. The liposome consists of an aqueous compartment surrounded by a phospholipid layer of two molecules thickness. Drugs can either be encapsulated in the aqueous space or accommodated in the bilayer membrane in case of a lipophilic drug or a lipophilic derivative of the drug. Targeting devices such as immunoglobulins can be attached to the liposomal surface.

Excellent reviews on the physiochemical characterization and preparation of liposomes using a wide variety of techniques are available[17,18,19]. An assortment of molecules, including peptides and proteins, has been incorporated in liposomes, which can then be administered by different routes[20,21,22]. In a number of instances, liposomal drug formulations have been shown to be markedly superior to conventional dosage forms, especially for intravenous and topical modes of administration of drugs. Egbaria & Weiner (1990) presented an excellent review on topically applied liposomal formulations with emphasis on the evaluation of liposomal systems in a wide variety of animal models and human skin using both *in vitro* and *in vivo* techniques[23].

When conventional liposomes are intravenously injected, they are quickly cleared from the blood by cells of the mononuclear phagocyte system (immune system). Organs rich in such cells (mainly macrophages) are liver and spleen. It is not surprising therefore that intravenously administered liposomes will localize primarily in these organs. With respect to the liver, it has been shown that not only the macrophages of this organ, also known as Kupffer cells participate in liposome uptake[24]. When the liposomes are small enough to pass the open fenestrations in the endothelial lining of the hepatic sinusoids, i.e. below approximately 100 nm in diameter, they will gain access to the liver parenchymal cells, the hepatocytes. Once there, they can be taken up by these cells quite efficiently, most likely by an endocytotic mechanism. Endocytosis is also the

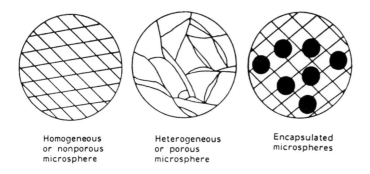

Homogeneous   Heterogeneous   Encapsulated
or nonporous   or porous   microspheres
microsphere   microsphere

Figure 7   Typical morphologies of microspheres employed for biomedical applications.

As an alternative to drug targeting with liposomes, Couvreur et al.[32] proposed biodegradable nanoparticles made by polymerization of various alkylcyanoacrylate monomers. A characteristic of cyanoacrylate polymers is that their rate of degradation is dependent on the length of their alkyl chains[33,34]. The current status of anticancer drug delivery by means of polyalkylcyanoacrylate nanoparticles has been reviewed by Couvreur et al.[35]. Their results obtained *in vitro* with multidrug-resistant cancer cells were impressive and indicate that nanoparticles may indeed be able to circumvent this resistance. Although they need to be confirmed in *in vivo* experimental models, the results open new perspectives in rendering multidrug-resistant cancers vulnerable to a carrier-mediated chemotherapy.

A challenging development in the treatment of preventing restenosis after angioplasty or vascular injury might be the application of biodegradable microspheres containing a drug that inhibits DNA synthesis in vascular smooth muscle cells. Coronary restenosis is a complex medical problem whose analysis has let to the development of a variety of conceptual models (paradigms)[36]. The term "restenosis" has thus been applied simultaneously to describe local neointimal hyperplasia at the site of catheter treatment (histologic restenosis), the development of significant luminal narrowing at the treatment site (angiographic restenosis), and the late recurrence of signs or symptoms of ischemia after an initially successful angioplasty (clinical restenosis) [37]. The choice of a paradigm dictates which associated analytical techniques should be used.

Smooth muscle cell proliferation plays a major role in the genesis of restenosis after angioplasty or vascular injury. Local application of agents capable of modulating vascular responses, including smooth muscle cell proliferation, has been achieved, but difficulty in maintaining active levels locally has been a factor limiting the efficacy of such approaches. Several types of catheters have been used for delivery of drug solutions or suspensions into the interior of the blood vessel wall by diffusion or as a result of an applied hydrostatic pressure[38,39]. These include modified angioplasty catheters with porous balloons. A difficulty with this approach is the possibility of rapid loss of efficacy because of the diffusion and convection within the wall and distribution into surrounding tissues or the lumen. One strategy addressing this issue of drug washout is the catheter-based administration of nondiffusible microparticles that contain the therapeutic agent, for example colchicine analoque that inhibits DNA synthesis in vascular smooth muscle cells[40]. These microparticles would preferably be composed of biocompatible and biodegradable materials so as to avoid inflammation and allow for gradual reabsorption. This approach is still in an experimental phase and the safety and efficacy of local drug delivery in preventing restenosis have not been proven yet.

7 Ports And Pumps

The theory behind pump development is the desirability of providing the diabetic with a continuous low basal level of insulin, boosted by a bolus delivery at mealtimes, in order to mimic more closely the action of the normal pancreas. By this basal-bolus delivery regime, control of blood glucose concentrations is much tighter, hence in the long term, overall metabolic control is better and the incidence of severe diabetic complications is reduced. In practice, following early reports on the effectiveness of pumps in the clinical setting, the devices proliferated outside the research setting and into the diabetic community too quickly, without adequate backing, so that much of the early promise has not been fulfilled, not least in terms of patient acceptability.

Blood glucose concentration control in diabetic subjects can theoretically be obtained by an insulin delivery device, which is directly controlled by a continuous measurement of the present glucose concentration of the body. Such a closed-loop system, often referred to as artificial ß cell or artificial pancreas, principally consists of three components: a glucose sensor, an insulin delivery pump and a microprocessor control unit regulating the insulin administration based on the measured amount of glucose. Such a system should ideally be implantable. However, the main difficulty for the realization of such a system is the development of a reliable long-term continuously measuring glucose sensor.

Various systems from single access ports to micropumps are now available for administration of repeated bolus doses, or continuous or programmable infusions. Three general situations exist where implantable drug delivery systems (pumps) are most effectively utilized:

- when continuous infusion or chronic exposure offers the greatest advantage for success pharmacokinetically and pharmacodynamically;
- when chronic site-specific delivery of regional (local) therapy is the preferred treatment strategy;
- when chronic bolus dosing or intermittent drug delivery is desired.

Access ports are small volume rigid chambers placed in contact with the compartment to be infused. They are designed to replace lumbar or intraventricular puncture or intravascular injection by a simple subcutaneous injection (Chapter "Port-catheter Systems: Design, Advantages and Applications").

8 Programmable Implantable Medication Systems(PIMS)

A first application of microcomputers implanted in humans is the PIMS, which has been developed at the Johns Hopkins University Applied Physics Laboratory[42]. This device has characteristics very similar to those of an orbiting spacecraft. The major similarities are: a command system, a telemetry system, a miniature, long-life power system, and very-large-scale integrated circuit chips. It also has been designed and developed using reliability and quality assurance techniques from the aerospace industry.

8.1 Command system

The first spacecraft (Sputnik, Explorer I) orbited the earth and merely emanated a steady radiofrequency output. The earlier invention of the transistor in 1947 made possible the first electronic device to be implanted in a human subject: an artificial fixed pulse rate cardiac pacemaker[43]. As spacecraft and implant technology developed, both spacecraft and implants required more flexibility and capability. Examples of control commands in modern pacemakers are those to change the stimulation pulse rate, pulse voltage, or pulse width; to enable or disable an electrical signal from the atrium; and to adjust the sensitivity for an electrical signal from the ventricle. The PIMS command system is used to change basal delivery rate, to turn the device on and off, and to set limits on medication usage. The command system permits the spacecraft to adapt to its environment and to the spacecraft's specific condition, thus enhancing its use for a variety of purposes. A command system for an implant allows it to adapt to the patient's changing needs.

8.2 Telemetry system

Telemetry involves the transmission of data from a remote location. For example, typical implant telemetered data might be the confirmation of the parameters that have been commanded into it, the battery voltage, and the rate of infusing medication, both in real time and as stored data.

8.3 Power system

Power systems must be small and long-lived. Both spacecraft and implant systems such as the pacemaker have used rechargeable nickel-cadmium cells to store energy and operate the system between recharges. These developments have resulted in reliabilities of more than 10 years of continuous operation of a pacemaker in humans[44,45].

8.4 Miniaturization of electronics

Until the age of the space shuttle with its huge payload capability, it was a struggle for spacecraft designers to meet payload size and weight limitations. Therefore, miniaturization or better microminiaturization of electronics became commonplace in spacecraft design. A spinn-off of these developments is the currently available multiprogrammable small-sized pacemaker.

8.5 Implantable microcomputer systems

Not many computer-controlled implantable medication systems exist. The implantable portion of PIMS, which contains the microcomputer subsystem is called the implantable program infusion pump (IPIP). The application of IPIP to diabetics by controlled release of insulin is one such obvious example[46]. The microcomputer-controlled delivery of medication by PIMS may provide improved treatment for several diseases such as Parkinson's disease, amyotrophic lateral sclerosis[47], chronic pain, Alzheimer's disease, etc. In this area of medication infusion, the current variation of PIMS is the sensor-actuated medication system (SAMS), a microcomputer-controlled, closed-loop system by which a physiological parameter will be sensed and medication will be released according to specific algorithms programmed into the IPIP. An example is to sense the electroencephalogram precursor of an epileptic seizure and to release antiseizure medication in the brain so as to better control epilepsy with a minimum amount of medication. Another candidate application for the system is sensing of glucose for proper insulin infusion. In the coming decade, the extensive use of PIMS in an open-loop mode should find applications to a large variety of diseases that have not been well treated in the past.

The implantable drug delivery devices have reservoirs of varying capacities (generally 12-50 ml) which can be refilled by injection through a self-sealing septum. The different systems can be classified in terms of their programmability, from simple pulsatile pumps to fully programmable, remote-controlled electric pumps. The pulsatile pump (Cordis Secor™), continuous flow pumps (Infusaid), and programmable pumps (Synchromed, Medtronic) are some of the examples on the current market.

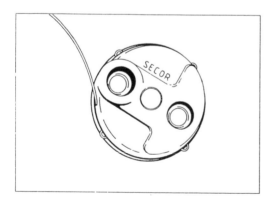

Figure 8
The SECOR implantable drug delivery device with left and right push button and in between the refill port.

All these pumps have their application mainly in the field of neurology (pain, spasticity) and cancer chemotherapy. With respect to costs of health care, the less-known, relatively cheap, totally implantable, patient-activated, drug delivery device (Secor™, developed by Cordis S.A., France) will be described here. This pulsatile pump system (Fig. 8), especially designed for intrathecal or epidural administration of opiates (morphine hydrochloride), consists of a patient-operated two-button dual pumping mechanism to transport boluses of 0.1 ml of medication from the reservoir through the integral system tubing and lumbar catheter to the selected space of the lumbar spinal canal. The volume of the pliable reservoir is 12 ml and refill is accomplished with a 25 G needle. The refill port consists of a self-sealing dome and safety valve to seal off the filling chamber or port from the reservoir. Suture rings are provided to allow fixation of the device in the implantation pocket. Accessories include a lumbar catheter, 14 G Tuohy needle, guide wire, suture sleeve, step-down connectors, 25 G needles, and luer connector. Accidental activation of one pump does not give a bolus. A bolus is delivered only when the pumps (buttons) are activated in the proper sequence. The disc-shaped pump has a diameter of 65 mm, a height of 15 mm and a weight of 44 grams. The function description of this device is explained in Figure 9.

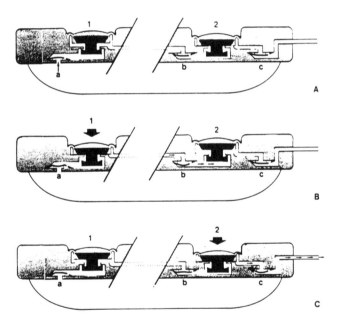

Figure 9   Position of the push buttons with the implantable drug delivery system completely filled with solution and in the inactive state (A). Button 1 is in the upper position, button 2 is in the lower position, valve a is open, and valves b and c are closed. To "prime" the system (B), button 1 is pushed and valve a closes while valve b opens, transporting solution from pump 1 to pump 2 and pushing button 2 to the upper position. (Valve c remains closed because of its high opening pressure). Pump 1 refills itself, once the button is released, via valve a and elastic forces push this button upward. To deliver a bolus to the catheter (C), button 2 is pressed down, closing valve b and opening valve a.

The technique of refilling the implantable drug delivery system is illustrated in Figure 10.

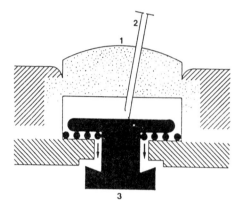

Figure 10
Technique of refilling the implantable drug delivery system. A standard 25 G needle (2) is inserted through the refill plug (1) and into the filling chamber to push down the safety valve. The injected solution flows along the valve (*arrows*) to fill the reservoir. When the needle is withdrawn, the valve returns to its upper position to seal off the filling chamber from the reservoir.

Clinical results demonstrated that this patient-controlled drug delivery system is safe and that it provides excellent relief of pain associated with terminal cancer[48]. A possible advantage of this drug delivery system over continuous infusion pumps is that patients can elect to have the morphine delivered only when they have pain. Thus pain relief would be maximized and tolerance build-up would be minimized[49]. In addition, the concept of dose intensity and circadian rhythms may be combined to improve treatment results. Treatments such as radiation, chemotherapy, and other medications have been shown to have markedly different efficacy and safety depending upon the pattern of administration within the day. The biologic knowledge has been incorporated into modern chronobiologic engineering, e.g. the homeostatic (equal-dose) versus chronobiologic (sinusoidal) treatment of specific cancers like leukemia.

This approach is also called chronotherapy. Both cell-cycle specificity and biologic rhythms determine the complexity of chemotherapeutic treatment schedules (chemotherapeutic strategy), but also the future pump design with different possibilities of modes of drug delivery (bolus, multi-step bolus, continuous, continuous complex, bolus-delay). Therefore computerized implantable devices may play a major role in future drug therapy.

References

[1]   White PF. Use of patient-controlled analgesia for management of acute pain. *JAMA*, 259:243,1988.
[2]   Scheving LE, Halberg F, Ehret CF (eds). *Chronobiotechnology and Chronobiological Engineering*. Martinus Nijhoff Publishers, Boston,1987.
[3]   Kost J, Langer R. Responsive polymeric delivery systems. *Adv Drug Deliv Rev* ,6:19,1991.
[4]   Kari B. Control of blood glucose levels in alloxan-diabetic rabbits by iontophoresis of insulin. *Diabetes,* 35:217,1986.
[5]   Meyer BR, Kreis W, Eschbach J, O'Mara V, Rosen S, Sibalis D. Successful transdermal administration of therapeutic doses of a polypeptide to normal human volunteers. *Clin Pharmacol Ther,* 44:607,1988.

[6]   Stephen R, Miotti D, Bettaglio R, Rossi C, Bonezzi C. Electromotive administration of a new morphine formulation: Morphine citrate. *Artif Organs,* 18:461,1994.

[7]   Halberg F. Physiologic 24-h periodicity; general and procedural considerations with reference to the adrenal cycle. *Z Vitamin, Hormon Fermentforsch,* 10:225,1959.

[8]   Lemmer B. Implications of chronopharmacokinetics for drug delivery: antiasthmatics, H_2-blockers and cardiovascular active drugs. *Adv Drug Deliv Rev,* 6:83,1991.

[9]   Floyer J. *A treatise of the asthma.* Wilkins, Innis, London,1698.

[10]   Barnes JP. Autonomic control of the airways and nocturnal asthma as a basis for drug treatment. In Lemmer B (ed.), *Chronopharmacology - Cellular and Biomedical Interactions.* Marcel Dekker, NY/Basel, pp.53,1989.

[11]   Dethlefsen U, Repges R. Ein neues Therapieprinzip bei nächtlichem Asthma. *Med Klin,* 80:44,1985.

[12]   Chien YW, Chien TY, Bagdon RE, Huang YC, Bierman RH. Transdermal dual-controlled delivery of contraceptive drugs: formulation development, in vitro and in vivo evaluations and clinical performance. *Pharm Res,*6:904,1989.

[13]   Bangham AD. In Knight G (ed.), *Liposomes from physical structure to therapeutic application.* Elsevier, North-Holland, NY,1981.

[14]   Bangham AD, Standish MM, Watkins JC. The action of steroids and streotolysins on the permeability of phospholipid structures to cations. *J Mol Biol,* 13:138,1965.

[15]   Gregoriadis G. The carrier potential of liposomes in biology and medicine. *New Engl J Med,* 295:704,1976.

[16]   Fendler JH, Romero A. Liposomes as drug carriers. *Life Sci,* 20:1109,1977.

[17]   Szoka F, Papahadjopoulos D. Liposomes: preparation and characterization. In Knight CG (ed.), *Liposomes from physical struture to therapeutic applications.* Elsevier, North-Holland, NY,1981.

[18]   Lichtenberg D, Barenholz Y. Liposomes, preparation characterization and preservation. In Glick D (ed.), *Methods of biological analysis.* John Wiley, NY, 33:337,1988.

[19]   Nassander UK, Storm G, Peeters PAM, Crommelin DJA. Liposomes. In Langer R, Chassin M (eds), *Biodegradable polymers as drug delivery systems.* Marcel Dekker, NY,1990.

[20]   Yatvin MB, Lelkes PI. Clinical prospects for liposomes. *Med Phys,* 9:149,1982.

[21]   Lopez-Berestein G, Fidler IJ (eds). *Liposomes in the therapy of infectious diseases and cancer.* Alan R. Liss, NY,1989.

[22]   Gregoriadis G (ed.). Liposomes as drug carriers. In *Recent Trends and Progress.* John Wiley, Chichester,1988.

[23]   Egbaria K, Weiner N. Liposomes as a topical drug delivery system. *Adv Drug Deliv Rev,* 5: 287,1990.

[24]   Scherphof GL. In vivo behavior of liposomes; interactions with the RES and implications of drug targeting. In *Handbook of Experimental Pharmacology.* Juliano RL (ed.), Springer Verlag, Heidelberg, pp.285,1991.

[25]   Maruyama KM, Kennel SJ, Van Borssum Waalkes M, Scherphof GL, Huang L. Drug delivery by organ specific liposomes. In *Polymeric drugs and drug delivery systems.* Dunn RL, Ottenbrik A (eds), pp.275,1991.

[26]   Daemen T, Dontje BHJ, Veninga A, Scherphof GL, Oosterhuis JW. Therapy of murine liver metastases by administration of MDP encapsulated in liposomes. *Selective Cancer Therapeutics,* 6:63,1990.

[27]   Forssen EA, Tökes ZA. Improved therapeutic benefits of doxorubicin by entrapment in anionic liposomes. *Cancer Res,* 43:546,1983.

[28]   Gabizon A, Dagau A, Coreu D, Barenholz I, Fuks Z. Liposomes as in vivo carriers of adriamycin: reduced cardiac uptake and preserved antitumor activity in mice. *Cancer Res,* 42:4734,1984.

[29]   Chang TMS. In Thomas C (ed.). *Artifical cells.* Springfield, IL,USA,1972.

[30]   Arshady R. Microspheres and microcapsules. A survey of manufacturing techniques: I. Suspension crosslinking. *Polym Eng Sci,* 29:1746,1989.

[31]   Arshady R. Microspheres for biomedical applications: preparation of reactive and labelled microspheres, *Biomaterials,* 14:5,1993.

[32]   Couvreur P, Kante B, Roland M, Guiot P, Baudhuin P, Speiser P. Polycyanoacrylates nanocapsules as potential lysosomotropic carriers: preparation, morphological and sorptive properties. *J Pharm Pharmacol,* 31:331,1979.

[33]   Vezin W, Florence A. In vitro degradation rates of biodegradable poly-N-alkylcyanoacrylates. *J Pharm Pharmacol,* 30:Suppl 5P,1978.

[34]   Leonard F, Kulkarni R, Brandes G, Nelson J, Cameron J. Synthesis and degradation of polyalkylcyanocrylates. *J Appl Polym Sci,* 10:259,1966.

[35]   Couvreur P, Roblot-Treupel L, Poupon MF, Braseur F, Puisieux F. Nanoparticles as microcarriers for anticancer drugs. *Adv Drug Deliv Rev,* 5:209,1990.

[36]   Kuhn TS. *The structure of scientific revolutions.* 2nd Ed. Chicago, University of Chicago Press,1970

[37]   Kuntz RE, Baim DS. Definning coronary restenosis; newer clinical and angiographic paradigms. *Circulation* 88:1310,1993.

[38]   Kerenyi T, Merkel V, Szabolcs Z, Pusztai P, Nadasy G. Local enzymatic treatment of artherosclerotic plaques. *Exp Mol Pathol,* 49:330,1988.

[39]   Wolinsky H, Thung SN. Use of a perforated balloon catheter to deliver concentrated heparin into the wall of the normal canine artery. *J Am Coll Cardiol,* 15:475,1990.

[40]   March KL, Mohanraj S, Ho PPK, Wilensky RL, Hathaway DR. Biodegradable microspheres containing a colchinine analogue inhibit DNA synthesis in vascular smooth muscle cells. *Circulation,* 89:1929,1994.

[41]   Fritschy WM. *Experimental studies on microencapsulation of pancreatic islets for transplantation in diabetes mellitus.* Thesis PhD, University of Groningen, Groningen, The Netherlands,1994.

[42]   Fischell RE. A programmable implantable medication system (PIMS) as a means for intracorporeal drug delivery. In Tyle P (ed.), *Drug Delivery Devices.* Marcel Dekker, NY, pp.261,1988.

[43]   Elmqvist R, Senning A. Implantable pacemaker for the heart. In *Medical Electronics.* (Proc 2nd Int Conf Med Electronics, Paris, June 1959), Smyth CN (ed.), Iliff & Son, London, pp.253,1960.

[44]   Fischell RE, Lewis KB, Schulman JH, Love JW. A long-lived reliable, rechargeable cardiac pacemaker. In Schaldach M, Furman S (Eds), *Advances in pacemaker technology.* Vol 1, Springer Verlag, NY,1975.

[45]   Love JW, Lewis KG, Fischell RE. The Johns Hopkins rechargeable pacemaker. *JAMA,* 234:64,1975.

[46]   Spencer WJ. For diabetes; an electronic pancreas. *IEEE Spectrum,* 15:38,1978.

[47]   Munsat TL, Reichlins S, Taft J, Andres P, Kaplan M, Kasdon D. Experience with long-term intrathecal infusion of TRH in ALS. *Muscle Nerve,* 9:103,1986.

[48]   Koning G, Feith F. A new implantable drug delivery system for patient-controlled analgesia. *Ann NY Acad Sci,* 531:48,1988.

[49]   Hoekstra A. Pain relief mediated by implantable drug delivery devices. *Int J Artif Organs,* 17:151,1994.

Port-Catheter Systems: Design, Advantages, and Applications

A. Hoekstra

1 Introduction

The concept of "total implantation" of vascular access devices, substituting for the "percutaneous exit" catheters, has added appreciably to the range of applications and the safe duration of catheterization. New, highly biocompatable materials have extended the catheter indwelling time from a few days to a year or longer. The parallel development of small, lightweight, compact, and reliable infusion pumps, carried or worn by the patients, have provided mobility and relatively unrestricted patient activity. These important benefits have reduced significantly the frequency, duration, and cost of hospitalization associated by extended drug therapy in the treatment of selected chronic and transient diseases[1,2].

Despite these advantages, there are still numerous complications; many of them are related to the design of the port systems[3,4]. Limitations of most access ports include size, profile, weight, use of metal which interferes with MRI and CT scans, and difficulties in locating the port septum for appropriate needle access. Design features which affect the ease of clinical use must be considered. Procedure-related complications are frequently underestimated[3].

In order to overcome such deficiences, two vascular access ports - the MULTIPURPOSE ACCESS PORT (MPAP) and MINIPORT systems - have been developed to address the above limitations. Access port design, its advantages and prevention of procedure-related complications are presented on the basis of animal studies and clinical results obtained with these port systems. Other applications such as treatment of infections, diabetes and pain in combination with drug infusion pumps are summarized in this paper.

2 Animal Studies

2.1 Materials and methods

2.1.1 Vascular access ports

The MPAP and MINIPORT systems, developed by Cordis S.A., France, are totally implantable, oval-shaped access ports designed to provide long-term, repeated access to the vascular system. The systems consist of a reservoir made of polysulfone. The chamber is accessed through a selfsealing silicone rubber septum using a 22G or smaller size standard bevel needle. Systems include a radiopaque silicone catheter, and an antikinking device (silicone) which is used to securely connect the (F6, F7, F8) catheter to the integrated connector of the reservoir. A flange with two suture holes is incorporated into the port base to facilitate anchoring the port to the underlying fascia. The technical specifications of these ports are summarized in Table 1, and a longitudinal section of the port is shown in Figure 1.

2.1.2 Animals and experimental procedure

Three healthy animals of different species, dog, pig, and goat, with a vascular system comparative to man were used for design evaluation or functionality testing of the MPAP-prototype. Each animal had normal haematology at the beginning of the study.

Parameters used for design evaluation or functionality testing were focussed on the use and handling of the port-catheter system such as easiness of flushing, injection, blood sampling, stiffness of the septum to prevent blood reflux after pushing the septum, and possible diffusion of blood into the saline present in the port-catheter system.

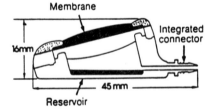

Figure 1
Longitudinal section of the port system. The black areas schematically correspond to the parts finally added to the membrane (septum) and reservoir volume, respectively.

For the purpose of this pilot experiment, the choice of animals was based on availability, overall size together with the human similarities between artery and vein sizes. In order to prevent any form of stress by the procedures involved, the experiments were conducted under general anaesthesia. Surgical procedures were performed under aseptical conditions for survival of the animals. The study is a combination of a *in vivo* and *ex vivo* design.

Table 1: Technical Specifications of MPAP and MINIPORT Systems

Technical Specifications	MPAP		MINIPORT
	Prototype	Final product	
Weight (g)	9.7	10.2	3.8
Length (mm)	44.0	45.0	32.0
Width (mm)	28.0	28.5	17.5
Height (mm)	14.5	16.0	12.0
Septum area (cm^2)	3.0	2.7	0.6
Septum thickness (mm)	4.0	7.0	6.0
Internal volume (ml)	0.9	1.5	0.25
Punctures*	1500	1500	500
Catheter Kits			
French Size	Length (cm)	External diameter (mm)	Internal diameter (mm)
F6	75	2.0	0.9
F7	75	2.3	1.0
F8	75	2.7	1.2

* Without leakage; using 22G needle with 3m H_2O back pressure

A male sheperd's dog weighing 27 kg was anaesthesized with pentobarbital (Nesdonal®, Rhône Merieux, Lyon, France) 30 mg/kg, i.v. and placed in the dorsal decubitus position. General anaesthesia was maintained with 1-2% halothane (Fluothane®, ICI, UK) in oxygen/nitrous oxide (1:2) by inhalation using a constant-volume respirator (Dräger Ventiloq®). An inguinal incision was made to expose the left femoral artery and vein. The vascular F6 catheter (50 cm) of the MPAP system, filled with heparinized saline (500 IU/ml), and clamped was introduced and positioned in the femoral artery, and subsequently connected to the access port filled with the same solution. A similar procedure was performed for venous access. The access ports were then placed on the operation-table in order to visualize the effects during and after the performance of different use and handling procedures. Parallel experiments were carried out with a young pig and a goat.

Table 2: Use and handling parameters of the MPAP-prototype tested *ex vivo*

Parameters	Dog i.a.	Dog i.v.	Pig i.a.	Pig i.v.	Goat i.a.	Goat i.v.
Downwards movement of septum during injection	++	+++	++	+++	++	++
Upwards movement of septum during needle withdrawal	+++	++	+++	++	++	++
Blood reflux after 2-4 times of pushing the septum	++	-	++	-	+	-
Blood reflux after 4-6 times of pushing the septum	++	+	++	+	++	+
Catheter-port patency at different time intervals	-	-	-	-	-	-
Ease of blood sampling (3 times 3 + 2 ml)	-	-	-	-	-	-
Flushing (replacement blood) of the system (3 times 5 ml)	-	-	-	-	-	-
End of use: injection of 5 ml heparinized saline (heparin lock)	-	-	-	-	-	-
Diffussion of blood into the catheter:						
- 1 hr after heparin lock	-	-	-	-	-	-
- 1.5 hrs after heparin lock	+	-	+	-	+	-
- 2 hrs after heparin lock	++	-	+	-	++	-
Diffussion of blood into the port reservoir:						
- 2 hrs after heparin lock	+	-	+	-	+	-
- 3 hrs after heparin lock	++	-	++	-	+	-

- : good result or no visible effect
+ : low grade
++ : moderate
+++ : intense or high degree of response

A female young pig weighing 25 kg was aneasthesized with 200 mg stresnil i.m. (Azaperon®, Janssen Pharmaceuticals, Beerse, Belgium), and after 0.5 hrs. with 200 mg hypnodil i.v. (Metomidaat®, Janssen Pharmaceuticals). Maintenance of anaesthesia was performed with i.v. (jugular vein) infusion of stresnil/hypnodil at a low concentration in saline by a double lumen catheter. The pig was intubated, and ventilated with oxygen/nitrous oxide (1:2) using a respirator (Dräger Ventiloq®). Catheters were placed in the femoral artery and femoral vein. Catheter-port connection was performed as mentioned before.

A similar procedure was performed with a female goat weighing 47 kg which was anaesthesized with valium (Diazepam®, Kabi-Pharmacia, Uppsala, Sweden) 1 mg/kg, i.v. and ketamine (Aescoket®, Aesculaap BV, Holland) 10 mg/kg, i.v. Salivation was prevented by glycopyrolate (Robinul®, A-H Robbins, UK), 0.4 mg/kg i.v. Anaesthesia maintenance was performed with ketamine infusion (10 mg/kg, 15 drops/min). The goat was intubated and ventilated with oxygen/nitrous oxide (1:2) with a respirator (Dräger Ventiloq®). Catheters were placed in the carotic artery and jugular vein, respectively. Catheter-port connection was performed as before. Flushing procedures and blood sampling were performed using a 22G needle connected via a 3-way stopcock to a 10 ml syringe. In all three cases the relative arterial pressure was measured with a Statham P23 dB transducer connected to the MPAP-reservoir by a 22 G needle, and registrated with a Hewlett Packard Polygraphic System. Venous pressures were obtained from a second transducer connected to the other MPAP-reservoir. Expired CO_2 by the anaesthesized animal was recorded (Dräger Capnoloq®) simultaneously. Measurements were performed during the whole period of experimentation (> 4 hrs) and monitored continuously in order to determine any possible obstruction or blood clot formation in the catheter-port system, and to control the status of anaesthesia.

2.2 Results

The results of the animal experiments are summarized in Table 2. During bolus injection, at the moment of needle penetration through the septum of the port into the chamber, a downwards movement of the septum (membrane) was seen in all cases. The result of this phenomenon was that the distance between septum inside the portal chamber and the bottom of the chamber (needle-stop) became relatively short, which easily could lead to uncertainty regarding the location of the needle tip before injection or infusion, especially in clinical practice where the port is implanted subcutaneously.

Withdrawing the needle from the access port showed that the septum could be lifted up for about 2 mm. Without maintaining a positive pressure in the syringe during needle withdrawal, reflux of blood into the catheter might appear. When applying pressure with a finger on the septum of the access port (compressing and releasing), reflux of blood into the catheter and port chamber appeared rapidly after pushing the septum 2-4 times (i.a. application) or 4-6 times (i.v. application). When the access ports were covered with a piece of cloth, simulating the skin, reflux of blood appeared much later after the same frequency of pushing the septum covered by the simulating skin.

Withdrawal of blood samples and flushing of the access port system, applied either i.a. or i.v., gave no problems when performed appropriately i.e. according to known clinical and pre-clinical procedures[5]. After leaving the access port systems filled with heparinized saline (500 IU/ml; heparin lock), diffusion of blood into the catheter was generally seen within 1.5 hrs (i.a. application) and entry of a very small amount of blood into the connected port chamber appeared in 2 hrs. This phenomenon was not seen in the venous application of the catheter-port system during the time of observation (> 4 hrs). During the whole time period of each experiment, the relative i.a. and i.v. pressures were recorded. There were no signs of obstruction or blood clot formation in the catheter-port system.

2.3 Discussion

Initially all samples of the MPAP-prototype tested in the different animals were patent during the experimental period of time. However, the results obtained from performing single use and handling procedures demonstrated a too high flexibility of the septum of the port system. Diffusion of blood into the catheter located i.a., and shortly after into the portal chamber might be attributed to a combination of both, the high flexibility of the septum and the i.a. pressure. Intravenous application of the catheter-port system did not show any problem of diffusion of blood into the system. The blood has no reason to enter the catheter as long as the port system is closed; only after prolonged time periods a diffusion of blood constituents might occur and so can cause obstruction of the catheter. This is also the reason why care and maintenance procedures are very important in clinical practice.

When there are extended periods between injections, infusions or blood sampling, the access port system must be flushed with saline and refilled with heparinized saline on a daily basis during the first 3 days post implantation, and after for i.a. application at least once a week, and for i.v. application at least every 3-4 weeks. Flushing of the system can be performed with saline only under the condition that the port system is left with a heparin lock, i.e. heparinized saline after each procedure. Thrombotic obstructions are often produced by faulty port handling[3]. Whether or not some little faults were made during this pilot experiment, it became clear that the septum of the MPAP-prototype was too flexible, which can easily result in unacceptable complications.

Therefore, it was concluded to redesign the prototype of the MPAP by compression of the silicone membrane, i.e. making the septum stiffer, and to increase the depth between inside septum and portal reservoir. The technical specifications of the prototype and redisigned final product (MPAP) are summarized in Table 1. The most important changes are a greater internal volume and thicker but little smaller septum area (Figure 1). This pilot experiment learned further that similar results possibly could be reached by *in vitro* or *ex vivo* methods, simulating clinical practice. However, before starting the earlier use of humans, i.e. clinical trials, one should at least confirm the functionality and safety of a new medical device by performing limited animal experiments. As reported earlier, whenever the experimental protocol allows, these catheter-port systems can also be applied to improve the well-being of the experimental animals, to facilitate the experimental work, to simplify serial blood sampling, and to reduce the problem of sepsis and malfunction of externalized catheters[6,7]. Good use, care and maintenance procedures of access port systems are prerequisites.

3 Clinical studies

3.1 Materials and methods

Patients were accepted if there was histological proof of malignant disease for which no other conventional therapy was available, i.e. repeated access to the vascular system was required, and a life expectancy preferably of more than 6 months was anticipated. Only patients who were willing to take an active part in an investigational study were accepted after informed consent was obtained. The involved evaluators received either MPAP's or MINIPORTS using the same protocol without any difference in patient selection criteria. Implantation of the system was performed under local or general anaesthesia using the percutaneous or open surgical technique. In all cases, the proximal end of the catheter was tunnelled subcutaneously and connected to the port previously positioned in the infraclavicular area or in the area of the orthoabdominal wall in cases where intrahepatic artery infusion was applied. The port system was flushed with heparinized saline prior to use and at the completion of injection or infusion. For continuous infusions standard procedures were performed to maintain a safe treatment.

The primary site of cancer (i.e., the most common diagnosis) of patients in this study was breast (35%; 27 cases), head and neck cancer (19%; 15), and colo-rectum cancer (22%; 17). In 13 cases (17%), the cancer was metastatic. Solutions administered included anticancer drug solutions (chemotherapeutic agents), antibiotics, electrolytes (1 case), nutrients (1 case), and blood product solutions (1 case). In some cases the port system was also used for blood sampling. Chemotherapeutic drugs included cisplatin, platinol (cisplatinanalog), 5-fluorouracil, cyclophosphamide, bleomycin, doxorubicin (Adriamycin), vincristine, vindesine, epirubicin, methylprednisolone, velban, thiotepa, epidophylotoxin (VP-16), mitomycin C, mitoxantrone, vepeside, dacarbazine and methotrexate.

The patients were evaluated for response to drug therapy including toxic reactions, for signs of local or systemic infection, and the implanted port system was monitored for clinical acceptance in terms of safety, reliability, and suitability/efficacy (patency). Data were recorded after the first medication, at 3 months, 6 months or longer post-implantation, or any time of any complication. At each follow-up evaluation, the evaluator assessed the course of treatment including dosage regimen by noting whether there have been any difficulties encountered with

Table 3: Number and type of complication during treatment

Type of complication	MPAP (40 cases: 228 treatment courses)			MINIPORT (38 cases: 141 treatment courses)		
	No. (Case No.)	%*	%**	No. (Case No.)	%*	%**
System-related:						
Difficulty in locating port	1 (36)	2.5	0.4	1 (16)	2.6	0.7
Difficulty in needle access	1 (21)	2.5	0.4	1 (16)	2.6	0.7
Needle disconnection	-	-	-	2 (14,16)	5.2	1.4
Drug extravasation	2 (21,31)	5	0.9	2 (14,16)	5.2	1.4
Catheter occlusion	2 (35,36)	5	0.9	1 (10)	2.6	0.7
Vein thrombosis	1 (31)	2.5	0.4	1 (11)	2.6	0.7
Skin erosion	1 (30)	2.5	n.a.	1 (14)	2.6	n.a.
Port-catheter explanted	4	10	n.a.	4	10.5	n.a.
Procedure-related:						
Difficulty in locating port	-	-	-	1 (13)	2.6	0.7
Difficulty in needle access	-	-	-	1 (13)	2.6	0.7
Needle disconnection	1 (33)	2.5	0.4	2 (6,13)	5.2	1.4
Drug extravasation	1 (33)	2.5	0.4	1 (13)	2.6	0.7
Port displacement	-	-	-	5 (24,27-30)	13.2	n.a.
Medication-related obstruction	4 (4,5,10,34)	10	1.8	-	-	-
Needle withdrawal accident	1 (3)	2.5	0.4	-	-	-
Spontaneous reopening port pocket	1 (12)	2.5	n.a.	-	-	-
Port-catheter explanted	7 (3,4,5,10, 12,33,34)	17.5	n.a.	6 (13,24, 27-30)	15.8	n.a.
Mortality	8	20	n.a.	8	21.1	n.a.
Ports functional at end of evaluation	29	87.9	n.a.	28	87.5	n.a.
Final complication rate	4	12.1	n.a.	4	12.5	n.a.

* Related to the number of cases i.e. ports
** Related to the number of treatment courses

the port system, such as difficulty in needle access, needle disconnection during infusion, catheter disconnection/migration, leakage, drug extravasation, vein/artery thrombosis, catheter occlusion, infection, skin erosion/necrosis over the port system.

3.2 Results

Multi-Purpose Access Port (MPAP) System: The MPAP system was implanted in 40 cancer patients, 36 for venous access and 4 for arterial access. The 22 women and 18 men ranged in age from 33 to 70 years. The implanted ports used for venous access functioned for an average of 212 days (range 33-371 days), and an average of 184 days (range 94-329 days) for arterial access. A total experience of 8376 implant days ($\cong$ 23 years), 228 treatment cycles and 870 infusion days ($\cong$ 2.3 years) was achieved. The system was used for more than 6 months (maximum of 371 days) in several cases.

MINIPORT System: The MINIPORT system was tested in 38 cancer patients, 35 for central venous access and 3 for arterial access. The 24 women and 14 men ranged in age from 32 to 72 years. The systems functioned for an average of 141 days (range 21-429 days) when used for central venous access and an average of 51 days (range 49-54 days) when used for arterial access, covering a total of 5109 implant days ($\cong$ 14 years), 141 treatment cycles and 500 infusion days ($\cong$ 1.4 years). The system was used in several cases for more than 6 months up to a maximum of 429 days.

No serious complications occurred during the implantation procedure when the catheter was placed intra-arterially or intra-venously by using an open surgical or percutaneous technique. Major complications encountered during treatment are summarized in Table 3. Procedure-related complications were not considered product-related and therefore were excluded from the analysis. Examples of procedure-related complications include the use of a needle too short for the port, difficulties in needle access due to instability of the MINIPORT system placed into an oversized pocket.

To address procedure-related complications, the manufacturer modified the instructions for use. The port was difficult to locate and access in one MPAP case at initial medication administration due to skin thickness. The same difficulty occurred with the MINIPORT system due to subcutaneous emphysema of one patient. This complication was considered unrelated to the port system. For the MPAP system (Table 3), there was one reported case of drug extravasation due to difficulty in needle access. This might indicate that the needle was not placed deep enough into the port reservoir.

Another case of drug extravasation occurred 347 days post implantation and is not clearly understood. In this case, vein thrombosis was also reported. Drug extravasation with the MINIPORT system occurred twice due to needle disconnection (Table 3). Vein thrombosis also occurred in one patient implanted with the MINIPORT. Catheter occlusions were reported for both port systems, however comparable to those reported in the literature. Some of these catheter occlusions were related to precipitation of drug solutions (cocktail crystallization) during treatment (Table 3).There were two cases of skin erosion, one due to drug extra-vasation, and another due to the poor nutritional status of the patient which died due to the disease. Spontaneous reopening of the MPAP pocket was reported in one case (Table 3). This phenomenon was related to the implant procedure and therefore not considered product-related. During the investigational period, no cases of infection or septicemia were observed. The system-related complication rate in terms of termination of treatment, i.e. explantation of the port-catheter system was 12.1 % for the MPAP and 12.5% for the MINIPORT. More than 87% of both port-catheter systems were still functional at the end of evaluation.

3.3 Discussion

As described in the literature[8,9], there are several factors that cause drug extravasation. One of these factors is, of course, the number or frequency of chemotherapy courses (cycles). Therefore, calculation of the incidence of complications should be related to the frequency of cycles and not to the number of ports (patients) as documented in the literature. Both approaches are given in Table 3. Nevertheless, without this approach, the incidence of drug extravasation (5%) due to needle dislocation and/or needle disconnection is comparable with that reported in the literature[8,10-14], being the relative risk of today's cancer chemotherapy. The incidence of vein thrombosis (2.5%) is low compared to the literature. Lokich et al.[11] reported the occurrance of subclavian vein thrombosis in 16% of 92 patients receiving chemotherapy and/or hyperalimentation by a totally implanted venous access port system. The relatively low number of implant days in patients with intra-arterial access is due to the fact that regional chemotherapy is typically completed or was stopped after 3-4 treatment courses (cycles).

As already mentioned before (results), there were some (procedure-related) complications which were considered unrelated to the port-catheter system. It is obvious that the use of a wrong (too short) needle, implantation of a small port into an oversized pocket, or implantation of a port too close to the breast, and difficulties in port location and needle access to a small port in a patient with subcutaneous emphysema, undoubtedly lead to complications. These results confirm earlier findings that the position of the port on the chest-wall is very important in this regard[2,9]. Port-catheter related problems such as obstructions due to precipitation of drug solutions (cocktail crystallization) during infusion were, with respect to the relatively large internal diameter (0.9 and 1.2 mm) of the catheters used (Table 1), not considered to be related to the port-system. Small lumen catheters (0.51 mm) become easily obstructed[10]. Greidanus et al.[2] reported that a catheter with an external diameter of less than 2.8 mm and a lumen of more than 0.51 mm may be advantageous with regard to the risk of thrombotic complications and irreversible obstruction of the lumen.

Most of these complications can be avoided by proper handling, and use of suitable drug combinations to prevent crystallization reactions within the port-catheter system. Without the procedure-related complications, the complication rate in terms of termination of treatment, i.e. explantation of the port-catheter system was 12.1% for the MPAP and 12.5% for the MINIPORT, which generally confirms the results of other groups[8,10-14]. During the investigational period no cases of infection or septicemia were observed, confirming one of the advantages of totally implantable access ports over external percutaneous central venous catheters[1]. Indications for using the MINIPORT system might be patients with little subcutaneous fat, children and, on the basis of clinical experience, patients with head and neck cancer which is frequently associated with malnutrition[15]. It needs to be emphasized that this report presents our preliminary observations in a relatively small number of patients. However, these findings may set the stage for a randomized study to compare these non-metallic ports with conventional ones.

4 Applications of access port systems

Various medical device manufactures now produce access ports. Pharmacia has developed access ports under the trade name Port-A-Cath®. They are constructed of stainless steel and titanium closed by a membrane of self-sealing compressed silicone stopper (septum). The access port is joined by a safety connector and an anti-kinking system to the catheter whose diameter depends on the actual site of infusion. The Port-A-Cath® system is now widely used in cancer chemotherapy. Other examples are the Life-Port ®, Infuse-Port®, Polysit®. The MPAP™ system developed by Cordis is of interest as it is equipped with a 7 to 8 mm thick septum with a larger surface area (2.7 cm^2) than the other systems. A miniature version, the Miniport™ with a silicone septum area of 0.6 cm^2 (6 mm thick) with a 8 mm deep reservoir is also available on the market.

Prior to the implantation of temporary or permanent drug delivery devices, the physician must evaluate the patient's requirements on an individual basis considering the following variables:
- Nature of the disease, the drugs involved, and the location of the drug discharge site
- Patient history: reliability, motivation, age, capacity for self-care, available family support and assistance
- Requirement and/or preference for the drug delivery mode of therapy
- Expected duration of therapy and anticipated number of drug infusion cycles
- Cost considerations, such as possible savings through outpatient therapy and earlier resumption or remurative work activities
- Proximity to a primary health care facility, emergency care, continuous drug supply, and other logistical considerations.

4.1 Direct access

At the time of port-catheter implantation, the drug discharge site is selected for the most advantageous delivery. Catheters are positioned to infuse drugs at, or near, most body compartments including relatively inaccessible locations such as the brain or spine. In particular, cerobrospinal fluid space, the peritoneal cavity, and pleural and pericardial spaces are treated more efficiently in terms of increased local drug concentrations (regional drug therapy) relative to systemic circulation. Ensminger stated[16]: "The future for regional chemotherapy appears exciting. There are many opportunities to apply therapeutic principles rationally with the potential of significant benefit to many patients".

4.2 Intravenous access

The most frequent applications for drug delivery via an access port system (catheter-port system) are: chemotherapy, intractable pain (cancer origin), infections, diabetes, postsurgical pain, osteomyelitis, thrombophlebitis, transfusion, cardiovascular diseases, clotting disorders, nutrition, hormonal imbalance. These selected applications have a common requirement for long-term, repetitious, frequent, or cyclical drug administration. These features justify the implantation of an access port system for safety, cost reduction, functional treatment requirement logistics, and medical advantages. Patients with serious refractory infections or recurrent infections usually requiring prolonged hospitalization are the prime cadidates for infusion therapy via implanted access port systems and external infusion pumps.

4.3 Intra-arterial access

The intra-arterial drug infusion route is used primarily in the treatment of selected localized malignant disease. This is based on the premise that in certain body compartments the concentration of drug in the target compartment will exceed significantly the concentration of drug going into systemic circulation, which provides a local "tumor kill" advantage.

4.4 Intraperitoneal access

With the development of access port systems and infusion technology, it is now possible to administer intraperitoneal chemotherapy on an outpatient basis to individuals with abdominal, gynecological, and other tumors or residual tumor fragments located in the peritoneal cavity. This mode of drug delivery is also used for insulin delivery in insulin-dependent patients. Infusion technology via access port systems in generally envisioned as a means for discharging into the body various drugs, fluids, nutrients, etc. Applications in this category include ascites, pleural fluid, spinal fluid, blood sampling. Totally implantable vascular access port systems have added considerably to the treatment and management of patients requiring protracted or frequent drug therapy. This technology has the following advantages:

- Safer and has lower infection rates than percutaneous catheterization
- Permits considerably longer indwelling catheter time
- Requires less frequent skilled nursing attention
- Much more convenient for the patient because it permits unrestricted patient mobility including swimming and bathing
- Easier to take care for in a home setting
- Readily available for a multiplicity of applications

References

[1] Schuman E, Brady A, Gross G, Hayes J. Vascular access options for outpatient cancer therapy. *Ann J Surgery* 153:487,1987.

[2] Greidanus J, De Vries EGE, Nieweg MB, De Langen ZJ, Willemse PHB. Evaluation of a totally implanted venous access port and portable pump in a continuous chemotherapy infusion schedule on an outpatient basis. *Eur J Cancer Clin Oncol* 23:1653 ,1987.

[3] Haindl H. Technical complications of port-catheter systems. *Regional Cancer Treatment* 2:238,1989.

[4] Sheen MC & Wang YW. Complications of port catheter systems in intra-arterial infusion chemotherapy. *Regional Cancer Treatment* 4:92,1991.

[5] Morton DB, Abbot D, Barclay R, Close BS, Ewbank R, Gask D, Heath M, Mattic S, Poole T, Seamer J, Southee J, Thompson A, Trussell B, West C, Jennings M. Removal of blood from laboratory mammals and birds. *Laboratory Animals* 27:1,1993.

[6] Bailie MB, Wixson SK & Landi MS. Vascular-access-port implantation for serial blood sampling in conscious swine. *Laboratory Animal Science* 36:431,1986.

[7] Grosze-Siestrup C & Lajous-Petter AM. Totally implantable catheter system in the dog. *Journal of Investigative Surgery* 3: 373,1990.

[8] Reed WP, Newman KA, Applefeld MM, Sutton FJ. Drug extravasation as a complication of venous access ports. *Ann Int Med* 102:788,1985.

[9] Kerr IG, Iscoe N, Sone M, Hanna S. Venous access ports (Letter). *Ann Int Med* 103:637,1985.

[10] Strum S, McDermed J, Korn A, Joseph C. Improved methods for venous access: the porth-a-cath, a totally implanted catheter system. *J Clin Oncol* 4:596,1986.

[11] Lokich JJ, Brothe A, Benotti P, Moore C. Complications and management of implanted venous access catheters. *J Clin Oncol* 3:710,1985.

[12] Moore CL, Erikson KA, Yanes LB, Franklin M, Gonzales L. Nursing care and management of venous access ports. *Oncol Nurs Forum* 13:35,1986.

[13] Coste JL, Donnadieu S, Perie AC, Miller B, Bassot V, Laccourreye H. Cathéters percutanés et cathéters implantables dans les cancers des voies aérodigestives supérieures. A propos de 600 cas. *Ann Oto-Laryng* (Paris) 105:97,1988.

[14] Brothers TE, Von Moll LK, Niederhuber JE, Roberts JA, Ensminger WD. Experience with subcutaneous infusion ports in three hundred patients. *Surg Gynecol Obstet* 166:295,1988.

[15] Santini J, Dassonville O, Milano G, Thyss A, Schneider M, Demard F. Loco-regional induction chemotherapy with Cis-DDP and 5-FU in head and neck cancer: a pilot study. Abstract of poster presented at the *Int Conf Adv Reg Cancer Ther*, Berchtesgaden, Germany, June 5,1989.

[16] Ensminger WE, Gyves JW. Regional chemotherapy of neoplastic diseases. *Pharm Ther* 21:277,1983.

Hydrocephalus Management With a Flow-Control Shunt: Overdrainage and Proximal Obstruction: Controllable Complications of Shunting

A. Hoekstra and M. Sussman

In collaboration with

G. Barrionuevo, M.D.[1], J.P. Castel, M.D.[2], M. Choux, M.D.[3], G. Costabile, M.D.[4], P. Dhellemmes, M.D.[5], C. Di Rocco, M.D.[6], J.A. Alvarez-Garijo, M.D.[7], J. Gilsbach, M.D.[8], A.E. MacKinnon, M.D.[9], D. Neuenfeldt, M.D.[10], S. Pezzotta, M.D.[11], L. Rabow, M.D.[12], B. Rilliet, M.D.[13], J. Sahuquillo, M.D.[14], W. Serlo M.D.[15], and L.G. Strömblad, M.D.[16]

[1]Hospital Virgen del Rocio, Sevilla, Spain, [2]Hospital Pellegrin-Tripode, Bordeaux, France, [3]Children's Hospital Timone, Marseille, France, [4]Kantonsspital Aarau, Switzerland, [5]University Hospital, Lille, France, [6]University Hospital Agostino Gemelli, Roma, Italy, [7]Hospital C.S.La Fe, Valencia, Spain, [8]Albert-Ludwigs-University Hospital, Freiburg, Germany, [9]Children's Hospital, Sheffield, U.K., [10]University Hospital Homburg/Saar, Germany, [11]University Hospital, Pavia, Italy, [12]University Hospital, Umea, Sweden, [13]University Hospital Vaudois, Genève, Switzerland, [14]Hospital Vall o'Hebron, Barcelona, Spain, [15]University Hospital, Oulu, Finland, and [16]University Hospital, Lund, Sweden.

1 Summary

A novel, variable-resistance valve has been developed to address the limitations that promote overdrainage in conventional differential pressure (DP) valves. When conditions favoring overdrainage occur, the Orbis-Sigma Valve (OSV) operates as a variable-resistance flow regulator within certain differential pressure values. Under other conditions, it functions as a low resistance DP valve. A total of 134 OSVs have been evaluated in 128 patients at 16 European centers. Most of the patients were younger than five years of age (mean 11.4 ± 7.6 years).

The major etiologies causing hydrocephalus were: aqueductal stenosis (21.6%); intracranial hemorrhage (17.2%); myelomeningocele (14.2%); infection (9%); and tumor (7.5%). One-third of the implants were performed to replace failed DP shunts. Two-thirds of all cases were followed for more than two years and nearly half have been followed over three years (mean 2.4 ± 1.2 years). Using actuarial statistics, 83.9% of the shunts continued to adequately manage hydrocephalus at three years.

Due to the low incidence of proximal obstruction (3; 2.25%), "obstruction" (19; 14.2%) was the major complication observed in the series. This number is remarkably low when compared with the incidence of obstruction in series of conventional DP shunts. Overdrainage occurred in 2 cases (1.5%) and insufficient drainage occurred in four cases (3%). Compared with literature for conventional DP shunts, the incidence of these phenomena is extremely low. Based on the results summarized in this paper and those already reported in the literature, the OSV appears to address the overdrainage limitations of conventional shunts.

Key Words: hydrocephalus, flow-control shunt, ventriculo-peritoneal shunt, ventriculoatrial shunt, variable-resistance valve

2 Introduction

2.1 History and complications

In 1952, unidirectional, differential pressure (DP) valves were successfully used to divert cerebrospinal fluid (CSF) from the cerebral ventricles[42]. This was the first universal method used to manage hydrocephalus. This therapy has been associated with a number of complications such as obstruction, overdrainage, infection[6,8,23,30,48,52,57,62]. During the past four decades, the main goal of shunting has been the prevention of shunt system complications[28,31,51].

Controlling complications: Potentially-controllable complications of shunting are poorly understood and therefore difficult to control. There are several reasons for this: (1) Some physicians accept complications as part of the therapy since some shunt revisions are considered to be normal (i.e., lengthening of distal catheters for growth in VA shunts, replacement of shunts because of time-related silicone elastomer deterioration, etc.) (2). Shunt complications are often multifactorial. For example, the development of one-piece systems to prevent disconnection make shunt placement more difficult; various ventricular catheter designs to prevent choroid plexus obstruction make removal more difficult; various surgical techniques for proximal and distal catheter placement, etc. Analysis of factors promoting successful long-term shunt insertion and, more importantly, the interaction of these factors, has never been addressed.

Infection: Because of the severity of its consequences, infection was the first complication to draw attention. Progressively, these efforts successfully led to lowered infection rates. Recently, reports with negligible infection rates have appeared[10]. The incidence of infection has been reduced because of improvements in shunt materials, design and careful implantation technique[19,33,43,50,58,64,65].

Connectors: Another example of a controllable complication of shunting is placement of a connector on the distal tubing. In adults, there is probably little influence of the connector on shunt function. However, in infants the placement of the connector in distal tubing has been associated with fracture and disconnection[52,53,54]. Infection and disconnection are now controllable complications[6,26,28,53,55].

Under-/Overdrainage complications: Recently, the relationship between various factors has been related to the complications of shunting. Subdural collections are complications related to overdrainage[52,53,54]. Only recently, the major risk of overdrainage has been indirectly related to proximal obstruction due to ventricular size reduction and associated choroid plexus and ependymal occlusion[52,53,54]. Another significant complication is the mismatch between the patient's requirements and the functional characteristics of DP valve mechanisms[4,24,26,27,38,44,47,51,52,53,54,56]. This may lead to under- or overdrainage complications[4,6,17,25,26,30,38,52,53,54,56] which are associated with complications such as proximal catheter obstruction. The control of overdrainage should be a primary concern in the management of hydrocephalus.

Surgeon's role: The surgeon, as the "final manufacturer" of a shunt system, determines the components to include, and the site of proximal and distal catheter placement (frontal or occipital ventricular catheter approach, peritoneum or atrium for distal catheter, etc.). These decisions greatly influence the outcome of the procedure.

Shunt complications such as obstruction, infection, or mechanical failure can often be traced to the implant technique, component selection or placement[53].

2.2 Conventional DP shunts

Limitations in DP valve mechanism design often lead to chronic overdrainage and slit-ventricle syndrome (SVS)[17,18,29,37,55]. Sainte-Rose *et al.* observed that the incidence of shunt failure is significantly higher when slit ventricles are present[52,53]. Overdrainage contributes to complications such as obstruction, shunt-dependency phenomena, slit ventricle syndrome (SVS) and subdural hematomas[17,18,29,37,51,52,53,55,56,62].

Mismatch of a valve system's operating characteristics to the patient's requirements results from one or more factors: inaccurate determination of the basal ICP level when selecting the system's opening/closing pressure;[6,18,26] and/or inadequate matching of the shunt's pressure/flow performance to the patient's needs;[6,14,15,46,50] inadequate surgical technique and/or strategies (i.e., overdrainage caused by shunt system placement - due primarily distal catheter placement)[18]. Correct ICP evaluation and appropriate valve mechanism selection and surgical technique may reduce the incidence and severity of overdrainage complications[1,7,11,20,21,22,34,35,36,46]. Of all these shunt complications, shunt obstruction is the major cause, by far, in the pediatric series[52,53,54]. In the adult series, overdrainage phenomenon is responsible for a number of complications such as subdural collections, orthostatic hypotension, etc.[16,48]. These OD phenomena, which occur in all shunts, are related to DP variations which are inevitable since shunt implantation opens a window to the atmosphere.

Several approaches have focused upon this problem: (1) From the first shunts, several generic opening pressure ranges have been produced. Physicians have developed a "comfort level" in selecting a particular generic range based upon experience and arbitrary criteria. If the selection proves incorrect, the system is revised with one in a higher or lower generic range. (2) Antisiphon devices have been designed to counteract the negative forces developed by the hydrostatic column in the distal catheter. (3) The Horizontal-Vertical Valve and programmable shunts offer the opportunity to vary resistance (automatically in the first device when the patient stands and via external programming in the second device).

Until recently, the available valve mechanisms for hydrocephalus treatment often limited the surgeon's ability to manage OD complications. Physicians and patients were left to deal with the complications caused by system mismatch/malfunction. Most patients with hydrocephalus have been treated by implantation of DP valves in which resistance to flow is fairly constant[51]. When opening pressure is exceeded, flow is determined by CSF input and output pressures[27,47]. With this type of valve, overdrainage is considered constant in standing patients, since the draining capacity of the shunt exceeds the ventricular secretion rate[51]. In this position, a DP increase or "siphon effect" occurs due to the negative pressure exerted by gravity on the hydrostatic column in the distal catheter[8,20,26].

Although these valves operate adequately at certain times, they cannot adjust to the wide range of physiological conditions caused by postural changes, coughing, etc. This reduces the effective opening pressure of the valve and/or keeps the valve open and draining continuously when conditions favoring "siphoning" are present. This chronic imbalance leads to overdrainage complications such as SVS, shunt dependency, subdural collections or hemorrhage, etc. To address this complication, antisiphon devices were developed[47]. Such devices stop shunt function when conditions favorable for siphoning occur even though CSF continues to be produced at the rate of approximately 21 ml/hr. They are affected by a number of factors such as implantation site and fibrous capsule formation[13].

2.3 Flow control shunts

Ten years ago, Cordis Corporation began investigation of a new direction for the control of hydrocephalus. The basic idea was to define a drainage system able to control CSF flow via the variation to resistance, thus indirectly restoring a normal intracranial pressure instead of trying directly to control ICP as conventional DP shunts do. As it was clear that drainage requirements vary from patient to patient, the initial requirement was to develop a system providing a drainage flow rate close to the average CSF secretion rate.

The *Orbis-Sigma™ valve (Cordis)* was designed to reduce overdrainage complications[51]. It is a variable-resistance (VR), flow regulator within certain DP values. When conditions favoring siphoning occur, the flow control stage (Stage II) of the mechanism maintains flow between 18 and 30 ml/hr (within the physiological ICP range), nominally the CSF production rate.

Several reports were recently published documenting the successful use the OSV in the treatment of the slit ventricle syndrome[45,59,60] tumoral hydrocephalus[61] triventricular hydrocephalus[49] and other etiologies of hydrocephalus[5]. This study was performed to evaluate the efficacy, safety and reliability of this new flow control shunt system.

3 Materials and methods

Clinical trials were performed in 16 European centers. Patients ranged in age from one day to 79 years. The majority of patients were less than five years of age (Figure 1). Authorization of hospital ethical committees and with written or oral consent of the patients, parents or representatives was obtained.

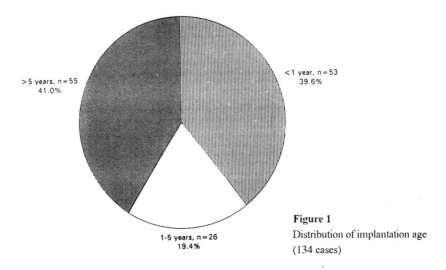

> 5 years, n = 55
41.0%

<1 year, n = 53
39.6%

1-5 years, n = 26
19.4%

Figure 1
Distribution of implantation age
(134 cases)

A total of 134 Orbis-Sigma valves in 128 patients were implanted for the treatment of hydrocephalus with various etiologies (Table 1). The major *etiologies* causing hydrocephalus were aqueductal stenosis (21.6%), intracranial hemorrhage (17.2%) myelomeningocele (14.2%), infection (9%) and tumor (7.5%). The etiology of hydrocephalus was unknown in 2.2% of the patients. Communicating hydrocephalus was present in 14.2% of the cases. The incidence of aqueductal stenosis in this series is higher than typically reported in hydrocephalus series because surgeons selected this shunt for their most difficult cases. Table 2 indicates the shunting history of the patients in this series. It should be noted that 67.9% of the cases were first implants while one-third were for replacement of failed DP shunts.

All patients were assessed for the degree to which hydrocephalus was controlled. *Evaluations* included measurement of ventricular size and configuration (X-ray, CT, MRI, Echoencephalography) prior to initial shunting. The ventricular sizes were classified as: slit, normal, large, severely enlarged (Figure 2). Whenever possible, long-duration pre- and postoperative ICP recordings were performed with the patient in the supine as well as in the erect position. The patients were regularly followed (see Figure 3 for evaluation forms) by clinical examinations (CT, MRI, etc.), at one week, two, six, 12 months, two years, and longer postoperatively following the same clinical evaluation protocol.

Table 1: Distribution of the Etiology of Hydrocephalus in 134 Cases

Etiology	No of cases	Percentage
Aqueductal stenosis	29	21.6
Intracranial hemorrhage	23	17.2
Myelomeningocele	19	14.2
Communicating hydrocephalus	19	14.2
Infection	12	9.0
Tumor	10	7.5
Dandy-Walker cyst	7	5.2
Encephalocele	4	3.0
IIIrd ventricle cyst	3	2.2
Intracranial hemorrhage + infection	1	0.75
Normal pressure hydrocephalus	2	1.5
Porencephaly	1	0.75
Fetal alcohol syndrome	1	0.75
Unknown	3	2.2

Table 2: Shunt History, Indications for Inserting the OSV in 134 Cases

Indication	No of cases	Percentage
Initial shunt implantation	91	67.9
CSF Overdrainage by previous shunt	19	14.2
Obstruction of previous shunt	14 - 2*	10.4 - 1.5
Infection of previous shunt	6 - 4*	4.5 - 3.0
Miscellaneous	4	3.0

* OSV's

The Manufacturer encouraged all explanted shunts to be returned for analysis. *Pressure-flow curve characteristics* of the shunt were measured *in vitro* using production test equipment. Shunts were considered to be within specification when the opening pressure ranged from 30 to 80 mm H_2O at a flow of 5 ml/hr; the shunt acted as a flow regulator at flows between 18 and 30 ml/hr, and as a low-resistance DP valve at DPs above 300 mm H_2O. When the pressure-flow curve was not within specification, the silicone elastomer boot surrounding the valve mechanism was opened to remove the core for further analysis of residues (blood, proteins, debris, etc.) that could possibly interfere with the operating characteristics of the shunt. Retesting was performed after flushing with a cleaning solution at 800 mm H_2O.

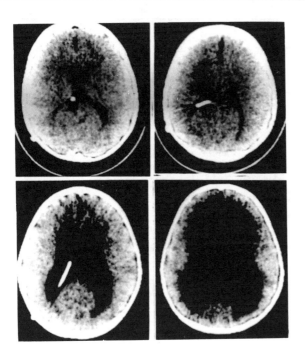

Figure 2
Ventricular sizes classified as normal (upper left), slit/small (upper right), large (lower left) and severely enlarged (lower right)

Particles in the shunt system were analyzed by various techniques such as hydrogen peroxide for blood detection; Fourier Transformed Infra-Red Spectroscopy (FTIRS) for fibers and proteins; and Scanning Electron Microscopy (SEM) for surface defects such as scratches, tears or other damage.

For the purpose of this study, a shunt complication was defined as any surgical procedure after initial implantation related for the treatment of hydrocephalus.

4 Results

The distribution of clinical follow-up time of the 134 cases is given in Table 3. Two-thirds were followed for more than two years and nearly half were followed for more than three years. The mean follow-up for all cases is 2.4 ± 1.2 years.

There were 11 cases (8.2%) of *shunt infection* (mainly *S. aureus)* (Table 4). In ten cases, the shunt was explanted. In one case infection was managed without shunt explantation. Proximal obstruction occurred in three cases (2.25%); in one, the ventricular catheter was revised, in the other two (1.5%), the shunt was explanted. Distal obstruction occurred in four cases (3.0%) with shunt explantation (1.5%) in two cases and distal catheter revision in two cases.

```
A

                        PATIENT HISTORY FORM


   Patient Name   :  ─────────────────────────    Hospital      : ──────────
   Date of birth  :  ──/──/──                      Neurosurgeon  : ──────────
   Sex            :  M  [ ]  F  [ ]

   Etiology
   Congenital   : [ ] if congenital indicate cause/syndrome ─────────────────
   Acquired     : [ ] if acquired [ ] Tumor         [ ] Trauma
                                  [ ] Hemorrhage    [ ] Infection
                                  [ ] Unknown       [ ] Other
   Type         : [ ] Communicating
                  [ ] Obstructive, if obstructive, obstruction location : ────
   What signs of hydrocephalus are evident ?
   [ ] Increased ICP ──────── mmH2O        [ ] Failing mental function
   [ ] Enlarged head ──────── cm               (test method: and score: ──────
   [ ] Open fontanelles                    ───────────────────────────────────)
   [ ] Setting sun sign                    [ ] Pyramidal tract signs
   [ ] Enlarged ventricles/CT scan Yes/NO  [ ] Ataxia
   [ ] Headaches                           [ ] Dementia
   [ ] Vomiting                            [ ] Gait disturbance
   [ ] Papilledema                         [ ] Urinary incontinence
                                           [ ] Other (indicate) ───────────────

   Are the following data available ? If available, please comment and add a copy.
   [ ] Skull X-ray           ─────────────────────────────────────────────────
   [ ] CT-scan  | either one of these must  ──────────────────────────────────
   [ ] NMR-scan | be made before implant !  ──────────────────────────────────
   [ ] Other                 ─────────────────────────────────────────────────

   If the patient receives the Orbis/Sigma Valve as a replacement for
   a valve of another type, please indicate :
   - Reason for revision : ────────────────────────────────────────────────────
   - Type of valve to be replaced : ──────────────────────────────────────────
   - Revision of total drainage system   [ ] Yes
                                         [ ] No
     If No, please indicate which part has not been revised :
   ─────────────────────────────────────────────────────────────────────────
```

Fig.3. Clinical evaluation forms: (A) Patient history form.

```
B
                          IMPLANT DATA FORM



Patient Name   :  ——————————————————————————
Implant date   :  ——————————————————————————
Valve no.      :  ——————————————————————————

CSF (sampled at implant) - Macroscopic aspect :  [ ] Clear
                                                 [ ] Xanthochromic
                                                 [ ] Hemorrhagic
  - Cells      ————————————— [   ] Specify unity used.
  - Protein    ————————————— [   ] Specify unity used.
  - Glucose    ————————————— [   ] Specify unity used.
  - Culture: pos./neg.
  Location of Valve unit : ————————————————————————————
  Ventricular catheter tip location: [ ] Frontal horn    [ ] Body lat. ventricle
                                      [ ] Temporal horn   [ ] Arachnoid cyst
                                      [ ] Occipital horn  [ ] Dandy Walker cyst
                                                          [ ] Isolated fourth ventricle
  Location of Drainage catheter   : [ ] Cardiac
                                     [ ] Peritoneal

Please comment on ease of implantation : ————————————————————
————————————————————————————————————————————————————————————

Complications / Difficulties encountered: ————————————————————
————————————————————————————————————————————————————————————

Were components of other shunts used with this system ? If so which ones ?
————————————————————————————————————————————————————————————



——/——/——                              ——————————————————————
Date                                   Physicians signature
```

Fig. 3. Clinical evaluation forms: (B) Implant data form

```
C1

                          FOLLOW-UP DATA FORM


  Patient Name    : ——————————————————  Hospital              : ————
  Date of birth   : ——/——/——             Implant performed by  : ————
  Implant date    : ——————————————————  Monitoring Physician  : ————
  Evaluation date : ——————————————————  Valve number          : ————
                    CT      CT    CT    CT    CT
  Follow-up : [1]  [2]  [3]  [4]  [5]  [6]  [7] (underline)
```

  - Valve functioning satisfactorily ?   [ ] Yes
                                                           [ ] No
  - Are signs of uncontrolled hydrocephalus evident ?   [ ] Yes
                                                           [ ] No
  [ ] Increased ICP ———————— mmH2O   [ ] Failing mental function
  [ ] Enlarged head ———————— cm   Score ——————————————————
  [ ] Open fontanelles   [ ] Pyramidal tract signs
  [ ] Setting sun sign   [ ] Ataxia
  [ ] Enlarged ventricles/CT scan Yes/NO [ ] Dementia
  [ ] Headaches   [ ] Gait disturbance
  [ ] Vomiting   [ ] Urinary incontinence
  [ ] Papilledema   [ ] Other (please indicate) ——————
  Please comment on above positive findings : (if ICP recorded please add copy)

  - If any supplementary data (x-ray, CT scan, etc.) is available, please send a copy
    and comments (see implant data)

  - Were complications encountered ?
    Infection: [ ] Wound   Breakage: [ ] Disconnection
                      [ ] Meningitis   [ ] Random site
                      [ ] Peritonitis   Please specify ——————
                      [ ] Septicemia
    Obstruction: [ ] Ventricular   [ ] Subdural collection
                      [ ] Valve   [ ] Orthostatic Hypotension
                      [ ] Distal   [ ] Slit ventricles
    Migration: Down / Up   [ ] Craniostenostosis
    Misplacement: [ ] Proximally   [ ] Other, Please specify : ——————
                      [ ] Distally   ——————————————————————————

Fig. 3. Clinical evaluation forms: (C1) Follow-up data form

```
C2
                        FOLLOW-UP DATA FORM (continued)


  - Please comment on encountered complications :
    _____

  - Please comment on action taken :  _____
    _____

  - Please comment on possible cause of the complication : _____
    _____

  - Was the shunt surgically revised ?      [ ] Yes   [ ] No
    If yes please indicate shunt manufacturer and model number : _____

  Please return non functional shunt components to :
  Cordis Europa N.V. for analysis
  with completion of the complication report and according to the prescribed shipping
  conditions (protocol page 8).
  Comments : _____
  _____
  _____
  _____


  ——/——/——                                    _____
  Date                                         Physicians signature
```

Fig. 3. Clinical evaluation forms: (C2) Follow-up data form (continued)

D

COMPLICATION REPORT

Report Date : ————————
(Please complete Part A only)
A. Valve Serial no. : ————————————
   Implant Date   : ——————————————— Explant Date : ————————
   Reason for removal : ————————————————————————

   Clinical consequences of incident : ————————————————————

   Hospital : ——————————— Reporting physician : ———————
   Patient : ——————————— Signature : —————————————

NOT TO BE COMPLETED BY MONITORING PHYSICIAN :

B. 1. Customer Service - Decontamination
   2. ——————————— For Analysis (Report to be attached)
   3. Disposition   [ ] Dead Storage
              [ ] Clinical Research
              [ ] Research and Development
              [ ] Other ————————————————

C. Action taken : ————————————————————————————
   ————————————————————————————————
   ————————————————————————————————

——/——/——   —————————
Date   Signature

Fig. 3. Clinical evaluation forms: (D) Complication report

Table 3: Distribution of Clinical Follow-Up Time of 134 Cases

Years	No of Cases	Percentage
0 - 1	22	16.4
1 - 2	29	21.6
2 - 3	25	18.7
3 - 4	47	35.1
4 - 5	11	8.2

Mean follow-up time: 2.42 ± 1.25 years

Obstruction was encountered in 19 cases (14.2%). In six (4.5%), revisions were performed, while in the remaining 13 cases (9.7%) the shunt was explanted. Valve obstruction occurred in 12 cases (9.0%). In three cases, gentle flushing of the shunt system with sterile saline (syringe method) cleared the obstruction. However, in nine cases (6.7%) the shunt was explanted. Valve infection at the beginning of the infection can be manifested as an obstruction; bacterial proliferation in the valve core can reduce the effective valve cross-section. Catheter fracture occurred in two cases (1.5%); in one case the shunt was explanted and in the other case the catheter was revised.

Improper distal catheter placement, with kinking or disconnection, occurred in four cases, (3.0%); all were revised. Overdrainage occurred in two cases (1.5%); in one case the shunt was explanted and in the other case an antisiphon device (ASD) was added to the OSV system*. There were four cases (3.0%) of insufficient (under) drainage resulting in shunt explantation. At last follow-up evaluation, the majority of patients (65; 62.5%) had normal ventricles; 21 (20.2%) had moderately enlarged ventricles, eight (7.7%) had severely enlarged ventricles and ten (9.6%) had small or slit ventricles one or more years post implantation. There were no signs of hydrocephalus reported after initial shunting (104; 77.6%).

Half of the explanted shunts were returned to the manufacturer for analysis (Table 5). Technical inspection and analysis showed no significant shunt-related malfunctions even in the presence of small amounts of blood. However, shunts obstructed with *debris or proteinaceous matter* were difficult to clean and were not in specification (pressure vs. flow characteristics).

* NOTE: The manufacturer of the OSV does not recommend using an ASD with this system.

Table 4:   Distribution of Type of Complication, Performed Revisions and Explantations in 134 Cases
(included relative frequency)

Type and No. of complications (%)	No. of revisions (%)	No. of shunt explantations (%)
Infection	11 (8.2)	1 (0.75)*	10 (7.5)
Proximal obstruction	3 (2.25)	1 (0.75)	2 (1.5)
Valve obstruction	12 (8.95)	3 (2.25)	9 (6.7)
Distal obstruction	4 (3.0)	2 (1.5)	2 (1.5)
Total obstructions	19 (14.2)	6 (4.5)	13 (9.7)
Catheter fracture	2 (1.5)	1 (0.75)	1 (0.75)
Improper catheter placement (kinking, disconnection)	4 (3.0)	4 (3.0)	- -
Overdrainage	2 (1.5)	1 (0.75)	1 (0.75)
Insufficient drainage	4 (3.0)	- -	4 (3.0)**
Miscellaneous	1 (0.75)	- -	1 (0.75)
Total	43 (32.1)***	13 (9.75)	30 (22.45)

*   This revision, i.e. corrective action, means drug treatment.

**   These four cases represent three young infants (2.1, 2.3 and 3.6 months old) and one disabled adult
(79 years old).

*** The total number of 43 complications occurred in 40 out of the 134 cases, i.e. in three cases (out of
40) there was a complication two times (case nos: 69, 78, 84).

Thus, 134-40 = <u>94 cases free of complications (70.1%).</u>
The overall success rate (compl. free) is:

$$\frac{94}{134-10 \text{ (inf. leading to explantation)}} \times 100 = 75.8\%$$

The overall success rate (without shunt revision, i.e. shunt explantation) =

$$\frac{134-30}{134-10 \text{ (expl. due to infection)}} \times 100 = 83.9\%$$

Table 5: Analysis of explanted OSV's

Case No	Complication	Technical Analysis
8	Infection	Debris in valve. After cleaning, valve in specification
55	Infection	Boot was damaged. Valve slightly out of specification
37	Infection	Valve in specification
15	Proximal obstruction	Valve in specification. Mislocation proximal catheter?
50	Shunt obstruction	Blood clots in valve. After cleaning, valve in specification
69	Shunt obstruction	Debris in valve verified. Valve not in specification
78	Shunt obstruction	Debris + blood in valve. Valve not in specification
85	Shunt obstruction	Debris in valve. Valve not in specification
95	Shunt obstruction	Deposits in valve. Valve not in specification
99	Shunt obstruction	Holes in boot. After cleaning, valve in specification
46	Overdrainage	Holes in boot of valve. After closing holes, valve in specification. Shunt damaged during implantation?
28	Underdrainage	Valve partly obstructed with proteinaceous matter, valve not in specification
113	Underdrainage	Valve partly obstructed with proteinaceous matter, valve not in specification
56	Catheter fracture	Valve in specification
17	Miscellaneous (no complaint)	Valve in specification

5 Discussion

Based upon the follow-up documentation and analysis, 104 out of 134 cases (77.6%) were adequately managed for hydrocephalus (Table 4) without shunt complications during the follow-up period ranging from 1.01 years to 4.57 years (mean = 2.3 ± 1.3 years). Since the incidence of shunt infection is not related to the shunt design, infectious complications were not included in the evaluation of indications for shunt revision. Actuarial analysis demonstrated that 83.9% of the shunts were functional for the specified follow-up period.

5.1 Shunt Complications

As with other shunt systems, *obstruction* is the major complication observed in the series. The incidence is significantly lower than that experienced in conventional DP shunt series due mainly to fewer cases of proximal obstruction. This can be attributed to the low incidence of slit ventricles following shunt insertion compared with conventional shunts.

It is well known that in cases with slit ventricles, the incidence of proximal shunt obstruction is high[52,53,54]. In this study, the obstruction rate was 14.2% (excluding some interventions resulting in explantation rate of 9.7%) which is remarkably low when compared to the number of cases which had an obstruction of the previous conventional shunt system (13 out of 38 cases: 34.2%) as well as compared to literature data[3,7,30,44,52,53,54]. For instance, Metzemaekers *et al.*[41] reported that shunt obstruction occurs in 60% of the cases during the first year after the shunt has been inserted.

There are several factors related to shunt system obstruction: patient-related factor (i.e. age, etiology of the hydrocephalus, degree of ventricular dilatation, etc.), and factors related to the physician and surgical technique (i.e. experience of the surgeon, therapeutic strategy, etc.)[9,23,52,53,54]. For instance, it has been reported that the ventricular catheter location (parietal or frontal) influences its functioning[2]. In addition, several patients with slit-like ventricles were successfully treated with the OSV[59,60]. This has also been confirmed in series reported by Pezzotta and Locatelli[45].

Because of the flow restriction in Stage II, the risk of *underdrainage* was a concern. There are very few cases of underdrainage, except in cases of choroid plexus tumors in which there was documented CSF overproduction that would overload the capabilities of the flow control Stage II. There is general agreement among neurosurgeons, however, to excise the tumor rather than implant a shunt. Underdrainage accounts for four cases in this study (3%). One of the reasons for underdrainage might be that the shunt system became partly obstructed. This was verified for two cases after technical analysis (Table 5).

There were only two cases where *overdrainage* was reported (1.5%). In one case, the problem was solved by addition of an antisiphon device. The other case of confirmed overdrainage cannot be explained by surgical manipulation or by disfunction of the shunt system, because the system was functionally within specification after closing the detected holes of the valve boot (technical inspection and analysis; Table 5). Whether or not this damage was related to overdrainage is unknown.

In fact, the incidence of overdrainage with the OSV (1.5%) is much lower than with conventional shunt systems (7%)[42,45]. Antisiphon mechanisms[66] have been designed to control differential pressure of shunt systems, but they have not eliminated the difficult problem of avoiding overdrainage assuring adequate drainage at the same time[32,38].

The OSV operates as *flow regulator* within defined pressure limits. It is characterized by a three-stage valve mechanism; each stage offers a different resistance to CSF flow, maintaining an almost constant flow rate close to the physiological rate of production when the patient stands erect[49]. The OSV's pressure-flow characteristics and its ability to stabilize CSF-flow (which reduces the danger of overdrainage) has also been confirmed by Czosnyka *et al.*[12] during control infusion tests.

The *remaining shunt complications* were related to catheter fracture, and improper catheter placement leading to kinking or disconnection. All could be repaired by intervention, except in one case of catheter fracture where the complete shunt system was explanted. The incidence of these complications, which appear to be related to surgical technique, is low when compared to literature[52,53,54].

The overall incidence of shunt complications requiring *surgical revision* in this study is low (9.75%). Anile *et a.l*[5] reported a revision rate of 28.65%, while Selman *et al.*[57] in a study of 130 lumboperitoneal shunted patients, with an average follow-up time of three years, reported a revision rate of 13.8% of the cases.

5.2 OSV Performance

On the basis of the results of the multi-center study as well of those already reported in the literature, the OSV is 1) effective in preventing post-shunt intracranial hypotension, 2) offers an adequate control of intracranial hypertension, 3) seems to be associated with a lower number of shunt failures as compared with the conventional systems, and 4) does not raise the problem of choosing the valve pressure range like in DP valves[49].

The *adaptability of the OSV* to the different physiopathological characteristics in hydrocephalus with different etiologies represents an advantage over conventional shunt systems. However, the OSV will never cover the entire spectrum of the physiopathological conditions of hydrocephalus. In cases of choroid plexus tumors and cases requiring drainage of extraventricular structures and collections, the use of the OSV is contraindicated.

The risk of overdrainage, particularly in patients greater than five years old, although still present, is lower than typically reported. This is not surprising since the characteristics is an average rate of flow (Stage II). It appears that the incidence of valve obstruction is no higher than with conventional valves.

Insufficient drainage can be observed in very young infants or disabled adults. It must be understood that in conventional shunts, the effects of the opening pressure are counterbalanced by a large overdrainage for each elevation of DP (leaving the supine position, REM sleep, etc.). This is limited in the OSV because of the flow limitation. Because of this, artificial elevation of DP by patient positioning is strongly advised.

One question remains open. For years, the goal of valve treatment was rapid reduction in *ventricular size*. There is no evidence in the literature that this forced reduction of the ventricular size should be a treatment goal. For instance, in tumoral hydrocephalus by tumor removal or in aqueductal stenosis treatment by third ventriculostomy, ventricular normalization frequently is not achieved without adverse consequences. Further investigations are necessary to clearly demonstrate this point and to avoid shunt removal as the ventricles did not normalize even though the patient was otherwise without complications.

5.3 New Shunts

During the past few years, several new shunts have been developed to address the problems of OD and UD.

Externally-Programmable Shunts: These do not solve the problem because the programming displaces the equilibrium point between OD and UD, but the valve continues to transition between OD and UD, with inadequate drainage in the recumbent position and excessive drainage when the patient stands.

ASDs: This alternative operates by cancellation of the "negative" output pressure of the shunt. There are several problems related to these devices: By construction, correct function is related to atmospheric pressure which is not exactly the case *in situ* under the skin. In fact, the function of the ASD, which is location dependent, depends upon location and skin thickness. Based upon its principle, an ASD will develop a "positive" pressure in the upright position which is not the case in normal physiology. In theory, it is possible to expect ventricular enlargement while venous pressure is normally becoming negative. In fact, this was reported earlier[39,40,63]. This device addresses the main cause of DP variations which are position related and not other possible causes of DP changes.

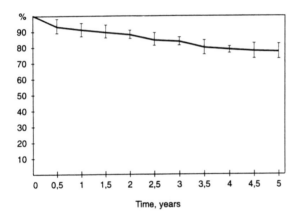

Figure 4

Actuarial probability of absence of shunt complications over a 5-year period after surgery for one series of patients treated with the OSV

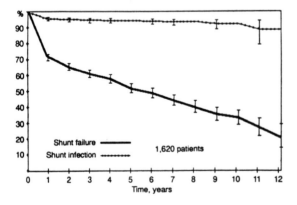

Figure 5

Actuarial probability of absence of shunt complications. Life-table analysis of mechanical complications in conventional DP shunts and shunt infections in all of the series over a 12-year follow-up period (After Ref.53: Permission S. Karger AG, Basel, Switserland)

5.4 OSV Shunt Survival

The final question is whether patients managed chronically with DP shunts can be subsequently managed with a flow-regulated device. The results indicate that 51.3% of the one-third of the patients chronically managed with DP shunts (revised because of OD complications) were successfully treated with the OSV.

For all cases (134) covering 128 patients treated with the OSV, the actuarial probability of the absence of shunt complications at one, two, three, four, and five years is 91%, 88%, 83%, 79%, and 77% respectively (Figure 4). The actuarial probability of absence of shunt complications is significantly higher than conventional DP shunts (Figure 5), and the relative frequency of complications such as obstructions, catheter disconnections, etc. is remarkably low.

In conclusion, the Orbis-Sigma Valve achieved an actuarial probability of survival significantly better than conventional DP shunts. The flow-control stage appears to reduce the incidence of overdrainage complications including the incidence of proximal obstructions.

References

[1]   Abbott R, Epstein F, Wisoff J. Chronic headache associated with a functioning shunt: Usefulness of pressure monitoring. *Neurosurg*, 28(1):72-77,1991.

[2]   Albright AL, Haines SJ, Taylor FH. Function of parietal and frontal shunt in childhood hydrocephalus. *J Neurosurg*, 69:883-886,1988.

[3]   Amacher AL, Wellington J. Infantile hydrocephalus: long-term results of surgical therapy. *Child's Brain*, 11:217-229,1984.

[4]   Ames RH. Ventriculo-peritoneal shunts in the management of hydrocephalus. *J Neurosurg*, 27:525-529,1967.

[5]   Anile C, Maira G, Puca A. Reliability and efficacy of different CSF-shunting devices in the treatment of CSF-dynamics disturbances. In Paoletti et al (eds), *Neuro-oncology*. The Netherlands: Kluwer Acad. Publ., pp 393-397,1991.

[6]   Basauri L, Zuleta A. Shunts and shunt problems. *Monogr Neural Sci*, 8:12-15,1982.

[7]   Bradley KC. Cerebrospinal fluid pressure. *J Neurol Neurosurg Psychiatry*, 33:387-397,1970.

[8]   Burcheit F, Maitrot D, Healy JL, et al. How to choose the best valve. *Monogr Neural Sci*, 8:184-187,1982.

[9]   Cardoso ER, Del Biggio MR. Age related changes of cerebral ventricular size. Part II *Acta Neurochir* (Wien), 97:135-138,1989.

[10]   Choux M, Lena G, Genitori L. Shunt implantation-Toward zero infection (meeting abstract). *Child's Nervous System*, 4(3):181,1988.

[11]   Cooper R, Hulme A. Intracranial pressure and related phenomena during sleep. *J Neurol Neurosurg Psychiatry*, 29:564-570,1986.

[12]   Czosnyka M, Maksymowicz W, Batorski L. Comparison between classic-differential and automatic shunt function on the basis of infusion tests. *Acta Neurochir* (Wien), 106:1-8,1990.

[13]   DaSilva MG, Drake JM. Effect of subcutaneous implantation of anti-siphon devices on CSF shunt function. *Pediatr Neurosurg*, 16:197-202,1990.

[14] Davson H. Formation and drainage of cerebrospinal fluid. In Shapiro K, Marmarou A, Portnoy H
      (eds), *Hydrocephalus*. New York: Raven Press, pp 3-40,1984.

[15] Drake JM, Sainte-Rose C, DaSilva M, et al. Cerebrospinal fluid flow dynamics in children with
      external ventricular drains. *Neurosurg*, 28:242-250,1991.

[16] Duplessis E, Decq Ph, Barat JL. Traitement de l'hydrocéphalie chronique de l'adulte par dérivation
      à débit régulé: A propos d'une série de 46 patients. *Neurochirurgie*, 37:40-43,1991.

[17] Epstein F, Lapras C, Wisoff JH. Slit-ventricle syndrome: Etiology and treatment. *Pediatr Neurosci*,
      14:5-10,1988.

[18] Faulhauer K, Schmitz P. Overdrainage phenomena in shunt treated hydrocephalus. *Acta Neurochir*,
      45:89-101,1978.

[19] Fitzgerald R, Conolly B. An operative technique to reduce valve colonisation. *Z Kinderchir* 39,
      (Suppl II):107-108,1984.

[20] Fox JL, McCullough DC, Green RC. Effect of spinal fluid shunts on intracranial pressure and
      cerebrospinal fluid dynamics. 2. A new technique of pressure measurements results and concepts.
      3. A concept of hydrocephalus. *J Neurol Neurosurg Psychiatry*, 36:302-312,1973.

[21] Fox JL, Portnoy HD, Schulte RR. Cerebrospinal fluid shunts: an experimental evaluation of the flow
      rates and pressure values in the antisiphon valve. *Surg Neurol*, 1:299-302,1973.

[22] Friden HG, Ekstedt J. Volume pressure relationship of cerebrospinal space in humans. *Neurosurg*,
      14:351-366,1983.

[23] Griebel R, Khan M, Tan L. CSF shunt complications: an analysis of contributory factors. *Child's
      Nerv Sys*, 1:77-80,1985.

[24] Gruber R. Should "normalisation" of the ventricles be the goal of hydrocephalus therapy. *Z
      Kinderchir* 38, (Suppl 2):80-83,1983.

[25] Hakim C. The physics and physiopathology of the hydraulic complex of the central nervous system.
      M.I.T. Thesis, 1985.

[26] Hakim S. Hydraulic and mechanical mis-matching of valve shunts used in the treatment of
      hydrocephalus: the need for a servo-valve shunt. *Dev Med Child Neurol*, 15:646-653,1973.

[27] Hakim S, de la Roche FD, Burton JD. A critical analysis of valve shunts used in the treatment of
      hydrocephalus. *Dev Med Child Neurol*, 15:230-255,1973.

[28] Hayden PW, Shurtleff DB, Stuntz TJ. A longitudinal study of shunt function in 360 patients with
      hydrocephalus. *Dev Med Child Neurol*, 25:334-337,1983.

[29] Hirayama A. Slit ventricle - a reluctant goal of ventriculo-peritoneal shunt. *Monogr Neural Sci*,
      8:101-111,1982.

[30] Hoffman HJ. Technical problems in shunts. *Monogr Neural Sci*, 8:158-169,1982.

[31] Hoffman HJ, Hendrick EB, Humphreys RP. Management of hydrocephalus. *Monogr Neural Sci*,
      8:21-25,1982.

[32] Horton D, Pollay M. Fluid flow performance of a new siphon-control device for ventricular shunts.
      J Neurosurg, 72:926-932,1990.

[33] Locatelli D, Bonfanti N, Sfogliarini R. CSF shunt infections in children. *Acts Int. Symposium,
      Control of hospital infections*, Rome, April 27-29, pp 198-200,1987.

[34] Lorenzo AV, Page LK, Watter GV. Relationship between cerebrospinal fluid formation, absorption
      and pressure in human hydrocephalus. *Brain*, 93:679-692,1970.

[35] Macnab GH. A comparison of the amount of cerebrospinal fluid formed in 24 hours with the amount
      delivered through the Holter valve. *Dev Med Child Neurol*, 4:293-294,1962.

[36] Magnaes B. Movement of cerebrospinal fluid within the craniospinal space when sitting up and lying down. *Surg Neurol*, 10:45-49,1978.

[37] Matsumoto S, Oi S. Slit ventricle syndrome. *Ann Rev Hydroceph*, 3:108-109,1985.

[38] McCullough DC, Fox JL. Negative intracranial pressure hydrocephalus in adults with shunts and its relationship to the production of subdural hematoma. *J Neurosurg*, 40:372-375,1974.

[39] McCullough DC, Wells M. Complications with antisiphon devices in hydrocephalics with ventriculoperitoneal shunts. *Concepts Pediatr Neurosurg* 2:63-75,1983.

[40] McCullough DC. Symptomatic progressive ventriculomegaly in hydrocephalics with patent shunts and antisiphon devices. *Neurosurg* 19:617-621,1986.

[41] Metzemaekers JDM, Beks JWF, Van Popta JS. Cerebrospinal fluid shunting for hydrocephalus: a retrospective analysis. *Acta Neurochir* (Wien), 88:75-78,1987.

[42] Nulsen FE, Spitz EB. Treatment of hydrocephalus by direct shunt from ventricle to jugular vein. *Surg Forum*, 2:399-403,1952.

[43] O'Brien M, Parent A, Davis B. Management of ventricular shunt infections. *Child's Brain*, 5:304-309,1979.

[44] Paraicz E. Mechanical problems in shunts. *Monogr Neural Sci*, 8:49-51,1982.

[45] Pezzotta S, Locattelli D. The treatment of hydrocephalus in pediatric patients with a variable resistance valve. In Paoletti et al (eds), *Neuro-oncology*, Kluwer Acad. Publ., The Netherlands, pp 373-377,1991.

[46] Pierre-Kahn A, Gabersek V, Hirsch JF. Intracranial pressure and rapid eye movement in hydrocephalus. *Child's Brain*, 2:156-166,1976.

[47] Portnoy HD. Hydrodynamics of shunts. *Monogr Neural Sci*, 8:179-183,1982.

[48] Puca A, Anile C, Maira G, et al. Cerebrospinal fluid shunting for hydrocephalus in the adult: factors related to shunt revision. *Neurosurg*, 29:822-826,1991.

[49] Rampini PM, Caroli M, Zavazone M. Advantages of the Orbis-Sigma valve in the treatment of triventricular hydrocephalus. In Paoletti et al (eds), *Neuro-oncology*, Kluwer Acad. Publ., The Netherlands, pp 387-391,1991.

[50] Renier D, Lacombe J, Pierre-Kahn A, et al. Factors causing shunt infection - computer analysis of 1174 operations. *J Neurosurg*, 61:1072-1078,1984.

[51] Sainte-Rose C, Hooven M, Hirsch JF. A new approach to the treatment of hydrocephalus. *J Neurosurg*, 66:213-226,1987.

[52] Sainte-Rose C, Hoffman HJ, Hirsch JF. Shunt failure. *Concepts in Pediatr Neurosurg*, 9:7-20,1989.

[53] Sainte-Rose C, Piatt JH, Renier D, et al. Mechanical complications. *Pediatr Neurosurg*, 17:2-9,1992.

[54] Sainte-Rose C. Shunt obstruction: A preventable complication? *Pediatr Neurosurg*, 19:156-164,1993.

[55] Salman JH. The collapsed ventricle: Management and prevention. *Surg Neurol*, 9:349-352,1978.

[56] Samuelson S, Long DM, Chou SN. Subdural hematoma as a complication of shunting procedures for normal pressure hydrocephalus. *J Neurosurg*, 37:548-551,1972.

[57] Selman R, Spetzler RF, Wilson WB. Percutaneous lumbo-peritoneal shunt: review of 130 cases. *Neurosurg*, 6:255-257,1980.

[58] Serlo W, Heikkinen ES, Von Wendt L. Infection prophylaxis in hydrocephalus shunt surgery. An evaluation of different regimes. *Z Kinderchir*, 34:121-127,1984.

[59]   Serlo W, Von Wendt L, Saukkonen AL, et al. The incidence and management of the slit ventricle syndrome. In Bhatia R, Kak VK, Sambasivan M(eds), Abstract no 107021 Frank Educational Aids Pvt Ltd, New Delhi, India, *the 9th International Congress of Neurological Survey,* New Dehli, India, Oct.8-13,1989.

[60]   Serlo W, Saukkonen AL, Keikkinen E, et al. The incidence and management of the slit ventricle syndrome. *Acta Neurochir* (Wien), 99:113-116,1989.

[61]   Spaziante R. The treatment of tumoral hydrocephalus in adults by means of a variable resistance valve. In Paoletti et al (eds), *Neuro-oncology,* The Netherlands, Kluwer Acad. Publ. pp 379-385,1991.

[62]   Steinbok P, Thompsom GB. Complications of ventriculo-vascular shunts: computer analysis of etiological factors. *Surg Neurol,* 5:31-35,1976.

[63]   Tokoro K, Chiba Y. Optimum Position of the antisiphon device. *Neurosurg,* 27:332(letter),1990.

[64]   Venes JL. Control of shunt infection. *J Neurosurg,* 45:311-314,1964.

[65]   Welch K. Residual shunt infection in a program aimed at its prevention. *Z Kinderchir,* 28(4):374-377,1978.

[66]   Yamada H. A flow-regulating device to control differential pressure in CSF-shunt system. *J Neurosurg,* 63:570-574,1982.

Introduction to the Microvita Section

R.F. Gauthier

There is a growing scientific interest in the relationships among matter, mind and consciousness. A rapidly increasing body of reliable experimental results exists in several areas of parapsychology and non-conventional medicine. Many of these results have not found a satisfactory scientific explanation or explanations in terms of conventional scientific theories. (See the section on the Role of Consciousness.) At the same time, the materialistic theoretical underpinnings of even the physical sciences are being increasingly challenged, for example in quantum physics, where a conscious observer may play a vital role in determining the result of an experiment.

In light of these growing challenges, new concepts and hypotheses are being proposed to build scientific thinking on a firmer though less materialistic foundation. These new concepts and hypotheses may help explain some of the currently unexplained scientific results, leading to a more integrated understanding of the nature of and relationships among matter, mind and consciousness.

One new concept is microvita, proposed in 1986 by Indian philosopher P.R. Sarkar. Microvita are related to the ancient tradition of Yoga, in which much importance is placed on the practical attainment of harmony of body and mind, in part through understanding their respective natures and interactions. In Yoga philosophy, consciousness is fundamental, while both mental and physical processes and structures are derived from a state of pure, unlimited consciousness. This view is in marked contrast to the materialistic view in modern science, where matter and energy are considered fundamental, while mind and consciousness are assumed to be derived or evolved from them, and are given secondary importance. So ideas from the Yogic tradition may have a vital role to play in creating a new scientific synthesis among matter, mind and consciousness.

Microvita are described as fundamental, sub-microscopic living entities which play a vital role in the structuring of matter, living beings, individual minds, and collective psychology. They are also partly responsible for both physical and mental health as well as illness, in individuals as well as social groups. If hypotheses about microvita can be elaborated and tested scientifically, this could lead to a complete revision of many areas of scientific knowledge, which now are based on purely materialistic foundations.

The first article, "The Origins of Mind", compares western and eastern ideas on the nature and evolution of complex structures. Growing levels of complexity on the physical level seems to correspond to higher levels of expressed mind. The existence of different hierarchal levels of mind in complex living beings is proposed to be related to and develop out of the organisation of microvita at the sub-atomic level. A systems dynamics approach sheds further light on the evolution and stability of complex physical and mental structures.

The second article, "Microvita: A New Approach to Matter, Life and Health", takes a closer and more quantitative look at how microvita may participate in the structuring of sub-atomic particles such as the photon and the electron, which are presently conceived as fundamental physical entities, each having both a wave and a particle nature. A quantum physics experiment is proposed to test the idea that perhaps a few million microvita compose a single electron. Some possible relationships of microvita to biology and medical treatment are also briefly discussed.

The third article, "The Quantum Field of the Healing Force", proposes microvita as the quanta of a healing force field that exists, according to eastern tradition, not only at the physical level but in a hierarchy of mental levels as well. The mathematics of quantum theory may be useful to describe the behaviour of microvita in these different levels. These microvita carry subtle psychological qualities, as well as disease or healing qualities, through physical and mental spaces.

The fourth article, "Is Another Physics Needed to Explain Fundamental Fluctuations in Physical Measurements?", proposes a new theoretical approach to explain the well established but as yet unexplained variability of rates of certain chemical reactions, even under constant experimental conditions. The author points to a need for new physical principles to explain the data. The unexplained variability of chemical reactions under constant experimental conditions is explained in the microvita hypothesis as possibly due to differences in the number and types of microvita contained within the atomic structure of different samples of the same chemical. Changes in the number and type of microvita over time in a chemical sample could also lead to corresponding changes in chemical reaction rates.

Microvita science is still in its infancy. The need for a radical new scientific hypothesis or hypotheses to explain the growing body of unexplained scientific data relating matter, mind and consciousness is becoming critical. Perhaps the microvita hypothesis, when subjected to further elaboration and testing by a new generation of scientific and medical researchers, may help to fill that need.

The Origins of Mind

M. Towsey and D.N. Ghista

1 Introduction

There is increasing concern that the strictly objective, quantitative and reductionist methodology of the natural sciences is inadequate to investigate the dynamics of mind and consciousness [2, 12]. By contrast the science of yoga has evolved as a system to expand mind and consciousness, and offers an appropriate methodology to investigate the so-called 'problem of consciousness'. We have discussed these epistemological issues in another chapter of this book (in the section on Consciousness [20]). In this paper, we focus on the question of the origins of mind from matter and the concomitant unfolding of consciousness during evolution.

No progress can be made in this field if one persists in the belief that mind and consciousness are mere epiphenomena of complex material processes or that they are merely modes of discourse or linguistic confusions. This position is succinctly expressed by Steven Rose, who defines mind as *"equivalent to the sum total of brain activity for discussions within the universe of discourse at a hierarchical level above that of the physiological description of the interaction of cells and below that of social analysis."* [11] Rose's definition stems from a commitment to materialism, which is currently the paradigm for all the natural sciences. But a paradigm is a set of working assumptions and not a set of proven facts. Even the 'neuro-philosopher', Patricia Churchland, who is completely committed to the methods of the natural sciences, admits that "We do our research as if materialism was a proven fact, but of course it isn't." [as quoted by Lewin, 8] And while she does not believe in a "nonphysical soul", she admits of Cartesian dualism that "we cannot claim to have ruled it out."

From the yogic perspective, mind is substantive, that is, it is some kind of substance or energy but different from the physical kind. This mental substance is known as *citta*, sometimes translated into English as *ectoplasm* to make a distinction from the physical protoplasm of living cells. An obvious corollary is that mental processes are primary or real phenomena as opposed to epiphenomena of material processes. In this paper we are more concerned with the investigation of mind as an objective form of non-physical energy and less so with the philosophical issues surrounding dualism. We begin with a classification of the categories of mind and propose that an understanding of each category requires its own notion of space and time. Next we discuss the evolution of mind and in particular the dynamics of complex systems as a model for the evolutionary mechanism. Finally we offer simple models for the first major evolutionary step, the emergence of mind from matter.

2 A Classification of Mind

In simplest terms, yogic ontology distinguishes three categories, matter, mind and consciousness, just as does western metaphysics (Figure 1). Mind can be further subdivided into three categories, the crude, subtle and causal minds or, to use Wilber's more revealing terminology, the prepersonal, personal and transpersonal minds [22]. The personal mind corresponds to the ordinary state of human consciousness, the prepersonal to plant and animal consciousness and on some occasions human consciousness, while the transpersonal mind is responsible for coordinating plants, animals and human beings on the larger scale. Finally, the transpersonal mind can be further divided into three levels which do not have English names, but which can be described as the 'intuitive-social' mind, the 'ethical-aesthetic' mind and the highest level of mind associated with the experience of universal consciousness. This gives a total of seven categories or steps on a spectrum of consciousness. In order to avoid difficulties of terminology, these seven levels will be subsequently referred to by numbers as in Figure 1. The five layers of mind are levels 2 to 6.

Consciousness	Consciousness	Consciousness	Atman		7
Mind	Transpersonal mind	Universal mind	Hiranmaya kosa		6
		Ethical, aesthetic mind	Vijinamaya kosa		5
		Socio-eco-, intuitive mind	Atimanas kosa		4
	Personal mind	Rational, senti-mental mind	Manomaya kosa		3
	Pre-personal mind	Sensory, instinctual mind	Kamamaya kosa		2
Matter	Matter	Matter	Annamaya Kosa		1

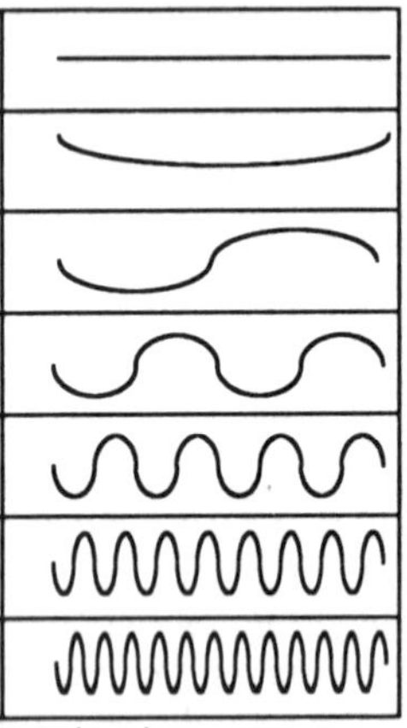

Figure 1 The different states or levels of consciousness as described by yogic science.

In addition to being a spectrum of consciousness, these categories also manifest objectively as a spectrum of waves. At the matter end of the spectrum, the wavelength is shortest but progressively increases towards the subtle end of the spectrum, until in the extreme, pure consciousness is described as having infinite wavelength. There is no movement or distortion within it. Just as in modern physics, waves are said to have a particle complement, so too in yoga, the notion of wave-particle duality is extended up to level 6. In descending the spectrum, inter-particle distance decreases just as does wavelength. Thus the matter end of the spectrum is said to be crude, dense, atomistic and bounded. Ascending the spectrum, the categories become increasingly subtle, rarefied, synthetic and unbounded.

Important concepts emerge from this yogic ontology. First, matter and mind are both substantive categories located on the same spectrum. Matter is at the dense end of the spectrum, which happens to be known to us through our sense organs. Indeed, matter is sometimes represented as the densest level of mind. In other words, the gap between mind and matter is not as great as is suggested by the simplistic trinity of matter, mind and consciousness. Thus we might expect matter and mind to display similar dynamics, because they are both wave phenomena. The main difference between the two is that mental phenomena are not directly accessible to us through the senses. Ironically, yogic science has been able to accommodate the recent trend in western post-modern philosophy to dissolve the sharp boundary between matter and mind.

Second, the spectrum of wave forms is continuous, but the boundaries between the different levels of consciousness are distinct. This hints at the physicist's notion of a phase change, as for example the temperature scale is continuous, but the transition from liquid to solid occurs at a distinct temperature.

Third, level 4 mind is, from a western perspective, the collective mind associated with social systems and eco-systems. In other words, collective minds are substantive real entities which endow a corporate identity. There has always been a debate in ecology, which has its parallels in sociology, as to whether ecosystems are real entities or just an artifice of the human desire to categorize and classify nature. James Lovelock's Gaia hypothesis is a recognition that the entire earth may be considered as a single homeostatic organism or corporate entity [8]. From a yogic perspective, the expanded level of mind which coordinates these corporate entities is the Intuitive or level 4 mind.

Fourth, level 7 labeled in Figure 1 as 'pure consciousness', does not involve consciousness of things or any objective experience as such. Rather it is a state which transcends experience. Hegel describes it as "the self-contained existence of spirit" [5]. He adds, "Two things must be distinguished in consciousness; first the fact *that I know*; secondly *what I know*. In *self*-consciousness these are merged in one; for spirit knows itself." As defined by Hegel and drawn in Figure 1, consciousness transcends mind. But this is an incomplete statement. Consciousness is also immanent in both mind and matter. Thus matter (level 1) is to some extent conscious or self-actualizing, while the experience of *universal consciousness* (level 6) is to some slight extent an objectivated form of energy.

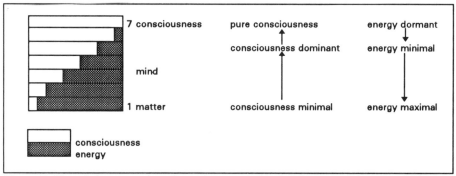

Figure 2 The relative expression of consciousness (subjectivity) compared to energy (objectivity) increases as one moves from level 1conscious state to level 7 conscious state.

In yogic science, the Consciousness principle and the Energy principle are inseparable. The analogy used is that of a piece of paper, which has two sides. Philosophically they are distinct but in practice they are inseparable. To the Consciousness principle we attribute our sense of Self, the witness (the 'I' of I experience), the will (the 'I' of I do) and the organizing principle. Out of it arises conscience. By contrast the Energy principle is pure motivity or the power to do work. In every other respect it is 'blind', that is lacking in awareness. Although practically speaking consciousness is never separated from energy (see Benor [1] who arrives at the same conclusion from his experience as a physician), the philosophical distinction is made because the two principles can express themselves in differing proportions (Figure 2). At the matter end of the spectrum, consciousness appears to be negligibly expressed while energy is maximally expressed. At level 6, the opposite is the case, while in the transcendent state of pure consciousness, energy is entirely dormant.

3 Mind as a Hierarchy of Space and Time

The levels of mind in yoga are described as *kosas*, meaning *shells*. The import is that the layers of mind are related to one another in a hierarchy of interpenetration which may be represented by a set of concentric circles (Figure 3). The inner-most circle represents matter (level 1). Level 7 is unbounded consciousness. Matter is interpenetrated by all levels of mind. The lowest level of mind (level 2) encompasses as well as infuses and penetrates matter, but is in turn encompassed, infused and penetrated by all higher levels of mind. And of course, Consciousness penetrates all levels of mind and matter. The converse cannot be said, because the scope of matter is limited compared to that of mind.

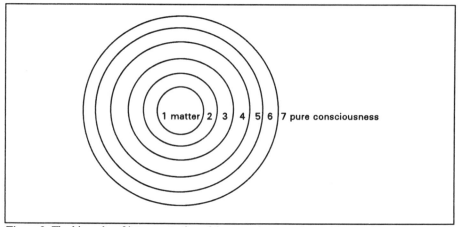

Figure 3 The hierarchy of interpenetration of the seven layers or kosas. The word kosa means shell. The 'higher' shells encompass and interpenetrate the lower shells.

This hierarchy of interpenetration brings us to the critical notions of space and time, as they apply to the spectrum of consciousness. Pure consciousness is all pervasive, without any limitation, parts or boundaries, while at the opposite end of the spectrum matter is subject to innumerable limitations and demarcations. One particular limitation is that the the field effects of matter particles are local, that is, there are spatial and temporal limitations to communication between particles. If pure consciousness (level 7) is beyond all notions of space and time, while physical (level 1) space and time are minutely divisible into measured parts, what can be said about space and time as they apply to the intermediate levels of mind (levels 2-6)? Mind must be subject to some limitations of space and time, otherwise the notions of wavelength and frequency would not be applicable. But from the perspective of the physical world, the mental worlds appear to be non-local and atemporal. This leads us to propose that each higher level on the spectrum of mind appears non-local and atemporal to the level below. Thus in the physical world, non-local phenomena (coordination between sub-atomic particles that cannot be explained in terms of information or field effects transmitted at or below the speed of light) are indicative of the action of mind level 2 impinging on matter level 1. Similarly, the phenomenon of 'synchronicity' as described by the Jungian school of psychology, is indicative of a transpersonal or collective mind (level 4) impinging on the personal mind (level 3).

In "Physics As Metaphor" [6], the physicist Roger Jones discusses time and space as we experience them physiologically and in dream states, that is without the prejudicial fore-structure imposed by the physicist's linear measure of space and time. Time, he says, sometimes speeds up sometimes slows down, distance appears more or less depending on our mood and so on. In some dreams, large distances can be covered in an instant, while in others it takes an age to sprint a few metres to safety from a wild animal. Time chops back and forth effortlessly. Jones considers physical space and time as a limiting case where a linear structure is imposed on something which is inherently fluid. In other words, he regards mental time as primary and physical time as derivative. In the same vein, Rudolph [13] draws a distinction between physical time, physiological time and experienced time.

Psychologists have developed a number of intriguing methods to map mental 'distance'. One is to measure the time it takes to classify objects or sensations into a number of simple abstract categories. The shorter the time, the closer an object is to the 'centre' of its category. In this way, one can build up a map of psychic 'distances' or 'similarities' between abstract categories and the objects in them. Shepard [17] constructs graphs of change in stimulus versus change in response for a wide variety of species and then mathematically transforms the curves so that they become similar for all species. The transformation required for a particular species describes its psychic space.

We consider the greatest challenge facing mind research is to find an adequate way to describe time and space as they pertain to the different levels of mind. If it is true that each level appears to operate nonlocally and atemporally to the layer below, then each level will require its own description of space and time. In Rudoph's terminology, perhaps physiological time applies to level 2 and experienced time to level 3.

4 The Evolution of Mind

From a materialistic perspective, evolution may broadly be described as the unfolding of structural and behavioural complexity in matter. Richard Dawkins [4] considers the origin of structural complexity as the central issue of evolutionary theory. From the point of view of physical structure, evolution begins with atoms, which join together to make cells, which in turn join to make organisms, which join to make eco- and social systems (see left side, Figure 4). At each step billions of parts combine to make a whole. (It is tempting to ask what might be the result of the coming together of billions of eco-social systems.)

From the yogic perspective, evolution is understood as the unfolding of higher layers of mind with increasingly subtle and expanded consciousness (right side of Figure 4). Obviously, these evolutionary steps represent big jumps. Between a cell and an organism, there are tissues, organs and organ systems. The sequences have been written this way to reveal the essential correspondence between the two schools of thought, but both admit substeps. To summarize 1) evolution is synthetic, involving the joining of numerous smaller parts to make larger wholes in a hierarchical sequence; 2) each whole is a complex system from the point of view of its parts; 3) evolution is as much expansion of consciousness as it is expansion of complexity; and 4) from a yogic perspective, evolution is driven by the tendency of consciousness to increase its scope and expression. This 'law of consciousness' has similar status in the yogic description of life as the 'law of entropy' has in the physical sciences description of matter. It expresses a fundamental behaviour of living systems which admits no further simplification. The interplay of these two laws (those of consciousness and entropy), lies at the heart of the evolution of mind.

That expansion of consciousness drives evolution has been expressed by such disparate western philosophers as Teilhard de Chardin and Hegel. Although the idea is not inconsistent with the scientific evidence, it is of course incompatible with neo-darwinism, the theory used to interpret that evidence. According to neo-darwinism, consciousness is an incidental adjunct that emerges as random mutation and natural selection conspire to mold extraordinary physical complexity in the brain. "Perhaps", says Dawkins [as quoted in 3], "consciousness arises when the brain's simulation of the world becomes so complete that it has to include a model of itself." However as Popp [10] argues, neo-darwinism has become a theory which cannot in principle be falsified, because "any predictable future in terms of an enveloped determinism is disapproved" and yet the theory can be used to contrive ad hoc after-the-fact explanations for any observed evolutionary process.

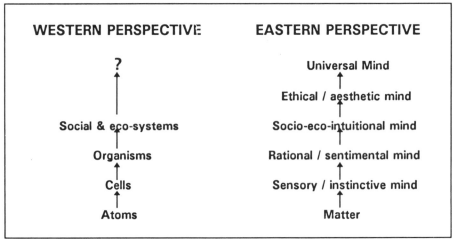

Figure 4 A comparison of the western and eastern descriptions of evolution. The former concentrates on an objective description of the physical structures, the latter concentrates on a subjective description of unfolding mental characteristics.

5 Synthesis and Self-organisation

The evolutionary sequence of Figure 4 can be written as a sequence of three synthetic 'reactions';

many atoms	⇒ cell	+	sensory-instinctive mind
many cells	⇒ organism	+	rational-sentimental mind
many organisms	⇒ eco-social system+		collective mind

In each case the first product on the right hand side is the 'sum of parts' and the second product is the 'emergent part'. The latter is substantive as opposed to being just an abstract emergent property. The sum of parts and the emergent part are inextricably meshed to constitute the whole. The generalized synthetic reaction is shown in Figure 5. It can go in the reverse analytic direction but not by the same route as the synthetic reaction, thus ruling out the possibility of attaining a stable equilibrium. This generalized synthetic reaction summarizes the fundamental process leading to both complexity of structure and emergence of mind and consciousness. Some of its characteristics are now discussed.

1) Critical Mass: In order for the emergent part to manifest, a huge number of parts are required, on the order of billions. Thus it appears as if there is a critical mass below which the emergent part does not manifest. For example in the reaction involving the synthesis of many cells, primitive multicellular organisms do not demonstrate rational or sentimental qualities because the sensory-instinctive qualities of their individual constituent cells still dominate. Only in mammals do we observe the emergence of sentimental behaviour and in humans rational behaviour (at least by the human definition of it!).

2) Diversity of Parts: A rock contains billions of atoms but it is not living, thus critical mass by itself is not enough. A diversity of component parts appears to be an essential feature of these synthetic reactions. For example, living cells can contain as many as 90 different kinds of atoms, most of them only required in trace amounts but nevertheless having functional importance. Human beings contain some 250 different kinds of cells and most eco-systems contain a great diversity of organisms.

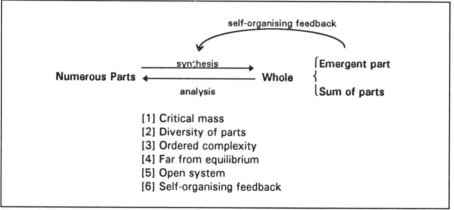

Figure 5 The emergence of the whole from its parts involves complex dynamics.

3) Ordered complexity: The wholes that arise from the synthetic reactions have a high degree of order but their order is complex. The ordered complexity of cellular structure or of a modern industrial state is self evident. By contrast, a rock crystal has a highly ordered structure but it is not complex while a turbulent flow of water has complexity, but its dynamics is chaotic as opposed to ordered. Recent advances in the study of complex dynamical systems reveal that the boundary between order and chaos is not clear cut. Computer modeling suggests that the 'ordered complexity' of living systems actually resides at the boundary between order and chaos, the so-called "edge of chaos" [7]. A biochemical consequence of living on the edge of chaos (as just one example) is that cell membranes are poised on the phase boundary between liquid and solid states. To maintain a delicately balanced position on a phase boundary far-from-equilibrium, is energetically expensive, so what are the advantages in doing so?

4) Far from equilibrium: From the perspective of the parts, the whole is far-from-equilibrium. It maintains a complex highly improbable structure despite the 'anarchistic tendencies' of its parts. In the language of yoga, life is said to be a never ending struggle to sustain an unstable equilibrium. The advantage is that such systems can respond non-linearly to a stimulus (i.e. small stimulus - rapid and large response) and they can be coded to process information [8]. Thus having parts far-from-equilibrium and poised on the edge of chaos would appear to be a necessary condition for a more subtle consciousness emerging in the whole.

5) Open systems: Living systems appear to defy the second law of thermodynamics (concerning entropy) because they develop complexity and maintain it far from equilibrium. The contradiction is resolved by recognizing that the second law is a description of isolated systems, whereas living systems are open. Open systems allow the transport of energy and matter across their boundaries (Figure 6) whereas isolated systems do not. The human body for example, is said to turn over all of its atoms about every seven years. A large corporation or school maintains its identity despite a yearly turnover of personnel. Prigogine has demonstrated that open systems far-from-equilibrium are able to maintain or increase their internal order by importing free energy (whether in the form of food, structured energy or information) and 'exporting' entropy (in the form of waste products, heat or noise). In other words, open, far-from-equilibrium systems use up or *dissipate* energy, and as a result they can spontaneously generate internal structure. They are self-organizing.

6) Self-organization: Self-organization implies that complex structure arises from the internal dynamics of the whole itself, in a deterministic but not easily predictable way. There are two possible mechanisms for self-organization; (1) macroscopic structure emerges spontaneously from the interactions between numerous parts (bottom-up dynamics), (2) an emergent coordinating factor simultaneously 'oversees' all the parts (top-down dynamics). The first option postulates only local communication between adjacent parts of complex dynamical systems, the second postulates global communication between parts. The two options are not exclusive however. Recent advances in neural networks, complexity theory and chaos theory have given tremendous insight into how complex dynamical systems can settle into stable patterns of behaviour, with only comparatively simple rules to define the local interactions between parts. These stable patterns are regarded as the outward manifestation of a deeper internal process which is described mathematically as a limit cycle in the phase space of the interacting parts. "There is order for free!", says Kauffman [as quoted in 8], one of the pioneers in the field of complexity. While the phase space of a huge number of parts would appear to offer unlimited possibilities for order, in fact complexity theory appears to permit only limited numbers of quite discrete ordered states. In other words, biological structure is limited by apparently abstract mathematical constraints. This is in contrast to neodarwinism which assumes that living organisms are plastic enough to be molded in any direction given an appropriate sequence of natural selection pressures.

The biophysicist, Fritz-Albert Popp argues for the second self-organizing option, an over-arching control mechanism using long range interactions. "Without supra-molecular, long-range interactions, evolution, i.e. the development of increasingly higher levels of organization, would be impossible and, of course, senseless." [10] He concludes on the basis of his investigation of very low-intensity long-range photon interactions between plants, that coordination in living organisms is established by photon exchange at specific wave-lengths. Both local and long range coordination between parts is optimized by using a spectrum of coherent wavelengths. Evolution, he argues, is driven by an optimization principle which causes an "expansion of coherent states", that is the use of longer and longer wavelengths to achieve greater spatio-temporal coordination. Although Popp's model of evolution is explained in physical terms, he states that an interconnectedness at the global level is necessary to account for the emergence of consciousness. "In a sense, the whole entity of living systems is a more or less considerably interlinked developing unit. Just as cells in an organism perform different tasks for the whole, different populations enfold themselves not only for their own, but principally more for the totality of all living systems, toward expanding a "consciousness" of the whole, and becoming more and more aware of it." [10]

6 The Microvita Hypothesis

From the perspective of the science of yoga, the evolution of complexity and the unfolding of mind are both spontaneous processes whose impetus comes from *within* the structures themselves. While the evolution of complexity may be described as self-organization, the unfolding of mind results from the inner impetus for self-realization. Self-organization and self-realization are complementary objective and subjective descriptions of the same process.

An important guiding principle is that 'something cannot come from nothing'. In other words, if the synthesis of billions of atoms produces recognizable qualities of mind, then those qualities must already lie latent within the constituent atoms. We will further draw upon several elements of Sarkar's *microvita* hypothesis [15]. Essentially the hypothesis states that sub-atomic particles such as electrons and protons are composed of innumerable minute 'particles' or microvita, some of which are physical in nature but some of which are psychic or mental in nature. These individual microvita express elemental psychic states or feelings such as joy, melancholy, the urge for survival and so on. In short, atoms have feelings! Far from being inviolable entities, electrons and protons in the microvita hypothesis are themselves open complex systems, in which there is a constant turnover of their constituent microvita. Within isolated atoms or inert physical structures, psychic microvita do not make themselves apparent. However when atoms come together in the ordered complex structure of biological cells, the psychic microvita manifest in collectivity, as evidenced by the sentience of the organism.

The idea that atoms themselves must contain the antecedents of mind and feeling has an impressive pedigree, as for example the mathematician Alfred Whitehead [21] and the evolutionist Sewell Wright [23 p14]. More recently, the Professor of Biology at University of New South Wales, Charles Birch, has summed up the idea; "There is but one theory, known to me, that casts any positive light on the ability of brain cells to furnish us with feelings. It is that brain cells can feel! What gives brain cells feelings? It is by the same logic that we may say - their molecules. And so on down the line to those individuals we call electrons, protons and the like. The theory is that things that feel are made of things that feel." [2 p32]

The idea that the universe is composed of innumerable 'particles' far smaller in size than electrons or protons is a central premise of *super-string* theory, which currently excites theoretical physicists. The particles, it postulates, are actually vibrating strings of energy about 10^{-35}m long. The theory promises to build up a complete description of the known sub-atomic particles and their interactions based on the mode of vibration of each string. The microvita hypothesis is also a 'grassroots' account of macroscopic diversity, but intended to account for mental phenomena as well as physical. The antecedents of mind and thus of consciousness are to be found latent within matter.

Finally we should note that the divide between the quantum level and the macroscopic level is beginning to break down. For example, in 1962 Brian Josephson predicted the occurrence of certain quantum effects at the macroscopic level which were subsequently verified experimentally. Biophysicists, Sitko and Gizhko [19] have suggested that the 'Weisskopf quantum ladder' can be extended from the sub-atomic, atomic and molecular levels to macro-molecules and cells. Experimental support for this idea comes from recent investigations of the way proteins change their conformation [9]. "We see that living systems can be described simultaneously in terms of synergetics [self-organization] as a hierarchy of dissipative structures and in terms of quantum mechanics as intact quantum systems." [19] Conversely they postulate that the synergetic (self-organisation) approach usually applied only to living systems, can be extended down to the atomic level. "The dynamic stability of an atom, as of a dissipative system, is created at the interchange between virtual photons with the physical vacuum." [19] We suggest that in terms of the general synthetic reaction of Figure 5, the sum of parts is best described by the dissipative system approach, while the emergent part is best described by the quantum approach. Sitko and Gizhko regard these as complementary descriptions of the same phenomenon.

The novel feature of the microvita hypothesis is the combination of all these ideas to produce an account of the emergence of mind from matter. We may note in passing, that the hypothesis offers as a testable consequence of sub-atomic particles being open systems, that certain chemical reaction rates will vary depending upon the constituent microvita of the reacting species in ways that cannot be explained by the traditional parameters of temperature, pressure, pH and so on. A paper by Shnol [18] describing anomalous biochemical reaction rates is interesting in this context.

In summary, the problem of explaining the emergence of mind from matter becomes one of explaining how the constituent psychic particles which remain unexpressed within individual atoms become recognizably expressed within a complex cellular structure. Traditional yogic science provides some clues to solve this problem, which we have used to develop the following models.

7 Models for the Emergence of Mind

1) Mind as the product of a phase change

The phase change phenomenon has an important place in yogic science just as it does in the physical sciences. In both cases it is used to account for the emergence of a new property by the coordination of innumerable parts. For example, in an unmagnetised iron bar, the individual crystal domains are magnetised in random directions, so that at the macroscopic level, no magnetic field is discernible. When the bar is magnetised, all the individual magnetic domains are oriented in the same direction, so that the sum of fields becomes recognizable at the macroscopic level.

In this model, we suppose that the psychic microvita within atomic matter are dipoles that give rise to psychic fields. Only when a large number of atoms are oriented correctly within a complex cellular structure, do the individual fields sum to a recognizable macroscopic psychic field. Furthermore, as the total psychic field becomes stronger, the field itself will act as a positive feedback mechanism to align other atoms and their microvita. The initial orientation of the atoms is achieved by the dynamics of open, far-from-equilibrium, dissipative, complex systems which have a spontaneous capacity to develop structure. In other words, the process begins with bottom-up dynamics, which in turn gives rise to top-down dynamics and the two become a self-reinforcing cycle.

2) Mind as the product of chemical reactions:

Chemical reactions involve the making and breaking of electron orbitals. This happens continually in protoplasm due to anabolic and catabolic reactions with an accompanying loss or gain of heat energy depending on the particular biochemical reaction. The production of heat energy results in the loss of an immeasurably small amount of mass (as described by Einstein's energy-mass equivalence). Some similar phenomenon is suggested in the following yogic description of the origin of mind from matter. "Whenever and wherever, as a result of chemical clash, a portion of the physical body gets powdered down, that is, transformed into factors subtler than [the physical], the effect is known as 'unit mind'. Unit mind is the product of a chemical reaction of physical clash." [14]. In this passage "chemical clash" most likely describes the making and breaking of chemical bonds in opposing anabolic and catabolic reactions. There is a consequent "powdering" which seems to describe a loss of mass in the form of much smaller particles, such as microvita, which in the released state manifest as mind. In this model, the emphasis is on cellular metabolism as the generator of mind whereas in the previous 'phase change' model, the emphasis was on cellular structure. The two models could be combined. Note that in this model mind can also be lost, that is absorbed within matter. Thus mind is fluid, in a constant state of production and reabsorption, dependent on an underlying substrate of physical metabolism.

3) Mind as a resonance phenomenon:

The resonance model is prompted by the yogic description of living organisms as having "psycho-physical parallelism" [14]. Resonance is a phenomenon whereby a vibrating structure can induce oscillations in another structure by imparting energy at a similar or harmonic frequency. The receiving system can absorb energy and thus vibrate only at specific frequencies. This model begins with the idea that cellular metabolism and structure represent a limit cycle in the phase space of a complex system of molecules. In some way the quasi-periodic cycling of a protoplasmic system is able to impart vibratory energy at the correct frequency to the microvita responsible for the expression of Level 2 mind.

10 References

[1]   Benor, D.J. 1993 *Spiritual healing and consciousness as energy* Proc 2nd Gauss Symposium, Munich 2-8 August
[2]   Birch, Charles 1990 *On Purpose* NSW University Press, Sydney
[3]   Blackmore, Susan 1989 *Consciousness: science tackles the self* New Scientist 1 April
[4]   Dawkins, Richard 1987 *The Blind Watchmaker* Norton
[5]   Hegel, G.W.F. 1959 *Lectures on the Philosophy of History* in "Treasury of World Philosophy" ed D.Runes, Littlefield, Adams & Co, Paterson, New Jersey
[6]   Jones, Roger 1983 *Physics as Metaphor* Abacus, London

[7]   Kauffman, Stuart 1991 *Antichaos and Adaption* Scientific American, August

[8]   Lewin, Roger 1992 *Complexity* Macmillan

[9]   New Scientist 1991 *Proteins change their shape in the blink of a laser flash* 16 March, p18

[10] Popp, Fritz-Albert 1992 *Evolution as the Expansion of Coherent States* in The interrelationship between mind and matter, Proceedings of a Conference hosted by the Centre for Frontier Science, ed. B. Rubik, Temple University Philadelphia.

[11] Rose, Steven 1976 *The Conscious Brain* Penguin Books/Pelican

[12] Rubik, Beverly 1992 *The interrelationship between mind and matter* Proceedings of a Conference hosted by the Centre for Frontier Science, Temple University Philadelphia.

[13] Rudolph, Hans-Joachim 1993 *Time patterns and the state of mind* Proc 2nd Gauss Symposium, Munich 2-8 August

[14] Sarkar, Prabhat R. 1978 *Idea and Ideology* Ananda Marga Publications, Calcutta 5th ed.

[15] Sarkar, Prabhat R. 1990 *Microvitum in a nutshell* Ananda Marga Publications, Calcutta, 3rd ed.

[16] Searle, John 1990 *Is the Brain's Mind a Computer Program?* Scientific American **262** (January 1990) p26-31

[17] Shepard, Roger N. 1987 *Toward a Universal Law of Generalization for Psychological Science* Science **237** (Sept 11), p1317-1323.

[18] Shnol, S.E. 1993 *Anomalous multipeaked structures of physical and biochemical rates of reaction* Proc 2nd Gauss Symposium, Munich 2-8 August

[19] Sitko, S.P. and Gizhko, V.V. 1991 *Towards a Quantum Physics of the Living State* J Biol Phys **18**, 1-10

[20] Towsey, Michael and Ghista, Dhanjoo N. 1993 *A Science of Consciousness* Proc 2nd Gauss Symposium, Munich 2-8 August

[21] Whitehead, Alfred 1966 *Modes of Thought* NY, Free Press.

[22] Wilber, Ken 1980 *Eye To Eye - the quest for a new paradigm* Double Day, Anchor Press

[23] Wright, Sewall 1953 *Gene and Organism* American Naturalist **87**, p5-18

Microvita: A New Approach to Matter, Life and Health

R.F. Gauthier

1 Scope

P.R. Sarkar's concept of microvita is proposed as the basis for a new scientific paradigm. Microvita are subtle, sub-microscopic living entities that organise energy to create forms, structures and processes in the universe. Millions of microvita may compose one electron. Quantitative microvita models for the form of an electron and a photon have been derived, using known experimental facts and a proposed spiral motion of microvita. The model explains the experimental value of the z-component of the electron's spin. A new interpretation of quantum theory is proposed, based on microvita. An experimental test of the microvita hypothesis in the area of quantum physics is suggested. Microvita are related to viruses, and may help explain the origin of life and evolution of species through the creation and addition of new genes. The interplay between positive and negative microvita in the human body and mind explains the cause of disease, and suggests a new approach to health.

2 Introduction

There is a search on for new, more holistic approaches to scientific and medical research.[4] One promising approach is the growing body of experimental evidence relating energy and form, particularly biological form.

It appears likely that biologically active or form-carrying information can be stored in water and transmitted through electromagnetic fields. There is the work of Popp [7] on biophotons, and the work of Reid [8] on the transmission of crystal structure information by an electric current. There is the hypothesis of Sheldrake [13] that morphic fields may transmit form and behavioural information without use of electromagnetic or other known physical fields. The work of Benveniste [3] on both in vitro and in vivo biochemical effects at very high dilutions shows that biologically active information can apparently be stored in water. The work of Endler et al [6] on the effect of highly diluted thyroxine in sealed ampules on amphibian development, shows that biochemical information from various substances can even be transmitted through glass with biological effects. However, none of this experimental evidence is supported by an accepted scientific theory, and the subject remains controversial.

3 Microvita

Indian philosopher P.R. Sarkar [12,11] proposed in 1986 a new concept that could lead to such a theory. It is the microvitum, or in plural, microvita. Some of the properties of microvita are described below.

Microvita are living, sub-microscopic, indivisible entities whose function is to organise energy into the structures and processes we call matter, living protoplasmic organisms, and minds. Microvita are themselves aware, having subjective experience, as well as having objective properties that may be determined through research, both empirical and mental. Microvita live, reproduce and die. They may be positive, negative or neutral. Microvita move unbarred throughout the universe, creating matter and individual minds, as well as destroying them.

Thousands of millions of microvita compose a carbon atom, according to Sarkar. So microvita may compose individual electrons as well. The number of microvita in a single electron could be in the millions, or perhaps even less than a million. Microvita can travel through various sensory modalities, as well as through mind, and can convey information from one mind to another. Positive microvita are responsible for the evolution of living protoplasmic beings and individual minds, and for maintaining good physical and mental health. Negative microvita evolve material structures such as sub-atomic particles out of energy, and direct the mind towards matter. Microvita may also be neutral in their effects.

3.1 The Formation of Matter

Energy is formed into material particles and fields by the action of microvita. A model for this for the photon and electron will be described in the section on microvita physics. Quantum phenomena such as the electron's spin and the paradoxical double-slit experiment may be better understood in the light of microvita structuring of physical particles.

3.2 The Evolution of Life

More complex physical and chemical structures are further evolved with the help of more subtle microvita. Protoplasmic organisms are formed with the help of positive microvita, and individual minds emerge at this stage. Bodies and minds evolve with the help of positive microvita, from unicellular organisms through plants and animals to human beings with highly evolved minds and bodies.

3.3 The Nature of Health

Disease and health can be understood in a new light through microvita. Health is a result of a proper balance or psycho-physical parallelism between the waves of the body and the mind. Physical or mental disease is caused by an excess of negative microvita. Positive microvita are therefore necessary to fight the disease-causing negative microvita and restore health. Medical treatment should supply positive microvita to the patient through various means to help the body's natural healing processes.

4 Microvita Physics

The concept of microvita can lead to useful quantitative descriptions in the area of sub-atomic physics by providing new models describing the size and structure of elementary particles such as the electron and photon. The microvita approach also leads to a new interpretation of quantum physics. Quantitative dynamic models of the electron and the photon have been developed based on the concept of microvita and combining known experimental facts about electrons and photons.[9] The models are summarised below.

According to Sarkar, each and every existence has its own peculiar wavelength and its peculiar rhythm. So this is applies to microvita as well. We may interpret rhythm here as vibrational frequency, measured in cycles per second. Wavelength might be measured in metres. So any particular microvitum will have its own frequency and wavelength. The photon or quantum of light energy is also described in terms of frequency and wavelength. Electrons and other sub-atomic particles with a "rest mass" also have a wave nature and can also be described in terms of a frequency (which depends on a particle's total energy) as well as a wavelength, called the de Broglie wavelength (which depends on a particle's velocity or its momentum.)

The frequency of a photon is directly proportional to its energy, while its wavelength is inversely proportional to its momentum. The wavelength of an electron is inversely proportional to its momentum, like the photon. The frequency of an electron is directly proportional to its total energy (its rest energy plus its kinetic energy or energy of motion.) So for both photons and particles with rest mass, the formulas for energy and momentum are

$$E=hf \quad and \quad P=h/L \tag{1}$$

where h is a fundamental physical constant called Planck's constant. These formulas are also true relativistically, i.e. for particles with velocities approaching the speed of light. Photons move at the speed of light, independent of their energy. In a vacuum their velocity is c = 300,000 kilometers per second, approximately.

For particles with rest mass, the energy contained in the particle is given according to Einstein by

$$E = mc^2 \tag{2}$$

where m is the total mass of the particle, which depends on its velocity. The momentum of a particle with mass is given by

$$P = mv \tag{3}$$

where m also depends on the velocity of the particle.

Since microvita are said to compose electrons and atoms, we can assume they compose photons as well. We can assume that microvita have a frequency and a wavelength, and that they carry energy and momentum related to their frequency and wavelength, respectively, similar to elementary particles. But the proportionality constant is not h but a constant much smaller than h, which may be called S.

$$h = NS \tag{4}$$

where N is the number of microvita in an electron, which may be in the millions or less. S is a fundamental constant that relates energy and momentum to the spacio-temporal pattern or form of movement of microvita. We may assume for now that the number of microvita in an electron and in a photon is the same, given by N.

So for microvita in a photon,

$$E = Sf \quad \text{and} \quad P = S/L \tag{5}$$

for the energy and forward momentum of a microvitum, respectively, where f is the frequency and L is the wavelength of a microvitum, and S = h/N .

5 The Microvita Structure of a Photon

A photon has a momentum given by

$$p = h/L \tag{6}$$

(where L is the wavelength) and angular momentum or spin is given by

$$s(photon) = h/(2\pi) \tag{7}$$

The spin of a photon may be positive (in the direction of motion) or negative (in the opposite direction.) Light that is right circularly polarised consists of photons with one type of spin, while left circularly polarised light consists of the other. With circularly polarised light, the electric field direction (perpendicular to the direction of propagation) rotates 360 degrees as the light moves ahead one wavelength. The tip of the electric field vector follows a helical path. So we can assume that the microvita making up a photon also move in a helical path which rotates 360 degrees clockwise or counterclockwise for each advance of one wavelength.

So the rotational movement of the microvita around the axis of the helix accounts for the spin of the photon, while their forward motion parallel to the axis of the helix gives the photon its linear momentum.

There will be a relationship between the radius of the microvita helix for a photon and the wavelength of the photon. (For a photon, a radius is not defined in current physics, only a wavelength.) This radius will tell us something of the microvita structure of a photon.Assume that the total momentum vector P for the photon is directed along the helical path of the microvita at each point on the helix. (See Figure 1.) Let A be the angle that this total momentum vector makes with the direction perpendicular to the photon's linear motion. Then the component of P (the vertical P vector in Figure 1) that contributes to the spin is P cosA.

The component of P (the horizontal P vector in Figure 1) that produces the linear momentum of the photon is P sinA.

So if r is the radius of the helix, then the angular momentum is

$$Angular\ momentum = rP\ cosA = h/(2\pi) \tag{8}$$

while the linear forward momentum is given by

$$Linear\ momentum = P\ sinA = h/L \tag{9}$$

where L is the wavelength (distance between turns of the helix.) Dividing (9) by (8), we get

$$(P\ sinA)/(rP\ cosA) = (h/L)/(h/2\pi) \tag{10}$$

or

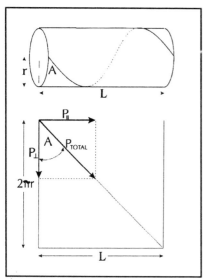

Figure 1
Determination of the radius of the photon helix.

$$(sinA)/(cosA) \ = (2\pi r) / L \tag{11}$$

$$tanA \ = (2\pi r) / L \tag{12}$$

When the microvita move around one circumference ($2\pi r$), they also move forward one wavelength L. So according to this geometrical relationship in Figure 1,

$$tanA = L / (2\pi r) \tag{13}$$

Comparing equations (12) and (13) we see that

$$(2\pi r) / L = L / (2\pi r) \tag{14}$$

This will only be true if

$$tan A = 1 \quad and \quad A = 45^{o} \tag{15}$$

for photons of any wavelength, and

$$L = 2\pi r \tag{16}$$

that is, the circumference of the microvita helix of a photon equals the wavelength of the photon.

Therefore, the radius of the microvita helix of a photon is

$$r = L/(2\pi) \tag{17}$$

where L is the photon's wavelength.

So if the wavelength of a photon increases or decreases, (corresponding to a decrease or increase in the energy and momentum of the photon, respectively,) the radius of its microvita helix changes proportionately.

Mathematically, the coordinates of position of one microvitum moving along its helix of radius R are

$$x = r\ cos(2\pi z/L) \tag{18}$$
$$= (L/2\pi)\ cos(2\pi z/L)\ =\ (1/k)\ cos(kz) \tag{19}$$

$$y = r\ sin(2\pi z/L) \tag{20}$$
$$= (L/2\pi)\ sin(2\pi z/L)\ =\ (1/k)\ sin(kz) \tag{21}$$

where k is the wave number ($k=2\pi/L$), z is the distance along the helix, and x and y are directions perpendicular to z. The maximum slope of these sine and cosine curves is 1, that is, 45 degrees, independent of the wavelength of the photon.

A photon whose energy equals the rest energy of an electron, has a wavelength equal to h/mc. So the radius of the microvita helix for this photon would be, from the above results,

$$r = L/2\pi = h/(2\pi mc) \tag{22}$$

The length h/mc is well known in physics and is called the Compton wavelength. It equals .0243 x 10^{-8} cm for an electron. A photon with this wavelength has an energy corresponding to the rest mass of an electron.

We may model a photon as a spiral of N microvita. All N microvita move along the same spiral path one after the other, like beads sliding along a string. The microvita may be considered to be equally spaced along the photon spiral. If the photon is many wavelengths long, the N microvita would be spread out evenly over the total length, while if the photon is only a few wavelengths long, the N microvita would be concentrated along this shorter total length. In each case the formula E = hf for total energy as a function of frequency would apply. A long and a short photon having the same wavelength and frequency have the same total energy. Let us see how this microvita model for a photon compares with that of an electron composed of microvita.

6 The Microvita Structure of an Electron

An electron is known experimentally to have its own spin or angular momentum. In quantum theory the electron is described as having "intrinsic" spin and no size, although it does have a measurable wavelength which varies inversely with its velocity. The spin is thought to be too small to arise from the orbital motion of particles. This leads to a lack of a visual model for an electron as well as for other particles in quantum theory. But the microvita model below shows that the electron's measured value of spin can arise from a closed spiraling movement of microvita. The microvita model of an electron accounts for the electron's spin, even though the moving microvita themselves are point entities. These microvita have no internal structure or spin, but carry momentum and energy due to their movement in space.

6.1 The Radius of a Microvita Electron

An electron can be first modeled as a closed double-looped circular path of vibrating energy, with microvita travelling around the circle in one rotational direction. Assume that all the microvita in the particle vibrate in parallel in a single circular pattern. The wave motion of the microvita moves around the double-looped circle at the speed of light.

Now assume that the circle's circumference, the distance around the vibrating circle of microvita, is the minimum length for a stable structure. That means the circumference of the circle is one wavelength. Assume that all the microvita in the particle vibrate together at a common frequency and common wavelength, so that the particle's wave pattern is a simple circular one. The radius of the particle can then be calculated by knowing the circle's circumference.

So to obtain a spin of 1/2 for an electron, the microvita structure requires a double circular loop of radius

$$r_e = h/4\pi mc = 1.9 \times 10^{-13} \; meters \qquad (23)$$

This calculated value for the radius of a microvita electron contrasts with the radius of an atom of about 10^{-10} meters and the radius of a small atomic nucleus of around 10^{-15} meters. That is, the electron's radius is calculated to be roughly 1/500 times the radius of an atom but around 200 times the size of a small nucleus.

In this preliminary model therefore, an electron consists of a quantum of energy (proportional to the mass of the electron) shared by perhaps millions of microvita circling at the speed of light on a double-looped path of total length equal to one Compton wavelength.

6.2 Detailed Microvita Structure of the Electron

The actual motion of microvita for the double looped particle above will not be a circular motion but a closed spiral motion with the double circular loop as the center line of the closed spiral. What is the amplitude of the microvita motion around the double loop? We can consider that the microvita structure of an electron is created when the helical path of microvita making a photon is curved so that the centre line of the microvita spiral follows a double-looped circular path and the spiral closes on itself after travelling a centre-line distance of one wavelength (the distance between two successive turns of the corresponding microvita photon spiral.) The circumference of the circle of each single loop is therefore one-half the wavelength of the corresponding microvita photon spiral.

The microvita electron structure is therefore generated by rotating the helix radius, whose length is $(1/2\pi)$ h/mc, around a point moving along a circle of radius $(1/4\pi)$ h/mc, rather than rotating it along a straight line, as with a microvita photon helix. The radius of the helix generating the microvita electron is therefore twice the radius of the double-looped circle. The point travelling along the circle moves twice around the circle, or 720°, while the radius of the helix rotates through 360° about the moving point.

The above operation creates a closed-loop three-dimensional path for microvita (see Figure 2) whose shape is uniquely specified (except for mirror image reversal) by the geometrical parameters of the photon (spin 1) and the electron (spin 1/2), and whose size is determined by the electron's mass (as well as by Planck's constant h and the speed of light c.)

We can assume for our microvita model that a number N of microvita move spread out along the three-dimensional pathway all in the same rotational direction. This microvita electron has several interesting features. First, if it is turned upside down (rotated 180 degrees about the x-axis) the same form is obtained, but with the microvita moving in the opposite direction. This corresponds to an electron whose spin is down (the original orientation is the spin up orientation.) Second, a mirror image of the form gives a "left-handed" version of the form, which cannot be made to coincide with the original form. This corresponds to a positron (the anti-particle of the electron) if the original form is that of an electron.

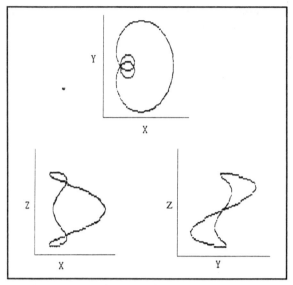

Figure 2
Two-dimensional projections of the
three-dimensional microvita
electron form.

The double-looped microvita model of an electron is consistent with an aspect of quantum theory of electrons related to spin. In the double-looped model, the helical path of microvita composing an electron closes on itself after one turn of the helix spread over two turns around a circle, i.e. after 2x360° or 720°. In quantum theory, the mathematical wave function describing the spin of the electron also has a rotational periodicity of 720°.

The coordinates of the above three-dimensional closed loop microvita structure for an electron can be derived straightforwardly from the above geometrical relationships and are given below. The xy, xz, and yz projections of the three dimensional form are shown in Figure 2.

$$x = r_e \, (1 + 2 \cos(\theta/2)) \, \cos(\theta) \tag{24}$$

$$y = r_e \, (1 + 2 \cos(\theta/2)) \, \sin(\theta) \tag{25}$$

$$z = 2 \, r_e \, \sin(\theta/2) \tag{26}$$

where θ (theta) is an angle running from 0 to 720° (once around each loop of the double-looped circle generating the helix) and

$$r_e = h/4\pi mc \tag{27}$$

the previously calculated microvita electron radius from equation (23).

A detailed calculation of the z component S_z of angular momentum of the microvita electron model above, for N equally spaced (in terms of angle) microvita gives

$$S_z = h/4\pi = \tfrac{1}{2} \, h/2\pi \tag{28}$$

which is exactly the experimentally measured spin of the electron.

7 A New Interpretation of Quantum Theory

The application of microvita concepts in physics goes beyond modelling sub-quantum structures of elementary particles. It introduces a new view of the dynamics of wave-particle systems, presently described by quantum theory. In the microvita model of the photon and the electron, a helical structure of microvita moves through space in different ways to form an electron or a photon. In these structures the total values of energy, momentum and spin are conserved for the electron or the photon composed of microvita. But momentum (in the case of the photon) or both momentum and energy (in the case of the electron whose microvita have variable speeds) are not conserved for each individual microvitum. Rather, the spiraling microvita in an electron exchange energy and momentum among themselves without directly contacting each other, since the microvita are point particles that maintain a certain physical separation. The spiraling microvita in a photon exchange momentum among themselves to keep each microvitum moving along its curving path.

It is the collective behaviour of microvita that makes them intriguing. In EPR-type experiments, so named for Einstein, Podolsky and Rosen [5], the measurement of the state of one particle, say spin orientation, apparently nearly instantaneously affects the state of spin orientation of a second particle which previously interacted with the first particle, and now is separated at a significant distance from it. So the states of the two particles remain correlated even at large separations. Note that in both the electron and the photon, the microvita are moving along their curved paths at speeds always greater than the speed of light, although a photon itself moves forward at the speed of light, and an electron moves always at less than the speed of light. This may be important since quantum systems appear to be able to communicate among themselves at speeds exceeding the speed of light.

In a similar way the states of the microvita in an electron or photon would also be correlated. *So the non-locality that characterizes quantum correlations of distant particles also applies to the microvita composing a single electron or photon.* Bell's theorem [2] and associated quantum experiments [1] established that if there are "hidden variables" underlying quantum phenomena, then the must be of a non-local nature. Microvita would satisfy this requirement for such hidden variables.

In terms of microvita, an electron or a photon (or other elementary physical particle) is a collective quantum system in itself, coordinating its physical component properties in a non-local manner. When two physical particles interact, the combined microvita of both particles form a larger quantum system, and the combined physical properties of this larger system are coordinated non-locally. What is the nature of this non-local coordination of physical properties in a quantum system, either within or among physical particles, in terms of the microvita concept?

In this interpretation of quantum theory, microvita make choices in their collective behaviour as electrons and other physical particles and quantum systems. These choices determine the distribution of energy and momentum among the microvita within a single particle or among different particles in a single quantum system. The choices of the collective microvita are constrained by the requirements of conservation of total energy, momentum, angular momentum, electric charge and any other relevant physical parameters of the combined quantum system. An atom as a combined quantum system would have a combined coordinating capacity corresponding to all its component particles.

How the quantum system actually functions is related to its capacity for choice, as determined by its collective structure of microvita.

8 Testing the Microvita Hypothesis

Certain "paradoxical" phenomena in quantum theory may be more readily understood with the microvita model, for example, the double-slit experiment. In it, individual electrons directed at two parallel slits produce a certain wavelike distribution of electrons detected on the other side of the slits. The pattern reflects the influence of both slits on each electron, and is difficult to explain if the electron is a single particle highly localised is space (it is currently explained in quantum theory as due to the electron's wavelike properties.) In the microvita model, the double slit result can be understood differently. It this view, some of the microvita in an electron could pass through one slit and some pass through the other, and recombine on the other side of the slits to form an electron that displays the observed double-slit behaviour. (The same explanation would apply to photons, composed of many microvita.)

The double-slit experiment suggests an experimental test of the microvita hypothesis that millions or possibly less than one million microvita compose an electron. If the microvita in an electron divide into two groups to pass through a double slit, they should divide into four groups when faced with four slits. If faced with a hundred slits, they should divide into a hundred groups of microvita. The microvita would then combine together on the other side of the screen to form a single electron which would be detected as usual.

Now if there were exactly a million microvita in an electron, then if faced with two million slits, the microvita could pass through no more than half of the slits. The position of the electron re-formed and detected on the other side would depend on which slits the microvita passed through. The microvita composing the next electron would pass through a different set of slits, etc. The statistical pattern of electrons detected on the other side would not be the same as the statistical pattern that would be created according to the predictions of quantum theory, which assumes that all slits, if they are open, contribute to the pattern of electrons observed at the detector on the other side.

However, as long as the number of slits is small, quantum theory would predict the correct distribution of electrons, as all slits would be used by large numbers of microvita. Quantum theory would thus describe the statistical behaviour of large numbers of microvita in their collective behaviour (as physical particles), but should start to break in its predictive power down when microvita are forced to show individual behaviour, for example when passing through many slits.

So the microvita hypothesis predicts a gradual breakdown in the accuracy of the predictions of quantum theory in a multiple slit experiment as the number of slits increases towards a million or more. If one million microvita compose an electron, the breakdown should be apparent with two million or more slits. Such an experiment could even yield a more precise estimate of the number of microvita composing an electron. (The fewer the microvita in an electron, the fewer the number of slits needed before the quantum theory predictions start to break down.) A grid of 1000 by 1000 small, closely spaced holes in a screen would provide one million "slits" conveniently, and the principle of the experiment is the same for slits or holes. So here is a clear, testable prediction of the hypothesis that an electron is composed of on the order of a million microvita or other sub-quantum particles. A positive result implies a breakdown in the accuracy of the predictions of quantum theory under certain conditions. Such a breakdown would have many implications for fundamental physics, as well as other scientific areas relying on physical understanding, such as biology and medicine.

9 Microvita Biology

Microvita can give insight into the nature of the origin and evolution of life.[10] Matter and energy have an inherent tendency to self-organise to form and evolve living beings, in a way which is not yet scientifically understood. Evolution can be seen as the development and expression of desires and longings in material systems, themselves composed of microvita, by the action of other microvita. Elementary particles first form themselves into atoms and molecules. Then simple molecules organise themselves into more complex molecules. Later, living protoplasmic beings can develop.

Microvita first form and then modify genetic molecules so that simple and then more advanced living structures are formed with desires and longings that can be expressed through their associated physical structures or bodies. The collective desires of a biological species to survive in a changing environment would attract microvita carrying information for needed genetic changes. These microvita would enter the reproductive cells of the species. They would form viruses and DNA molecules carrying the new genes necessary for evolving desired changes. In this way genetic changes may be implemented during evolution and spread throughout a group of members of a species. These species members would undergo evolutionary transformations that give them greater psycho-physical expressional capacity in new environments, and a new species or sub-species would be formed.

With microvita, genetic changes would not be random (as in neo-Darwinism) or based only on use or disuse of certain organs (as in Lamarckism). Changes would be in directions desired or needed by organisms in order to survive or develop. Microvita-created evolutionary viruses and DNA originally inserted in the genes during evolution, would become activated in the developmental process of an organism to cause the organism to develop according to its evolutionary pattern.

10 Microvita Medicine

Microvita are described as positive, negative or neutral in relation to physical and mental health.[12] Medicines containing positive microvita from different sources may be developed to help control cancer and other diseases in the future. Already a technique has been developed in Orissa, India for making microvita medicines for treat various diseases. Positive microvita are purportedly attracted from the environment by a mental concentration technique and trapped in sealed glass containers of water. The water is then "potentised" as in homeopathy by diluting with alcohol and shaking, and then used as a medicine. Some encouraging results have been found so far in the treatment of cancer and several other serious diseases.

There are many types of alternative medical practices such as homeopathy, acupuncture, distant healing, etc. which can be effective but whose mechanism of action is not explained in scientific terms. It may be that the action of microvita underlies some of these techniques.

11 Conclusion

Microvita are proposed as a basis for a new, holistic paradigm for physics, biology and medicine. In physics, a microvita model for the dynamic structure of the electron explains its experimentally measured z-component of spin as due to a group of point microvita circulating faster than light. A proposed multi-slit electron experiment provides a direct test of the hypothesis that an electron is composed of millions of microvita.

References

[1] A. Aspect and P. Grangier. Experiments on Einstein-Podolsky-Rosen-type correlations with pairs of visible photons. In R. Penrose and C. J. Isham, editors, *Quantum Concepts in Space and Time*. Oxford University Press, 1986.

[2] J.S. Bell. *Speakable and Unspeakable in Quantum Mechanics*. Cambridge University Press, Cambridge,1987.

[3] J. Benveniste. Transfer of biological activity by electromagnetic fields, *Frontier Perspectives*, Vol. 3, No. 2, Fall, 1993.

[4] J.P. Briggs and F.D. Peat. *Looking Glass Universe*. Fontana, London, 1985

[5] A. Einstein, B. Podolsky, and N. Rosen, Can quantum-mechanical description of physical reality be considered complete?, *Phys. Rev.*, 47:777-780, 1935.

[6] P.C. Endler, W. Pongratz, and C.W. Smith. Effects of highly diluted successed thyroxine on amphibian development, *Frontier Perspectives*, Vol.3, No. 2, Fall, 1993.

[7] F.A. Popp, K.H. Li, and Q. Gu., editors. *Recent Advances in Biophoton Research and its Applications*. World Scientific Publishing Co., London, 1992.

[8] B.L. Reid. The ability of an electric current to carry information for crystal growth patterns. *Journal of Biological Physics*, 15:33-36, 1987.

[9] A. Rudreshananda. Do microvita form sub-atomic particles?, *New Renaissance*, 3:2:21-25, Mainz, 1992.

[10] A. Rudreshananda. *Microvita: Cosmic Seeds of Life*. Dharma Verlag, Mainz, 1988.

[11] A. Rudreshananda. P.R. Sarkar's concept of microvita. in *Shrii P.R. Sarkar and His Mission*, pages 28-40, Ananda Marga Publications, Tiljala, Calcutta 700039, 1992.

[12] P.R. Sarkar. *Microvitum in a Nutshell*. Ananda Marga Publications, Calcutta, 1991.

[13] R. Sheldrake. A New Science of Life: the Hypothesis of Formative Causation. Blond and Briggs, London, 1981.

The Quantum Field of the Healing Force

N.N. Vasilyev

1 Scope

A preliminary scientific approach is suggested to describe the healing process. The necessity arises because of difficulties in curing many present-day diseases by modern medicine. The description is based on the recognition of the existence of a healing force field, which is responsible for the healing of any disease. This force field is represented using quantum electrodynamics. *The quanta of the healing force field are considered to be microvita.* They can be positive or negative with respect to human health and surroundings. Healing occurs as a result of the attraction of positive microvita, which compensate a predominant influence of illness-creating negative microvita. In this context a new paradigm for gaining scientific knowledge is suggested.

2 Introduction

The Eastern tradition considers that a medical effect depends not so much on external treatments but on some supreme Force [2] that creates a healing force field. External treatments may include different procedures and remedies which might be described as objective. On the other hand, the action of the healing force field might be supra-sensory, and may arise as the result of an internal appeal. Sometimes that is fulfilled instantly, avoiding external treatments and going beyond known medical concepts.

Nevertheless, as a rule, healing requires some time to be completed. In a healing process, a moment may come when the assurance of overcoming the crisis arises. The recovery then seems to be the result of the external treatments and time. Physical objects or living beings may be transmitters of the healing force. Their action may extend over some distance. Herein we will recall certain previously known approaches to healing and to suggest how these may relate to a new scientific approach to understanding the nature of the healing process.

3 General Considerations

The concept of a healing force field requires a short background description. In accordance with the Indian tradition [7,5], the highest being, i.e. Cosmic consciousness (or, simply, Consciousness) is pure and unexpressed. In principle it cannot be described intellectually, but can be realized through direct experience. This Consciousness expresses itself through its Operative Principle, which through its Force emanates microvita [8], the fundamental quanta of the Force's field. These microvita organise energy to form all physical and mental objects, events and actions at all scales. This microvita field is defined in six spaces, giving rise to worlds [7] which can be described with the help of some mathematical functions. The pure Consciousness is the highest, 7th space, which is qualitatively unlimited and featureless; there is no function defined in it. One can use a parameter $s = 1,...,6$ corresponding to the six spaces of expressed Consciousness, with the lowest space having $s = 1$ and corresponding to our ordinary three-dimensional space developing in time. The spaces U_s form a complex hierarchical structure in which the higher spaces enclose the lower ones:

$$U_{s-1} \in U_s, \quad s = 2,...,7$$

The 6th world, the highest mental level, is usually considered to be outside of space and time. In western philosophy, it is known as Plato's world of ideas. This world can be experienced after certain intuitional practices, but it is also known to creative thinkers at moments of creativity. At these times the Force has been described as breaking through eternity into time. But from a mathematical point of view, one should not speak of an absence of space-time, but rather of radically distinct space-time properties. The spaces U_6 and U_7 are inconceivable for an ordinary mind, just as for a "flat" being of a two-dimensional surface it is impossible to imagine a three-dimensional space which encloses the surface. The space-time properties and dimensions of the spaces differ, but are assumed to be related to those of Riemann spaces in relativistic quantum mechanics. This is undoubtedly true for the first and lowest space, which includes only the physical world evolving in time. It is likely to be true for the second space, which is the world of desires and actions related to the physical world, and the third space which is the world of thinking and remembering [7]; the higher spaces are of a more complex nature and are only partially reflected in the lower spaces.

At the individual level, these six levels of expressed Consciousness correspond to the physical body ($s=1$) and five levels of individual mind ($s=2,3,4,5,6$), with the levels differing by their expressions--the physical sensations (hearing, touch, vision, taste and smell) in level 1, and various subtler mental propensities (desires, feelings, thoughts, and ideas) in levels 2-6.

Microvita are postulated to be unstructured, i.e. qualitatively elemental entities [8]. The idea is close to that of Leibniz's monads. According to their subtlety, microvita can be perceived through their actions and reflected mentally in concepts. Thus, they are mostly investigated in one's internal mental laboratory. Since microvita are the agents or operators in the force field's substrata, (which include propensities and sensations), both are media for their movements in the individual mind.

From the point of view of human health and quality of existence, microvita may be positive, neutral and negative, the first ones being closer to the higher and subtler worlds, the latter to the cruder. Human health is based upon the harmonic balance of microvita at every individual level. When diseases develop, the balance is disturbed by a surplus of negative microvita. The healing force field, i.e. a flow of positive microvita, restores the harmony and thus destroys the source of disease.

Although the field of microvita exists irrespective of human awareness of it, one can consciously and intelligently utilize microvita. [8]. This implies an active internal and external action. People have been always doing this, with variable success. But the conditions of high efficiency of the conscious interference is not still clearly known. Nevertheless, these have been given by the Eastern tradition, and are personal peace and proper concentration on the effect intended. This can be a kind of an appeal to the Force to descend into the place to be cured: an imagination of light, warmth, etc. But even peacefulness in all the mind's levels and of the physical body, i.e. relaxation, creates conditions for Force's released activity in the most harmonizing way. The highest efficiency of conscious action is achieved by persons who succeed in intuitional practices [2]. In the limit, these persons open a supramental vision, which means taking possession of the supreme Force. The supramental vision is inalienably active and creative, knowing a true causality. The physical body is seen as an expression of this causality [4].

This supramental transformation is not easy to accomplish as it requires patient work to calm all internal processes, beginning with the intellect and lower mind's activity, then going through feelings and sensations down to a "cellular" mind. That also means finding peace at every level of the human being, from the top down to the bottom. The supramental transforming force field comes down to the cells of the physical body, causing a sympathetic resonance at all levels. Once this is done, any disease becomes the object of the influence of the supramental healing force and supramental law. An application of the supramental force for healing causes an immense curing effect. In human history are many examples of this integral transformation: mental, vital and physical. This has also been the highest aim of alchemy of different peoples and times, with the conversion of the base metals into gold only being a symbolic expression or part of the process. [1]

While the manifestations of the healing force field corresponding to physical sensations are often considered to be objective, those corresponding to propensities and ideas are thought to subjective or supra-sensory. The Eastern tradition has always included all human sensations and propensities in a united scheme, differing from the Western approach. The former has worked out an approach to investigate these subtle supra- and extra-sensory realms. It is to become a witness [7,5] so that one can observe internal objects as separate from oneself. The level of the observing position should be higher than that of the object. Indeed, this rule of investigation of reality does not differ fundamentally from that of Western science. But while Eastern science has explored the supra- and extra-sensory realms, Western science has emphasized the sensory realms. To research feelings by means of the mind is common to both traditions, but to research lower mental realms with the help of higher mental realms is the speciality of the Eastern approach. This research yielded its own objective laws, which do not depend on the person involved, although personal experiences of some 'internal objects' may differ. This is a more subtle approach to the problem of measurement. And as in the physical sciences, different persons usually get somewhat different experimental results, without infringing on the laws, so it is with experiments in the supra- and extra-sensory realms.

4 An Approach to a Theoretical Description

The field of the universal Force in visual sensation is described by the known theory of photon fields. Quantum electrodynamical photon annihilation and creation operators at the state α, $a_\alpha^\pm$, may describe microvita action. According to the definition of microvita as agents of the overall Operative Principle [8], photons (as the quanta of the electromagnetic field) are suggested to be elemental microvita and the quanta of the field in visual sensation. The eigenvalues of those operators, defined at coherent states, are described by their eigenvectors [6].

As the above-mentioned spaces were suggested to have Reimann-like geometro-physical properties, the laws of quantum statistics would be applicable. Nevertheless, higher spaces are presumed to be more deterministic and less statistical in nature, so that greater causality exists at higher levels. As an approximation, one could extend the above approach over the remaining physical sensations and all the subtler mental propensities. Then *microvita would be considered to be quantum objects and quanta of the fields in the corresponding spaces*, with microvita annihilation and creation operators being introduced. The "quantum object" mentioned is commonly accepted to mean that which can only be "measured" with a fundamental and significant change of its state taking place. The human mind, the most subtle and important instrument for investigating microvita, influences their states in every case, except the case when a quite high level of mind is reached. The latter marks the next, higher stage of cognition of reality, where the laws of reality are inseparable from those of cognition and extend beyond known physical laws.

The number of microvita is postulated to be quantitatively unlimited in the α-state [8]. This permits one to consider them to be quasi-bosons. They have an integer spin-like parameter, quasi-spin, and are described by Bose-Einstein statistics for an ideal boson gas, i.e. without interaction. The set of parameters characterizing the α-state, $\{\alpha\}$ in the phase spaces includes: an index of microvita type (sensations or various propensities), an index of space s where these microvita move, a characteristic frequency ν, a quasi-spin q, and a wave vector k.

Heard sounds are connected with physical vibrations only in the spaces with $s = 1,2$ (the physical world and the world of physical desires). Remembered sounds are related to the spaces with $s = 3$. Developing increased sensitivity, one can hear subtle sounds of higher spaces with $s = 4-6$, which are not related to acoustic physical vibrations but are the result of higher mental development.

As microvita are considered to be massless [8], they cannot have zero projection of quasi-spin but only two non-zero projections along the propagation vector, according to quantum electrodynamics. The quasi-spin is suggested to be the parameter related to individual influences of microvita. (An alternative is to introduce a microvita charge inherent in all types of microvita and which can be positive or negative.) Let us then propose these two projections to be related to positive and negative microvita; one can accept the anti-parallel projection as related to negative microvita and the parallel projection to positive microvita. However, the general effect of microvita on people in similar circumstances is often different, as their reactions are different. For example, the attitude to the same case of rapid recovery can be gratitude, unbelief or curiosity. People are influenced by different types of microvita beyond those used in healing, and this produces such variable results.

Each type of microvita can be associated with its Hilbert space, built on the basis of the coherent states' eigenvectors $/\alpha >$. Points of this phase space correspond to the states in the representation of the coherent states. Then, the mean number of microvita is given by

$$< N_\alpha > \; = \; < \alpha \, / \, a_\alpha^+ a_\alpha^- \, / \, \alpha > \; = \; / \, \alpha_- \, / ^2 ,$$

where α_- is the eigenvalue of the operator $| a_\alpha^- >$.

Generally, a boson system is described by mixed states, and a density matrix ρ. The entropy of the system gives a measure of chaos and is maximum if $\rho_{\alpha\alpha} = const(\alpha)$. On the contrary, complete order and the maximum of information can only be reached if the system is kept in a coherent (pure) state.

In the healing process, it is important that the coherent state mentioned would be one of positive microvita. Fixed negative states would create physical and mental pathologies like cancer and obsession. They can be described in terms of generation of a coherent radiation of corresponding negative microvita with a positive feedback loop, as in laser theory. On the other hand, considering this mechanism to be coherent generation of microvita, one could build a theory of positive psychological states of mind based on an amplification of positive microvita flow. Indeed, this result can be obtained by already known mental practices. Valuable results are reached when certain types of negative microvita are completely absent in a person's physical and mental structures, i.e. at local zero temperature for those negative microvita boson systems. This corresponds to purified states of mind, which are reached with the help of mental training. Negative microvita systems remaining would be associated only with necessary functioning of the physical body. This is the condition of health.

5 A New Scientific Paradigm

A complex idea, visual images, music, psychological complexes and formations can be reflected by the movement of collective bodies of elemental microvita in different spaces. These bodies can also be considered to be entities described by the laws of ideal boson systems. In particular, an image of a certain person may carry subtle vibrations which can be mixed with a whole set of feelings, thoughts and ideas. Then this visual symbol may be a conductor of different kinds of collective bodies of microvita to be manifested in several spaces at once. So, the transfer of an image's properties from one place to another is not just due to visual photons, but due to simultaneous complex movement of microvita in different spaces.

The theoretical approach described above is limited, first of all, by its statistical nature. This reveals a temporary scientific weakness, and it should be definitely supplemented by a deeper description. Besides, dialectics teaches that there cannot be the same laws applicable to different levels of existence: physical inanimate matter, life and mind. The laws of life and mind were discovered mostly in the East and ancient Egypt, and are partially studied by modern occultism and esoteric science. Those laws were expressed in their special languages and terms. It seems reasonable to look for contact points with those expressions of this knowledge, and to convert the mosaic of our knowledge into a unified vision. A physical field model seemed to have been a way to a synthesis. But efforts in the 20th century have not appeared very fruitful, despite the efforts of the most outstanding scientists. The main reason is the intellect was trying to solve an essentially, on its own level, insoluble problem. For the intellect manifests only one, and not the highest, power in the cluster of universal forces, and is incapable either of describing them completely or radically changing them. A higher entity should be included into an integral picture of the universe.

Such a unifying entity has been known since Vedic times: "A child of waters, a child of woods, a child of things moving and unmoving, even a stone has it", says the ancient Rig Veda [3]. This is a psychic being [5], (Purusha Antaratman, in Sanksrit), which carries the original Consciousness and is manifested in its Force, called Agni, as a causal background of the material and psychic worlds. Vedic wisdom saw Consciousness to be behind all manifested things, events and processes. The universe is only a symbolization of the Agni-Force, and is not a self-existing and self-sustaining entity.

Microvita are seen to be quanta of a force field in a descending movement of the supramental Force from Consciousness down to matter. They represent forms and forces of descending psychic beings manifested in the universal spaces, from the 6th down to the 1st.

The acceptance of the unifying idea of psychic beings, and Agni as their original and supreme Force, would not mean a radical shift of the scientific method of obtaining knowledge. Some reflection of this idea was always retained by some thinkers during the evolution of science, especially in philosophy (atomists, neo-platonists, Leibniz's monadology etc.) In this century, the idea has persisted in the natural sciences, mostly in the quantum theory of physics. It may be time *to change the scientific paradigm*, as was done at the beginning of the century in physics.

First, science in its aspiration to integrate, to make an order and synthesis, should accept a unifying entity as a basic element of the all-pervading Consciousness, whatever name will it be called: *psychic being, Antaratman, or monad*. It is notable the term "atom" originally meant just that elemental, basic entity of the universe, in Greek philosophy. Obviously, integral knowledge would be impossible without this fundamental acceptance based on millennial human wisdom.

Then science can overcome its present limitations by acknowledging *the superiority of the psychic being over the intellect*. This is a point of much resistance. The intellectual, civilized human being is considered by many to be the highest and final creature of universal evolution. But if that were true, there would not have been the disastrous global crisis which has been guided by the reasoning intellect; there would not have been grief, sorrow and depression in human minds; there would not have been the aversions towards and perversions of so many brilliant ideas. The intellect is not capable of changing life radically. This unfortunate situation has its solution: to accept a limitation on the intellect and subject it to the superior psychic being.

Once that is done, a direct way is opened to get essential knowledge through an identification of the psychic being with the object researched. This process is spontaneous and natural since a psychic being is released from the many-leveled fetters of the mind via the calming intuitional practices mentioned. Indeed, this identification has always been used by outstanding scientists, but it has been usually applied to get knowledge of the physical world, grounded in materialistic and positivistic or neo-positivistic conceptions. The material reality of the universe was thus discovered. By means of this identification, other realities have also been investigated in other Eastern and Western traditions, such as: vital energy supporting life; astral forms and forces supporting feelings and emotions; mental forms grounding thoughts and ideas; and supramental realms. There is no use insisting that one of these realities is more real or proper than the rest. They are all real and objective if realised through identification. Nevertheless, there is a hierarchy of knowledge and realities, of the spaces described above. The laws of higher spaces, with larger parameter *s*, predominate over the lower ones. For example a human being is not bound to all the laws of ants or sponges, but is able to intervene and change the expressions of these laws.

This identification is a crucial part of the proposed new paradigm. On the other hand, intellectual expression should be a secondary source of meanings as it is often insufficient and inefficient, and sometimes harmful, for alone it may misguide. Sometimes other symbolic languages may be more adequate to express and convey knowledge about the force fields of some spaces; for instance, higher realisations may be attained by mental incantation of mantras. Various component levels of the total human structure may be realised: the physical, vital, emotional, mental and supramental levels. If this is so, these other expressions do not have to be less valuable than intellectual meanings in their description of material and higher realities. Many consider mathematics to be the highest language, but mathematics is just a kind of identification; the realities discovered relate to some mental level or to the fields in higher, though not the highest, spaces. On the other hand, recognising the principle of identification as a main tool of science, opens not only wider realms of knowledge, but also higher states of being for human beings. The latter is surely the most important role of knowledge.

While peace of mind is being developed via intuitional practices, a person's external surroundings progressively reflect a more subtle experience. This creative reflection follows the same law of identification of the psychic being in manifesting its descent or elevation through microvita activity. Creativity is a natural consequence of this identification, at all levels, in all spaces of the universe. In the physical world, it is the creation of new forms of matter. The most impressive example has been officially documented alchemical conversions of base metals into gold [1], and the artificial conversions of nuclei in physics. It is a science of the future to create necessary external conditions and to transform matter for the needs of progressing psychic beings, not by external activity but through an internal process of identification.

Creativity as an inalienable law of universal development is manifested in a double movement of descending psychic beings (devolution [5]) and of elevating psychic beings (evolution), which is mediated via microvita activity. This process develops in accordance with an individual law of psychic beings--a causal law of Karma [5] which is called fate. Applied to human beings, this law admits a conscious process of identification with more elevated states of internal being, and thus permits a great widening of individual consciousness into the higher spaces up to the Supreme Consciousness, the origin of all-pervading creativity. This opens a way to extremely fast progress up to a supramental being. The latter implies an evolution of humanity into a supramental race [4]. But this is a topic which goes beyond this work.

6 Conclusion

A preliminary approach is proposed to describe healing. It is based on a proposed new scientific paradigm and on a concept of a healing force field applied by a healer (living or material) to a patient. The new paradigm proposed here is as follows:
 - the recognition of the psychic being as a unifying entity of the universe;
 - the acknowledgement of the psychic being's superiority over the intellect and other expressions of knowledge;
 - the acceptance of acquiring knowledge through identification of the psychic being with the object researched, as a general method of science; this expands the sphere of science into the vast realms of the supra- and extra-sensory universe which have been previously considered subjective realms;
 - admitting other languages or expressional forms as having equal rights with intellectual language, to adequately express different levels of reality; the hierarchy of knowledge is based on the levels of realities expressed--not on the means or the languages of their expression.

This identification is creative. It can create any form, or master any process in the universe visible and invisible. Therefore it is also an origin and essence of universal development, which is of double movement: elevation (evolution) and descent (devolution) of psychic beings.

Based on this proposed paradigm, on the identification experienced and described in the Indian tradition, a new cosmology and cosmogony emerge from this double movement: continuing devolution from and elevation to the original Consciousness, of which psychic beings are parts and at the same time complete representatives. The universe is considered to be a permanent manifestation of a field of the Force of Consciousness or of psychic beings' forces in a hierarchy of six spaces (one physical and five mental, from the viewpoint of individual minds.)

The known theoretical approach of secondary quantization, previously applied only in physical space, is proposed to be applied to describe this field in the five mental spaces. The quanta of the field in all these spaces of sensations, feelings, emotions, thoughts and ideas are unified by a general term "microvita". These are quasi-bosons of α-states which are defined by: a specific index of microvita type, an index of the space where they move, a frequency band, a quasi-spin and a wave vector. *Quasi-spin is considered to be the most important parameter for healing.* Healing is qualitatively described in terms of coherent generation or transmission of curing (positive) microvita to overcome a surplus of illness-creating (negative) microvita.

This short consideration of healing action is inevitably limited both theoretically and in concrete models and methods. Undoubtedly however, the topic will be extended and deepened on the basis of a general approach to Eastern and Western cultural and linguistic traditions.

References

[1] For example, W. Ganzenmüller, *Die Alchemie im Mittelalter*. - Paderbon, 1938; N. Sivin, *Chinese Alchemy: Preliminary Studies*. - Cambridge, Mass., 1968; V.B. Dash, *Alchemy and Metallic Medicines in Avurveda*. - New Delhi, 1986.

[2] A. Ghose, *The Basis of Yoga*. - Pondicherry, 14th ed., 1981.

[3] A. Ghose, *Hymns to the Mystic Fire* - Pondicherry, 6th imp., 1991.

[4] A. Ghose, *The Supramental Manifestation on the Earth*. - Pondicherry, 2nd imp., 1980.

[5] A. Ghose, *The Synthesis of Yoga*. - Pondicherry, 9th ed., 1980.

[6] J.R.Klauder, *Coherent States: Applications in Physics and Mathematical Physics*. - World Scientific Publications, 1985.

[7] P.R.Sarkar, *Ananda Sutram*. Ananda Marga Publications, Calcutta, 1989.

[8] P.R. Sarkar, *Microvitum in a Nutshell*. Ananda Marga Publications, Calcutta, 1991.

Is Another Physics Needed to Explain Fundamental Fluctuations in Physical Measurements?

V.A. Kolombet

1 Scope

In the course of empirical investigations of spontaneous alterations in the measured rates of biochemical reactions, new quantum effects were found. Theoretical attempts to interpret these effects lead to a new physical idea of the existence of a "non-Bornian" physics based on a modified link between the wave function and probability. Herein we deal with new "non-Bornian" effects in physics and biology.

2 Introduction

The essential part of the theory of measurements is the analysis of distributions of values measured. This is a reason why during our investigations we first focused our attention on the surprising discreteness observed in the amplitude of deviations of measurement results [4]. Instead of smooth unimodal distributions, there are often spectra of thin peaks or states. The main result of our work is a law describing the distance between these peaks. It appears that this law in its main features is the same for different physical and chemical phenomena, and is characterized by a new dimensionless constant D. Because the value of D appears to be closely related to fundamental physical constants, we can consider the states as "fundamental states" and transitions between fundamental states as "fundamental transitions". The transitions appear during the course of repetitions of measurements - the discreteness of these states leads to the appearance of reversible "fundamental transitions", called "fundamental fluctuations".

3 Fundamental Fluctuations in Biochemistry

The phenomenon of "fundamental fluctuations" was first discovered in the course of biochemical measurements, as apparently random noise in the results of repetitions of measurements of ATP-ase activity of actomyosin solutions [4]. Subsequent samples of the same enzyme solution lead to sharply differing values of the rate of respective biochemical reactions. Controlled experiments showed that experimental errors were small. Moreover sometimes -- during a long set of repetitions (in other words, during monitoring of the rate of a biochemical reaction) -- one could observe nonreversible discrete transitions from an old mean rate of reaction to a new one. Mean magnitudes of reaction rates on these "macrostates" form a system of equidistant levels - first, second, third ...

Fundamental fluctuations did not depend on the type of enzyme used, the experimenter, the equipment and other variables. For instance, the same discrete states were discovered during the investigation of the rate of the creatine kinase reaction [5].

A very interesting feature of fundamental fluctuations is that when the mean value of the rate decreases from one discrete level to another then the variance of fluctuations increases proportionally. When this decrease becomes 1/3 (and respectively 2/3) of the distance between "macrostates" then the linear link between the decrease and variance vanishes.

The phenomenon of "fundamental fluctuations" has been under study for nearly thirty-five years (the first paper is Ref.4), but only at the present moment can we say something essential about its inner nature. It appears that the values of the relative variances (dispersions) of fundamental fluctuations represent equidistant spectra of states. They are also quantized. The quantum of relative variance -- the dimensionless constant D -- was calculated from experimental data whose accuracy was one and a half percent. This precision is sufficient for the reconstruction of the physical nature of states and transitions observed.

4 Fundamental Fluctuations in Physics

Because all quantum processes of usual quantum physics are based on Planck's constant "h", we concentrate our attention on the probable link between the dimensionless constant D and Planck's constant "h". If the link really exists, one should find a way to transform "h" to a dimensionless constant too. Since Einstein's special theory of relativity, physicists know that unification of temporal and spatial coordinates gives a simple relation between a meter and a second: 1sec equals nearly 300,000,000m. In this way the dimension of "h" becomes simpler and coincides with the dimension of the square of electric charge. In other words, one of the simplest dimensionless representations of Planck's constant is the value of the so-called fine structure constant alpha, because alpha is proportional to the square of electric charge, too.

The link under question appears to be the following: D with high precision coincides with alpha to the power of 3/2. Because alpha is a probability and the square of the respective psi-function, then the constant D represents the third power of a psi-function. (Naturally, the electric charge is a very special case of a psi-function but nevertheless it is really a psi-function, see for instance Ref.1). Fundamental fluctuations this way are probably described by psi-cubed expressions. This result is new one because all observable values usually represent even powers of the psi-function (as a consequence of transition from amplitudes to probabilities).

If we are right, then Max Born's interpretation of the statistical nature of wave functions is true, but another truth exists too: there are properties of matter described by some other functions, and in order to transform this function to probability, one must raise it not to second power as usual but to the power of two-thirds as for example in the case of physics of mass [2] (see below also) and maybe to some others in the case of some other phenomena. For example, in the case of fundamental fluctuations, one must raise D to the power of two-thirds to get alpha, belonging to a probability level.

If so then there are some interesting consequences, in particular,

1. there is no explanation for fundamental fluctuations in framework of traditional physics, and

2. the existence of at least some unsolved physical problems could be the result of accepting only Born's interpretation of the link between wave function and probability.

In order to check this hypothesis, one should consider the most fundamental unsolved physical problems such as the question of the physical nature of mass and electric charge, the question of the mass-spectrum of elementary particles, the question "Why 137?" (an integer nearly equal to the reciprocal of the fine structure constant alpha) and so on.

It appears that using our extended Born's interpretation may lead to simple and effective steps to the solution of some of these questions. In particular, the ratios of the masses of elementary particles to the mass of the electron appear to be the values described by psi-cubed expressions; these ratios raised to power of two-thirds represent almost integer numbers [2]. See Figure 1.

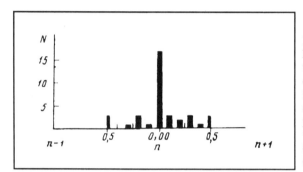

Figure 1
Distribution of the fractional parts of the magnitudes $(m_x/m_e)^{2/3}$ calculated from mean mass m_x of isotopic multiplets of light elementary particles. All sufficiently precisely (errors < 0.5) determined values $(m_x/m_e)^{2/3}$ except values $(m_x/m_e)^{2/3}$ of all heavy particles (psi, K, B, upsilons,...) were used.

Travelling through this "terra incognito", we found another example of fundamental fluctuations. The fluctuating value in this case is the value of measured mass of an object. The first hint of the existence of fundamental fluctuations of mass was the following. After describing in Ref.2 some strongly determined corrections of the masses of elementary particles, the familiar equidistant spectrum of states (levels) with abnormalities at 1/3 (and respectively at 2/3) of distance between neighboring levels was suspected.

Do fundamental fluctuations in mass occur? In order to answer this question one should consider experimental deviations in the results of mass measurements. Do discrete states exist? Does the value D characterizing the distance between peaks exist?

We analyzed published data concerning measurements of different masses, from the mass measurements of the electron neutrino to weighing macroscopic objects by means of precise balances. All cases led to the result (more precisely, to the strong suspicion) that fundamental fluctuations of mass measurements really exist, and the transitions between discrete states are mainly described by the value D. Nevertheless fundamental fluctuations of mass measurements and of the speed of biochemical reactions differ sharply from each other by some integer parameter. The parameter for reaction rate measurements is 1 and for mass measurements is 3. This difference could explain why fundamental fluctuations were first found in biochemistry -- for in biochemistry they are near thousand times greater. This "one-to-three ratio" of parameters seems to be very promising for the development of the theory of fundamental fluctuations.

5 Consequences of Fundamental Fluctuations

There are many practical consequences of our research results. Some of them are in the medical and biological areas. For example, our phenomenological ideas lead to prediction of a unique resonant mass at the vicinity of near 28,000 proton masses. Masses of light-weight proteins are situated in this region, and we can test this mass region by observing the masses of precisely measured proteins. Because proteins are modified during biological evolution, they could change their masses to or from the predicted value. Distribution of protein masses really does have a peak at vicinity of the predicted value [3]. See Figure 2.

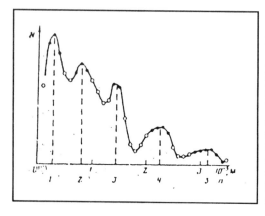

Figure 2
    Smoothed distribution of masses of proteins appears to be the system of broad peaks at M_1, M_2, M_3 ... . Values $(M_n /M_1)^{2/3}$ represent the set of numbers 1, 2, 3, ... The picture can be easy distorted by different factors (measurement errors, estimated values instead of precise measurement results, multiprotein complexes existence and so on). The value of M_1 appears in course of investigation of the fundamental fluctuations as a unique resonant mass.

Another resonant mass is 2.1 billion proton masses. This is the region of mass for example of the genetic material of light-weight viruses. Are viruses sensitive to this special mass? Yes, our preliminary investigation shows that the masses of single-chain DNA and RNA of viruses tend to be close to this predicted mass.

6 Conclusion

The main theoretical result of this investigation of fundamental fluctuations is their probable link with a new region of physics missed by other scientists. In some features the situation may resemble the discoveries of non-euclidian geometries by changing the famous fifth postulate of euclidian geometry (parallel lines never meet). Born's interpretation is a postulate of physics and we are trying to explore new areas of physics by varying this postulate. This article predicts the existence of a new region of physics, in which many old problems of physics may have simple solutions. Many striking phenomena are in this new region. Macroscopic fluctuations are only one way of penetrating this new physics in our usual world. Others are through the values of masses of elementary particles, abnormalities in protein distributions, etc. When this physics is more developed the set of phenomena explained may be much larger.

    Maybe the proposed "microvita" phenomenon is not even the strangest one in this field of "non-Bornian" physics.

Acknowledgements

The author is grateful to Prof. S.E.Shnoll for support and helpful discussions and to Dr. Richard Gauthier for editing the paper.

References

[1] R.P. Feynman. *QED. The Strange Theory of Light and Matter.* Princeton University Press, 1985.

[2] V.A. Kolombet. Macroscopic fluctuations, masses of particles and discrete space-time. *Biophysics*, 37:402-409, 1992.

[3] V.A. Kolombet. On the probable existence of a system of special values of protein masses. *Biophyzika*, 31:426-429, 1986.

[4] S.E. Shnoll, On spontaneous synchronous transitions of actomyosin molecules from one state to another. *Voprosy Meditsinskoi Khimii.* 4:443-454, 1957.

[5] S.E. Shnoll and E.P. Chetverikova. Synchronous reversible alterations in enzymatic activity in actomyosin and creatine kinase preparations. *Biophys. Biochem. Acta*, 403:89-97, 1975.

Scientific Research into Consciousness

B.D. Josephson

The following few chapters are concerned with the problem of consciousness. Their authors do not subscribe to the conventional view that consciousness is 'just another problem', to be tackled in a somewhat similar way to the way that science has tackled problems in the past. They believe, and argue in the following, that including conscious experience within the scope of science will involve much more radical changes. Science so far has been based on 'objectifying' the world, and conscious experience has also a 'subjective' aspect. As Beverly Rubik observes in the first of these chapters, conventional science has explored only certain aspects of consciousness, and in so doing has asked only limited questions, leading to limited understanding. It excludes for example from its considerations experiences such as those of a spiritual, mystical or psychic nature, and the so-called 'near-death experiences', all of which "have the power to change people's lives" Conventional science (as Towsey and Ghista note in their chapter) does not handle the *meaning* of information, whilst our consciousness does. In the section entitled 'Obstacles to a Scientific Exploration of Consciousness', Rubik discusses the factors (such as emphasis on reductionist methods of analysis and investigation, and adherence to Descartes' unjustifiable assumption of a fundamental split between consciousness and the material world) which lead current science to have difficulty with consciousness, and sketches in its place proposed outlines of a scientific epistemology that will not exclude consciousness. She notes that 'mind-body medicine' (a field of study concerned with details of the influence of consciousness on the body) provides already an instance of a field of scientific study where the conventional assumptions break down.

My own chapter which follows next focuses on the fact that conventional thinking has led certain scientists and philosophers to make remarkable assertions regarding everyday beliefs, such as the assertion that such beliefs are best characterised as works of fiction, only science having the ability to tell us of the truth. I argue a case that the modes of thinking involved in drawing such conclusions are pathological consequences of effectively denying the reality of processes of ordinary life and retreating thereby into a world of abstractions, and conclude with the question, who is really dabbling with fiction, the scientist-philosopher in his world of abstractions, or the common man?

There follow three chapters discussing specific kinds of direct evidence that argue in favour of deviations from the conventional points of view. David Lorimer focuses on the 'near-death experiences' that have been experienced by many critically ill people. These experiences indicate that a person who to ordinary observation is clinically dead may in reality be enjoying clear experiences of his surroundings, albeit from a viewpoint that is located outside his body. From this and other observations Lorimer argues that "consciousness itself is non-physical and extrasensory", and that we have "non-physical as well as physical senses or organs of perception". He argues also, as Rubik does, that some nominally 'subjective' experiences have such transformative power as to deserve to be categorised as encounters with something *real*.

Next Deborah Delanoy discusses experimental evidence provided by parapsychology suggestive of the same conclusion, i.e. that consciousness can act independently of the body. Recent methods of data analysis indicate a degree of consistency between the results of experiments on a given category of paranormal phenomenon (e.g. psychokinesis or precognition) which would be surprising if the measured effects were purely the result of experimental defects, which latter one would expect to vary considerably from one experiment on a given category of phenomenon to another. These analyses cast doubt on commonly made claims to the effect that psychic phenomena are irreparably irreproducible. Delanoy suggests that it appears "increasingly likely" that what is involved in these phenomena is a "genuinely new, hitherto unrecognised characteristic of mind or consciousness". Following this, Daniel Benor reviews the body of research on spiritual healing, which suggests that healing has objectively measurable effects, some occurring at a considerable distance. His speculations of the mechanisms involved lead him to propose that some subtle form of energy, perhaps bridging the gap between consciousness and matter, is involved in these processes.

In the concluding chapter of this section, entitled "The Science of Consciousness", Michael Towsey and Dhanjoo Ghista describe the approach of Yoga Psychology. They observe (following Wilber) that in explorations of conscious experience the experimenter's mind serves as the experimental apparatus, whose scope can be increased by suitable preparatory disciplines which increase its sensitivity to subtle experiences. Merging one's consciousness with objects, as opposed to becoming separate or distant from them, is the essential key to success, and those who have followed the path before can show the way to others by communicating their own experiences to them, thereby attesting to the essential intersubjectivity or universality of conscious experience. The authors argue that these methods allow conscious experience to be studied in a scientific manner. They conclude with speculations concerning how experiences in consciousness might be related to the 'objective' phenomena of conventional science.

The Challenge of Consciousness Study

B. Rubik

1 Introduction

Those unfamiliar with consciousness study may profess that science has not encountered any anomalous phenomena resulting from consciousness actively interacting with the physical realm. As a result, they may be puzzled about this paper or the entire section on consciousness in this book. However, conventional science has only explored certain aspects of consciousness such as cognition and perception, and in so doing has asked only limited questions. Science is bounded by the present dominant paradigm, which maintains that consciousness is a localised epiphenomenon of complex physical brain processes that manifests only passively in the physical world. Thus, the research questions posed by the scientific community typically reflect this view.

In relation to this, all that we know in science about nature is gleaned through the types of questions that we ask. Not only should scientists carefully consider their questions, they should also consider the limitations of their questions and how to enlarge their scope of inquiry. Indeed, some do pose a different set of questions that go beyond or even challenge the dominant paradigm. These novel questions take them into entirely new terrain, to the furthest frontiers of science, but it is not without opposition.

The history of science has shown repeatedly that novel ideas or questions that challenge the status quo are rejected, and their proponents suffer extraordinary obstacles. This is true regardless of the stature of these scientists in the scientific community. It is virtually impossible to publish the resulting ideas or data in respectable, widely read scientific journals, and one may also face loss of funding and camaraderie. Occasionally such maverick questions lead to significant breakthroughs as seen throughout the history of science. They have also led to some notable failures. On the other hand, entirely new fields of science have dawned in this manner, and this has greatly enhanced scientific progress. The present is no exception to this; there are several "frontier sciences", that is, areas of science not yet mainstream. One example of an emerging frontier science is bioelectromagnetics, the interaction with biological systems of low intensity, extremely low frequency electromagnetic fields.

The subject of this paper is the challenge of consciousness study, the frontier science exploring the interrelationship between mind and matter. It is an emerging interdisciplinary field of inquiry that has drawn an increasing number of noteworthy scientists and other scholars in recent years. However, some of the research in this area challenges not only the dominant paradigm, but also many of the assumptions at the foundation of science.

2 Evidence from Experiments on Consciousness Interactions with the Physical Realm

One important sphere of evidence for the mind-matter interrelationship that is now rapidly growing is the interface of the mind-body system in biomedicine. Different states of consciousness, beliefs, attitudes, emotions, and intentions have been shown to play active roles in bodily functions, health, and the healing process. A new field of medicine has sprung forth called mind-body medicine [1], which goes well beyond the older psychosomatic medicine. In addition to being possible causal factors in disease aetiology, mental and emotional states can be mobilised to help cure disease, even serious chronic degenerative diseases such as cancer. Among the positive interventions of consciousness, there is the well known placebo response that occurs in over 30% of all patients, whereby inert substances such as sugar pills or injections of saline reduce pain and may also cure disease, if the patient believes they will. There is now a plethora of mind-body techniques used in healing, such as biofeedback, yoga, dance therapy, affirmation and prayer, visualisation and relaxation techniques [2]. Changes in the immune system function correlated with changes in thought, belief, and other states of mind are well documented in the new field of psychoneuroimmunology, and these form the basis of medical applications to improve immune function [3]. Research from psychiatry on multiple personality disorders indicates that the patient's physiology shifts measurably with each personality. These studies show that changes in allergy profiles, disease states, visual acuity, and EEG occur instantaneously upon shifts in personality [4]. The evidence suggests that the mind-body is a fundamental indivisible unit, that consciousness interacts with the body to yield significant physical effects. There is no known mechanism for these phenomena that begins with a shift in consciousness and ends with a physiological response. However, conventional science that poses questions of the underlying biological mechanism has gleaned some insight into central nervous system activity. This conventional biological approach lies outside of the scope of this paper and will not be discussed here.

Numerous experiments have been done that address the question of whether conscious intentionality can interact with or ascertain information concerning a physical process beyond the level of the individual. It is statistically well documented (see for example the chapter by Delanoy in this book) that the conscious intention of operators can apparently influence the outcome of sensitive electronic or mechanical devices such as random event generators [5, 6]. Remote perception, the capability to ascertain information nonlocally without any known sensory cues has also been demonstrated [7]. Dozens of studies have shown that prayers and meditations of widely separated individuals correlate with a significant improvement in the health and well being of others [8].

In 1989 an international roundtable meeting to explore various approaches to consciousness research was convened at Temple University. Composed of sixteen pioneering scientists and scholars working at the frontiers of biology, engineering science, medicine, physics, psychology, and the philosophy of science for over a decade, these investigators have been asking novel questions on the interrelationship between mind and matter, studying phenomena that challenge presently accepted scientific concepts. Their results are reported in a proceedings of the meeting [9]. Although these findings may seem to defy conventional science, it must be pointed out that there is no official body of science that applies to conscious beings. We have only a science of the material realm that has no clear position on the possibility of an actively participating consciousness.

A large number of case studies on extraordinary inner experiences can be found in the psychological and anthropological literature. Reported phenomena include near-death experiences, dreams, spiritual or mystical insights, creative discoveries, and psychic experiences, all of which are presently unaccepted by science. Of course, these reports are by no means equivalent to controlled experiments. However, they are much too prevalent to ignore. Moreover, anecdotes regarding such phenomena have been reported in every culture to date. Although these phenomena are similar in nature across many cultures, they remain outside of orthodox science.

3 Obstacles to a Scientific Exploration of Consciousness

Besides the political difficulties in studying unconventional topics in science mentioned earlier, consciousness study poses several intellectual problems that challenge science. As already noted, some experiments for example show correlations well beyond chance expectations between different states of consciousness or conscious intention, and specific measurable changes in the physical world. First of all, such study on the interaction of consciousness and the physical world challenges the conventional notion of mind as a local epiphenomenon of brain processes and nothing more. Secondly, it challenges the long standing separation between the realms of mind and matter known as Cartesian dualism. Third, it challenges the conventional notion of causality which is physical. Fourth, some of the phenomena in this area challenge our usual notions of space and time, as physical effects associated with conscious intent may occur nonlocally and either before, simultaneously, or after the time period of applied intentionality.

The issue of causality warrants further discussion. Causality in physical science is ascribed to physical events, not mental states. Consciousness has conventionally been conceived as being caused by brain processes, not causal in itself. However, one of the interpretations of quantum theory, which is by no means widely accepted, is that the collapse of the wave function requires consciousness in the act of measurement. Causality in conventional biology and psychology is "upward"; for example, changes in consciousness are explained in terms of chemical and physical changes in the brain. This is the basis of molecular reductionism, the dominant paradigm in biology and medicine. On the other hand, the effect of states of consciousness to change brain activity has been termed "downward causation" by Roger Sperry [10]. Although this term exists in the neuroscience literature, it is considered unconventional by most biologists and physicists. Nonetheless, an extension of the notion of "downward" causation may be invoked to describe changes in the physical world correlated to shifts in conscious intent. A third possible position consistent with the results from consciousness study is that consciousness is neither caused nor causal, but acausal; that is, merely a synchronous, meaningful correlation with observed physical events.

The exclusion of inner subjective experience from our science has allowed us to admit only certain limited features of consciousness into scientific inquiry such as perception and cognition, features which presumably could be measured for participants in the laboratory. However, it has excluded much larger dimensions of consciousness. Inner experiences such as dreams and creative insights have the power to change people's lives, the face of science itself, and even the world, yet tend to be dismissed when serious scientific investigation is proposed.

Perhaps one of the most important challenges to science raised by consciousness study concerns the assumed separability of the observer from the observed in the act of scientific observation. Because consciousness is highly interactive and displays nonlocality, how can we draw a line between where our consciousness stops and that of another being begins?

There is no suitable scientific epistemology from which to explore the whole of consciousness. The lack of an adequate epistemological framework has truly hampered its scientific study. Understanding consciousness, which is the basis of all our experience, including science itself, should be part of a complete science that addresses the physical, biological, and human realms.

A deep exploration of consciousness is important not just for consciousness study, but for all of science. Consider that experiments in the psychophysics of perception show that vision, despite scientific training, has elements of subjectivity. For example, research on visual perception reveals that the persistence of visual illusions and the experience of virtual reality indicate that the "external" world is largely a projection of expectations. The history of science shows that cultural as well as individual beliefs have been "projected" into what scientists actually "see." Hence, science is context dependent. A notable example is that of Van Leewanhoek, an early microscopist, who proclaimed that he saw tiny babies in the heads of human sperm, which was the cultural belief then and for 2,000 years prior.

What we term as "external" may indeed be intertwined with our mental orientation, attitude, expectations, and beliefs. The experienced so-called "external" world is actually private to each human. Thus, no pure "objectivity" exists. Only consensus reality is possible, and science may be considered the consensus product of the scientific community. This also means that what we refer to as the "physical" world becomes a subset of the full range of experience.

To solve all these dilemmas, I propose an integral role of consciousness within science and its epistemology. The question is, then, what types of conceptual frameworks can be used for the scientific exploration of consciousness, both ordinary and extraordinary experiences? How must science be revised to integrate consciousness?

4 Toward a Scientific Epistemology that Embraces Consciousness

An international roundtable meeting of fifteen key scientists and philosophers interested in this field was held recently to explore this question. It was convened by Willis Harman of the Institute of Noetic Sciences of Sausalito, CA, who has pursued this question for a number of years and has published similar ideas on this topic [11, 12]. The remainder of this paper is a summary of the work of this group.

Consider the epistemology of reductionist science, which is the basis of classical physics, biology, psychology, and medicine; in short, the ordinary world of human experience. Reductionist science offers a third-person perspective pertaining to those phenomena whose observations are completely public. It is maintained that such a science is objective, that is, refers to the external world and is also unbiased. Observations are expected to be repeatable by all observers, regardless of their beliefs, attitudes, or persuasions. Of course, reductionist science suffers from Cartesian dualism, in that it omits consciousness from the onset; reality and scientific laws are assumed to exist independently of conscious beings. Consequently, the subjective domain goes unstudied. Quantitative methodology is the ideal in reductionist science; that which cannot be measured is not taken very seriously. Reductionist science ignores the persona of the scientist, who is assumed to be completely irrelevant to the scientific method. Beyond the use of blinded and double-blinded controls in experiments, possible subtle influences by experimenters are ignored. Reductionist science aims at causal understanding, prediction, and control. It maintains that the world may be dissected into discrete parts for analysis, although the ontology, methodology, and epistemology are typically confused in practice. Causality is assumed to be physical and reductionistic. In reductionist science, the observer is considered to be separate from the observed, and in effect, this constitutes a particular stance or posture toward the observed, although this is considered "objective."

Consider the following proposed characteristics for science that accommodates consciousness and also regards it as a fundamental that cannot be ignored.

1. The rift between the first- and third- person is considerably reduced. In addressing consciousness, the experimenter and the subject are in a more similar position with respect to the data. They can merely offer their own private experiencing of their phenomenal worlds and thus become collaborators or partners in exploring consciousness.

2. The subjective domain is acknowledged to be "real" although not reducible to physical mechanism. Thus, public observations of the "external" world and inwardly-directed observations are both appropriate for scientific inquiry. However, it is acknowledged that the subjective will provide a different level of knowing. First-person inquiry will provide intelligibility, whereas third-person inquiry will provide explanation; these are complementary.

3. Objectivity in the sense of unbiased inquiry remains a goal. It is recognised that open inquiry in consciousness study can lead to intersubjective consensus for third-person inquiry, and this is the most that science can achieve. It is acknowledged that different levels of spiritual or psychic development will lead to differences in first-person inquiry of consciousness, but a commitment to unbiased inquiry can, nonetheless, remain a goal.

4. Although the phenomena of the subjective domain are not strictly repeatable, they are repeatable in terms of similarity. For example, there is a considerable similarity in near-death experiences and in the types of achievements demonstrated worldwide in research studies on extraordinary human abilities, such as in psi research (as described in the chapter of Delanoy).

5. Interaction is expected between the observer with the observed in all observations, both inner- and outer-directed. This is true of the experimenter consciously exploring his or her own inner states as well as the experimenter exploring the consciousness of another being.

6. Many years ago, the psychologist William James advocated a "radical empiricism" [13] that addresses the totality of human experience, including sensory data, the broad spectrum of inner subjective experiences, and the experience of the physical world. It is timely to launch a new science that encompasses this totality, recognising that in all science, even in conventional science that has tried to ignore or explain away consciousness, the consciousness of the human being is central.

7. A qualitative methodology is most appropriate to study consciousness in most cases. Such heuristic methods have already been developed for social and humanistic psychology. It is acknowledged that quantitative approaches used in conventional science will, in most cases, not be useful to study consciousness.

8. The scientist's persona is critical in the observation process. It is acknowledged that each scientist brings a unique genetic profile, experience, and personal way of seeing. A corollary of this is that creative discovery in science is not just an accident that may happen to anyone, but a unique process of exploration. In addition, the scientist may also be transformed in the exploration process through a shift in consciousness, especially in the personal exploration of consciousness.

9. There is a need to reassess science in the light of unconscious influences. For example, scientific creativity and discovery are related to flashes of intuition that may emerge from the unconscious in response to conscious, active work on a particular scientific puzzle.

10. There is an emphasis on the unity of experience, as represented in people's stories and the "maps" of conventional science, recognising that these two facets, anecdotal and map-like evidence, together constitute all human knowledge. The use of reductionism as a method is not excluded, but its limitations are noted. Reductionism *per se* is not part of the ontology or epistemology.

11. The concept of causality is extended to include "downward" as well as the conventional "upward" notion. Furthermore, it is admitted that causality itself is a limited concept useful in depicting the dynamics of simple physical systems, but it becomes meaningless for very complex living systems that are conscious. Acausal synchronicity may be a useful concept here.

12. Science becomes participatory, whereby the observer cooperates, experiences, or identifies with the observed. Thus, a partnership is envisioned between researcher and the observed. It is recognised that the consciousness of the scientist and of the scientific community are both intimately part of science, which is a human construct.

5 Conclusions

The problems with the present scientific epistemology have been examined in the light of new evidence emerging from the frontier science of consciousness study, and applications in mind-body medicine. A revised epistemology is necessary in order to embrace the full potential of human consciousness. What is proposed is an epistemology centred around the primacy of consciousness. The characteristics of this epistemology are only sketched here, offering a provisional epistemology that may need further refinement as guided by a more extensive exploration of consciousness.

A radically new outlook in science as is proposed here would have extensive repercussions in the world. In this regard, conventional reductionist science presently constrains us to assume a certain posture toward the observed, nature, from which we feel separate. Although we may consider this "objective", it has created a scenario in which we are alienated from nature, inventing technology to fight, control, and manipulate it. It is expected that the "participatory" outlook in which scientists enter into relationship with the observed, imbued with consciousness from the start, will produce a different outcome: a science in which we feel part of nature, and a technology whereby we take greater responsibility for balancing, maintaining, and restoring natural processes.

Finally, if scientists have not had or cannot admit to having exceptional inner experiences presently unaccepted in science, the data from consciousness study will hardly be convincing. Significant change in human endeavours begins with an inner transformation that is usually much deeper than the tip-of-the-iceberg of consciousness known as intellect. I concede that it would take a considerable shift in the consciousness of individual scientists and the scientific community to adopt this type of proposal.

References

[1] J. Achterberg, L. Dossey, J. S. Gordon, et al. *Mind/Body Interventions*. In B. Rubik, B. Berman, R. Pavek, et al. (eds.) Alternative Medicine: Expanding Medical Horizons. U.S. Government Printing Office, Washington, D.C., in press, 1994.

[2] H. Benson, J. Beary, et al. *The relaxation response*. Psychiatry **37** (1974), 37–46,.

[3] R. Ader. *Developmental psychoneuroimmunology*. Dev. Psychobiol. **16** (1983), 251–267.

[4] F. Putnam. *Differential autonomic nervous systems in multiple personality disorders*. Psychiatric Research J. **31** (1990), 251–260.

[5] R. G. Jahn, B. Dunne, and R.D. Nelson. *Engineering anomalies research*. J. Scientific Exploration **1** (1987), 21–50.

[6] H. Schmidt. *Collapse of the state vector and PK effect*. Foundations of Physics **12** (1992), 565–581.

[7] H. E. Puthoff and R. Targ. *A perceptual channel for information transferred over a conglomerate disturbance: historical perspective and research*. Proc. IEEE **64** (1976), 329–354.

[8] R. C. Byrd. *Positive therapeutic effects of intercessory prayer in a coronary care unit population*. Southern Med. J. **81** (1988), 826–29.

[9] B. Rubik (ed.). *The Interrelationship Between Mind and Matter*. Center for Frontier Sciences at Temple University, Philadelphia, 1994.

[10] R. Sperry. *Mind/brain interaction: mentalism, yes; dualism, no*. Neuroscience **2** (1980), 195–206.

[11] W. W. Harman. *Towards an adequate epistemology for the scientific exploration of consciousness*. J. Scientific Exploration **7** (1993), 133–143.

[12] W. W. Harman. *Toward an adequate science of consciousness*. Noetic Sciences Review **27** (Autumn 1993), 77–78,

[13] W. James. *Essays in Radical Empiricism*. Longmans, Green, and Co., New York, 1912.

First-Person Experience and the Scientific Exploration of Consciousness

B.D. Josephson

1 Introduction

What makes conscious experience a difficult or confusing subject for science to deal with is its personal or individualistic character (that is to say the fact that a given experience is an experience apparently tied to a particular individual). It is in this respect very different from the other phenomena studied by science, where while the phenomena may be observed by a particular individual they are considered to be in principle independent of that individual. To say that an individual's experience is merely the functioning of that individual's brain does not fully resolve the problem, since experiences are so very different in nature from brain processes.

The temptation for science is to ignore this first-person aspect of consciousness altogether, and treat it as no different in principle from any other natural phenomenon. A number of workers on consciousness would claim, furthermore, that as far as science is concerned, the presence of consciousness adds nothing new in principle: all the *behaviour* observed in conscious beings such as ourselves can be understood entirely in terms of brain function, without invoking any new laws or principles relating specifically to consciousness. There is a clear opposition between the point of view just stated, and the viewpoint taken in these chapters on consciousness, which presuppose the validity of a first-person point of view, and take it for granted that study of conscious experience can provide valuable input to our understanding of the mind.

Some such scientists seem to feel almost compelled to take a negative attitude in regard to everyday knowledge of consciousness and of mind (that utilising everyday concepts such as believing, intending and desiring, and referred to in the scientific literature as 'folk psychology'). The psychologist Mandler [1], for example, fairly typically dismisses folk psychology as 'convenient fictions that help our limited understanding', while speaking of science on the other hand as a 'productive game', namely the creation of models that enable one to understand and to predict; while Dennett [2] puts forth strongly the idea that our everyday thinking is suspect, citing circumstances in which people are seriously wrong about what they are doing and why they are doing it. Specifically, they 'fill in gaps in what they know, guess, speculate and mistake theorising for observing'. Although sincere in what they say, they are nevertheless 'unwitting creators of fiction'. Even when a person says he is (say) imagining a purple cow, or that he feels puzzled, we should not take these assertions as factual in nature, even if the speaker believes it to be true. The line followed by Dennett and Mandler appears to be this: only scientific method (as conventionally understood) can lead us to certainty, or to the facts of a situation; hence ordinary thinking leads us to something other than the facts; and if it leads to something other than the facts then what it leads us to must be non-fact, that is to say fiction.

2 An alternative perspective

In what has been described above, one kind of thought is being used to cast doubt upon the validity of another, an interesting situation indeed. Should we trust such kinds of argument? Might a person not be equally entitled to argue that a kind of thinking whose conclusion is to doubt commonly accepted ways of thinking should itself be doubted?

To clear our minds in regard to such questions, it may be helpful to consider how it came about that scientific method attained to such an exalted status as it currently does obtain. The answer would seem to be a combination of two things, the great achievements of the scientific method, and the fact that, not infrequently in the past, commonly held opinions have been disproved by application of the methods of science. These facts would be evidentially more compelling if it were not for the converse facts that great achievements have come about in the past on the basis of methods other than science, while there have also been occasions where scientific beliefs have proved incorrect because faulty scientific assumptions were made. Thus both science and everyday opinion can be wrong as well as right, and it is up to the practitioners of each to minimise errors as far as they can.

The assumption may remain, however (as the above quotations indicate), that science, even if admittedly sometimes in error, has a unique ability to get at the truth that everyday analysis does not. I cannot myself help thinking that this assumption represents nothing more than a narcissistic attitude that the scientist has towards the products of himself and his companions in the scientific enterprise. In the context of understanding the mind, it entails the dogmatic assertion that people do not really understand their minds while scientific psychologists do. Is this actually so? The scientist's claim rests on the fact that models or theories are produced which can *predict in advance* the results that new experiments will produce. Can these theories, however, allow the scientist to cope well with the flow of real life situations in the way that people do? The answer here is no. Mandler's characterisation of folk psychology as 'convenient fictions that help our limited understanding' can very well be countered by the assertion that the scientist's understanding of the mind, however good it may be in the realm of carefully chosen experiments, is extremely limited and fragmentary when applied to the context of ordinary life.

In conclusion, then, ordinary knowledge derived on the basis of introspection is good, if we look in an *appropriate* context, and is often better in such situations than is scientific knowledge. The question may be asked, what is meant by good knowledge, and how do we acquire it? In short, the answer is that knowledge is something that we *use*, and the good or bad outcomes of its use determine whether we judge the knowledge as being good or valueless. We may become expert at assessing knowledge by paying careful attention to the outcomes of acting upon various types of knowledge.

Studying the arguments of Dennett in depth one sees that they amount to the following: that the concepts that seem so clear in everyday use have no clear meaning *in science*, and they must therefore be dismissed as nonsense, or at least as unreliable. We should rather see this situation as implying a judgement on *conventional science*'s inadequacies. More discussion of this point is given in Sec. 4.

There are many alternative frameworks of thought to those of science, each with the potential to arrive at its own meaningful truths. In reference 3 many examples were listed: the humanities and the arts in general; spiritual or religious practices, creativity; the use of symbol, myth, and metaphor; the role of the feminine; the historical perspective; and cross-cultural aspects of knowledge.

3 Concrete experience vs. abstractions

There are significant differences between the first-person, experiential approaches to the study of conscious experience and superficially similar approaches such as philosophy, differences that parallel the differences between science itself and the philosophy of science. These differences can be characterised as differences along the abstract/concrete dimension, or as differences in immediacy. There is a subtle difference between stating what one experiences and philosophising about it, related to the difference between the idea of an experience and the experience itself, and the chapters of this book on consciousness focus on experience in a way that philosophy generally does not. In short, *experience is capable of teaching us things that words or symbolisms cannot.*

In connection with this last point, observe that in scientific discussions of consciousness it is commonly stated that the goal of investigations into consciousness is fully achieved if they can provide a full *abstract description* of the experiences concerned. One must question the terms on which such an understanding would be satisfactory. From the standpoint of abstractions itself (the usual scientific perspective), abstractions may be felt to give satisfactory accounts of experience, and yet from the point of view of experience itself or of life, abstractions seem inadequate substitutes.

We thus see the unbalanced consequences of excessive attachment to seeing things purely from a scientific perspective. There seems to be a genuine necessity, not addressed by the scientific, third-person perspective, for non-abstract ways of appreciating consciousness. Possibly everyday experience provides us with such non-abstract aids to interpretation, giving us a kind of knowing that is not quite an abstraction, but may nevertheless function as a source of the abstractions that we subsequently make. It may also be that literature, with its emphasis on metaphorical representation, can provide better contact with this kind of knowing than can the highly formal representations of knowledge that science utilises.

In the case of investigations centred around a first-person perspective, one may admittedly be unable to accede to some of the usual demands of science such as the ability to attach an instrument and take measurements, and to obtain results independent of the individuality of the person whose experiences are being investigated; but rigour, and the evaluation of the significance of any hypotheses that may be proposed, need not be compromised because of this, and openness to the need to change one's ideas if contrary evidence should emerge is in no way excluded, any more than it is in the case of ordinary life. But the mere fact that a scientific argument says one thing and other approaches say something else does not provide sufficient grounds *in itself* for the abandonment of the non-scientific belief, as this conflict may merely represent alternative ways of thinking about the same thing. The case of the problem of the self, to be discussed in Sec. 6, provides a good illustration of this theme.

4 Working with the new perspective

We now examine some of the ramifications of what has been discussed above, especially in regard to some of the claims of Dennett to which reference has been made previously. As we have seen, one of Dennett's claims is that since there is much confusion in people's minds concerning conscious experience then we should treat all people's utterances as if they were works of fiction. Note at work here a characteristic of the scientific approach, namely the demand for a uniform approach to all situations of a given kind (contrasting with the commonplace approach of forming one's own judgements with regard to the circumstances under which a given type of statement may reasonably be accepted). The alternative, developed in Sec. 2 above, is to recognise that people's utterances or beliefs have practical uses, and to judge them in accord with how effective they are in such a context.

A simple example to illustrate this point is provided by the letters on the visual display units of most computers. Such letters are in reality not letter shaped at all, but are made up out of a number of conjoined rectangles (pixels). From one point of view (i.e. that of characterising the geometrical shapes on the screen) it is incorrect, or a piece of fiction, to talk of there being letters, words, and sentences on the screen. But from another point of view (the wider context in which the screen is providing us with information that we require), our perceptions do not lie: they are providing us with the precise information that we need.

An interesting feature of everyday descriptions and everyday language is that by taking into account some of the subtleties of the way our minds work they are able to make effective distinctions of a kind that science currently does not make. An example is the case, discussed at considerable length by Dennett, of optical illusions. Dennett's discussion follows by now familiar lines: the illusory pink ring displayed on the jacket of the book is not actually there (as demonstrated by colorimetric measurements) and merely 'seems' to be there. It is not there as a fact, and must therefore be deemed fiction ("there's no such thing as a ring that merely *seems* to be there"). But ordinary thought has ways of transcending such apparent contradictions, for example by introducing the *concept* of illusion, clearly distinct from that of fiction (a ring which appeared to exist, but under closer examination was seen to be something else, would be an illusory one; while a ring that was said to exist but could not be discovered by a person who looked for it in the appropriate manner would be a fictional one). The holograms used for example on credit cards to deter forgery provide an interesting case of an illusory image that is effectively 'objective', and has a very definite practical value.

The distinction to be made can be expressed in this way: science relies to a high degree on making its judgements mechanical, whilst everyday life makes freer use of 'judgements of experience' that cannot be readily mechanised. If we exclude 'judgements of experience' we are left with a void in our knowledge. Dennett and others erroneously deny the possibilities of developing sound judgements of experience, and optimistically assert that science can in the course of time fill the void that is left. But why deny ourselves the use and fruits of using the powerful tool for studying consciousness that we already have?

I must close by commenting that none of the above should be taken as an unconditional attack on being scientific. When science can show convincingly that opinions need to be adjusted since they do not hold up under close examination (concerning which kind of situation Dennett gives a variety of examples), the corresponding views should indeed be adjusted. It is however necessary to take into account, as previous examples have shown, that different descriptions suit different purposes, and that the localised purposes of scientific description may be different from those of the totality of life.

5 The reality of the field of consciousness

A place where Dennett's arguments and the common-sense view of reality appear to be in collision lies in his claim to have proved that there is no specific field of consciousness (called by him 'Cartesian Theatre'), observed or possessed by a specific observer. Examination of his arguments shows that they depend on refuting particular formulations of the idea of an observer observing a field of consciousness, generally deriving from Descartes' proposal that the conscious self has a particular location in the brain. Once we extricate ourselves from such models and adopt a genuinely phenomenological/experiential approach, the alleged contradictions largely disappear. For example, there is no contradiction, as Dennett appears to believe, between his 'multiple drafts' picture (which does appear to be a necessary assumption in order to account for the results of certain experiments) and the detailed aspects of conscious experience, the latter of which may very well include such elements of a fleeting or provisional character as the multiple drafts picture requires. Contradictions come only from insisting on the metaphor of a theatre stage, on to which conscious images enter and depart like actors, requiring that a given entity be either 'in consciousness' (on the 'stage'), or not.

We do of course make judgements as to whether we are aware of something or not, and appear to arrive at definitive yes/no answers when we do so, apparently confirming the Cartesian Theatre idea. Going into the issue more deeply however, we recognise that in such cases our perceptions are subtly influenced by our judgements. Once we have decided the answer to a question concerning awareness of something (e.g. footsteps) is 'yes' we seem to be rather clearly aware of the percept of interest, while if on the other hand our judgement is that it is not there, our perception changes into one wherein the relevant percept is felt to be definitively absent. This discreteness in judged percepts is thus a consequence of our having decided to impose upon our perceptions a *theory* about those perceptions, thereby essentially vetoing perceptions of an intermediate nature, ones conforming less with our predetermined preconceptions. The mode of perception, in conformity with Dennett's ideas, where what we experience can be identified precisely with available descriptions of our experience, is not the totality of possibilities. There can be alternatively a receptive, non-judgmental state, that is open to experiences of a fundamentally (as far as one is concerned at the time of the initial encounter) elusive kind. Of such experiences Dennett argues that there is nothing we can do other than point to them saying 'that experience'. This is only partly the case. A new experience can gradually become linked to other experiences, and thereby acquire its own significance. Further, the significance of a given experience does not have to be closed or definite; it may be possible for it to augment itself indefinitely through accumulated experience of it.

6 The concept of the self

Dennett's theory of the self, conceived as an agent that perceives and acts, is formulated on the basis of the third-person perspective. He sees it as a creative abstraction that helps survival by designating what should be given favourable attention. From the first-person point of view these can be perceived as interesting philosophical ideas that are not however particularly helpful from the point of view of getting on with life and are perhaps best left for the philosopher to enjoy himself with. For the purposes of life, the actual concrete experience essential. Typically, I may have a recognisable experience of causing something to happen (e.g. making a car move faster by pressing down on the accelerator), or of programming myself to do certain things (such as to turn right at a particular place). Whether my 'I' is concrete in nature or an abstraction is generally not a matter of moment for me as a philosopher might think it ought to be: what really matters as far as I am concerned in ordinary life is simply the need to keep track by any means available of might be happening to the 'I', so that experiences may be avoided of a kind that I wish to avoid. Other people (scientists: philosophers) may have other models for what is happening under such conditions; if for their own purposes they wish to think in these ways I may be content to let them do it, but my personal interest is (say) to drive to the beginning of a walk that I wish to go on, and experience indicates that it is more productive (to borrow Mandler's word) simply to use the 'myself' model than to analyse what state my brain might be need to be in for the achievement of my goal.

The actual *experiences* associated with the concept of myself or my soul are also of value in connection with personal development; however, this topic lies outside the scope of this chapter (see, however, the chapter by Towsey and Ghista).

In this last section we have gained a glimpse into the pathology that is liable to attend over-reliance on the principles of the scientific, third-person approach. The concepts of 'myself' and 'I' that underlie our ordinary actions are simple ones, closely linked to actual experiences and thus to our reality. Science constructs its own, different version of reality that is extremely useful for its own purposes. However, when it starts to create explanatory models that purport to account for our experiences but do not fully fit with the totality of those experiences one may not unreasonably suggest that if the models are to be taken seriously than they should be adjusted to conform better to the latter. Otherwise, who is dabbling with fiction?

7 Conclusion

This article was written primarily as a counter to attacks made by a number of scientists on experiential approaches to studying consciousness. Its main tenets are that there is nothing sacred or special about the scientific approach to knowledge, and that all knowledge-claims need to be judged in terms that are appropriate to the claims that might be made in any given case. With this understanding, many of the criticisms of the experiential approaches are seen to be inappropriate.

Acknowledgements

The ideas contained in this article have been considerably influenced by discussions carried out over the course of a number of years with Prof. Steven Rosen and Dr. Beverly Rubik. I should like to thank them for their inspiration.

References

[1] G. Mandler, *Mind and Body*, Norton, 1984.
[2] D. Dennett, *Consciousness Explained*, Allen Lane, 1992.

[3] B. D. Josephson and B. Rubik, *The Challenge of Consciousness Research*, Frontier Perspectives (Center for Frontier Sciences at Temple University) **3**:1 (1992), 15–19 (electronic version available in the IPPE preprint archive: World Wide Web at URL http://phil-preprints.L.chiba-u.ac.jp/IPPE.html; ftp site Phil-Preprints.L.Chiba-U.ac.jp: directories Philosophy of Mind and /pub/preprints/Phil_of_Mind respectively).

The Near-Death Experience and the Nature of Consciousness

D. Lorimer

Since the publication of Raymond Moody's popular book 'Life after Life' in 1975, a good deal of academic research [1] has been conducted into the nature and effects of the near-death experience (NDE), a characteristic type of experience that has been reported by many individuals who have been for a time in a state close to that of death, but survived to tell of their experience, often involving an apparent entry into other realms of existence. Many investigators begin with a sceptical position, but invariably come to the conclusion that the NDE exists as a phenomenon, although they differ in their interpretations, which range from the physiological and pharmacological to the psychological and the spiritual. These interpretations reflect the number of levels at which the NDE can be analysed across a range of disciplines; it is quite possible that they are complementary rather than mutually exclusive. The basic issue discussed in this paper is whether the NDE can be satisfactorily accounted for by materialist theories of mind.

Two theories about the nature of consciousness can be stated very succinctly and were advanced at the end of the last century by the American philosopher and pioneering psychologist William James and the Oxford philosopher F. C. S. Schiller. James started from the statement 'Thought is a function of the brain', asking whether acceptance of this doctrine logically compels us to disbelieve in immortality [2]. He argues that the dependence of thought on the brain can be interpreted in two ways: either thought (and consciousness) are actually produced by the brain, or else thought (and consciousness) is transmitted by or through the brain. I call these respectively the 'productive' and 'transmissive' theories of consciousness. The productive theory entails that NDEs are in some sense produced by the brain, and that consciousness is extinguished at bodily death; the transmissive theory admits possible correlations with brain processes, but leaves open the possibility that death is not so much an extinction as a transformation of consciousness.

Schiller puts the argument in the following form as his 'final answer to materialism', proposing an inversion of the materialistic understanding of consciousness: 'Matter is not that which produces consciousness, but that which limits and confines its intensity within certain limits: material organisation does not construct consciousness out of arrangements of atoms, but contracts its manifestation within the sphere which it permits' [3].

He goes on to argue that this explanation does not involve either the denial of the facts of materialism with its insistence on the unity of life and continuity of all existence; it admits the connection of matter and consciousness, but denies that matter actually gives rise to consciousness. And, as James points out as well, 'it fits the facts alleged in favour of materialism equally well, besides enabling us to understand facts which materialism rejected as "supernatural". It explains the lower by the higher, Matter by Spirit, instead of vice versa, and thereby attains to an explanation which is ultimately tenable instead of one which is ultimately absurd'.

In modern terms, one can use the analogy of a television set. Is the programme produced by or transmitted through the set? A person ignorant of the facts of transmission might mistakenly conclude that the programme was actually produced by the set: when it is unplugged or parts are removed, the programme does not come through. In his own example, Schiller contends that a loss of consciousness following a brain injury can be as well explained by saying that the injury to the brain destroyed the mechanism by which the manifestation of consciousness was rendered possible, as to say that it destroyed the seat of consciousness.

Later in the book Schiller speaks of death in terms that are relevant to an understanding of the NDE. He writes that our view of death is necessarily imperfect and one-sided, 'for we contemplate it only from the point of view of the survivors, never from that of the dying. To us it is a catastrophic change, whereby a complex of phenomenal appearances, which we call the body of the dead, ceases to suggest to us the presence of the ulterior existence which we call his spirit'. The spirit, however, is inferred even during physical life, so that 'this does not prove, or even tend to prove, that the spirit of the dead has ceased to exist. It merely shows that he has ceased to form part of our little world, to interact, at least in the way to which we had been accustomed, with our spirits. But it is at least as probable that the result is to be ascribed to his having been promoted or removed as to his having been destroyed'.

In modern psychological parlance one would distinguish here [4] between the first- and the third-person perspective. Looking at someone who is clinically dead from the third-person perspective, one would never guess that there was any conscious experience going on at all. The NDE, however, is a first-person experience which can only have a partial correlation and interface with events in the physical world. Perhaps the most significant interface is the so-called veridical out-of-the body experience (OBE). OBEs do not necessarily occur only when a person is near death, but they are recognised as one of the more common features of the NDE. Phenomenologically, the person's consciousness comes apart from the physical body, which it can 'see'. There is also a corresponding feeling of emotional detachment, as from an old coat which has been discarded.

Once again, interpretations vary according to whether one posits that something leaves the body or not. Susan Blackmore [5] is one of the leading proponents of theories which insist that nothing leaves the body and that apparent OBEs are reconstructions or memories of cognitive maps from past experience. This theory would be falsified by cases reported 'in real time' which correspond to the experience of people in the physical world. I shall discuss two cases of this kind below.

One of the challenges for any science of consciousness is the nature of subjective experience or qualia. Gerald Edelman points out that 'we cannot construct a phenomenal psychology that can be shared in the same way as physics can be shared. What is directly experienced as qualia by one individual cannot be fully shared by another individual as an observer' [6]. Qualia are essentially first-person experiences, as is the NDE. Edelman suggests that it is 'our ability to report and correlate while individually experiencing qualia that opens up the possibility of a scientific investigation of consciousness'. This is precisely the procedure followed in NDE research. Individual reports are correlated so as to reveal common patterns. Only certain aspects of the experience, such as the OBE, can be correlated with external observers. Other patterns of experience — the tunnel and the light — can only be correlated internally between experiencers. Here again the question arises: are such commonalites a reflection of the unity of brain structure, or intimations of an independently existing spiritual world?

I will now give a contemporary, and an older example of a near-death OBE. The first case comes from a Dutch cardiologist, Dr Pim van Lommel [7]. A man suffered a cardiac arrest and was brought unconscious into the emergency unit, where his vital functions were restored. At no time, however, did he show any signs of consciousness. He was then transferred into intensive care, and came out of the ward some ten days later. When a nurse came in, he recognised her as one of those involved in his resuscitation and asked what she had done with his false teeth, which he had seen her place in a glass of water! He then accurately described the resuscitation procedure as if he had physically witnessed the scene from a point above his physical body, which he could see. He had not previously met this nurse, and yet was able to identify her correctly apparently as a result of his OBE.

The second case is a classic one involving Lord Geddes, a former professor of medicine and British government minister. He had such a severe case of gastro-enteritis that he found himself unable to ring for any assistance. He then found that his consciousness was separating from another consciousness which was also himself; these he called his A-consciousness and his B-consciousness, the latter remaining identified with the body while the former 'which was now me, seemed to be altogether outside my body, which it could see'. He gradually realised that he was 'free in a time-dimension of space, wherein "now" was in some way equivalent to "here" in the ordinary three-dimensional space of everyday life'. After a while his wife comes into the room and calls his doctor. He sees (at a distance, as it were) his doctor leave his patients and come very quickly. Revealingly he claims that he hears the doctor say, 'or sees him think' that he is nearly dead. The doctor subsequently gives Geddes an injection, and the A-consciousness is gradually drawn back into the body so that 'all the clarity of vision of anything and everything disappeared and I was possessed of a glimmer of consciousness suffused with pain' [8].

What is striking here is the clarity of the A-consciousness when it is ostensibly operating independently of the physical body, and the fact that it can apparently project through space-time as if totally free from normal restrictions. No less striking is the diminution of consciousness when the A-consciousness rejoins the physical body. This experience is inexplicable in terms of the productive theory but makes a great deal of sense in terms of the transmissive or filter theory of consciousness outlined above. The third person perspective of an unconscious physical body would give no hint of the intensity of first-person experience in either of these cases. The perceptions can hardly be of the normal sensory kind, since the physical senses cannot be said to have been functioning in any meaningful way. They can therefore be deemed extra-sensory. It is important to note that the first experience cannot have been reconstructed from memory as the subject had never before seen the nurse he recognised.

The implication here is that the mind cannot be wholly equated or identified with the brain, as is assumed by the majority of contemporary philosophers and neuroscientists, few of whom are aware of the kind of evidence I have outlined. Does this mean that one is forced into a form of dualism, so dreaded by those who have spent their lives trying to reduce mind to matter? At one level, any form of the survival of consciousness implies a dualism in so far as the mind is separable from the physical body. At another level, though, both mind and matter may be aspects of some higher order. The physical body is the vehicle of manifestation in the material world, but this does not preclude the possibility that we may have other bodies which correspond to different degrees of manifestation. If, for instance, we lived in a world of light, we might manifest in a body of light.

Many of the conceptual difficulties raised in connection with dualism are based on an antiquated, mechanistic view of matter which does not recognise it as a dynamic form of energy which is shaped by fields. Apart from the work by Sir John Eccles [9], very few people have advanced theories of interaction because all their thinking has been done within frameworks which assume the identity of mind and brain and the emergence of consciousness from matter. The radically empirical approach advocated here demands that such experiences as outlined above be considered when formulating models of mind/brain relations.

From the above illustrations, it can be argued that both the OBE and the NDE fall into the category of extra-sensory phenomena. The question then arises: what or who is perceiving? What kind of light is perceived? Experiencers sometimes describe themselves as being in a duplicate body, but on other occasions they experience themselves as spheres of light able to perceive in all directions at once. The sharpness of perception and thinking processes has already been noted. The fact that any perception occurs at all suggests that the basic capacity to perceive belongs to the non-sensory mind rather than the physical body, which simply permits a certain range of perception and experience.

The corollary of this is that the nature of consciousness itself is non-physical and extra-sensory, and that we have non-physical as well as physical senses or organs of perception. It is naturally hard to investigate these capacities by physical means which tend to bring one back within the limitations of the reductionist and materialistic model, but OBEs and NDEs can provide a starting-point for such a process through careful analysis of their phenomenology.

Another component of the NDE, common in accounts of mystical experience, is the encounter with the light. Some languages such as Bulgarian have separate words for physical and spiritual light, this recognising a distinction between the experience of inner and outer senses. In religious art, saintly figures are frequently portrayed surrounded by light, while the goal of spiritual practice is enlightenment. In the NDE the encounter is generally with a 'being of light', variously interpreted in accordance with the cultural background of the subject. The experience is a unitive one, where no separation is sensed between the subject and the light: the feeling is that of being immersed in love and illuminated with the light of cosmic knowledge.

There can clearly be no independent corroboration of an experience with the light, but Melvin Morse [10] has correlated the transformation of NDE subjects directly with this component. The work of Sir Alister Hardy and his centre in Oxford also witnesses to the transformative nature of the spiritual light [11]. None of these subjects, or indeed any of the great mystics, interprets this experience as a purely subjective emanation of the mind. They all insist that it indicates a spiritual reality in which human beings participate and of which physical reality is a limited aspect.

The phenomenon of the NDE may help researchers realise that a complete account of the nature of consciousness cannot be derived from normal or even abnormal experience. They may need to look at exceptional human experiences in order to achieve a full understanding. The veridical OBE poses a fundamental challenge to the idea that perception can only be mediated by the physical senses. It also hints at the possibility that human consciousness may be capable of surviving bodily death: if the mind can function independently from normal brain processes during physical life, as proposed by the transmissive theory of consciousness, then perhaps we do not need brains in order to be conscious. The theory opens up a whole new avenue of research questions.

References

[1] See Kenneth Ring, *Life at Death*, Coward, McCann and Geoghagan, New York 1980; Michael Sabom, *Recollections of Death*, Collins, London, 1982; Kenneth Ring, *Heading Toward Omega*, William Morrow, New York, 1984; Scott Rogo, *The Return from Silence*, Aquarian, London, 1989; David Lorimer, *Whole in One*, Arkana, London, 1990; Susan Blackmore, *Dying to Live*, Grafton, London 1993; also issues of the Journal of Near-Death Studies.

[2] See James, William, *Human Immortality*, London, Constable, 1899.

[3] F. C. S. Schiller, *Riddles of the Sphinx*, Swan Sonnenschein, New York 1892, 289.

[4] See Max Velmans, *A Reflexive Science of Consciousness*, In: Gregory Bock (ed.), Experimental and Theoretical Studies of Consciousness, Wiley, London 1993.

[5] See Susan Blackmore, *Beyond the Body*, Paladin, London, 1982 and *Dying to Live*, Grafton, London, 1993.

[6] Edelman, Gerald, *Bright Air, Brilliant Fire*, Penguin, London and New York, 1993, 114–115.

[7] Personal communication

[8] Quoted in Lorimer, David, *Survival?*, Routledge and Kegan Paul, London, 1984, 249–250.

[9] Eccles, Sir John, *Evolution of the Brain, Creation of the Self*, Routledge, London, 1989; *The Mystery of Being Human* In: Mott, Sir Nevill, Can Scientists Believe? James and James, 1991, 79–99.

[10] See Morse, Melvin, *Transformed by the Light*, Piatkus, London, 1993.

[11] See Hardy, Sir Alister, *The Spiritual Nature of Man*, OUP, Oxford, 1976.

Experimental Evidence Suggestive of Anomalous Consciousness Interactions

D.L. Delanoy

1 Introduction

The so-called "mind-body problem" is arguably humankind's most enduring question. The crux of this question is whether mind can exist independently of the body. Or, to re-phrase it, is mind an epiphenomenon of brain functioning, or is it, to some degree, independent of the mechanistic properties of our physical brains? Throughout written history, the greatest philosophical thinkers have pondered this matter. However, outside of the field of parapsychology, there has been very little experimental research exploring whether consciousness can interact with its environment independently of the physical body. This paper will address these issues by exploring patterns found in experimental parapsychological research which suggest that mind or consciousness can interact directly with its environment without mediation by known physical mechanisms, e.g. senses, motor activity, physiological output. If the patterns emerging from this experimental work are as they appear, they may help shed some light on the ability of consciousness to act independently of the physical body.

Parapsychological research can broadly be conceptualised as addressing two main areas. The first of these, extrasensory perception (ESP), refers to the apparent obtaining of information by the mind without recourse to currently understood sensory means of gaining such information. The second area, psychokinesis (PK), refers to changes in physical systems apparently brought about by an act of conscious intention, without recourse to currently understood means of effecting such changes. Both ESP and PK can be conceptualised as anomalous interactions between mind and its environment, apparently not mediated by any currently understood physical, sensory means. Psi is a term used to refer to both ESP and PK phenomena.

This paper will present seven major *meta-analyses* carried out on various parapsychological databases. These seven were chosen as they demonstrate both ESP and PK research, and highlight the wide scope of psi experimentation which has been conducted over the last 60 years. Meta-analysis is a term which refers to a group of statistical procedures that are used to summarise and describe bodies of research. They provide a systematic means of combining results from groups of related individual studies to assess overall consistency of results, and can assist in identifying variables within the database that appear to affect outcomes, known as "moderating variables". Meta-analytic techniques provide *quantitative*, as opposed to qualitative, reviews of bodies of research. The term "meta-analysis" was first coined by Glass in 1976[1], although the basic procedures had been known for several decades (Snedecor [2]; Mosteller and Bush [3]). More recently, many books have been published detailing methods, procedures and theoretical considerations for conducting meta-analyses (e.g. Glass, McGaw and Smith [4]; Hedges and Olkin [5]; Wolf [6]; Hunter and Schmidt [7]; and Rosenthal [8]); these references will provide further details of the procedures and statistical formulae described generally below.

For readers who are unfamiliar with meta-analytic techniques, a brief summary of the basic components of meta-analysis will be given. After identifying a domain of study, all relevant studies are gathered together. The characteristics of those which are of interest are then coded, e.g. procedural variables and constants, study quality, etc. Ideally, this coding should be performed by one or more individuals who are not closely involved with the research topic, to avoid investigators' biases influencing any coding decisions. The statistical measures generated for each study (commonly referred to as "test statistics", e.g. z, t, chi-square, etc.) are converted into effect sizes. An effect size is a measure of the degree to which a phenomenon is present in the population (i.e., of how large the effect is). As noted by Rosenthal [8], commonly used statistical measures are usually a product how large the effect is and some function of the size of the study, often the square root of the number of trials or individuals. He expresses this (p. 20) as:

$$Test\ of\ significance = Size\ of\ effect \times Size\ of\ Study$$

Most effect size measures are defined so that they are zero when the null hypothesis is true. Unlike "test" statistical measures, effect sizes do not grow in magnitude with the size of the study. Thus they provide a more accurate picture of replication across studies than can be provided by the standard statistical measures alone. This can be especially important when dealing with small effects and thus with studies with relatively low power, such as those commonly found in parapsychological research. Effect sizes allow studies to be assessed with a continuous measure, rather than the dichotomous measure to which test statistics are often reduced (i.e., statistically significant or not).

Replication across different studies is measured in terms of the consistency, or the "homogeneity" of the magnitude of the observed effect sizes. Again, this differs from the more traditional approach using test statistics, in which replication is defined by whether or not the null hypothesis is rejected in each study. Of course, when evaluating replication across a group of studies, confidence in any estimate of overall effect will be increased as the amount of confirming data increases. Using test statistics, outcomes from a group of studies can be combined and/or summarised to give an overall outcome for the database, weighting each study according to its size. In the following meta-analyses of parapsychological studies, the overall likelihood of observing the results if the null hypothesis is always true can be assessed by finding a combined z-score for all studies. This is simply a weighted average of the number of standard deviations the results deviated from chance, and its likelihood can be assessed using the standard statistical tables. One method of combining studies in this way is with a "Stouffer z" [3, 6, 9], and if the null hypotheses are always true, this statistic follows a standard normal curve. Stouffer's z provides a measure of how many standard deviations from chance the combined results of all of the studies fell. Using Stouffer's z, we can compute a "p-value" which gives us the probability of observing such extreme results if chance alone is the explanation. As we will see, the p-values for the meta-analyses in parapsychology are extremely low, thus effectively ruling out chance as an explanation for the data.

Using meta-analytic techniques, the impact of flaws upon study outcome and of various moderating variables can be quantitatively assessed, leading to improvements in study design and identification of factors associated with optimal outcomes. Possible relationships between variables can be recognised and tested in future experiments.

One problem that plagues all literature reviews is the tendency to report/publish only significant findings, commonly referred to as the "file drawer problem". However, there are a variety of methods available to estimate the size of the file drawer effect (i.e., the number of non-significant studies which would be required to nullify the outcome of a meta-analysis). For example, Rosenthal [8] provides a statistical measure, referred to as the "Fail-Safe N", to determine how many unpublished, null studies would be needed to negate an observed effect in any size of database, with the general guideline that a 5:1 ratio of null, unpublished studies to each published study should be obtained before the possibility of a negating file drawer effect can be safely eliminated.

In the following sections the findings of the selected meta-analyses will be presented. Consideration of possible interpretations, explanations and implications of this work will be found in the concluding Discussion section.

2 The ganzfeld debate

One group of parapsychological studies, the ganzfeld studies, have received more recent publicity, in terms of published articles examining the overall effect of the database, than has any other area of psi research. This attention is the result of detailed meta-analyses of the ganzfeld studies by a leading ganzfeld researcher, Honorton [9], and a critic of this work, Hyman [10]. Following the publication of these two meta-analyses, many other "pro" and "con" evaluations and commentaries have been published [11, 12, 13, 14]. The ganzfeld debate, often referred to as the "Honorton/Hyman debate", will be summarised below, but first a brief description of a ganzfeld study will be presented.

The ganzfeld technique consists of presenting a relaxed percipient with homogenous, unpatterned visual and auditory stimuli, which assists in increasing the mental imagery experienced by the percipient. While receiving this stimulus, the percipients verbalise all their experiences, their goal being to gain impressions which will relate to a sensorially isolated and remote target picture or short video clip. The "target" is being watched frequently by another person (a "sender" or "agent") who is attempting mentally to convey impressions of the target to the percipient or "receiver". These studies utilise a "free-response" methodology, in which the contents of the target material are unknown to the receiver (i.e., the percipient is "free" to respond with whatever impressions they generate, as he or she has no information regarding the specific contents of the possible target). The most common method of analysis used in ganzfeld studies is for the percipient or an independent judge(s) to compare the obtained impressions to four different target pictures/video clips, one of which is a duplicate of the actual target, looking for similarities. Using blind procedures, the judge has a one in four chance of correctly identifying the actual target (i.e., mean chance expectancy = 25 per cent "hit" rate). Study outcome is based upon whether similarities between the percipient's impressions and the actual target enabled the target to be correctly identified significantly more often than chance would allow. For further information regarding this experimental technique, procedural details and methods of analysis, see Honorton [9], and Honorton et al. [15].

A meta-analysis of twenty-eight ganzfeld studies was performed by Honorton (1985) [9], in response to a flaw analysis of the ganzfeld database conducted by Hyman (1985) [10].

Hyman found a highly significant overall effect in the database, but concluded that this effect was negated as he found a significant relationship between the study outcomes and procedural and statistical flaws contained in the studies. However, Hyman's flaw categorisations were severely criticised by Honorton, and a psychometrician, Saunders [16], found faults in Hyman's statistical analyses.

Honorton's meta-analysis found there were no significant relationships between study outcomes and quality. The overall composite (Stouffer) z score for the 28 ganzfeld studies included in the Honorton meta-analysis was highly significant ($z = 6.6$, $p < 10^{-9}$, two-tailed). The effect sizes were homogeneous, overall and across experimenters. The discrepancy between the Honorton and Hyman analyses of the ganzfeld studies prompted a further meta-analysis by Rosenthal [17], an independent specialist in meta-analysis. Like Honorton, Rosenthal found an overall composite z score of 6.60 for the twenty-eight ganzfeld studies. His file drawer estimate agreed with that of Honorton, requiring 423 unreported, null studies to negate the significance of the database. Here it is worth noting that another critic, Blackmore [18] conducted a survey to discover the number of unreported ganzfeld studies in 1980, prior to the Honorton/Hyman debate. Her survey found 32 unreported studies, of which 12 were never completed, and one could not be analysed. Of the remaining 19 studies, 14 were judged by Blackmore to have adequate methodology, with 5 of these (36 percent) reporting significant results. She concluded that "the bias introduced by selective reporting of ESP ganzfeld studies is not a major contributor to the overall proportion of significant results" (p. 217). Rosenthal, after considering the possible influence of various flaws upon study outcome, concluded that the overall hit rate of the studies could be estimated to be 33 percent, whereas chance expectancy was 25 percent.

In 1986 Honorton and Hyman published a "Joint Communiqué" [19] in which they agreed that there was an overall effect in the database, but continued to disagree as to what extent this effect may have been influenced by methodological flaws. In their communiqué they outlined the necessary methodological precautions that should be taken to avoid the possibility of future studies giving rise to the same level of debate that had surrounded the previous ones. They concluded that more studies needed to be conducted, using the controls they had documented, before any final verdict about the database could be reached.

Honorton and his research team proceeded to design a new ganzfeld system which met the criteria he and Hyman had specified in their communiqué. This system, and studies using it, are referred to as "autoganzfeld studies", as much of the procedure is under automated computer control in order to avoid the problems found in some of the earlier studies. Before Honorton's lab closed in 1989, 11 experimental series, representing 355 sessions, conducted by eight experimenters, had been collected using the autoganzfeld. Honorton et al. [15] published a summary of the autoganzfeld studies and compared them with his earlier meta-analysis. The autoganzfeld sessions yielded overall significant results ($z = 3.89$, $p = 0.00005$), with an obtained hit rate of 34.4 percent (with 25 percent being chance expectancy). The effect sizes by series and by experimenter were both homogeneous. Comparing the autoganzfeld outcomes to those of the 28 studies of the earlier meta-analysis revealed very similar outcomes, with the autoganzfeld showing slightly better ESP scoring than that obtained in the earlier studies (autoganzfeld results by series: effect size or $es = .29$, earlier 28 meta-analysis studies by experiment: $es = .28$).

Hyman, in 1991 [20] commenting upon a presentation of these results by the statistician, Utts [12], concluded that "Honorton's experiments have produced intriguing results. If, as Utts suggests, independent laboratories can produce similar results with the same relationships and with the same attention to rigorous methodology, then parapsychology may indeed have finally captured its elusive quarry." (p. 392). Replications are currently being undertaken at various labs; the only replication using a full autoganzfeld environment which has been reported to date was conducted at the University of Edinburgh [21], where the obtained significant, overall hit rate was 33 percent ($z = 1.67$, $p < 0.05$). This outcome is consistent with Honorton's autoganzfeld scoring rate of 34.4 percent, and replicates Rosenthal's hit rate estimate based on the earlier ganzfeld studies. The procedure for the Edinburgh study incorporated additional safeguards against subject and experimenter fraud.

3 Looking into the future: Meta-analysis of precognition ESP studies

Folklore and many anecdotal stories have relayed how some individuals have claimed to be able to "foretell" the future, or have experienced premonitions of events before they actually occurred. While much of this information is likely due to misinterpretation, misrepresentation or other flaws of human perception, memory and reasoning, there are experimental findings which suggest that precognition may occur (see Wiseman and Morris [22] for an overview of ways we can be deceived, or can deceive ourselves into interpreting a normal incident as being paranormal).

Honorton and Ferrari [23] conducted a meta-analysis of 309 precognition studies conducted between 1935 and 1987. These studies all used a "forced-choice" methodology, in which the subject is aware of the possible target choices, and is asked to choose one of them as his answer (as opposed to "free-response" methodologies, such as ganzfeld studies). In all of these studies, the subject made their choice as to the target identity prior to the target identity actually being randomly generated. Thus the subjects' responses were to targets which did not exist at the time of their response. These studies are thought by some to be methodologically superior to other ESP studies as there is little possibility of the subject "cheating", or receiving any subtle cues about the target identity, as the target does not exist when their response is made.

The studies included in this meta-analysis were conducted by 62 different senior investigators, and included nearly two million individual trials contributed by over 50,000 subjects. While the mean effect size per trial is small ($es = .02$), it is sufficiently consistent for the overall effect from these studies to be highly significant (combined $z = 11.41$, $p = 6.3 \times 10^{-25}$). Using eight different measures of study quality, no systematic relationship was found between study outcome and study quality. A "fail-safe N" estimate would require 14,268 unreported, null studies to reduce the significance of the database to chance levels. Given the wide diversity of study methods and procedures found in this database, it is not surprising that the study outcomes were extremely heterogeneous. The authors eliminated outliers by discarding those studies with z scores falling within the top and bottom 10 percent of the distribution, leaving 248 studies. It should be noted that the elimination of outlier studies to obtain homogeneity is a common practice, and in other, non-parapsychological reviews "it is sometimes necessary to discard as many as 45% of the studies to achieve a homogeneous effect size distribution" (p. 1507) [24]. The resulting mean trial effect size was .012, and the combined z still highly significant ($z = 6.06$, $p = 1.1 \times 10^{-9}$). While it was found that study quality improved significantly over the 55 year period during which these studies were conducted (correlation coefficient $r[246$ degrees of freedom$] = .282$, $p = 2 \times 10^{-7}$), study effect sizes did not significantly co-vary with the year of publication. Study effect sizes are homogenous across the 57 investigators contributing to the trimmed database. The rest of the analyses conducted were all performed upon this smaller database.

The authors identified four "moderating" variables that appeared to relate systematically to study outcome. The first variable involved the subject population. It was found that studies using subjects who were selected on the basis of good ESP performance in previous experimental sessions obtained significantly better ESP effects than those studies using unselected subjects (a t test with 246 degrees of freedom [df] giving $t = 3.16$, $p = 0.001$). Another variable which covaried with study effect size was whether the subjects were tested individually or in groups, with individual testing studies obtaining significantly higher outcomes than those using group testing methods ($t[200$ df$] = 1.89$, $p = 0.03$).

A further moderating variable involved the type of feedback subjects received about the accuracy of their responses. There were four feedback categories, including no feedback, delayed feedback (usually via mail), feedback given after a sequence of responses (often after 25 responses), and feedback given after each response. Of the 104 studies which supplied the necessary information, there was a linear and significant correlation between the precognition effect and feedback level ($r[102$ df$] = .231$, $p = 0.009$), with effect sizes increasing with level of feedback. A related finding involves the time interval between the subject's responses and the target selection. This finding is confounded by the feedback level, as time duration between the response and target generation may co-vary with feedback level (i.e., when feedback was given after every response, the time interval between response and target selection would have to be shorter than was necessarily the case when feedback was given after a sequence of calls, or a month after the responses had been made). There were seven different time interval categories, varying from a millisecond to months. There was found to be a significant decline in precognition effect sizes as the time interval between response and target selection increased ($r[142$ df$] = -.199$, $p = 0.017$). The significant temporal decline/study effect size relationship is due entirely to those studies which used unselected subjects, with the studies that tested selected subjects showing a small, non-significant increase in precognition scoring as the time interval increased (the difference between these groups was not significant).

It should be noted that there was no significant difference in quality between studies using selected and unselected subjects. Also, studies which tested subjects individually did show significantly higher study quality than those utilising group testing procedures ($t[137$ df$] = 3.08$, $p = 0.003$). A correlation between feedback level and research quality was positive, but not significant ($r [103] = .173$, $p = 0.82$).

In summarising the precognition findings, Honorton and Ferrari concluded "the forced-choice precognition experiments confirm the existence of a small but highly significant precognition effect." (p. 300). Furthermore, they concluded that the most important outcome of the meta-analysis was the identification of moderating variables, which not only provides guidelines for future research, but may also help expand our understanding of the phenomena.

4 Influencing randomness in physical systems: Two meta-analyses

"Mind over matter" is a frequently used phrase, but is there any evidence suggesting that mind can exert some influence over the behaviour of physical, material systems? Two meta-analyses dealing with such effects will be reviewed, both of which suggest that mind can directly interact with matter. Both these databases involve participants attempting to make a random system behave in a non-random manner.

The first of these databases involves studies in which people tried to influence the outcome of falling dice. This work was initially suggested by claims of gamblers that they were able to influence the outcome in dice throwing situations in gaming casinos. Radin and Ferrari [25] conducted a meta-analysis of 148 dice studies conducted between 1935 and 1987. This database also included 31 control studies in which no conscious influence of outcome was attempted. The results showed a significant overall effect for the experimental influence studies (es = .012, Stouffer z = 18.2, $p < 10^{-70}$), and chance results for the control studies (Stouffer z = 0.18). To obtain a homogeneous distribution of effect sizes, 53 studies (35 per cent) of the database had to be deleted. Of these deleted studies, 33 had positive and 19 had negative effect sizes. Eleven study quality measures were considered. While the relationship was not significant, the authors did find that effect size decreased as study quality increased.

Another methodological problem affecting this database is that the probability of obtaining a specific outcome is not necessarily equally distributed across all the die faces (e.g., if using pipped dice, the six typically has the least mass and is thus most likely to come up). To examine the possible influence of this "non-random" aspect of dice throwing, the results for a subset of 69 studies, in which targets were balanced equally across the six die faces, were examined. A significant overall effect was still obtained (Stouffer z = 7.617, $p < 10^{-11}$). For these 69 studies, the effect size was relatively constant across the different measures of study quality, and a file drawer analysis revealed that a 20:1 ratio of unreported, nonsignificant studies for each reported study would be required to reduce the database to chance expectations.

The second "mind over matter" meta-analysis involves studies in which a person attempts to influence a microelectronic random number generator (RNG) to behave in a non-random manner. This meta-analysis, conducted by Radin and Nelson [24], involves the largest parapsychological database to date, with 832 series, of which 597 were experimental series and 235 control series. The general protocol of these studies involves having a RNG drive a visual display, which an observer tries to influence, by means of mental intention, in accordance with prespecified instructions. The randomness of the RNG is usually provided by radioactive decay, electronic noise or pseudorandom number sequence seeded with true random sources; the RNG's are frequently monitored to ensure true random output in these studies. The observer initiates a "trial" by means of a button push, which starts the collection of a fixed length sequence of data. For each data sequence, a z score may then be computed. The mean effect size per trial for the experimental series was very small, but very robust (es = .0003, combined z = 15.58, $p = 1.8 \times 10^{-35}$) and significantly higher (z = 4.1, p = 0.00004) than the effect size for the control series (es = –.00004). Sixteen study quality measures were investigated; effect size did not significantly co-vary with study quality. The file drawer estimate for this data base is enormous, requiring 54,000 null, unreported studies to reduce the observed effect to chance levels. Given these findings, Radin and Nelson concluded that "it is difficult to avoid the conclusion that under certain circumstances, consciousness interacts with random physical systems" (p. 1512, [24]).

5 Direct mental interactions with living systems (DMILS)

Direct mental interactions with living systems (DMILS) research involves testing procedures where a person (an "agent") is trying to interact with a biological target system, e.g., another person's physiological responses or the behaviour of small animals or fish. In DMILS studies the biological target is located in a sensorially shielded room, providing isolation from any physical contact with the agent. The target's spontaneously fluctuating activity is monitored continuously while the agent, during randomly interspersed influence and noninfluence (control) periods, tries to influence mentally the target's activity in a pre-specified manner. The target system is unaware of timing or goal orientation (i.e., influence or non-influence) of the agent's mental intentions. When human physiological responses are the target system, the target person's only goal during the experimental session is to remain passively alert and to wish mentally that their physiology will unconsciously respond appropriately to the agent's intentions. The mental strategies used by the agent to interact with the remote, shielded target includes wishing and willing the desired changes to manifest in the target, mental imaging of the desired outcome, and in some instances simply paying attention to the target system. The randomised order of the influence or non-influence period is usually conveyed to the agent by a message on a computer monitor; the monitor may also convey to the agent the actual recordings of the target's activity, thereby providing on-going feedback about the effects of their mental intentions upon the remote target system. The experimental design eliminates possible confounding factors such as recording errors, placebo effects, confounding internal rhythms and chance correspondences.

The majority of the recent DMILS research has been conducted by Braud and his colleagues, who published a meta-analytic summary of 37 of their experiments (Braud and Schlitz [26]). This work involved 13 different experimenters and 655 sessions. These 37 studies examined seven different target systems, including electrodermal activity (EDA) with the agent trying to influence the subject's EDA to increase or decrease (i.e., trying to "calm" or "activate" the subject), blood pressure, fish orientation, mammal locomotion, and the rate of haemolysis of human red blood cells. The overall results from this work have been highly significant (per session overall $es = .33$, Stouffer $z = 7.72$, $p = 2.58 \times 10^{-14}$).

While this work was conducted by 13 different experimenters, it was all performed at the same laboratory. Other laboratories are now attempting to replicate this work, with the initial results generally conforming to those obtained by Braud et al. For example, Delanoy and Sah (1994 [27]) compared EDA responses to conscious responses in a DMILS environment, in which the agent was either remembering and trying to re-experience a very positive, exhilarating emotion ("activate" condition) or was thinking of an emotionally neutral object ("control" condition). The subject's EDA showed significantly greater activity during the activate periods than during the control periods ($es = .31$, $t[31] = 1.77$, $p = 0.04$). However, the subject's conscious responses (i.e., their guesses as to whether the agent was trying to activate or calm them) did not differ from chance expectancy. The finding of a significant physiological effect, with no corresponding effect shown by a conscious response measure, supports similar findings from Tart [28] and Targ and Puthoff [29], and suggests that subtle psi interactions may occur without any conscious recognition on the part of the subject.

6 Relating ESP to personality traits: Two meta-analyses

Parapsychological researchers have long been interested in exploring if there are any factors which might relate to why some people report having more psi experiences in their everyday life than do others. Similarly, while most experimental work is done with volunteer subjects who have not been chosen on the basis of their supposed psi ability, it has been observed that some people appear to do better in experimental psi tests than others. One approach to examining possible reasons for these observed differences has involved exploring the relationship between various personality factors and psi ability.

Two meta-analyses of studies which have looked for correlations between performance on a psi task and different personality traits will be discussed here. One of these involved studies which looked for a relationship between a person's opinion of psi and their own psi abilities with their psi test performance. Research examining what has come to be known as the sheep/goat effect, supported the hypothesis that in experimental psi tests those with positive attitudes ("sheep") tend to score above chance, and those with negative attitudes ("goats") below chance.

Lawrence [30] conducted a meta-analysis of the 73 published studies examining the sheep/goat effect. These studies were conducted by 37 principal investigators, and involved over 4,500 subjects who completed over 685,000 trials. The overall effect size per trial is small ($r = 0.029$), but highly significant over these studies which involved a large number of procedural manipulations and potential modifying variables. The combined Stouffer $z = 8.17$, $p = 1.33 \times 10^{-16}$. Using seven different measure of study quality, Lawrence found that effect size did not covary with study quality. A file-drawer estimate (Rosenthal's "fail-safe N") revealed that 1726 unreported studies with null results (i.e., 23 unreported studies for each of the 73 reported ones) would be required to reduce the significance of the database to chance expectancy.

This database has used a wide range of different sheep/goat scales, ranging from single questions to more lengthy questionnaires. The means of determining belief have also varied, with most focusing upon previous personal psi experiences, self-evaluation of personal psi ability, opinions regarding one's ability to display psi ability in the specific testing situation and/or one's general attitudes towards such phenomena. Lawrence found there was no overall relationship between effect size and the type of measure used, from which he concluded that the sheep/goat effect was quite robust regardless of how it was measured.

Another personality trait that has been studied in relation to psi performance is extraversion/introversion. Honorton, Ferrari and Bem [31] conducted a meta-analysis on the 60 published studies examining this relationship. Prior to this meta-analysis, descriptive reviews of this database had concluded that extraverts performed better than introverts on psi tasks (Eysenck, [32], Palmer [33], Sargent [34]). However, the ability of meta-analysis to identify flaws and modifying variables led to a different finding in the meta-analysis. While the meta-analysis did find a significant overall effect ($r = .09$, combined $z = 4.63$, $p = 0.000004$), the effect sizes were non-homogeneous. The studies were divided into smaller groups according to various procedural variables in order to discover the source of the non-homogeneity. The authors separated the 45 studies using forced-choice procedures from the 14 studies using free-response methods. Once again, significant but non-homogeneous, effects were found (forced choice: $r = .06$, combined $z = 2.86$, $p = 0.0042$; free-response: $r = .20$, combined $z = 4.82$, $p = 0.0000015$). A further division of these two groups of studies examined whether testing subjects individually or in groups had any impact on the outcomes.

This analysis revealed that of the forced-choice studies, 21 studies had tested subjects individually, resulting in a significant, but non-homogeneous effect ($r = .15$, combined $z = 4.54$, $p = 0.000006$). In the 24 forced-choice studies where participants were tested in groups, there was no significant effect ($r = .00$, $z = -0.02$), although there was homogeneity .

A flaw analysis showed that the significant effect in the forced-choice database was entirely due to 18 studies in which the extraversion measure had been given after the ESP test, the significance of this correlation being due to 9 of these studies in which the subjects knew how they had performed on their psi task before they completed the extraversion questionnaire. This finding raises the strong possibility that the correlation was due to psychological, as opposed to paranormal, factors. Thus the previous descriptive reviews which had found a significant, positive relationship between extraversion and psi-scoring had failed to uncover the inconsistency in the degree to which this effect was present in these studies, and the flaw which lead Honorton, Ferrari and Bem to conclude that the relationship in forced-choice studies would appear to be artifactual.

In the subset of 14 free-response extraversion studies, a significant ($r = .20$, combined $z = 4.82$, $p = 0000015$) but non-homogeneous effect was obtained. Dividing the studies according to individual or group testing procedures revealed that the 2 studies employing group testing were responsible for the non-homogeneity. The results for the 12 studies which testing subjects individually show homogeneity and a significant correlation ($r = .20$, combined $z = 4.46$, $p = 0.0000083$). Eleven of the studies documented the presentation order of the psi test and extraversion questionnaire. In all of these studies, the extraversion questionnaire was given prior to the ESP test, thereby avoiding the potential problem of subject's knowledge of their ESP results influencing the way that they completed their extraversion questionnaire. These 11 studies show a significant and homogeneous extraversion/ESP correlation ($r = .21$, combined $z = 4.57$, $p = .000005$).

After completing the extraversion/ESP meta-analysis, Honorton et al. examined the autoganzfeld database to see if they could confirm the relationship. For the 221 autoganzfeld trials for which they had extraversion data, they obtained a significant ESP/extraversion correlation ($r = .18$, $t[219 \text{ df}] = 2.67$, $p = 0.008$). This finding is consistent with those from the free-response extraversion meta-analysis.

7 Discussion

The above seven meta-analyses represent a cross-section of the meta-analyses that have been performed on parapsychological research. They were not chosen to illustrate the greatest effects or to "paint the rosiest picture", but rather to provide a window into the range of effects and variety of methodologies found in psi experimentation. The effect sizes in these studies tend to be very small (RNG–PK) to moderate (i.e., DMILS) in size. However, even the smaller effect sizes appear to be reliably found in the databases. Furthermore, the size of an effect does not provide a good indication of its potential meaningfulness or applicability. For example, a recent medical study investigating whether aspirin could help prevent heart attacks was ended prematurely because the effectiveness of the treatment was so clearly demonstrated after six months of trials that the investigators thought it would be unethical to withhold the treatment further from the control group. Indeed, the findings from the study were heralded as a major medical breakthrough. While the findings from this study were highly significant ($\chi^2 = 25.01$, $p = 0.00001$), the effect size is .068, considerably smaller than.

some of the effect sizes found in the psi literature [12, 13]. It should be noted that small effects have low statistical power [8, 12, 13]. For example, the aspirin study involved over 22,000 subjects. If there had only been 3,000 subjects, the investigators would have had less than a 50 percent chance of finding a conventionally significant effect [13]. Given the small effect sizes which are typical in psi experiments, low replicability is to be expected. Rosenthal [17] notes that "even though controversial research areas are characterised by small effects, that does not mean that the effects are of no practical importance." (p. 324). Indeed, in an article addressing behavioural research in general, Rosenthal [35] warned: "Given the levels of statistical power at which we normally operate, we have no right to expect the proportion of significant results that we typically do expect, even if in nature there is a very real and very important effect" (p. 16).

What can these findings tell us about the functioning of apparent psi abilities? The conceptualisation of ESP as the anomalous input of information into consciousness and PK as the anomalous output of influence are "working" models, which help convey possible interpretations of the obtained phenomena. However, the distinction between ESP and PK is often blurred. If one accepts the precognition database as suggesting that information about an event can be obtained before the occurrence of the event, many of the psi results could be interpreted as representing acts of precognition. For example, while most ganzfeld and DMILS studies are "real-time" and involve an "agent", it is possible that the actual mechanism at work may be the subject obtaining information about the target by "looking" into the future to gain relevant target information and then generating appropriate impressions in the case of ganzfeld studies, or producing the appropriate self-regulatory responses in the case of the DMILS studies. In this context, it should be mentioned that the role of the agent is unclear. No DMILS studies have yet been reported which have not used an agent, but in the case of ganzfeld studies, the recent Edinburgh study [21] found equally significant outcomes in sender and no sender conditions. Similarly, other ESP studies have obtained significant, positive outcomes without using an agent (for a review of this work see Palmer [36]). Findings such as these indicate that a sender appears not to be a necessary component in anomalous information transfer studies, although they may still have a beneficial psychological impact upon the study outcomes [36]. The RNG–PK work has been traditionally conceptualised as representing "influencing" effects, as has the DMILS work which was initially known as "bio-PK". However, alternative interpretations of these apparent effects may involve ESP. May et al. [37] have proposed that apparent RNG–PK effects could be the result of the observer, via precognition, knowing what would be the right moment to initiate a sequence of random event (i.e., when to push the button) to get the desired outcome, thereby making use of the random fluctuations found in RNG systems to create a non-random outcome. Others have questioned the validity of a precognition interpretation of psi data. For example, Morris [38] discusses models based on "real-time" psi effects as possible alternative explanations of precognition. For example, using PK a subject or investigator could influence the random source used to choose the target in precognition studies to obtain a selection consistent with the subject's response.

As the above comments make apparent, the mechanisms which may be involved in the producing the effects found in these databases are still unknown. Process-oriented research is ongoing in parapsychology. In future studies, correlations such as those found in the precognition database may help us better differentiate between the differing theoretical interpretations of these anomalous effects. While this paper has focused upon presenting summaries of experimental data, there are a variety of theoretical models which address these findings. Although it is outside the scope of this paper to review these models, a thorough presentation of theoretical parapsychology is provided by Stokes [39].

In conclusion, the findings from these meta-analyses suggest that consistent trends and patterns are to be found in the database. The consistency of outcomes found in the ganzfeld research, the robust PK effects, the modifying variables revealed by the precognition database, the variety of target systems displaying DMILS effects and the correlations found with personality traits are all indicative of lawful relationships. Given these relationships it is difficult to dismiss the findings as "merely an unexplained departure from a theoretical chance baseline" p. 301 [23]. Whether these effects will prove to represent some combination of currently unrecognised statistical problems, undetected methodological artefacts, or, as seems increasingly likely, a genuinely new, hitherto unrecognised characteristic of mind or consciousness remains to be seen.

Acknowledgements: I would like to thank Robert L. Morris and Jessica Utts for their helpful comments on earlier drafts of this paper.

References

[1] G. V. Glass, *Primary, secondary, and meta-analysis of research,* Educational Researcher **5** (1976), 5–8.

[2] G. W. Snedecor, *Statistical Methods,* Iowa State College Press, 1946.

[3] F. M. Mosteller, and R. R. Bush, *Selected quantitative techniques,* In: G. Lindzey (ed.), Handbook of Social Psychology: Vol. 1. Theory and Method, Addison-Wesley, 1954, 289–334.

[4] G. V. Glass, B. McGaw and M. L. Smith, *Meta-Analysis in Social Research,* Sage Publications Inc., 1981.

[5] L. V. Hedges and I. Olkin, *Statistical Methods for Meta-Analysis,* Academic Press, Ltd., 1984.

[6] F. M. Wolf, *Meta-Analysis: Quantitative Methods for Research Synthesis,* Sage Publications, Inc., 1986.

[7] J. E. Hunter and F. L. Schmidt, *Methods of Meta-Analysis,* Sage Publications, Inc., 1990.

[8] R. Rosenthal, *Meta-Analytic Procedures for Social Research.,* Revised edition, Sage Publications, Inc., 1991.

[9] C. Honorton, *Meta-analysis of psi ganzfeld research: A response to Hyman,* J. of Parapsychology **49**:1 (1985), 51–91.

[10] R. Hyman, *The ganzfeld psi experiment: A critical appraisal,* J. of Parapsychology **49**:1 (1985), 3–50.

[11] J. of Parapsychology **50**:4, 1986.

[12] J. Utts, *Replication and meta-analysis in parapsychology,* Statistical Science **6**:4 (1991), 363–403.

[13] D. J. Bem and C. Honorton, *Does psi exist? Replicable evidence for an anomalous process of information transfer,* Psych. Bulletin **115** (1994),4–18.

[14] R. Hyman, *Anomaly or artifact? Comments on Bem and Honorton,* Psych. Bulletin **115** (1994), 19–24.

[15] C. Honorton, R. E. Berger, M. P. Varvoglis, M. Quant, P. Derr, E. I. Schechter and D. C. Ferrari, *Psi communication in the ganzfeld,* J. of Parapsychology **54**:2 (1990), 99–139.

[16] D. R. Saunders, *On Hyman's factor analysis,* J. of Parapsychology **49**:1 (1985), 86–88.

[17] R. Rosenthal, *Meta-analytic procedures and the nature of replication: The ganzfeld debate,* J. of Parapsychology **50**:4 (1986), 315–336.

[18] S. J. Blackmore, *The extent of selective reporting of ESP ganzfeld studies*, European J. of Parapsychology 3:3 (1980), 213–220.

[19] R. Hyman and C. Honorton, *A joint communiqué: The psi ganzfeld controversy*, J. of Parapsychology 50:4 (1986), 351–164.

[20] R. Hyman, *Comment*, Statistical Science 6:4 (1991), 389–392.

[21] K. S. Dalton, R. L. Morris, D. L. Delanoy, D. Radin, R. Taylor and R. Wiseman, *Security measures in an automated ganzfeld system*, In: Proceedings of the 37th Annual Convention of the Parapsychological Association, Parapsychological Association, 1994, 114–123.

[22] R. Wiseman and R. L. Morris, *Guidelines for testing psychic claimants*, University of Hertfordshire Press, 1995.

[23] C. Honorton and D. C. Ferrari, *"Future telling": A meta-analysis of forced-choice precognition experiments, 1935–1987*, J. of Parapsychology 35 (1989), 281–308.

[24] D. I. Radin and R. D. Nelson, *Evidence for consciousness-related anomalies in random physical systems*, Foundations of Physics 19:12 (1989), 1499–1514.

[25] D. I. Radin and D. C. Ferrari, *Effects of consciousness on the fall of dice: A meta-analysis*, J. of Scientific Exploration 5:1 (1991), 61–85.

[26] W. G. Braud and M. J. Schlitz, *Consciousness interactions with remote biological systems: Anomalous intentionality effects*, Subtle Energies 2:1 (1991), 1–46.

[27] D. L. Delanoy and S. Sah, *Cognitive and physiological psi responses to remote positive and neutral emotional states*, In: Proceedings of the 37th Annual Convention of the Parapsychological Association, Parapsychological Association, 1994, 128–138.

[28] C. T. Tart, *Physiological correlates of psi cognition*, Int'l. J. of Parapsychology 5 (1963), 357–386.

[29] R. Targ and H. Puthoff, *Information transmission under conditions of sensory shielding*, Nature 252 (1974), 602–607.

[30] A. R. Lawrence, *Gathering in the sheep and goats... A meta-analysis of forced-choice sheep-goat ESP studies, 1947–1993*, In: Proceedings of the 36th Annual Convention of the Parapsychological Association, Parapsychological Association, 1993, 75–86.

[31] C. Honorton, D. C. Ferrari and D. J. Bem, *Extraversion and ESP performance: A meta-analysis and a new confirmation*, In: L. A Henkel and G. R Schmeidler (eds.) Research in Parapsychology 1990, Scarecrow Press, 1992, 35–38.

[32] H. J. Eysenck, *Personality and extra-sensory perception*, J. of the Society for Psychical Research 44 (1967), 55–70.

[33] J. Palmer, *Attitudes and personality traits in experimental ESP research*. In B B. Wolman (Ed.) Handbook of parapsychology. Van Nostrand Reinhold, 1977, 175–201.

[34] C. L. Sargent, *Extraversion and performance in 'extra-sensory perception' tasks*, J. Personality and Individual Differences 3 (1981), 137–143.

[35] R. Rosenthal, *Replication in behavioral research*, Journal of Social Behavior and Personality 5 (1990), 1–30.

[36] J. Palmer, *Extrasensory perception: Research findings*, In: S. Krippner (ed.), Advances in Parapsychological Research, 2 Extrasensory Perception, Plenum Press, 1978, 59–244.

[37] E. C. May, D. I. Radin, G. S. Hubbard, B. S. Humphrey and J. M. Utts, *Psi experiments with random number generators: An informational model*, In: D. H. Weiner and D. I. Radin (eds.), Research in Parapsychology 1985, Scarecrow Press, 1986, 119–120.

[38] R. L. Morris, *Assessing experimental support for true precognition*, J. of Parapsychology 46 (1982), 321–336.

[39] D. M. Stokes, *Theoretical parapsychology*, In: S. Krippner (ed.), Advances in Parapsychological Research 5, McFarland, 1987, 77–189.

[40] R. Rosenthal, *Replication in behavioral research*, J. of Social Behavior and Personality **5** (1990), 1–30.

Theoretical Speculations and Energy Phenomena Associated with Spiritual Healing

D.J. Benor

1 Introduction

Modern physics describes the world in terms of energetic fields, matter being a secondary phenomenon. While for some phenomena it may be an adequate approximation to think purely in terms of the individual particles involved, each with their own characteristic energies, this approximation is not good enough in general.

We have no way of knowing what residual energies (some not even known to present day science) may be relevant in biosystems. Newtonian medicine ignores such possibilities, continuing to describe human and animal bodies almost exclusively in terms of the component material parts, and ignoring any possible residual effects such as those mentioned. Diseases are conceptualised in terms of biochemical, cellular, tissue, organ and organismic *material* abnormalities. In the instances of physical trauma, infections, hormonal dysfunctions and genetic abnormalities, conventional medical approaches have borne great fruit in explaining and treating the *physical* problems. However, many ailments remain for which conventional medicine has only limited means of treatment, including arthritis, neurological diseases such as multiple sclerosis, cancers, degenerative changes of old age and many more. Numerous unpleasant and even dangerous side effects which may accompany conventional therapies (e.g. drowsiness, obtunded consciousness, habituation or addiction, constipation, allergies, nausea, vomiting, hair loss, and more) may force the patient to have to choose between symptom management and disease control, and diminished quality of life — and this at a time when the clock approaching the end of physical existence may be ticking towards an early end. Both patients and doctors are increasingly turning to healing for help with such problems, as healing is both effective and free of negative side effects.

You may be surprised to know that 155 controlled studies of healing have been published. About a dozen of these are doctoral dissertations. More than half of these studies demonstrate statistically significant effects — on water, enzymes, yeasts, bacteria, plants, animals and humans [1]. These support the claims of healers that they can influence states of health and illness through focused intent, which may be expressed through the laying-on of hands, prayer, meditation and the like.

Healing is just one example of a process where the mind is a relevant factor in states of physical health and illness, an aspect that has been elucidated over the better part of this century. It is presumed by many that the mind is the product of brain electrochemical activities and that its influence on the body is achieved through neuronal and neurohormonal mechanisms. The new and rapidly growing field of psychoneuroimmunology (PNI) has highlighted the contributions of mind and emotions to physical well-being and illness [2]. PNI demonstrates that mental exercises such as visualisations, meditations, breathing and other relaxation techniques may enhance the functions of the immune system. We are beginning to identify dozens of neuropeptides in white cells which are identical to those in neurones, suggesting that the brain may extend into the body and the influence of the mind may be expressed via the immune system as well [3].

2 Healers' accounts of healing: subtle energies

Spiritual healers claim that they are addressing the *subtle energy* aspects of health and illness. They claim that they can perceive energies which are different from conventional electromagnetic energies. Because of the biases of conventional, Newtonian medicine towards theories and interpretations of health and illness based purely on the material constituents and ignoring subtler aspects related to energy, the reports and theories of healers have been largely ignored or dismissed. This is a gross oversight.

It has been difficult for Newtonian medicine to even consider the possibility that such a subtle energy might exist, that it might influence states of health and illness, or that its effects might be demonstrable in scientific studies. This negative bias has also led to a reluctance to accept studies of healing for publication in conventional medical journals until the last few years. You will find most of the studies published in parapsychology journals (which, it should be noted, have a peer review system).

What mechanisms might be relevant in explaining self-healing and healer assisted healings? Healers report that several interpenetrating, subtle energy fields surround the physical body [4, 5, 6]. They claim that the physical body is an expression of the states of these energy fields, each of which is distinctly related to an aspect of being (physical, emotional, mental and spiritual). The fields are said to be hierarchically organised. The emotional and mental fields can therefore also influence the physical body. The emotional field is governed by the emotions, the mental field by the mind (with apologies to our German colleagues, whose language does not contain a precise translation for the word *mind*).

These subtle energy fields extend to various distances from the physical body. Though only a few sensitive people are able to perceive these subtle energies as visual halos or auras of colour around the bodies of living organisms and inanimate objects [5, 7, 8], many people can sense them with their hands [9]. One has only to hold one's hands near one another and then move the hands slowly apart and back together again to sense them. About 90 percent of doctors and nurses I have instructed in workshops on healing are able to do this, and 25–75 percent of general audiences can do so as well. (The lower percents are among academics, and others who tend towards scepticism.)

The human body is a very sensitive instrument for the perception of subtle energies. This presents problems of 'noise' in perceiving and interpreting human reports. There is also a wide variability in healers' subjective experiences of perceiving and directing energies. Most people who are gifted with healing are not academically or research oriented. These problems have contributed to the difficulties of science in accepting reports of healers.

Giving and receiving healing are perceived subjectively as very powerful experiences, with a sense that they are more 'real' than everyday, material reality [10]. When coloured by sensitives' cultural and religious beliefs, these experiences tend to be taken as verifications of the beliefs. It becomes difficult to separate the subjective from whatever underlying objective realities, or at least common denominators, might be discernible through a survey of a spectrum of healers [1, 11, 12].

Healers claim, and controlled studies substantiate, that they are able to send healing from any distance without diminution of effect, whether they are some metres [1, 13, 14] or even many kilometres away from the people or other living organisms they are treating [1, 15].

At the time of giving healing by laying-on of hands or by transmitting from a distance, healers may also be able to diagnose states of health and illness of healees. Evidence in support of these claims may be found in the similar studies of the Stanford Research Institute [16] and Princeton Engineering Laboratories [17] on remote viewing. Subjects were able to report accurate details of locations at great distances from these laboratories.

My own impression (from a dozen years of researching healing, reviewing the literature and learning to be a healer) is that consciousness is an energy phenomenon which can interact with and shape physical, material reality. Healers report they 'turn on' and 'focus' the healing energies through specific mental intentions. Preliminary research confirms that there may be varying healing effects with varied mental focus, and/or intents, of healers [18, 19]. Healers have been able to 'turn off' healing by deliberately focusing on mental exercises unrelated to healing — in order to serve as controls in healing research. In the experimental conditions (identical in outward appearances to the control conditions) the same healers demonstrated healing effects when their mental intention was to give healing [20].

3 Theoretical speculations

Several theoretical concepts are proposed to explain the observed healing phenomena. Principal amongst these are the following:

1. Biological energy fields are said to surround and interpenetrate the physical body of living systems. Healers claim that they are able to use their own biological fields to influence the fields of other living systems. [1, 5, 6, 9] Healers presume that the energy field acts as a template for the physical body, and that influences upon the energy field of an organism will subsequently be reflected in the physical body.

This theory could explain healings by the laying-on of hands. Typically, healers' hands are held either lightly touching the healee's body or 100–200 mm. away from the body. Numbers of controlled studies have demonstrated that healers can bring about changes in living systems with their hands near to but not touching them. [1]

2. Information may be transmitted from the mind of the healer to the mind of the healee, with the subsequent changes actually being produced by the healee as a 'self-healing'. This theory may explain placebo effects. This theory is extended by some healers, who suggest that telepathic or clairvoyant communications [1] may convey the requisite healing information. The information might influence the mind of the 'target' organism, its energy field, or its material substance by direct, psychokinetic ('mind over matter') influence [12]. Early research indicates that particular states of mind or intent in the healer may produce discrete healing effects [18, 19].

3. Spectrophotometric shifts in the infrared (2.5–3.0 microns) and ultraviolet (188.8 millimicrons) have been reported in some studies [18, 21, 22] when the water is given healing by healers. It is generally suggested that these are related to hydrogen bonding, as this infrared peak corresponds to that of the hydrogen bond.

4. Glen Rein speculates from a series of experiments that non-Hertzian fields may impart a patterning in water which conveys healing properties. He also demonstrated that Leonard Laskow, a medical doctor who is a gifted healer, could produce varying effects upon the growth of tumour cells, depending on his particular state of mind [18]. Non-Hertzian fields are said to be present around wire coils which are counter-wound so that the ordinary EM field of one coil precisely cancels that of the second coil. Rein demonstrated that non-Hertzian fields can have significant effects directly on biological systems [23, 24, 25], on water, and on biological systems via the water as a vehicle for the effect [18].

4 Avenues for future study

The dearth of research directed at the *mechanisms* of healing has contributed to the limited acceptance of healing by the scientific community. Early exploratory research suggests several avenues for the study of possible mechanisms to explain the mind/matter interactions of spiritual healing:

1. Some healers report that they are able to give healing to water, which is then said to transmit healing effects. Laboratory research has confirmed that healer-treated water enhances plant growth [1, 26, 27, 28, 29, 31].

Water appears to be capable of storing energy patterns, as suggested by extensive research on homeopathic remedies. [4] Remedies have demonstrated potency when they are diluted past the point that even a single molecule of the original medicinal substance is present in the administered solution. It is suggested that the therapeutic energetic patterning in the medicinal substance is transferred to the water, which, in turn, transfers the patterning to the energy field of the treated organism.

2. The qualitative reports of healers and healees engaged in healing may help to clarify the roles of consciousness in organising the matter of the body and may provide a starting point for understanding mind-matter interactions [1, 9, 10, 11, 12]

3. Light-shielded photographic film held between the hands of a healer [32] and between the hand of a healer and the body of a healee [33] have shown various patterns of exposure when developed. The production of images on photographic film through mental intent, unassociated with spiritual healing, has also been reported by several investigators [34, 35, 36]. These studies should be replicated and extended, as they might suggest quantum relationships in interactions between mind and matter. They might also clarify aspects of the nature of subtle energies.

4. The reports of healers [1, 10, 12, 37] and mystics [38] may extend our understanding of energies to include dimensions postulated in quantum physics. It appears possible that the mind may be able to perceive and interact with energies described by quantum physics, and/or to transcend the ordinary, three or four dimensional world. For instance, theoretical constructs in quantum physics postulate that it may be possible to move backwards as well as forwards in time. Healers I know are able to identify traumatic experiences which occurred years earlier and contributed to physical problems of healees, and to predict (in some cases) when illnesses will be cured or not. Precognition, retrocognition and retroactive psychokinetic influence on random number generators have been demonstrated in numbers of studies in parapsychology [1, 39].

5. The reports of healers and others who have explored inner worlds which we label *mystical* may help us comprehend the spiritual dimensions [1, 10, 12, 37, 38].

5 Conclusion

It has been suggested in numerous popular books that principles and research in quantum physics may contribute to the understanding of healing phenomena. This paper suggests conversely that research in healing may contribute to understanding of aspects of quantum physics.

References

[1] Benor, D. J. *Healing Research: Holistic Energy Medicine and Spirituality. Volume I. Research in Healing.* Helix Verlag GmbH 1993.

[2] Solomon, G. F. *The emerging field of psychoneuroimmunology with a special note on Aids,* Advances **2**(1) (1985), 6–19.

[3] Booth, R. J. and Ashbridge, K. R. *Is the mind part of the immune system?* Advances, **9**:2 (1993), 4–23, plus comments 24–65.

[4] Benor, D. J. *Healing Research: Holistic Energy Medicine and Spirituality. Volume II. Holistic Energy Medicine and the Energy Body.* Helix Verlag GmbH.

[5] Brennan, B. *Hands of Light.* London/New York: Bantam 1987.

[6] Brennan, B. *Light Emerging.* London/New York: Bantam 1993.

[7] Kunz, Dora van Gelder, *The Personal Aura,* Wheaton, IL: Quest/Theosophical 1991.

[8] Leadbeater, C. W., *Man Visible and Invisible,* Wheaton, IL: Quest/Theosophical 1969 (Orig. 1902).

[9] Krieger, D. *Accepting Your Power to Heal: The Personal Practice of Therapeutic Touch,* Santa Fe, NM: Bear & Co. 1993.

[10] LeShan, L. *The Medium, the Mystic and the Physicist,* New York: Ballantine 1974 (UK edition: Clairvoyant Reality, Wellingborough, Thorsons 1974).

[11] Benor, D. J. *Intuitive diagnosis,* Subtle Energies **3**:2 (1992), 41–64.

[12] Benor, D. J. *Healing Research: Holistic Energy Medicine and Spirituality. Volume IV: Healing in the Light of Recent Research,* Helix Verlag GmbH (in press).

[13] Watkins, G. K. and Watkins, A. M. *Possible PK influence on the resuscitation of anaesthetised mice,* Journal of Parapsychology **35**:4 (1971), 257–272.

[14] Braud, W. and Schlitz, M. *A methodology for the objective study of transpersonal imagery,* Journal of Scientific Exploration **3**:1 (1989), 43–63.

[15] Byrd, R. C. *Positive therapeutic effects of intercessory prayer in a coronary care population,* Southern Medical Journal **81**:7 (1988), 826–829.

[16] Targ, R. Puthoff, H. & May, E. *Direct perception of remote geographical locations.* In: Tart, C. T., Puthoff, H. E. & Targ, R. (Eds.), Mind at Large. New York: Praeger 1979.

[17] Jahn, R. and Dunne, B. *The Margins of Reality.* San Diego: Harcourt, Brace Jovanovich 1987.

[18] Rein, G. *Quantum Biology: Healing with Subtle Energy,* Institute of HeartMath, PO Box 1463, 14700 W. Park Avenue, Boulder Creek, CA 95006.

[19] Spindrift, Inc., *The Spindrift Papers,* Century Plaza Building, 100 W. Main Street, Lansdale, PA 19446, USA.

[20] Heidt, Patricia, *Effect of Therapeutic Touch on Anxiety Level of Hospitalised Patients,* PhD dissertation, New York University 1979.

[21] Schwartz, S. et al. *Infrared spectra alterations in water proximate to the palms of therapeutic practitioners.* Subtle Energies **1** (1991), 43–72.

[22] Dean, Douglas, *An examination of infra-red and ultra-violet techniques for changes in water following the laying-on of hands*, Unpublished Doctoral Dissertation, Saybrook Institute, CA 1983.

[23] Rein, G. *Biological interactions with scalar energy: Cellular mechanisms of action*. In: Proceedings of the 7th International Association for Psychotronics Research, Georgia, USA 1988.

[24] Rein, G. *Effect of non-Hertzian scalar waves on the immune system*, Journal of the U.S. Psychotronics Association (1989), 15–17.

[25] Rein, G. *Utilisation of a cell culture bioassay for measuring quantum fields generated from a modified caduceus coil*, Proceedings of the 26th International Energy Conversion Engineering Conference, Boston, MA 1991.

[26] Grad, B., *A telekinetic effect on plant growth. I*. International Journal of Parapsychology **5**(2) (1963), 117–134.

[27] Grad, B., *A telekinetic effect on plant growth II. Experiments Involving Treatment of Saline in Stoppered Bottles*, International Journal of Parapsychology **6** (1964(a)), 473–498.

[28] Grad, B. R., *Some biological effects of laying-on of hands: A review of experiments with animals and plants*, Journal of the American Society for Psychical Research , **59** (1965(a)), 95–127 (Also reproduced In: Schmeidler, Gerturde (Ed.) Parapsychology: Its Relation to Physics, Biology, Psychology and Psychiatry, Metuchen, NJ: Scarecrow 1976.)

[29] Miller, R. *Methods of detecting and measuring healing energies*, In: White, John and Krippner, Stanley, Future Science. Garden City, NY: Anchor/Doubleday 1977.

[30] Macdonald, R. G.; Hickman, J. L. & Dakin, H. S., *Preliminary physical measurements of psychophysical effects associated with three alleged psychic healers*, Research Brief, July 1, 1976; Summary in Research in Parapsychology 1976, Metuchen NJ/London: Scarecrow 1977.

[31] Saklani, A. *Psi-ability in shamans of Garhwal Himalaya*: Preliminary tests, Journal of the Society for Psychical Research **55** (1988), 60–70.

[32] Moss, T. *The Body Electric*, New York: St. Martin's 1979.

[33] Turner, G. *What power is transmitted in treatment?* (Part 1 of 4-Part Series), Two Worlds (July 1969), 199–201.

[34] Eisenbud, J. *The World of Ted Serios*, New York; Pocket Books 1967.

[35] Eisenbud, J. and Stillings, D. *Paranormal Film Forms and Palaeolithic Rock Engravings*, Archaeus **2**:1 (1984), 9–18; 18–26.

[36] Fukurai, T, *Clairvoyance and Thoughtography*, New York: Arno Press 1975.

[37] Benor, D. J. *Healing Research: Holistic Energy Medicine and Spirituality. Volume III, Research in Spiritual Healing*. Munich: Helix Verlag GmbH (in press).

[38] Capra, F. *The Tao of Physics*, Boulder, CO: Shambala 1975.

[39] Edge, H. L. et al., *Foundations of Parapsychology: Exploring the Boundaries of Human Capability*, Boston & London: Routledge & Kegan Paul 1986.

Towards a Science of Consciousness

M. Towsey and D.N. Ghista

1 Introduction

Reading Paul Davies recent book *The New Physics* [6], one cannot help but marvel at the immense success of modern physics in describing the physical world. Yet that very success leads one reviewer [9] to pause and consider where amongst the successes "may lurk some tiny murkiness destined to become the seed of the next scientific revolution" just as the quantum behaviour of light turned out to be the seed which overthrew classical physics. Davies himself notes that the role of *consciousness* in a quantum observation still remains an unresolved issue. "..it may be that this frontier - the interface of mind and matter - will turn out to be the most challenging legacy of the New Physics." [6, p6]

The European Community has declared the 1990's to be the 'Decade of the Brain' and no doubt many researchers are hoping that with the concentrated attention of scientists around the world, they will conquer the problem of consciousness, just as in the 1960's a determined effort put a human on the moon. The assumption is of course, that consciousness is an epiphenomenon of physical processes and must eventually succumb to traditional scientific method aided by the latest technology. Physicists tend to approach consciousness from the perspective of information theory, while biologists think in terms of an evolutionary mechanism, as for example Blackmore who asks "So why did consciousness evolve? Evolution must have had some reason for making us conscious - mustn't it?" [3]

Consciousness may be defined as the faculty of being aware or of experiencing or of knowing. It is the subjective factor which lies behind any type of experience and indeed behind the very activity of scientific inquiry itself. From the point of view of the natural sciences, consciousness constitutes a 'problem' because it is not at all clear how inert matter which makes up the living organisms studied by natural science, can give rise to the faculty of awareness. The present day science of material processes is faced with a challenge, because being premised on materialism (the belief that matter alone exists and is devoid of any kind of awareness, intent, desire or purpose), it can only formulate models of the world which are, by definition, devoid of consciousness. But the awareness that 'I exist and am having particular experiences' is such an integral part of being human that it can no longer be ignored.

The hope that consciousness will eventually be explained in terms of coding and information theory ignores the limitations of the theory. Even one of its founders warns, "Information theory *does not* handle the *meaning* of information, it treats only the amount of information." [8] But it is precisely the meaning of information which lies at the heart of the 'problem' of consciousness. Meaning may be defined as the relation or mapping between events in the external objective world and events in our internal subjective world.

Information theory can endow the physical processes giving rise to an experience with a number but it is our consciousness which endows the experience with meaning.

Because of the limitations which materialism places on traditional scientific thought and methodology, there has been discussion about the need for a new kind of science which can investigate consciousness on its own terms [7, 12]. Its raw data would be qualitative (rather than quantitative) descriptions of subjective internal experience. It would set up hypotheses concerning relationships between categories of subjective experience, test those hypotheses according to some internally consistent logic and reformulate hypotheses according to new data. We would expect the laws of this science to change according to the level of consciousness, just as the laws of physics change as one goes from the subatomic to the macroscopic world [7]. Wilbur [14] develops this line of thought in much detail, his major contribution being to demonstrate that eastern schools of philosophy offer a paradigm which can place western scentific method in a broader context.

It is our contention that the yogic tradition offers an appropriate scientific methodology to investigate consciousness. The term yogic practice is used here in a very general sense to include various schools of spiritual and intuitional practice which have evolved in the Asian sub-continent and whose purpose is to investigate the origins of consciousness. The emphasis here is on the word *practice*. The term yoga has the advantage of being immediately recognizable as a systematic discipline or practice, as opposed to being just a school of philosophy. Yoga may correctly be described as a science in the sense that it has, and continues, to give rise to evolving bodies of knowledge grounded in practical experience. Its worth, like that of any science, can be judged pragmatically by its ability to expand the horizons of human understanding and to improve the quality of life.

This paper will discuss 1) the epistemology of yoga as a science of consciousness; 2) yoga's account of the physical world; and 3) yoga psychology. A convenient place to begin is with yoga's classification of the different kinds of conscious experience.

2 A Classification of Conscious Experience

Typically, yoga distinguishes five broad categories of conscious experience (Fig 1). The consciousness of most people in every day life belongs to the third level, the rational-sentimental. A more restricted level of consciousness is the instinctive-sensory, typical of plants and animals but also of human beings in certain circumstances. This level of experience concerns survival and maintenance of the physical body. The fourth level is the supra-mental or transpersonal, that is a state of consciousness which spans smaller or larger groups of people. We typically experience this state for short periods when we get intuitional insights into other people's states of mind or when we 'see' the solution to a problem which has been bothering us for some time. The fifth level is described as the level of discrimination, detachment or wisdom [13] or as the level of archetypes [14]. This uncommon state of consciousness involves the direct and pure apprehension of notions such as truth and falsehood, good and bad. Finally, the sixth level of consciousness is the unitary experience of universal consciousness sometimes described by saints and mystics. The word yoga itself means *union*, that is the union of one's individual consciousness with universal consciousness [1].

These levels of consciousness lie on a spectrum from least conscious to most conscious because in comparing them, we sense different degrees of increased or reduced awareness.

For example, rational-sentimental consciousness gives a sense of greater awareness or of 'seeing' more than the instinctive-sensory state, but in comparison to the transpersonal mind it is a more contracted, atomistic state of consciousness. In general, higher (more expanded) levels of consciousness encompass lower levels.

In addition, yoga distinguishes two more levels of consciousness. At the bottom end of the spectrum (level 1), physical matter is said to have some small degree of consciousness, that is, matter is 'self-actualizing' in some primitive sense. At the top end of the spectrum, there is a state of consciousness which is beyond the experience of universal oneness. In this state there is no longer any objective experience as such and so it is not accessible to any form of scientific inquiry. It is said to be a state of pure, self-reflected being.

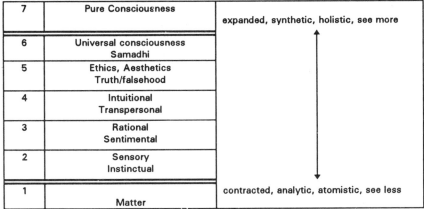

Figure 1 The seven levels or states of consciousness according to yogic science. Levels 2 to 6 constitute the levels of mind. Matter (level 1), although apparently pure objectivity, is said to exhibit some minimal degree of consciousness. Level 7 is pure subjectivity, where objectivity or energy is entirely unmanifest.

In common English usage, the word 'consciousness' is used loosely and therefore the term 'a science of consciousness' embraces the investigation a number of quite different phenomena, whose only feature in common is their non-materiality. Subjective emotional feelings and thought processes, the sense of 'self', para-psychic phenomena such as psycho-kinesis and psychic healing, and spiritual experience all seem to come within the field of 'consciousness research' as evidenced by the literature. In yogic philosophy, the term consciousness refers specifically to pure subjectivity. This discrepancy between eastern and western usage can lead to confusion.

3 Scientific Method

Now we ask the epistemological question, what is the justification for the above classification of consciousness? The short answer is that it arises from a long tradition of scientific inquiry into subjective experience. The important word here is *scientific*, which we are using in its broadest sense as elucidated by Wilber [14]. Wilber defines scientific method in terms of its three fundamental strands, *injunction, illumination* and *consensual validation*. The injunction is a recipe or set of instructions for both the preparation of the experimenter's mind and the experimental apparatus. The second step is the actual performing of the experiment which gives rise to a particular experience or state of consciousness. It is important to note that the experience itself is the product of three inputs, the observer, the observed and the process of observation which links the two. In other words, the experience is not purely an objective result of that which is being observed as is sometimes naively supposed. Finally, the experience must be shared and validated communally, which is usually done through written descriptions in books or journals. Here the emphasis is on science as a social endeavour.

According to Wilber, these three strands of the scientific method apply whether the object of the science is to investigate the physical world, to investigate mental states through psychoanalysis or to experience the most expanded states of consciousness. However the three strands will take different forms appropriate to the object of study. To investigate the physical world, sensory observation, measurement and the usual methods of the natural sciences have proved to be exceptionally powerful. In case of exploring the rational-sentimental mind, phenomenological methods are appropriate, such as psychoanalysis. However for exploring more expanded levels of consciousness, meditative techniques are required. Thus, says Wilber, the yogic investigation of mind involves all the three strands of scientific investigation, injunction, illumination and consensual validation. The experimental apparatus is the experimenter's own mind which must be prepared and trained in a disciplined way, just as is true for the injunctive step in the natural sciences. The illuminative or experiential step is the performance of the meditative technique itself which leads to a certain quality of conscious experience. And finally the conscious experience, even that of universal oneness, requires validation if it is to come within the scope of systematic investigation. In this case however, validation is in form of corroboration by a teacher.

It is important to note that measurement, enumeration and calculation are not the defining criteria for scientific method. Of course they are essential for the investigation of the physical world but where they are not appropriate, as in the study of higher states of consciousness, their absence does not disqualify a particular methodology from scientific status.

Meditative techniques to investigate higher states of mind are necessarily introspective and therefore involve withdrawing one's attention from the external sensory world. There are three steps to such withdrawal; withdrawal from external environmental sensation, withdrawal from body sensation and withdrawal from internal 'auditory' and 'visual' distractions that bubble up from the subconscious when one tries to still the mind.

The withdrawal stage is followed by concentrated attention to some auto-suggestion and/or visualization. The exact nature of these depends on the purpose of the technique. Success, as in any endeavour, requires practice but it is important to note that the experiences so obtained are not purely personal or without meaning to others. On the contrary, all six levels of consciousness correspond to objectively existing categories or worlds and are accessible by anyone with the appropriate training and technique. Some yogic meditations can be considered as the equivalent of scales for a musician. They are not an end in themselves but practice of them builds technique and mental stamina.

An essential characteristic of yoga as a scientific method is that the experimental instrument, the mind, must be deployed using a *holistic approach*. The levels of consciousness described earlier correspond to objectively existing worlds in a hierarchical relationship of wholes and parts (Fig 2). At the top of the hierarchy, the biggest whole corresponds to the experience of *universal consciousness*. As we descend the hierarchy, the worlds become more atomistic and divided. If one adopts the *reductionist* or analytic viewpoint of traditional science, then one necessarily stands at the bottom of the hierarchy where, blinded by the profusion of parts, one cannot see the whole. In yoga, one is traditionally interested in the experience of universal consciousness and thus one adopts a holistic stance by placing oneself at the top of the hierarchy. The parts are recognized but in the context of the whole.

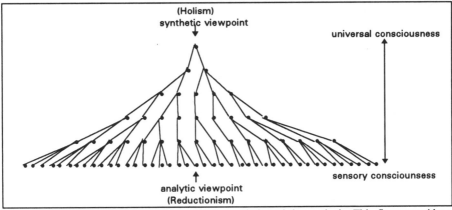

Figure 2 Two ways of looking at the world, the analytic and the synthetic. This figure provides an intuitive explanation for the difficulties involved in reconciling accounts of the world obtained from the two different points of view.

While the difference between reductionism and holism is certainly important, it does not convey the fundamental epistemological difference between the natural sciences and yogic science. Any experience is a unity of the observer, the observed and the process of observation which links the two. Experiences differ in their sense of 'distance' between the observer and the observed. Where the distance is great, the experience is said to be more objective and in the case of the natural sciences, more reliable. Where the distance is small, the experience is said to be more subjective. In yogic practice, unlike the natural sciences, a deliberate attempt is made to *reduce* the distance between the observer and the observed. Indeed, one attempts to merge one's identity with that observed entity and thereby acquire an intuitive knowledge of it. It is an intuitional approach.

Of course *intuition* has an important role to play in the natural sciences and even more so in the psychological sciences, where the goal is to understand the workings of another person's mind. And while the role of intuition in these sciences has not yet been formalized, it is increasingly recognized as a valuable ingredient which can be enhanced through life-experience, appropriate schooling and so on. In the case of investigating the more expanded levels of consciousness (levels 4, 5 and 6 of Fig 1) however, that which we wish to understand is *transpersonal*, so the methodologies of the physical and phenomenological sciences are no longer adequate. Here intuition has to be abstracted from its material and personal context and employed directly as the method of knowing. For this reason, yogic methodology is also called *"intuitional science"* [1]. Just as one's power of sensory observation can be improved through training, so too, can one's power of intuitive perception.

To summarize, the distinctive features of yogic scientific method are; 1) the experimental tool is the experimenter's *mind,*. 2) the mind is directed *introspectively*, 3) the mind adopts a *holistic* perspective, 4) the observer comes to know the observed through a gradually developed *intuition* which is abstracted from its material and personal context, 5) consciousness is considered to be primary and energy its derivation, as compared to the natural sciences where energy is primary and consciousness its derivation.

Rubik [12] describes three obstacles to progress in consciousness studies; 1) the dictum that subjective experience is an unacceptable basis for knowledge, 2) that consciousness studies require us to look at the world in a different way and 3) lack of a suitable epistemology. She advances twelve characteristics of a suitable epistemology which include; inwardly directed as well as outwardly directed observations, inter subjective validation as opposed to purely objective validation, acceptability of qualitative data as well as quantitative data, recognition of the critical role of the scientist's persona or mental orientation, a holistic approach as opposed to purely analytic and reducing the distance between 'first person' and 'third person' experience. Yogic scientific method as we have described it, meets these epistemological requirements.

Nevertheless, it is very difficult for a scientist trained in the orthodox tradition, to overcome a parochial bias that only objective, quantitative, reductionist science yields useful models of reality and that all else is mythology. Thus it would be of interest to know how the two methodologies compare in describing the same phenomena. Let us take the challenging case of comparing their descriptions of the physical universe, since modern physics gives us no insights into higher states of consciousness. Fortunately yogic science does give an account of the physical universe, but a comparison would have been difficult before the advent of quantum mechanics and relativity.

4 Consciousness and the Physical World

The pioneers of quantum mechanics were quick to perceive connections between their *new physics* and eastern cosmology [10] and these were popularized by Fritjof Capra in *The Tao of Physics* [4]. However these connections were primarily philosophical, in particular the notion that some kind of conscious mind lies behind the dance of matter and gives rise to matter. Our contribution to this enterprise has been to discover a surprising and far deeper degree of correspondence between the yogic and the new physics descriptions of the physical world.

The first historical record we have of any philosophically 'abstracted' description of the physical world goes back 7000 years and more to the Vedas, where the diversity of physical phenomena is explained in terms of five simple elements or *fundamental factors*, ethereal, aerial, luminous, liquid and solid factors. Many chemistry texts acknowledge this as the beginnings of chemical thought, but omit to mention that this simple idea acquired considerable sophistication over the ensuing millennia.

By the 1st century BC, the five fundamental factors were understood as a spectrum of wave forms, ethereal factor having the longest wavelength down to solid factor having the shortest wavelength. By the 10th century AD, the five factors had become associated with sensory properties called *tanmatras* [2, 11] as shown in Fig 3. Tanmatras are described as having a particle like character and yet are intimately associated with the underlying wave forms of the fundamental factors. In other words, the dual wave-particle nature of the physical world is also to be found in the yogic tradition, as is the notion of continuity versus discreteness. The sensory properties or tanmatras associated with a wave form increase in diversity as wavelength decreases. Continuing to the present century, the yogic description of the physical world has acquired more detail. Of relevance here is the idea that the energy associated with a wave increases as wavelength decreases across the spectrum [13].

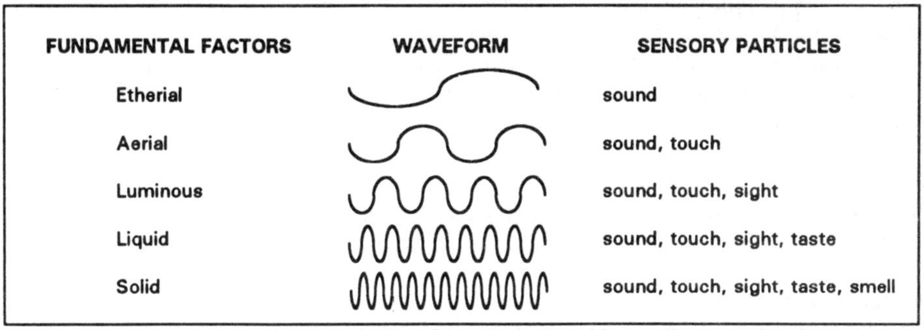

FUNDAMENTAL FACTORS	WAVEFORM	SENSORY PARTICLES
Etherial		sound
Aerial		sound, touch
Luminous		sound, touch, sight
Liquid		sound, touch, sight, taste
Solid		sound, touch, sight, taste, smell

Figure 3 The dual wave-particle description of the physical world according to yogic science. There is a spectrum of wavelengths but discrete sensory boundaries.

Now the question arises as to the relationship between this description of the physical world and that of the 'new physics'. In Fig 4, the two descriptions are placed side by side and, in order to make comparisons, we look for common patterns. The most obvious is that there are five levels of structure in the new physics, whose decreasing size corresponds to the decreasing wavelengths of the five fundamental factors.

Yoga's Description			Physics' Description		
Fundamental factor	Waveform	Sensory Attributes	Particle Type	Particle Charges	Physical Structure
Etherial		sound	vacuum state particles	?	space/time
Aerial		sound, touch	dark matter particles	flavour	galactic
Luminous		sound, touch, sight	?	?	stellar system & planets
Liquid		sound, touch, sight, taste	electron	flavour, electric	atomic
Solid		sound, touch, sight, taste, smell	quark	flavour, electric, colour	nuclear

Figure 4 A comparison of yoga's description of the physical world with that of modern physics. Although the language and methodologies are very different there are patterns of similarity which suggest that they are indeed describing the same world.

Secondly there is an increase in the mass of particle types from vacuum state particles to quarks, that corresponds with the increasing energy and decreasing wavelength of the five fundamental factors. And most significantly, the increase in particle interactions as we descend the table corresponds to the increase in sensory properties associated with each fundamental factor. So compelling are these comparisons that they suggest the following five correspondences. We propose that; 1) ethereal factor corresponds to the vacuum state and space-time; 2) aerial factor corresponds to dark matter and determines structure at the galactic level; 3) luminous factor determines structure at the stellar level, but as yet there appears to be no suitable particle type corresponding to it (the photon seems an obvious choice but it is not a matter particle); 4) liquid factor corresponds to electron particle-waves and determines structure at the atomic level; and 5) solid factor corresponds to quark particle-waves and determines structure at the nuclear level.

The correspondence between electron particle-waves and liquid factor illustrates why the holistic approach of yogic methodology conveys an apparently unfamiliar picture of the world. Yogic cosmology does not 'see' individual electron clouds around innumerable atoms and molecules but rather 'sees' the totality of electron clouds in the universe as a single identity. In fact, the quantum principle of *indistinguishability*, which states that individual electrons cannot be tagged or labelled, is not in conflict with this perspective. The state of a two electron system with electron *1* at location *a* and electron *2* at location *b* is indistinguishable from the state where electron *1* is at *b* and electron *2* is at *a*. The same principle can be extended to the entire universe with an indefinite number of electrons, with the consequence that all the electrons in the universe can be treated as a conceptual unity. It is in this sense that liquid factor may be considered as an equivalent description to innumerable electrons.

According to yogic cosmology, the five fundamental factors are continually changing one into the other, just as in physics, particles are colliding, merging and disintegrating. In particular, yogic cosmology describes a process of compression [13], whereby ethereal factor gives rise to aerial factor, aerial factor to luminous factor and so on. In the case of solid factor, if it is compressed beyond a certain critical point, it explodes with tremendous force giving rise to the four subtler factors. (Shown in fig 5 as the asterix below solid factor with arrows returning upwards, representing the recycling of matter and energy within the physical universe.) This appears to be the equivalent description of a supernova, where highly compressed matter within a contracting star causes it to explode. The universe of yogic cosmology is in tumult, not at all like the 'music of the spheres' of ancient Greek cosmology.

Figure 5 illustrates two further features of yogic cosmology. The first is that the physical universe is not a closed system. There is a continual input of energy coming from the universal mind and there is a continual output of energy by way of the individual minds of living organisms. The implication is that the physical universe will not suffer 'thermal death' as implied by the second law of thermodynamics [13] because it is not a closed system. It should be noted that the notion of a universal mind residing behind the physical universe has also been suggested by some western physicists. "Today ... the stream of knowledge is heading towards a non-mechanical reality; the universe begins to look more like a great thought than like a great machine. Mind no longer appears as an accidental intruder into the realm of matter; we are beginning to suspect that we ought to hail it as the creator and governor of the realm of matter..." [10]

Secondly, Fig 5 shows a barrier to the reverse flow from matter back to universal mind. When solid factor explodes violently and is converted back to liquid, luminous, aerial and ethereal factors, the point is specifically emphasized that no subtler mental factors are formed. Universal mind does not reabsorb matter or withdraw its projection. What might this correspond to in the 'new physics'? We suggest it corresponds to the speed of light as a limiting maximum. It is the speed of light which limits interactions between far-flung parts of the physical universe and thus our knowledge of it through physical means. Non-locality as observed in certain quantum systems (coordination between particles that cannot be explained in terms of information or field effects transmitted at or below the speed of light) can be interpreted as the action of mind in the physical world.

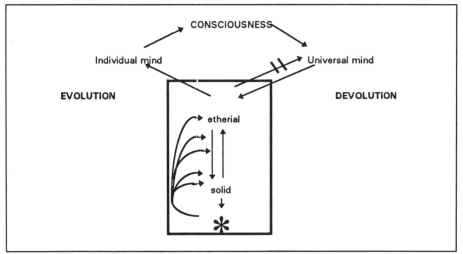

Figure 5 The cycle of creation according to yogic science consists of two phases. In the first phase, mind devolves from consciousness and matter devolves from mind. In the second phase, mind evolves from matter and consciousness evolves from mind. Note that there are cyclical flows of energy within the different levels of matter. In addition the backward flow from matter to universal mind is blocked (shown as the double slashed arrow). This block manifests to the physical observer as the limiting speed of light.

Apart from the philosophical interest in finding correspondences between two superficially dissimilar descriptions of the physical world, another purpose has been served. We began with the inadequacy of the natural sciences to investigate subjective experience and consciousness. We agree with Rubik that the need for "an extended science with revised epistemological foundations and methodologies appropriate for consciousness studies is indicated." [12] We have shown that yogic methodology shares those characteristics which Rubik identifies as necessary for the study of consciousness. Our confidence in yogic methodology is however much increased when we find that it arrives at a description of the physical world which has a strong correspondence to that of western science. We now turn to a brief comparison of yoga psychology with the western schools of psychology.

5 YOGA PSYCHOLOGY

Several major theoretical viewpoints have evolved within western psychology. These include behavioural psychology, cognitive psychology, psychodynamics, humanistic psychology and the socio-cultural schools of psychology. Furthermore, various interpretations have emerged within each of these schools. For example Jung and Adler both founded schools which are derivations of Freudian psychodynamics. While it is possible for the clinical psychologist to take an eclectic approach, Carson et al believe there is a real need "to tackle the theoretical clutter and develop a unified viewpoint that is both comprehensive and internally consistent and that actually reflects what we empirically know about abnormal behaviour." [5, p88] They regard this as the challenge for the next generation of theorists in human psychology.

We suggest that the spectrum of conscious states as shown in the left column of Fig 6 offers a useful unifying paradigm. An important feature of the spectrum is that the dynamics of any one level simultaneously impinge on the level above and on the level below, so that it is not possible to isolate the schools of thought into rigid categories. Generally speaking however, Level 1 corresponds to the biological school of thought which focuses on the organic component of psychological disorders. Level 2 corresponds to the behavioural school which focuses on objectively observable behaviour and thus is best suited for an account of sensory-motor and instinctive behaviours that are not significantly mediated by cognitive processes. Level 3 corresponds to the broad range of psychodynamic schools, such as the cognitive, psychoanalytic and humanistic schools. Level 4 corresponds to the transpersonal and socio-cultural schools of thought. Levels 5 and 6 have not generally come within the scope of 20th century western psychology but are hinted at by a variety of altered states of consciousness that have received some attention in recent years. Thus we do not perceive yoga psychology as a competing theoretical viewpoint but rather one which can both accommodate and complement the existing schools and so systematize the spectrum of psychological theory.

The Spectrum of Consciousness	Relevent School of Psychology
Universal consciousness	Yoga psychology
Ethical, Aesthetic, Truth consciousness	Yoga psychology
Intuitional, eco-social consciousness	Yoga psychology Transpersonal and psycho-social schools
Rational, sentimental consciousness	Humanist, psycho-analytic, and cognitive psychologies
Sensory, instinctual consciousness	Behaviourism
Matter	Organic disorders

Figure 6 The five levels of mind in yogic science provide a unifying paradigm to accomodate the various schools of psychology.

One useful consequence of this systematization is that it generates a theoretical framework (or paradigm) within which altered states of consciousness, such as near-death experiences, religious experiences, past-life experiences, out-of-body experiences and so on can be legitimately studied on their own terms. Because these altered states are not easily understood in terms of existing physical or psychological theories, there has been a tendency to deny their legitimacy or to regard them as pathological. For example, DSM III (the internationally accepted listing of recognised psychiatric disorders) has a class of 'depersonalization disorders' which includes any experience of the separation of self from the physical body. From the yogic point of view, altered states of consciousness such as those listed above are not necessarily pathological. Rather they are a consequence of the transition to other modalities of awareness not reliant upon the senses. These modalities are usually removed from awareness because the senses dominate our attention, in the same way that one is not aware of the fine texture of a cinema screen until the projected image is stopped. Rather than being viewed as pathological, expanded states of consciousness can be used as a therapeutic intervention to deal with psychiatric problems arising in the rational-sentimental mind.

The study of altered states of consciousness seems to demand a new epistemological and methodological approach. Yogic psychology has much to offer here but the benefits will not be one way. Yoga psychology is itself an incompletely developed science (otherwise it would not be a science). For example although the spectrum of conscious experience as understood by yoga is expansive (hence its ability to accommodate the many schools of psychological thought), the five divisions on the spectrum (six including the physical-organic level) are very broad and clearly require more analytical study. Wilber [14] demonstrates that the work of Maslow and Piaget points to several sub-levels within the rational-sentimental mind (level 3) not described in the yogic literature. A more comprehensive study of altered states of consciousness will help to fill in the detailed structure of the entire spectrum of conscious experience.

We noted earlier the traditional interest of yoga in exploring the more expanded (transpersonal) levels of consciousness. However it has always been recognized that such exploration depends upon healthy development of the lower levels of mind as well as the physical body. Thus yogic science has developed a range of techniques which have relevance to the repertoire of western psychology. For example from the biological-organic viewpoint, yoga emphasizes diet and asanas (yogic postures) as measures which directly affect and stabilize the endocrine system. Endocrine disorders, according to yoga psychology, underlie mood disorders and developmental personality disorders. Relaxation and breathing exercises are useful interventions at the behavioural level, while at the cognitive level, yoga has long used visualizations and autosuggestions which are now accepted techniques in behavioural and cognitive therapies.

The dominant themes in yoga psychology are *transcendence* and *harmony*, themes which are temperamentally in accord with the humanistic and transpersonal schools of western psychology. Transcendence is the innate desire in human beings for expansion of consciousness, that is, the desire to go beyond what is currently normal and to realize one's full potential. In both yogic and humanistic psychology, the denial, frustration or distortion of this fundamental urge for self-realization is regarded as the root cause of psychopathology. Maslow even went so far as to express concern about the 'psychopathology of the normal', that is, the disappointing waste of human potential caused by inhibiting social pressures to conform. In yogic psychology, self-realization is understood to mean the unfolding of higher and higher levels of mind. It occurs both during the development of an individual organism and in the evolutionary context of a biological species or social system. A limitation of humanistic psychology is that it seldom clearly defines where self-realization will lead. In yogic psychology, the pursuit of self-realization must eventually lead to a culminating point, the permanent condition of *universal consciousness*.

We have already referred to the importance in yoga psychology of a balanced development of all levels of mind. The importance of *harmony* between the different levels of mind is related to Maslow's well known hierarchy of needs. However, a conflict can arise between *transcendence* and *harmony,* because the urge to explore new mental horizons inevitably demands a change in one's existing view of the world. In the language of developmental psychology, growing-up requires a letting go of past attachments in order to face the challenges of a larger world. Wilber [14] gives an elegant exposition of the conflict between transcendence and harmony in human development and regards its unsatisfactory resolution as a major source of psychopathology. By contrast, a satisfactory resolution is what makes 'growing up' so rewarding! A final resolution of this conflict is possible only in the state of universal consciousness. This state can be described as a fulfillment so complete, that the question of going elsewhere does not arise.

6   REFERENCES

[1]   Anandamurti. Intuitional Science of the Vedas, in *Subhasita Samgraha*, Ananda Marga, Calcutta, 1956

[2]   Bhoskari and Abhinavagupta (commentators). *Ishvara Pratyabhijna Vimarsini*, Volume III, ed. K.C. Pandy, pub. Princess of Wales Sarasvati Bhavan.

[3]   S. Blackmore. Consciousness: science tackles the self. *New Scientist,* 1 April 1989.

[4]   F. Capra. *The Tao Of Physics,* Bantam NY, 1977.

[5]   R.C. Carson, J.N. Butcher and J.C. Coleman. *Abnormal Psychology and Modern Life,* 8th edition Scott, Foresman and Coy, USA. 1988.

[6]   P. Davies. *The New Physics,* ed P. Davies, Cambridge University Press, 1989.

[7]   P. Fenwick and D. Lorimer. Can Brains Be Conscious? *New Scientist,* 5 Aug 1989.

[8]   R.W. Hamming. *Coding and Information Theory,* Prentice-Hall 1980.

[9]   N. Herbert. The end of physics. *New Scientist,* 24 June 1989.

[10]   J. Jeans. *The Mysterious Universe,* pub E.P. Dutton, NY. 1932.

[11]   Rajanaka Ananda (commentator) *Sattrimsattattva Samdoh,* translator Debrata Sen Sharma, Kuruckshetra University.

[12]   B. Rubik. The Challenge of Consciousness Studies, *Proceedings 2nd Gauss Symposium,* Munich 2-8 August, 1993

[13]   P.R. Sarkar. *Idea and Ideology.* Ananda Marga, Calcutta 5th ed. 1978

[14]   K. Wilber. *Eye To Eye - the quest for a new paradigm.* Double Day, Anchor Press, 1980

Glucose Metabolism and Insulin Delivery

E. Sarti

Feedback controlled insulin infusing devices were introduced in the therapeutic practice since mid 70's. Biostator *and* Betalike *were cumbersome, bedside devices detecting glucose, and infusing insulin or glucose directly in vein. They could not be used outside of hospital environment and, moreover, the need of preventing hyperglycemia, whatever be the insulin sensitivity of the individual patient, led in most cases to insulin hyperinfusion. The large scale integration of electronic devices, together with the availability of small size insulin infusers, suggested, in the last decade, the study of the so called* wearable artificial pancreas. *The main problems to overcome concern with glucose detection, that must be performed subcutaneously, thus introducing a dynamic, scarcely known link between the* controlled variable *(i.e. blood glucose concentration) and the* measured output *(i.e. subcutaneous concentration). Moreover, a stable and biocompatible glucose sensor must be produced: this is by far the most arduous target: large efforts are devoted to it, as proved also by some contributions which hold a considerable part of this chapter. Additional problems arise when the needs of saving insulin and performing a more physiological control lead to the synthesis of* self-tuning controllers, *able to match themselves to the individual patient response and, if necessary, to its changes in time.* Minimum variance controllers and neural networks, *developed on the basis of accurate models of insulin and glucose metabolism, have been suggested to this purpose.*

After an introductory section of C. Cobelli *and* E. Sarti, *which presents a general overview on the problems involved in producing suitable feedback controllers of blood glucose and on the perspectives opened by new technological tools and methods, two sections are devoted to some theoretical questions to be kept in mind while facing the technical problems.* Issam El Mugamer et al. *expose an overview of the main long term complicances following prolonged hyperglycemia in diabetes mellitus, that should be avoided or, at least, weakened by automatic insulin infusion.* E. Biermann, *on his side, develops a mathematical model of non-insulin dependent diabetes, that can be used as an interactive tool to obtain more insight in the physiopathology and to develop therapeutic strategies.*

The following sections are shared between the two main problems above mentioned: sensors *and* control strategies. G. Urban *et al. describe a miniaturized sensor for* in vivo *applications, stable and well reproducible, obtained by a thin film photolitographic coating process, The following section, by* E. Wilkins *and* P. Atanasov, *relates on an implantable sensor, where the enzyme is immobilized on carbon powder dispersed on a liquid suspension, thus allowing replacement of spent enzyme by fresh one. A different approach is due to* M. Casolaro, *who describes a complete feedback control system based on a* chemical valve, *i. e. a porous membrane enclosing a pH-sensitive polymer, which allows insulin transit accordingly to the acidity changes induced by enzymatic glucose oxidation.*

The last two sections concern with self-adaptive controllers: P.G. Fabietti *et al. describe a minimum variance auto-tuning controller implemented on a micro-controller portable unit.* R. Bafunno *and* C. Coltelli, *finally, present a pioneer application of neural networks to the self-adaptive explorative control of blood glucose.*

Recent Developments and Open Problems in Feedback Control of the Glucose System in Diabetes

E. Sarti and C. Cobelli

1 Feedback control of blood glucose

The glycemia regulation in healthy people is entrusted to a feedback loop in the organism, based on the production and release of insulin by the pancreas. This loop was extensively studied and described through various mathematical models [1,2,3]. In type I diabetes this pancreas function is destroyed and has to be replaced by a therapeutic action of insulin administration. Very soon, it was thought to automatize the administration by submitting it to a measure of the glucose amount present in the organism, that is to say to close artificially the feedback loop that the disease had opened. Two main classes of such feedback devices are now under development: micro-encapsulated implanted β-cells, and artificial infusers. This lecture deals with the second class, i.e. the artificial pancreas (AP). We shall examine them from the point of view of the system engineer, that means we shall consider the performances of their components only as they affect the performances of the whole system.

An artificial pancreas was firstly described and tested by ALBISSER in 1974 [4, 5] while CLEMENS developed a similar device, thus made commercially available as Biostator GCIS [4]; another commercial AP, Betalike, was produced in 1984 [7]. Biostator and Betalike are the first industrial applications of feedback control in humans. They are cumbersome, bedside devices, which do not allow the patient to do any appreciable physical activity, and have to be used under direct clinician's surveillance. So, their use is confined to the hospital. Nevertheless, we shall describe them briefly, as they have been precious therapy and research tools in the past years, and are the basis of any further development.

Biostator includes a dual-lumen catheter, which draws heparinized blood out of a vein. Glucose concentration therein is measured through an electro-enzymatic sensor; then, the concentration signal is processed by a nonlinear dynamic controller, which is programmed in a small digital computing unit. The controller output drives two pumps, which infuse insulin and glucose again into a vein. Therefore, Biostator performs a typical feedback control system (Fig.1): the patient is the controlled system, or plant; blood glucose concentration is the controlled variable, which must be kept close to the reference value (say 95-100 mg/dl) in spite of disturbances, i.e. glucose inputs, (e.g. meals) and unexpected consumption (e.g. physical activity). Therefore, the system acts as a regulator device. Finally, insulin and glucose infusion are

---

* This text has been partially published, in Italian, in: D. DE ROSSI, M. GRATTAROLA, *Eds.: Bioelettronica e nanotecnologie per la bioingegneria*. Pàtron Editore, Bologna 1992. By kind permission of the publisher.

the control variables. More precisely, it should be said that they are, together, one control variable, as they are alternatively active, in case of hyper- and hypo-glycemia, respectively.

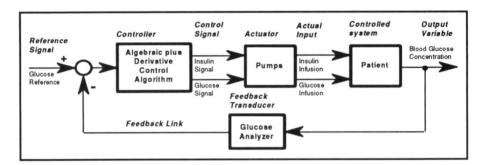

Fig. 1
Feedback control system

The Biostator control algorithms were set up empirically, based on a wide clinical experience – and not on the basis of a mathematical model, as we shall see in the next section. As an example, the algorithm ruling the insulin infusion is shown here. If glucose concentration GY is higher than reference value BI, then insulin is infused at a rate

$$IR = RI [K_g (GY - BI) + 1]^2 + RI K_d (GY - BI) m$$

where RI is the basal infusion rate, needed to keep glucose concentration at the basal value in absence of disturbances, m is the slope of the glucose response, and K_g, K_d are the gain coefficients of the static and dynamic term respectively. Otherwise, if GY is lower than BI, insulin infusion is stopped and glucose infusion can take place. A similar algorithm is provided for glucose infusion, but the fourth power of the difference between desired and actual glycemia is included there, meaning that hypoglycemia is much more feared than hyperglycemia. To smooth the measurement noise, blood glucose level and slope are averaged over 5 blood samples, taken at intervals of 1 minute. It can be noticed that both terms in the algorithm are nonlinear: the static term is proportional to the square of glucose blood level, and the dynamic one to the product of glucose level and slope. Therefore, the Biostator controller may be qualified as a nonlinear algebraic-plus-derivative controller. We shall see that both these features, nonlinearity and derivative control, are used, or at least desirable, in the feedback regulation of biological functions.

After building such hospital devices, most of the research was devoted to wearable artificial pancreas, i.e. tools able to be used in normal life, out of medical surveillance. This target implies to satisfy some requirements:

-   size and weight of the control unit must be drastically reduced: this result is easily reached, now, by large scale integration of electronic components;
-   similarly, power consumption of both controller and infusion devices must be strongly reduced: today's electronics solves also this problem, by providing low rating devices;

- the portable device is intrinsically unable to recover hypoglycemia, e.g. by glucose infusion: therefore, hypoglycemia must be accurately avoided by an appropriate design of the whole unit, that must include control software and some alarm devices;
- glucose detection and insulin infusion must follow a less invasive route than the intravenous one: this need generates the most complex problems in the system design; they will be discussed below;
- the control software should be optimized for the individual patient response; in fact, the Biostator calibration for an "average" patient can lead to an insulin hyperinfusion (and, therefore, to the danger of hypoglycemia), that should be avoided; this result can be accomplished by employing self-adaptive controllers;
- more generally, the portable device should ensure that the patient has the same safety level he benefits from the direct medical assistance in the hospital. This means that all parts of the system should be intrinsically safe, i.e., at least, guarantee the basal insulin infusion in case of fault.

In the past years the research has followed different lines. The most advanced results have been obtained, probably, in the field of infusing devices: miniaturized industrial infusers, both portable and implantable, have been realized by different manufacturers (the portable Minimed by Hoechst, and the implantable Infusaid by Siemens), and are now in use for open-loop administration, i.e. predictive control [8]. The need for a self-adaptive control was realized since 1972, when PAGUREK proposed his pioneer controller [9]; now, our group is working on the subject [10], and some results are described in chapters 5.6 and 5.7. An electronic portable unit (but not yet really wearable: its weight was about 400 g) was described by SHICHIRI in 1982 [11]: it employed a standard PID regulator and a needle replaceable sensor, which was tested on animals and humans [12]. Our group developed in 1991 a much smaller unit, using an integrated microcontroller and C-MOS technology [13]. Finally, the widest effort was devoted to develop long-term stable and accurate sensing devices: this is, undoubtedly, the most challenging research field, as tested by the number of realizations which followed the above mentioned sensor of SHICHIRI. We shall give in Section 4 some more data thereon.

Later on we shall discuss the open problems rising from the above listed requirements. But we point out here that the most difficult ones concern with the development of a safe enough control software and the realization of a stable and reliable sensor.

2 A mathematical model of the diabetic patient

The development of the control algorithms, whatever they are, requires an accurate model of the controlled system, i.e. of the patient. This requirement can be removed, as in the case of Biostator, if the controller structure is rather simple, and mathematical computations are replaced by a large clinical experience performed on the device itself. Anyway, a mathematical model is strictly needed to develop more sophisticated controllers, such as self-adaptive. A number of mathematical models of the glycemia regulation system in healthy and diabetic subjects has been developed, starting from the early works of BOLIE [1], ACKERMAN [2] and, mainly, from the basic contribution of CERASI and LUFT [3]. Most of them are compartmental and nonlinear, thus reflecting the main peculiarities of the phenomenon. We shall describe the model used by our group, mainly with the purpose of pointing out some of its features, which are relevant for the structure and accuracy of the control system.

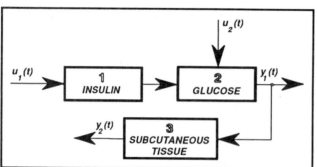

Fig. 2
Block diagram of the diabetic patient model:
$u_1(t)$ = insulin infusion rate (input variable);
$u_2(t)$ = glucose input (disturbance);
$y_1(t)$ = blood glucose concentration (controlled variable);
$y_2(t)$ = subcutaneous glucose concentration (measured variable).

The whole model is shown in the block diagram of Fig.2. It comes from a preceding model, which was developed to study the physiology of the glucose-insulin system and was largely validated by clinical tests [16]. It must be warned that the modifications introduced to represent type I diabetes, and the external connections to the sensor and to the actuator, are not likewise supported by experimental tests. Yet, they were tested through many computer simulations, performed while varying the values of the parameters in the entire foreseeable range. These simulations did not check the intrinsic accuracy of the model, but only that the control system preserves the requested performance.

The model has the inputs already seen in Fig.1: u_1 represents the *control variable*, which is the rate of infusion of exogenous insulin; u_2 is the *disturbance*, i.e. the gastrointestinal absorption of food: simulations tests were performed with oral glucose loads. The control variable enters block **1**, that describes the insulin metabolism. The structure of this block is shown in Fig.3: it is a linear three compartment model. Exogenous insulin is infused in a *plasma compartment*: this hypothesis, which reflects the most common infusion route, was adopted in the simulations till this time, but has to be modified for portable devices, as it is nearly impossible for a patient to live normally with a needle in a vein. Insulin moves from plasma to the *hepatic* and *peripheral compartment*. The latter represents fundamentally the insulin in interstitial fluids, which controls the insulin-dependent glucose utilization taking place especially in muscles. Two of the three state variables in block **1**, that is to say the insulin concentrations x_1 and x_2 in hepatic and peripheral compartments, influence the metabolism of glucose which is represented in block **2**. Block **2** is a compartmental model as well (Fig.4): it includes two compartments, representing the tissues that exchange glucose with blood rapidly and slowly, respectively. The model is nonlinear: as a matter of fact, glucose release and removal are nonlinear functions of glucose concentrations, and also of insulin concentrations in the hepatic and peripheral compartments.

The nonlinearities included in the model, as those of almost all biological processes, are of the *gradual threshold* and *saturation* type. Moreover, in insulin-dependent glucose utilization, an effect depending on the *product* of the concentrations of glucose and insulin is supposed. This means that glucose metabolism is *controlled* by insulin. Using the terminology of automatic control, it could be said (as long as saturation does not take place) that the glucose metabolism model is *bilinear* in its state variables x_4 and x_5, and in its input variables (coming from the insulin model) x_2 and x_3. Therefore, the model of Fig.4 exhibits almost all the nonlinearities that can be met in control systems. This is nothing but the application of a principle well-known to control engineers: *nature is intrinsically nonlinear*. As a consequence, *an adequate controller should be nonlinear*, to compensate the nonlinearities of the plant. This principle was applied in the experimentally derived Biostator algorithms.

The model in Fig.2 has two outputs as well: the first one, y_1, represents the blood glucose concentration and is the *controlled variable*: to keep its value at a fixed level is the control *aim*. The second output variable y_2, leaving block **3**, is present whenever y_1 is not measurable, and, therefore, the measurement has to be performed on a quantity in some way derived from it. In our case, y_1 is not measurable because a patient cannot live normally with a sensor in a vein, so it is necessary to find a less invasive location. The simulation tests were carried out supposing a subcutaneous sensor, and, therefore, block **3** represents glucose kinetics through subcutaneous tissues. The operation of moving the measuring point has to be done very carefully, because a part of the controlled system is now between the two variables, the controlled one and the actually measured one, and, therefore, belongs to the *feedback link*. So, an inaccuracy in the knowledge of block **3** immediately affects the precision of y_1. By the way, it is interesting to notice that this problem did not exist in glycemia industrial controllers, because their sensor measured glucose in blood drawn from venous circulation. Of course, the structure of block **3** depends on the sensor position, that may vary from patient to patient, and also in one patient, when, for testing or maintenance need, the sensor is removed and introduced again in a nearby – but not the same – position.

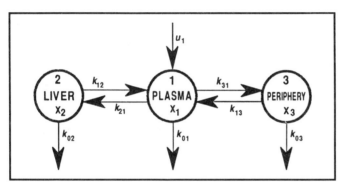

Fig. 3
Compartmental model of insulin metabolism

In other words: while the set of biological (and mathematical) functions that determine the output y_1 is known enough to be described with a model reflecting the physiological structure, the subsequent block **3** can only be of the input-output type, identified through a difficult *calibrating* operation: as a matter of fact, the whole structure or at least the parameters of the model may vary according to the location of the sensor. So the structure and the parameters of the controller, and in principle also the performance of the control system may vary as well. If a subcutaneous sensor location is hypothesized, block **3** represents the signal attenuation and delay due to the transportation of glucose from the vessels to the subcutaneous tissue; it can be schematized with a transfer function of the first order without zeros

$$G(s) = y_2(s)/y_1(s) = K/(1+\tau s)$$

where K and τ were varied in a wide range to simulate various locations of the sensor and various individual patient responses.

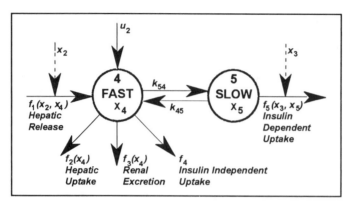

Fig. 4
Compartmental model of glucose metabolism.

3 Controllers

The controllers already used for biological systems have a quite simple dynamical structure: if linear, they are standard regulators of PD or PID type, or sometimes purely algebraic ones [17]. The reasons for this structural simplicity lie in the basic characteristic of the biological systems, that have a response which is normally *stable* (an instability should be considered as an *illness*) and nearly always *aperiodic*; in addition the systems are never requested of too strict specifications, neither for precision nor for time response.

On the contrary, controllers are normally *nonlinear*. As the controlled systems, they exhibit the entire variety of the nonlinear control theory. They may include *thresholds* or *saturations* of the output variable, which are introduced for patient safety, to avoid the infusion of the drugs being completely interrupted or exceeding the biological tolerance. Sometimes, the controller presents a *dead zone* for the input variable: the controller output does not change as long as the difference between the controlled variable and its desired value remains inside a definite range. Conceptually, this means that not a single desired value but a *welfare band* needs to be determined for the controlled variable. No control action, or change in the previously set up parameters, is needed within this band. In practice, this reduces the number and the duration of the operations on the patient.

Among the nonlinearities of these controllers we must mention the intrinsic *asymmetry* of the control actions, if they are carried out with the administration of one single drug – and not with two antagonist administrations, as it happens with insulin and glucose in the industrial artificial pancreas. It may happen that wide variations of the controlled variable – in such a direction that an increase in the administration is necessary – are easily compensated by the control system; while variations in the opposite direction, which would need a *negative* control action, can only interrupt the administration: then, the only thing to do is to wait for a *spontaneous* normalization of the situation, with the control loop open between controller and patient, and the patient left to his own natural evolution. This happens, typically, in the portable artificial pancreas, which cannot infuse glucose, and, therefore, must perform all effective control actions only by insulin infusion. Finally, nonlinearities can be deliberately introduced to *compensate* those of the controlled system. This practice, rather common in industrial plants, is adopted also in bioengineering: a nonlinear controller is used to *linearize* the whole system response. A typical example are the already seen algorithms of the Biostator.

Persons who widely used Biostator in past years complain of an insulin overinfusion and a postprandial insulinemia quite superior to physiological values, in particular when Biostator parameters are set to the values suggested by the manufacturer. Such a drawback was attributed to many reasons. The infusion route does not correspond to the physiological entry site of insulin produced by pancreas in healthy people. The extracorporeal path of the blood sample towards the sensor causes a finite delay. A further delay of the artificial control system, in comparison with the natural one, is due to a physiological process that cannot be reproduced by the artificial organ: food assumption and its first absorption cause a response of the endocrine and autonomous nervous systems, that in some way *anticipate* glucose appearance in blood. Finally, the controller parameters are not *individualized*: they do not consider interindividual and intraindividual (in time) variations; they are chosen to get an acceptable response also from less insulin-sensitive patients, at the price of overloading with insulin the more sensitive ones.

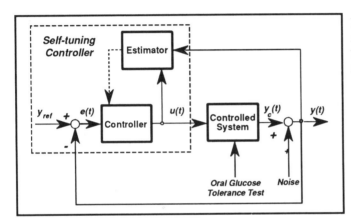

Fig. 5
A scheme of self-tuning control of blood glucose concentration

Self-tuning control

This overloading of insulin, which tends to produce hypoglycemia in the insulin-sensitive subjects, can be compensated, in the industrial devices, by glucose infusion. Portable devices, which cannot infuse glucose, must avoid it. Any delay must be reduced by minimizing or suppressing, if possible, the blood extracorporeal path. And the last cause, i.e. non-personalized control (which is probably the most important one) is faced with *adaptive controllers*.

The application of adaptive control techniques – more exactly called *self-tuning* – to biological systems has two concurrent reasons. The first one is the interindividual variability: unlike the industrial controllers, that are applied to a fixed process and are calibrated on its own characteristics, our instruments need to be used by more people with different responses: so it is advisable to avoid a new calibration on every new patient. The second reason is the slow variability of the controlled system parameters: it is not fast and large enough to elicit dynamic effects on the control system performances; but certainly a controller calibration, which was optimized at a certain time for the therapeutic aim, will be no more optimal in the future. To these two reasons we can add a third one, hopefully only temporary: the short maintenance of characteristics of the so far experimented sensors. They need periodic calibrating, which implies a readaptation of the controller, possibly automatic.

The block diagram of a self-tuning controller is shown in Fig.5 [14]. The apparatus consists, at least conceptually, of two parts: the *estimator* and the real *controller*. The estimator has in itself a discrete-time model of the controlled system. In the most common devices the model is linear, of a rather low order, with a finite delay. Its structure is chosen on the basis of the physiological knowledge of the process under control. Its parameters are estimated automatically through a comparison of the input and output signals in the controlled system; the estimation is generally performed by a least-square method. Once the *current* model of the process is obtained, the estimator calculates the controller parameters and transmits them to the controller itself. In this calculus of the parameters, some typical procedures for the synthesis of a control system are applied. We will briefly comment on the most interesting ones.

In the *minimum-variance regulators* the structure and the finite delay of the model are previously fixed. Once the parameters of the model have been obtained by a least-square method, the controller parameters are calculated according to an optimal control strategy, to minimize a *performance index*

$$J_k = (y_{k+h} - y_b)^2 + Q\, u_k^2$$

The index k represents here the generic sample; h is the controller finite delay; u and y indicate the variables u_1 and y_2 of Fig.2, that is to say respectively the infusion rate of insulin and the glucose concentration in the sensing point; y_b is the basal value of the concentration itself. The weight coefficient Q in the performance index is of great importance because it determines the control policy, i.e. the therapeutic strategy. As a matter of fact, raising its value means increasing the *weight* of infused insulin in the index J, and therefore reducing the amount of administered substance at the price of a larger tolerance in the glycemia variations above the basal value. The parameter Q can also be used to compensate changes in the controlled system due, for instance, to changes of the feedback block 3 in Fig.2.

Pole assignment is used as well. Really, with this name a *cancellation technique* is meant, rather than the pole displacement obtained through state variable feedback. Once the model of the controlled system in the estimator is identified, the controller cancels the model poles and zeros with its own zeros and poles, and replaces them with a new set of poles and zeros that, after closure of the feedback loop, should give the desired response. Also in this case some parameters are available to which the control strategy can be assigned. For example, the modulus of the dominant poles can be used: by increasing it the response is made quicker, therefore the postprandial glycemia increase is reduced at the price of a larger insulin consumption. Finally, some *model matching* controls are suggested in bibliography, but only little is so far known about their usefulness.

Neural Networks

A severe drawback of these controllers is their intrinsic linearity in the estimator as well as in the controller, while the plant, i.e. the patient, is, as already seen, strongly nonlinear. Certainly, nonlinearities can be introduced in any controller, even self-adaptive, to compensate the plant nonlinearities; and, in self-tuning controllers, they can belong both to the estimator and to the controller. However, they make the plant identification more laborious; and, above all, the plant nonlinearity should be fully known *a priori*, thus contradicting the assumption that an unknown plant has to be identified. From this viewpoint, a completely different adaptive control strategy looks very attractive: the *neural networks*.

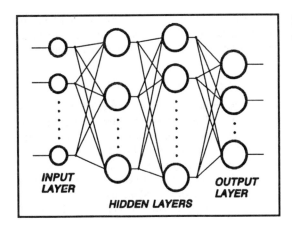

Fig. 6
A scheme of a generic neural network

A neural network (commonly abbreviated in NN) is shown in Fig.6. It is made of at least three layers of *neurons*, i.e. devices operating like biologic nervous cells. Neurons can be carried out by hardware circuits or by software algorithms. Analog neurons have been proposed, but, for control purposes, digital devices appear preferable. The *input layer* is made of as many neurons as required by the number of signals entering it: dynamical systems require more than one neuron for each signal. The input layer is followed by one or more *hidden layers*: the number of neurons therein depends on the accuracy which the NN has to operate with. Finally, an *output layer* has as many neurons as the output signals. In a *fully connected* NN each neuron is connected to all neurons of the following layer: this means that the number of connections is rapidly increasing with the number of neurons. As in the central nervous system of living beings, information is memorized in the state of connections: a *training procedure* activates some connections and inhibits some others. The neuron response can be easily made nonlinear: so, a NN is particularly able to represent a nonlinear system.

The application of neural networks to the adaptive control is rather recent, and still bounded to the predictive control; nevertheless, their extension to the feedback control is quite promising. In principle, a NN can be used in self-tuning control both as estimator and controller. As estimator, the NN is fed, together with the controlled system, with an appropriate input signal, and is trained to minimize the error between its own output and the output of the controlled system. Similarly, the NN can be connected and trained to realize also the controller function. The training signal is generally a stochastic sequence. Training a fully connected network can require many thousands of samples, an unacceptable duration for the patient. However, a NN can be pre-trained off-line, on the basis of recorded input-output sequences, obtained, in case, by computer simulation. On the other hand, when, as in our case, the plant structure is at least partially known, a *structure-oriented* network can be used, with a reduced number of neurons and connections, and, therefore, a much shorter training time.

The behavior of these machines is rather surprising: for instance, it can happen that their training, after a rather long period, during which it seems that nothing happens, suddenly reduces the error to a low level, which remains nearly unchanged even if the training goes on. An old scholar, Japanese and Buddhist, talks about "illumination". The word reveals the attitude of the researcher towards these machines: *something* happens, but nobody knows *how* it happens. One of the reasons of the interest of this topic is the return to good heuristic methods after a long period of refined mathematical speculations. Another reason is that controllers which imitate the central nervous system in its structure and presumably also in its functions, are introduced in bioengineering.

Open problems

The use of self-tuning controllers is very interesting, and in the future will probably be generalized to all biological systems in which the variability between individuals and in time of a single individual cannot be either ignored or correlated to an easily measurable quantity, as the body weight. Anyhow, it presents some relevant problems. Firstly, the choice of the signal activating the parameter estimation. Of course, the dynamic response of a system can be recognized only while the system is in a transient condition due to *known* inputs – even not intentionally applied, as disturbances are, but, anyway, known. Many biological processes, and glucose metabolism in particular, are most of the time in stationary conditions; but if they leave these conditions, it is due to *unknown* disturbances – in our case, meals and physical activity. Therefore it is necessary to *stimulate* the dynamic response in order to start the estimation mechanism. In industrial plants this is usually done by adding an appropriate signal to the controller output. A *white noise* or a *square wave* is commonly used. The energy and frequency range of these signals must be sufficient to produce in the sensor a response distinguishable from the one produced by other stimuli. But the patient may not tolerate such energy and frequency, especially if they are continuously applied. Therefore, a compromise must be found between patient and estimator needs: estimation should be performed only when needed, and kept active for the time strictly required for parameter tuning. This compromise conditions the choice of the controller and of its algorithms.

A second problem is generated by non-measurable disturbances. If a nuisance enters the controlled system while its identification is active, the estimator registers a change in the plant output not correlated to any variation of its input. Such change cannot be interpreted but as due to a variation *inside* the system, and produces a wrong estimate. Therefore, whenever possible the estimator is provided with a measure or a forecast of the disturbances; when it is not possible – and this is nearly always the case in the biological systems – the estimation should be interrupted for the whole length of the nuisances.

But the most serious problem is stability. An adaptive controller may become unstable for various reasons. It may happen that the model inside the estimator is too simplified, so that some parts of the controlled process are ignored, as not relevant, in its identification, while on the contrary they greatly influence the feedback stability. Or interference may occur between the two loops in Fig.5, the control loop and the estimation one. The effects of this interference cannot be easily foreseen because the whole controlled system is in common between the two loops. A third cause of instability is typical of the pole assignment: the presence of pure delays in the controlled system can induce the estimator to identify a non-minimal phase model, with zeros outside the circle of stability. The controller deletes them with unstable poles; however the cancellation does not eliminate the unstable poles but at the most renders them non-observable from the output: the instability shows itself with macroscopic oscillations, mostly at the controller output.

In systems with slowly varying parameters, the stability problems due to the interference between estimation and control loop are reduced, if not entirely removed, by alternating estimation and feedback control, so that the loops of Fig. 5 are never simultaneously closed. During normal operation the control loop is closed, while the link between estimator and controller is open; the controller works with fixed parameters. When the controller parameters need to be updated, the control loop is open, the control signal is kept constant by a memory device, and, therefore, an open-loop control replaces the feedback one; thereafter, the estimation signal is superimposed to the control one, and the new parameter values are computed and fed to the controller; finally, the estimator-controller link is broken, the control loop is closed again on the controller equipped with the new parameters, and the feedback control is restored. Such alternation is the rule in digital controllers, which have a sequential functioning. Yet much attention should be paid to two questions: on one side, the two operations should be separated by intervals long enough that no interference takes place between them, as a consequence of time delays due to the system dynamics. On the other side, parameter changes should be slow enough, as compared both to the response of the control system and to the duration of the estimating process, that no appreciable transient phenomena take place when they are updated by re-tuning. This happens in glycemia control, so that problems in developing self-adaptive controllers are considerably reduced.

Some algorithms must be inserted in the controller, to ensure patient safety. In particular:
– a minimal insulin infusion is maintained, no matter how the measured glycemia lowers,
– postprandial infusion is reduced to basal level, whatever be the residual hyperglycemia, after infusing a predetermined amount of insulin within a predetermined time.

All these features (intrinsically robust self-tuning algorithms, compensation of nonlinearities, periodical open-loop updating of controller parameters, estimation interrupted when disturbances arise, safety algorithms) concur in producing a controller reliable and accurate enough, to be used out of medical assistance. Therefore we may say that the problem of synthesizing the controller for portable units is solved, at least theoretically. We cannot say the same for the sensing devices.

4 Sensors

Sensors are, as we noticed, the most problematic part of the system. Their interest is witnessed by an impressive number of publications: the reader can see, e.g., the review papers of WILKINS [18] and PFEIFFER [19]. Always, in control technology, the physical characteristics of the sensor determine the accuracy and reliability of the whole system. The signal emitted by any sensor is affected by *noise* and *drift*; its *sensitivity* can vary in time; the chain formed by the sensor and the signal conditioners, that follow it, can introduce non negligible *delays* in comparison with the control system time response. Because the sensor is placed in the retour branch of the feedback loop, all these peculiarities can generate serious errors, as it was told about block **3** in Fig.2. Therefore, it should not surprise that the sensor has still most unsolved problems.

In principle, sensing noise reduces the control accuracy. Therefore, it is generally smoothed by digital filtering, as mentioned for the biostator algorithm. But, if its amplitude is small enough and its frequency range is high enough with respect to the system response (so that it is filtered by the system itself), its presence can be even advantageous, as it speeds up the estimation process by keeping it active when the tuning signal is constant.

Similarly, the collection of glucose samples, their transfer to the sensing device, the development of chemical reactions and the conversion of their effect into an input signal for the controller, introduce a finite delay. This delay is reduced with needle type sensors, which introduce the sensing element directly in the tissue to be tested. Anyway, if its length is comparable with the time constants of the closed-loop system, it reduces its stability margin and deteriorates its transient response: but its value is generally well-known, and, therefore, it can be easily compensated, e.g. by a *Smith predictor* [20].

Sensor drift and sensitivity variation affect in a substantial way its stability (the word *stability* has here a different meaning than we used for dynamical systems in the preceding section), and, therefore, the accuracy and reliability of the control. Then, the sensor must undergo periodical *re-calibrations* to guarantee the required steady-state control performance. Most of the research effort on the subject is devoted to increase the interval between subsequent calibrations. Soon in 1983 SHICHIRI tested his needle sensor on pancreatomized dogs, replacing it – that is equivalent to a re-calibration – every four days [11]. The sensor was tested, with similar results, also on diabetic humans [12]. This encouraging result seems unsurpassed in the following years, notwithstanding a great number of efforts done in refining the chemical process as well as the physical structure. Besides needle sensors [21, 22], which minimize the measurement delay but must be replaced for calibration, microdialysis sensors look attractive [23]. They employ a hollow fiber, inserted under skin, to collect biological fluid and carry it to the chemical reactor. Their main interest lies in the easy calibration – a calibrated glucose solution is sent to the reactor in place of the biological fluid –, even if the dependance of sensor output from glucose concentration is of quite problematic evaluation. Anyway, the choice between implanted needles and microdialysis sensors is still an open question [24, 25]. A different way has been tried, once more, by the SHICHIRI group, which detected the changes in the infra-red absorbance of oral mucosa [26]. On the other side, two chapters in this sections, due to WILKINS and ATANASOV and to CASOLARO, seem to open some new perspectives both in glucose sensing and insulin administration.

One more requirement, to be satisfied for the routine use of sensors, is the uniformity between different specimens. This means that industrial production should be developed on rather different principles than experimental prototypes. The work of URBAN, in this section, is devoted to this target; and Eli Lilly Laboratories are producing sensitive microchips through a bioelectronics technique derived from the semiconductor production [25]. As far as we know, they have been tested *in vitro* with encouraging results. If such results are confirmed by *in vivo* experiments, a feasible solution could come out from inserting them in microdialysis devices. So, advanced technologies such as large scale integration of electronic components enter also this field: they witness, together with so large efforts documented also by some contributions of this section, to the hardness and interest of this problem.

5 Conclusions

To conclude, we may resume the advantages but also the limits of these technologies. Certainly the equipments already realized on industrial scale and applied in clinical *routine*, produced remarkable benefits from the practical point of view – they improved the conditions of patients who could not be effectively treated with conventional therapies – and also from the viewpoint of knowledge both of physiological processes and of therapeutical strategies. But the *completely automatic* management of feedback glycemia control, seems to be a target still far from realization and perhaps too ambitious. On one side a sensor reliable enough, that a diabetic patient can be fully entrusted to it, has not yet been produced. Sensor calibration compels to periodical interventions, that at the moment can be only manually performed.

This is the main obstacle on the way to the wearable artificial pancreas. On the other side, even if good controllers, from the engineering viewpoint, are already available, yet the artificial control can imitate the natural one of a healthy person only partially and roughly. As already mentioned, the autonomous nervous system intervenes in glycemia control; and the endocrine system itself starts working during the intestinal absorption, that is to say *before* the glucose appears in the blood. Moreover, the location of the artificial sensor is anyway close to the body surface, that means far from that of the biological glucose receptors. Finally, the infusion route is by no means physiological: intravenous route was used in the industrial controllers, and the problems connected to it are well known. The intraperitoneal route was proposed, and experimented with some success both on animals and on humans [26]: it is closer to the physiological one, and, in addition, it is more suitable for a portable system; but it doesn't exactly correspond to the physiological route of insulin secreted by pancreas either.

This has to be added to the uncertainty and arbitrariness in the choice of biological variables to be measured and controlled. We don't know, yet, where a sensor should be placed to give a correct representation of the state of a patient; and, probably, some other variables should be measured as well. These lacks of knowledge show the limits of a purely technical solution of therapeutic problems. Even if the researcher has correctly chosen one or more variables to be controlled, has defined a control strategy and has realized a very refined controller from the engineer's point of view, we cannot say that he has produced a therapy, but only an automatic therapeutic instrument. The clinician remains the only responsible of its use. He has to interpret equally, on the basis of clinical experience, the patient symptoms and the signals coming from the automatic system, and he has to program the use of the automatic device as a fundamental part of his therapeutic project. This applies to all bioengineering applications of feedback control, and in particular to glycemia. Within these bounds, it is reasonable to wish and foresee that portable glycemia controllers are developed quickly, as soon as the technological difficulties we described are overcome.

References

[1] V.W. BOLIE: Coefficients of normal blood glucose regulation. *J. Appl. Physiol.*, 16 (5): 783-8, 1961.

[2] E. ACKERMAN, J.W. ROSEVEAR, W.F. McGUCKIN: A mathematical model of the glucose-tolerance test. *Phys. Med. Biol.*, 9: 203-13, 1964.

[3] R. LUFT, E. CERASI: An analogue computer model for the glucose-insulin interrelationship. *Acta Diab. Latina*, 5 (Suppl. 1): 264-76, 1968.

[4] A.H. CLEMENS, P.H. CHANC, R.W. METERS: The development of Biostator, a glucose controlled insulin infusion system (GCIS). *Horm. Metab. Res.*, Suppl. 7: 22-33, 1977.

[5] – *Betalike Operating Manual.* Eseaote Biomedica, Genova (*s.d.*).

[6] J.L. SELAM, M.A. CHARLES: Devices for insulin administration. *Diabetes Care*, 13: 955-79, 1990.

[7] D. PAGUREK, J.S. RIORDON, S. MAHAMOUND: Adaptive control of the human glucose regulatory system. *Med. & Biol. Eng. & Comput.*, 10: 752-61, 1972.

[8] P. BRUNETTI, C. COBELLI, P. CRUCIANI, P.G. FABIETTI, F. FILIPPUCCI, F. SANTEUSANIO, E. SARTI: A simulation study on a self-tuning portable controller of blood glucose. *Int. J. Artif. Organs*, 16: 51-57, 1993.

[9] M. SHICHIRI, R. KAWAMORI, Y. YAMASAKI, N. HAKUI, H. ABE: Wearable-type artificial endocrine pancreas with needle-type glucose sensor. *Lancet* II: 1129-31, 1982.

[10] M. SHICHIRI, R. KAWAMORI, Y. YAMASAKI, N. HAKUI, H. ABE: Closed-loop glycemic control with a wearable endocrine pancreas. *Diabetes*, 33: 1200-2, 1984.

[11] P.G. FABIETTI, M. MASSI BENEDETTI, F. BRONZO, G.P. REBOLDI, E. SARTI, P. BRUNETTI: Wearable system for acquisition, processing and storage of the signal from amperometric glucose sensors. *Int. J. Artif. Organs*, 14: 175-8, 1991.

[12] C. COBELLI, A. MARI: Validation of the mathematical models of complex endocrine-metabolic systems. A case study of glucose regulation. *Med. & Biol. Eng. & Comput.*, 21: 390-9, 1983.

[13] H.J. CHIZEK, P.G. KATONA: Closed loop control. *In*: E.R. CARSON, D.G. CRAMP, *Eds.*: *Computers and control in clinical medicine*, p. 95-151. Plenum Press, New York 1985.

[14] P.E. WELLSTEAD, M.B. ZARROP: *Self-tuning systems, control and signal processing*. John Wiley, Chichester 1991.

[15] K. WARWICK, G.W. IRWIN, K.J. HUNT: *Neural networks for control and systems*. Peter Peregrinus, London 1992.

[16] E. WILKINS: Towards implantable sensors: a review. *J. Biomed. Eng.*, 11: 354-61, 1989.

[17] E.F. PFEIFFER: Artificial pancreas, glucose sensors and the impact upon diabetology. *Int. J. Artif. Organs* 16: 636-44, 1993.

[18] G.P. REBOLDI, P.D. HOME, G. CALABRESE, P.G. FABIETTI, P. BRUNETTI, M. MASSI BENEDETTI: Time delay compensation for closed-loop insulin delivery systems: a simulation study. *Int. J. Artif. Organs*, 14: 350-8, 1991.

[19] P. ABEL, A. MULLER, U. FISCHER: Experience with an implantable glucose sensor as a prerequisite of an implantable beta cell. *Biomed. Biochem. Acta*, 43: 577-84, 1984.

[20] J.S. CHURCHOUSE, W. H. MULLER, F.H. KEEDY, C. M. BALTERSBY, P.M. VADGAMA: Studies on needle glucose sensors. *Anal. Proc.*, 23: 146-8, 1986.

[21] A.L. AALDERS, F.J. SCHMIDT, A.J.M. SCHOONEN, I.R. BROEK, A.G.F.M. MAESSEN, H. DOORENBOS: Development of a wearable glucose sensor; studies in healthy volunteers and in diabetic patients. *Int. J. Artif. Organs*, 14: 102-8, 1991.

[22] V. POITOUT, D. MOATTI-SIRAT, G. REACH, Y. ZHANG, G.S. WILSON, F. LEMONNIER, J.C. KLEIN: A glucose monitoring system for on line estimation in man of blood glucose concentration using a miniaturized glucose sensor implanted in the subcutaneous tissue and a wearable control unit. *Diabetologia* 36: 658-63, 1993.

[23] J. BOLINDER, U. UNGESTEDT, P. ARNER: Long-term continuous glucose monitoring with microdialysis in ambulatory insulin-dependent diabetic patients. *Lancet* 342: 1080-8, 1993.

[24] K. KAJIWARA, T. UEMURA, H. KISHIKAWA, K. NISHIDA, Y. HASHIGUCHI, M. UEHARA, M. SAKAKIDA, K. ICHINOSE, M. SHICHIRI: Noninvasive measurement of blood glucose concentration by analysing Fourier transform infra-red absorbance spectra through oral mucosa. *Med. & Biol. Eng. Comput* 31: S17-22, 1993.

[25] J.J. MASTROTOTARO, K.W. JOHNSON, R.J. MORFF, D. LIPSON, C.A. ANDREW, D.J. ALLEN: An electroenzymatic glucose sensor fabricated on a flexible substrate. *Sensors and Actuators*, B. 5: 139-44, 1991.

[26] J.L. SELAM: Insulin therapy of diabetes with implantable infusion pumps: clinical aspects. *Int. J. Artif. Organs*, 13: 261-6, 1990.

Autonomic Dysfunction in Diabetes Mellitus

I.T. El Mugamer, K.D. Desai, M. Towsey, and D.N. Ghista

I.   BACKGROUND AND PURPOSE

1.   DIABETES MELLITUS

Diabetes Mellitus (DM) is a heterogeneous clinical syndrome characterized by hyperglycaemia and long term specific complications; neuropathy, retinopathy, and nephropathy.   Diabetics have an increased incidence of ischaemic heart disease and peripheral vascular disease. A triad of neuropathy, micro and macro- angiopathy and infection operate synergistically to produce foot disease. Infection is more common in uncontrolled diabetes, especially urinary tract and genital infections. Loss of dentition is accelerated due to gingivitis and odontopalpitis. Autonomic neuropathy leads to visceral denervations producing a variety of clinical abnormalities, cardiac and respiratory dysarrhythmias, gastrointestinal motility disorders, urinary bladder dysfunction   and impotence.

Gestational and diabetic pregnancies are associated with an increased risk to the baby and the mother. DM is a chronic, debilitating, costly disease, attended by severe complications.[1] DM is a leading cause of blindness, renal failure and limb amputation all over the world. The delineation of the main risk factors for diabetes and its devastating complications may have an important bearing on reducing both the impact and cost of diabetes in many societies, particularly in countries where it is rapidly becoming the major cause of morbidity and mortality.[2]

2.   CORONARY HEART DISEASE IN DIABETES MELLITUS

Several population studies have shown that the incidence of CHD events is from 1.5 to more than 3 times in middle age and in elderly diabetic patients than in non-diabetic subjects of the same age. Autopsy studies have shown an increase in both fatty streaks and raised atherosclerotic lesions in diabetic patients than in non diabetic subjects.[3] This is even true in populations of low level of CHD. The majority of these patients had non-insulin dependent diabetes mellitus (NIDDM), although the distinction between the main two types of diabetes was not made in most of the studies. The excessive risk of CHD in diabetes is more marked in women than men. CHD occurs at an increased rate in subjects who have impaired glucose tolerance (IGT).[3] There is a positive non-linear relationship between glucose levels after an oral glucose load and CHD mortality and morbidity, with a sharp increase of CHD rate at the upper end of blood glucose distribution. Hyperglycaemia or factors associated with diabetes may have a direct effect on athrogenesis and/or thrombogenesis.

Diabetes mellitus causes a specific heart disease apart from CHD.[4] Clinical studies using haemodynamic and non-invasive techniques have shown subtle changes in systolic and diastolic left ventricular function in diabetic patients in the absence of any detectable clinical abnormalities in left ventricular function. Cardiac changes In Insulin-dependent diabetes mellitus (IDDM) are related to the presence of nephropathy and duration of the diabetes. However, NIDDM patients may have CHD associated changes in the left ventricle without relation to duration of the disease or presence of other complications. The morphological changes are microangiopathy, perivascular and interstitial fibrosis and deposition of glycoproteins and lipids.[4] The frequency of these changes and their relation to metabolic control are still poorly understood.

The clinical course of CHD in diabetics is more severe than in non diabetics. Mortality and morbidity is 2-3 times non diabetic subjects. Access to specialized cardiac centers in the majority of the developing countries is limited, leading to a higher mortality and morbidity of CHD in diabetics in these countries. Early detection of CHD in diabetics prior to clinical events, coupled with implementations of preventive management strategies, can delay the progressive down-hill course of the disease or can avert the development of myocardial infarction. Yet, there is no diagnostic test short of an Exercise ECG, which if positive will be followed by Coronary Angiography to diagnose CHD.

3.   DIABETIC AUTONOMIC NEUROPATHY

Diabetic autonomic neuropathy (DAN), as defined by the 1988 San Antonio (Texas) consensus meeting, is a descriptive term restricted to disorders in the autonomic nervous system, manifesting as dysfunction of several organ systems either clinical or subclinical.[5] Thus DAN encompasses disturbances in reflex arcs involving one or more sensors, an afferent branch, a central processing unit, an efferent branch and neuromuscular junctions inervating end organs.[5] The definition leaves the diagnostic method unspecified. Thus DAN is defined the way it is measured. Although the consensus meeting has recommended a battery of five tests, its recommendation is not followed in many studies.[6] More over, these recommendations have been criticized in 1991, by the Diabetic Study Group of the European Association for the Study of Diabetes.[6] Both the quality and practical appropriateness in the clinical setting of these tests have been questioned.

Symptoms of DAN are vague and non specific. Signs are difficult to detect, except in advanced cases. Prevalence studies based on symptoms detection give ranges from 6% to 42%. In a community study using standard non invasive cardiovascular tests, Neil *et al* reported a prevalence rate of 20% and 15% for DAN in IDDM and NIDDM respectively. Hospital clinics studies show a prevalence rate ranging from 17% to 55%m due to the selected nature of these patients.[5] Prevalence rates underestimate the true risk of DAN especially when mortality rates is incised by DAN. There is no available data on the incidence of DAN. DAN is associated with increased morbidity and mortality. Nephropathy, cardiac arrhythmia's, myocardial ischaemia and sudden cardiac death are associated with DAN.[3]

4.   CONVENTIONAL AUTONOMIC FUNCTION TESTS

The Autonomic Function Tests,[6,7] recommended by the San Antonio (Texas 1988) consensus meeting are:

❑ Heart (HR) and Arterial Blood Pressure (ABP) responses to standing up;
❑ HR response to deep breathing;
❑ ABP response to sustained hand grip;
❑ HR response to Valsava maneuver;
❑ Spectral analysis of HR and ABP variability.

Success of a test in clinical practice depend on many factors which usually differ from research accepted tests. The following points are important:
1. Simplicity: It has a non complex methodology and could be carried out by one and the same person
2. Safety: entails no unnecessary risk to the patient
3. Non-time consuming
4. Cost effective
5. Requires minimal patient co-operation
6. Sensitive, reproducible and specific
7. Interpretation is easy

In the HR response to sustained hand grip using a dynamometer, a hand grip is maintained at 30% of maximum voluntary contraction up to 5 minutes. ABP is measured every minute, the difference between the diastolic blood pressure before the release of the hand grip and the start of the test being taken as the measure of the response. However, the response is related to the strength of contraction, has a strong gender influence; moreover adequate performance is beyond the ability of many diabetic and elder subjects.[5] The test is cumbersome and of limited use in a practical clinical setup.

HR and ABP responses to Valsava maneuver are complex and difficult to perform in a clinical environment. The procedure consists of forced expiration to a set pressure against a forced expiratory route. There is an initial rise in the ABP and a baroreflex mediated decrease in HR; next, a fall in ABP occurs, with an increase in HR and peripheral resistance (PR); this is followed by a recovery of ABP. Immediately after the termination of the test, there is a transient increase in ABP, leading to a fall in HR. It is difficult to index this complex response. Both the tests, Valsava maneuver and ABP response to sustained hand grip, will not be considered further in this study.

Heart Rate Response to Standing
Change from horizontal to vertical position brings the influence of gravity on the circulation in to action. The test is easily performed by asking the subject to lie quietly on a couch for at least five minutes, then stand up unaided as quick as possible, and remain standing thereafter for at least one minute. After standing, there is an immediate increase in HR, then a subsequent decrease is observed in normal people. The HR increase is mediated by the vagal nerves since it is abolished by atropine and not by propranolol. The decrease in HR is due to a vagal reflex to a sympathetically mediated vasoconstriction and ABP overshoot. The increase in the HR is quantified by measuring the maximum increase during the first 15 seconds compared to the HR in the supine position. The decrease in HR is quantified by measuring the RR interval ratio in the 15 and 30 seconds after standing (RR max./RR min.).[5,8,9] Unfortunately different values for this ratio are used by several investigators.

ABP Response to Standing
In normal subjects, the initial ABP response to standing is an immediate blood pressure rise followed by a fall, recovery and sometimes an overshoot of the arterial blood pressure. This change is related to a reduction in cardiac vagal activity and a rise in **sympathetic vasomotor tone**. A greater loss of the major part of the sympathetic outflow is necessary before orthostatic hypotension occurs. Thus orthostsatic hypotension is believed to reflect sympathetic dysfunction. Consensus is lacking on the time between supine and standing blood pressure measurements. Systolic blood pressure fall of greater than 30 mmHg is generally taken to be abnormal, although disagreement still exists on the level. The influence of age on systolic blood pressure fall is reported to be significant.[5,6,7,9]

HR Response to Deep Breathing
The origin of the respiratory sinus arrhythmia (RSA) is still not clear. It generally agreed that it is mainly mediated by the **parasympathetic** nervous system since atropine greatly reduces its degree.[5,9,10] The sympathetic nervous system has a minor role to play in RSA, since blocking it with propranolol leads to a slight reduction of the RSA, which is more evident in standing than in the supine position. There is still no consensus on the method or on the way the response is quantified or on the limits of the reference values.

In most tests, RSA is measured when the subject lies quietly and breathes deeply at a rate of six breaths per minute; this rate produces maximum variations in HR.[6] In most studies RSA is quantified as the mean difference between the maximum and minimum HR during consecutive deep breaths(Δ IE), which can be expressed either in beats per minutes IEB or in seconds IES. Another way of expressing RSA is by IEratio (IER) or by the mean value of the ratio of the six longest RR intervals to the six shortest intervals. Further modification is the measurement of successive maximum and minimum intervals during a period of deep breathing. IES measures the RR variations without bias, since unlike IEB or IRE it is not influenced by the mean HR or the mean RR interval length.[5-10]

II.   HEART RATE (BEAT-TO-BEAT) VARIATION (HRV)

1.   CONCEPT
Fluctuations in heart rate (HR) and arterial blood pressure (ABP) are quasi-periodical, repeating with a period which is not strictly constant showing changes in morphology, amplitude and phase from one beat to the other. In steady state conditions, these oscillations are maintained around a certain mean value. The variability is the result of complex neural control mechanisms by the Autonomic Nervous System (ANS) whose function is to maintain the controlled parameters around set physiological values. The ANS responds to changing conditions according to certain temporal dynamics that set the controlled process to a new working point.[11]

The study of these changes on a beat-to-beat basis provides important information about such mechanisms in both physiological and pathological conditions. The beat-to-beat variations in HR and ABP contain (non-invasively monitorable) information about the sympathetic and parasympathetic cardiovascular control mechanisms. Regulation of beat-to-beat changes in ABP by the ANS involves dynamic changes in HR, stroke volume (SV) and peripheral resistance (PR).[12,13]

Modulations in these three main levers by the ANS is achieved by autonomic reflex arcs, the most relevant of which is the baroreceptor reflex (BRF). The BRF consist of stretch sensors in the carotid sinus and aortic arch reacting to changes in blood pressure with afferents running in the glossopharyngeal and vagus nerve synapsing in the nucleus tractus solitarii (NTS). The NTS is an integration centre receiving impulses from other brain stem nucleii and higher centres. The heart rate and blood pressure control systemcentral wiring diagram or loops involved are still obscure, but nevertheless displayed in figure 1. The efferents are from the vasomotor centre via the sympathetic nerves to the heart and peripheral blood vessels. The parasympathetic efferents from the nucleus ambiguous descend to the heart via the vagus nerves. The BRF activity is assessed from HR and ABP signals.[6]

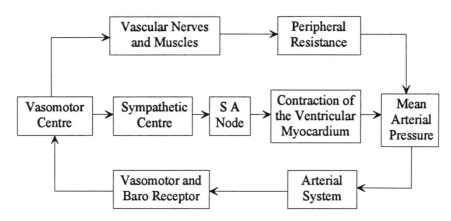

Fig. 1. Controlled system of the Heart Rate and Blood Pressure regulation

From the electrocardiogram (ECG), the RR waves are determined (with the interval lengths between consecutive R-waves) along with the HR. The variations in the interbeat interval series are not random, but show certain frequency-specific properties. The power spectrum (PS) consist of three major frequency bands ranging from 0 to 0.5 Hz. The boundaries of the bands are defined differently by different authors. The bands with most commonly employed boundaries are the following:[14-16]

(a)   **Low Frequency Band (LF) 0.02 - 0.05 Hz**
       In this band, variations are related to temperature regulation of the body, the vasomotor control and the renin-angiotensin system, with the centre-of-frequency at 0.04 Hz. This is an ill defined band modulated with influence of both parasympathetic and sympathetic systems. A very low frequency band (VLF) 0.01 to 0.0.04 is described with influence mainly by the sympathetic system. Recently an ultra low frequency band (ULF) at 0.002 to 0.005 is reported, but its physiological significance is not yet known.

(b)   **Mid Frequency Band (MF) 0.05 - 0.2 Hz**
        This band consists of variations related to the ABP control system, with the centre-of-frequency at 0.1. It is also called the Mayer band, corresponding to oscillatios in blood pressure described by Mayer in 1874. It is influenced by parasypmpathetic and sympathetic systems. Increase in vagal activity augments peak MF power, or energy content of the HRV wave form in this frequency range. Conversely parasympathetic blockade diminishes MF power, especially in the supine position.

        Interventions which increase sympathetic activity (e.g., passive tilting, standing, mental and physical stress, sympathomimetic agents, baroreceptor unloading with nitroglycerine infusion and coronary occlusion) are known to increase the MF component (figure 2).

        The magnitude of the power in this band decreases monotonically with age, it is increased by standing when the sympathetic system is activated. It is not influenced by breathing when the respiratory rate is above 9 breath per min. In supine position its magnitude is mainly dependent on parasympathetic system.

(c)   **High Frequency Band (HF) 0.14 - 0.5 Hz**
        Variations related to respiration is associated with parasypmathetic activity with centre of frequency at 0.25 Hz which varies with the respiratory rate. This band is mediated solely by the parasympathetic system. The magnitude of the power in this band is more in the supine than the standing position (figure 3). There is a linear decline in the power of this band up to the age of 30 years, and does not change thereafter.[17]

NORMAL SUBJECT-(SUPINE POSITION)
POWER SPECTRUM (AR)

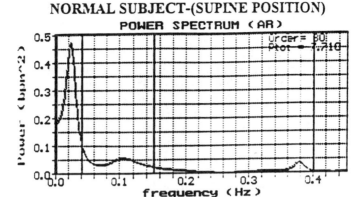

NORMAL SUBJECT-(STANDING)
POWER SPECTRUM (AR)

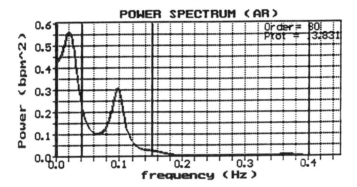

NORMAL SUBJECT-(DEEP BREATHING)
POWER SPECTRUM (AR)

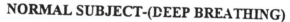

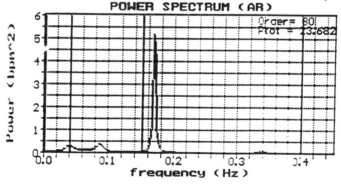

Figure 2 shows HRVPS plots of a normal subject in supine, standing and deep-breathing modes. The power statistics on the right side shows the power under low and high requency bands of the HRVPS. There is increase in mid frequency power in standing position compared to supine position and increase in high frequency power in deep breathing test. Note changes in the scale of y-axis.

III.   HRV AND POWER SPECTRAL ANALYSIS (PSA)
CHANGES IN DISEASE

1.   HRV Changes in DM
The frequency of abnormal heart rate variability (HRV) is almost the same in IDDM and NIDDM without clinical evidence of peripheral or autonomic neuropathy.   Normal individuals show a progressive decrease of HRV with age with a loss of 4.6 beats per each decade of life. In IDDM, HRV is inversely related to age and to the duration of the disease; there is no nonlinearity in the effect of each variable. In NIDDM, HRV is inversely related to age; however, the relation to the duration of the disease did not reach statistically significant level, and there was no nonlinear effect.[5,16-18]

2.   HRV Changes in DM with Neuropathy
Almost two-thirds of patients with clinically detectable peripheral neuropathy (having parathesiae, burning feet, pains, decreased knee and ankle reflexes, decreased vibration pinprick and position sense) have significant changes in HRV. All patients with clinically detectable autonomic neuropathy (having gastroparesis, neurogenic bladder or orthostatic hypotension) have abnormal HRV.[19]

3.   HRV-PSA Changes in DM
There is greater reduction in the magnitude of the power of all the frequency components of the PS with age, in diabetics than in normal subjects, in both supine and standing posture, suggesting a combined decrease in sympathetic and parasympthetic out flow. The decrease is more pronounced in the MF power in the standing position.[30]

In the case of diabetics, our studies have demonstrated a shift in the mid-fequency band peak from 0.1 Hz to 0.005 Hz, while the amplitude of this peak is not augmented in response to orthostatic stress, in contrast to its augmentation in normals (Figure 3).

4.   PSA Changes in CHD
There is a significant reduction of the HF power in CHD. The reduction correlates very well with the angiograhpic severity but not with other CHD features, including the presence of a previous myocardial infarcion, location of diseased coronary arteries, indices of left ventricular function and New York Heart Association (NYHA) class.   There is also reduction in the MF power which is attributed to the decrease in vagal out-flow, since there is no change in the HF/LF ratio, which suggests that the sympathetic cardiac function is not affected in CHD.[18]

DIABETIC SUBJECT-(SUPINE POSITION)

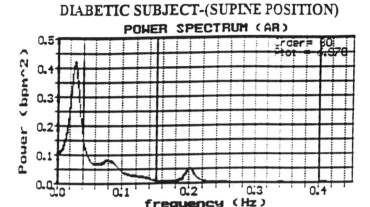

DIABETIC SUBJECT-(STANDING)

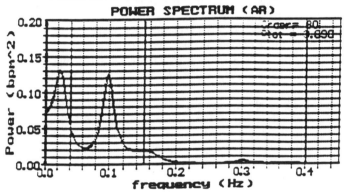

DIABETIC SUBJECT-(DEEP BREATHING)

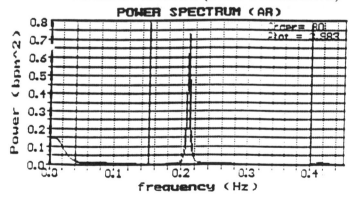

Figure 3 shows HRVPS plots of a diabetic subject in supine, standing and deep-breathing modes. The power statistics on the right side shows the power under low and high requency bands of the HRVPS. There is decrease in mid frequency power in standing position and in high frequency power in deep breathing-mode compared to corresponding power levels of normal subject.

IV.   REFERENCES

[1]   Report of a WHO Study Group, Prevention of diabetes mellitus, WHO Technical Report Series, 844, Geneva, 1994.

[2]   King, H., Global estimates for prevalence of diabetes mellitus and impaired glucose tolerance in adults, Diabetes Care, Vol. 16, No. 1, January 1993, pp. 157-177.

[3]   Pyörälä, K., "Diabetes and Heart Disease," in <u>Prevention and treatment of diabetic late complications</u>, edited by Mogensen, C.E. and Standl, E., Walterde Gruyter, Berlin, 1989, pp. 226.

[4]   Fein, S., Sonnebuck, E.H., Diabetic cardiomyopathy, Progr Cardiovasc Dis, 1985, No. 27, pp. 225-270.

[5]   Faes, T.J.C., Assessment of cardiovascular autonomic function: An inquiry into measurement, Dessertation thesis, Department of Medical Physics, Faculty of Medicine 'Vrije Universiteit Amsterdam,' Amsterdam, The Netherlands, 1992.

[6]   Kahn, R., Proceedings of a consensus development conference on standardized measures in diabetic neuropathy, Autonomic nervous sytem testing, Diabetes Care, 1992; No. 15, pp. 1095-1103.

[7]   Ewing, D.J., Clarke, B.F., Diagnosis and management of diabetic autonomic neuropathy, British Medical Journal, Vol. 285, 2 October 1982, pp. 916-918.

[8]   Ewing, D.J., et al, The value of cardiovascular autonomic function tests: 10 years experience in diabetes, Diabetes Care, Vol. 8, No. 5, September-October 1985, pp. 491-498.

[9]   Ewing, D.J., et al, Assessment of cardiovascular effects in diabetic autonomic neuropathy and prognostic implications, Annals of Internal Medicine,   1980:92 (Part 2), pp. 308-311.

[10]   Eckberg, D.L., Parasympathetic cardiovascular control in human disease: a critical review of methods and results, American Physiological Society, 1980, H581-H593.

[11]   Bianchi, A.M., et al, Time-variant power spectrum analysis for the detection of transient episodes in HRV signal, IEEE Transactions on Biomedical Engineering, Vol. 40, No. 2, February 1993, pp. 136-144.

[12]   Akselrod, S., et al, Power spectrum analysis of heart rate fluctuation: A quantitative probe of beat-to-beat cardiovascular control, Science, Vol. 213, 10 July 1981, pp. 220-222.

[13]   Rimoldi, O., et al, Analysis of short-term oscillations of R-R and arterial pressure in conscious dogs, American Physiological Society, 1990, H-967-H976.

[14]   Fallen, E.L., et al, Power spectrum of heart rate variability: A non-invasive test of integrated neurocardiac function, Clinical and Investigative Medicine, Vol. 11, No. 5, 1988, pp. 331-340.

[15]   van Steenis, H.G., et al, Heart rate variability spectral based on non-equidistant sampling: the spectrum of counts and the instataneous heart rate spectrum, Medical Engineering Physics, Vol. 16, September 1994, pp. 355-362.

[16]   Ori, Z., et al, Heart rate variability: frequency domain analysis, Cardiology Clinics, Vol. 10, No. 3, August 1992, pp. 499-537.

[17]   Rottman, J.N., et al, Efficient estimation of the heart period power spectrum suitable for physiologic or pharmacologic studies, The American Journal of Cardiology, Vol. 66, 1990, pp. 1522-1524.

[18] Eckberg, D.L., et al, Defective cardiac parasympathetic control in patients with heart disease, New England Journal of Medicine, Vol. 285, No. 16, October 1974, pp. 877-873.

[19] van den Akker, T.J., et al, Heart rate variability and blood pressure oscillations in diabetics with autonomic neuropathy, Automedica, Vol. 4, 1983, pp. 201-208.

[20] Bianchi, A., et al, Spectral analysis of heart rate variability signal and respiration in diabetic subjects, Medical & Biological Engineering & Computing, Vol. 28, 1990, pp. 205-211.

Physiologic Modelling of Type-2-Diabetes for Medical Education

E. Biermann

1 Introduction

Non-Insulin-Dependent-Diabetes Mellitus (NIDDM) is a disease that is frequent in the elderly. Elevated serum glucose levels in the preprandial and postprandial state are well known signs. Research in recent decades has demonstrated that insulin resistance of the glucose-consuming tissue and a disorder of insulin secretion are both responsible. However, the interplay between glucose appearance and disappearance and its controlling agents, mainly insulin, is complex and the contributions of the single mechanisms or organs to hyperglycemia are still controversial [14].

In order to make these interplays more transparent, a mathematical model of the main physiologic steps has been set up and introduced in a computer program named DIACATOR. This program is designed to be used by medical students, physicians and researchers in experimental diabetology on an interactive basis to obtain more insight in the physiopathology and therapeutic approaches to NIDDM.

2 The model of insulin-glucose dynamics

Figure 1 displays the basic structure of glucose-insulin system. The equations of the mathematical model have been taken from known dependences available from the literature. The basic model equations are shown in the appendix. It must be pointed out that the model is not an organ model, with exception of the kidney. All organs contributing to glucose uptake are lumped together either to non-insulin dependent or insulin dependent glucose uptake. The equations form a framework of differential equations to be solved numerically.

2.1 Modelling the pathological state of NIDDM and its precursors

The degree of glucose intolerance is characterized in this model by four parameters that can vary according to the stage of the disease :

a) the pancreatic sensitivity for glucose to excrete insulin
b) the presence of a first insulin secretion phase
c) the insulin sensitivity of the insulin dependent tissue
d) the non-suppressible part of the hepatic glucose production

Type-2-diabetes has two mayor abnormalities to be introduced in the physiological
model of the healthy state: a defect in insulin secretion (a+b) and tissue insulin resistance
(c+d) [7,14,22,24,29].

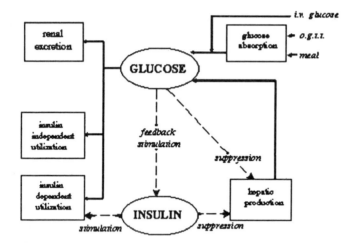

Fig1:
Illustration of the
metabolic interplay of
all relevant steps of
the model for the
glucose-insulin sub-
system

The entire spectrum of observations can only be decribed by assuming a heterogeneity of
metabolic abnormalities where the characteristics above are differently pronounced.

Table 1: seven different states of the disease

stage of disease	tissue sensitivity	pancreatic responsiveness	non-suppress. production	first phase
normal	100%	100%	0	yes
pre-diabetic	70%	100%	1.0 mg/kg/min	yes
pre-diabetic	42%	100%	1.9 mg/kg/min	no
diabetic	42%	70%	3.0 mg/kg/min	no
diabetic	42%	50%	3.8 mg/kg/min	no
diabetic	30%	25%	4.0 mg/kg/min	no
diabetic	35%	13%	3.5 mg/kg/min	no

The stages indicate either the progression of the disease or independently developing
states [14,16]. Seven stages of metabolic abnormalities (including a healthy and two prediabetic individuals) can be simulated using seven different sets of parameters according to
the table 1. These parameter sets were arbitrarily chosen for educational purposes.

2.2 Modelling therapy

Therapy of NIDDM is based on diet, three different types of oral hypoglycemic drugs and insulin. *Diet* is modeled by using a delayed absorption function for glucose and by distribution of the carbohydrates to six instead of three meals per day. *Sulfonylureas* are simulated by adding a constant amount of insulin to the prevailing insulin concentration [18,21]. *Biguanides* enhance tissue sensitivity for glucose [2]. *Glucosidase inhibitors* are modeled by delaying glucose absorption from the intestine [19]. Insulin therapy is introduced by modelling subcutaneous absorption according to reference [3,4].

3 Results of the modelling process: simulated curves

The model is used for simulation of curves that are either accessible to intuition or direct observation like 24-hour glucose profiles or the glucose tolerance test or the curves show more basic dependences that serve for comprehension. These type of curves are similar to the *characteristics* of an electronic device and can thus be considered as the "characteristical curves of an individual patient" or precisely of a certain stage of the disease under a certain therapy. Essential for the learning process is the interactive use by which curves either with different disease stages od different therapies can be compared.

3.1 Glucose utilization, -clearance, -production and insulin concentration

Glucose production is represented by a set of glucose dependent curves ("liver-characteristics"), each for a distinct insulin concentration according to Eq. 4 (Appendix). Two graphical representations can be created to demonstrate the law of mass action and the action of insulin on glucose utilization.

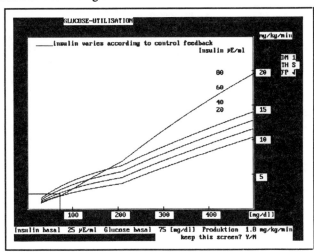

Fig.2:
Characteristical curves for glucose utilisation of a prediabetic individual with fixed insulin concentrations. The steeper curve represents variable insulin concentration according to the feedback. The slope of the curves represent insulin sensitivity.

The first demonstrates the different methods of utilization, i.e. insulin dependent- , insulin independent- and renal elimination (representation of Eq 1-3 from appendix, graph not shown).

The second graph (Fig. 2) shows the effect of different insulin concentrations: four curves each for one fixed insulin level, and one curve where the insulin concentration varies according to the feedback with actual glucose values. The curves for the glucose clearance are generated by dividing the disposal rate by the actual glucose concentration. These graphs are useful in understanding the effect of mass action on glucose utilization and to see the additional effect of insulin in one picture. The serum insulin concentration as a function of the glucose concentration under equilibrium conditions (see Appendix, Eq.6) can be shown in a similar graph (not shown). This represents the activity of the feedback in equilibrium (gain) between glucose and insulin [5]. However the human endocrine pancreas exhibits a fast dynamic control behavior according to Eq. 7 (Appendix) in order to properly control a carbohydrate laod.

3.2 Glucose profiles and glucose tolerance test

Profiles of glucose- and insulin concentrations along with glucose appearance rate from the liver and from the intestines (after meals) can be simulated using three normal meals or a glucose infusion over 18 hours. Figure 3 shows two profiles for the same diabetic individual, one with and the other without sulfonylurea therapy.

The oral glucose tolerance test is simulated by applying the formula for exogenous glucose absorption. Futhermore, the appearance and the disappearance rate as a function of the glucose concentration are displayed forming the typical hysteresis curve that represents a dynamic state of balances: as long as the input is higher than the output the glucose rises and vice versa. The *ogtt* is the classical example, where all the above described functions act in concert and the entire model is challenged.

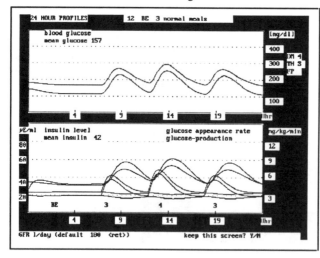

Fig.3:
24 hour profiles of glucose and insulin concentrations, appearance rate from hepatic glucose production and intestinal absorption of glucose. The curves represent a diabetic patient with and without *sulfonylurea* treatment.

3.3 Glucose clamp (Experimental diabetology)

This technique is used in experimental diabetology for estimating the degree of insulin resistance by forcing glucose (and insulin) concentration to a predetermined value (clamp) [15]. This is performed by frequent sampling of the glucose concentration and infusion of an amount of glucose solution, which is determined by a clamp algorithm. This is a classical problem of control theory and several clamp algorithms have been described in the literature. The clamp-algorithm used here has been designed by the author and is shown in the appendix. The user can enter the desired glucose concentration and, if applicable, the desired insulin concentration. Such a clamp experiment is simulated on the screen over a period of 400 minutes. The amount of glucose that has been infused per unit time is displayed at the end of the experiment (Figure 4).

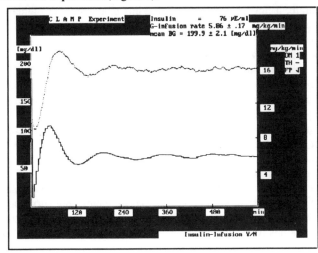

Fig.4:
Feedback controlled infusion of glucose to maintain the glucose concentration at a defined level (glucose clamp).Simulated glucose concentration (upper curve)and glucose infusion rate (lower curve, right axis). Calculations are based on the underlying model for a prediabetic induvidual and Eq.8 from Appendix.

In this program, the clamp-technique can be used to demonstrate how the tissue sensitivity for insulin S_i can be determined and thus insulin resistance (reverse of S_i) can be quantified. The definitions of S_i and of the glucose effectivness E can be taken from the literature [9]:

$$E = \frac{\delta(-dG/dt)}{\delta G} \qquad S_i = \frac{\delta E}{\delta I}$$

To derive proper values for E and S_i, four clamp experiments need to be carried out with two insulin concentrations and two glucose concentrations respectively. S_i is then equal to the insulin sensitivity F_i introduced in the equation for glucose uptake (Eq. 2, see appendix). The values from table 1 would result.

4 Discussion

Learning with the computer (CAI= computer aided instruction, CBT= computer based training) is becoming more and more frequent with spreading of PC's among students and medical doctors [20]. Functional dependences are well suited for learning with the computer if simulation of curves, interactive use and intuition can be combined. Mathematical models of glucose insulin interaction have been designed by other authors for different purposes [4,9,10,23] such as the artificial endocrine pancreas or for quantification of insulin resistance under non-equilibibrium conditions.

The program DIACATOR has been designed in a way to be used for instruction, particularely to demonstrate the different facets of non-insulin-dependent diabetes. The program was evaluated by intuition and plausibility considerations and - if available - by curves taken from the literature. It must be pointed out, that this model and the simulated curves do not pretend to generate "real" curves of a given patient nor do they serve as an expert system for therapy planning.

The model and the simulation program considers NIDDM and its pre-stages not as a homogeneous disease that can be handled with one set of parameters. According to DeFronzo and coworkers [14] the glucose intolerance is considered as result of a gradually developing pathological state where insulin resistance and later on insulin secretion are impaired. This is realized by the programm by seven sets of parameters (table 1).

The glucose clamp technique has been introduced into the program to demonstrate a fundamental experiment of diabetology: the quantification of insulin resistance. With a proper selection of experiments, the user can find the model-parameters according to 3.3 and thereby close the circle between the author and the user.

First studies with medical students and physicians have shown good acceptance and demonstrate the preferential mode of use of such programs: experimenting with curves and comparing progression of the disease under different therapies in an interactive way. Future work is neccessary to prove that this type of model and simulation program is equivalent or even superior to learning from textbooks or lectures.

Adress of the author:
Dr.Eberhard Biermann
Institut für Diabetesforschung
Kölnerplatz 1
80804 München   Germany

Appendix

A. Model equations

The following equation have been taken from the literatur and were modified if neccessary.

Glucose elimination is based on a saturation kinetics and for first approximation, it is proportional to insulin concentration :[8,11,13,17,26,27,30]

$$U_i = \frac{5.5 \times G}{G + K_M} \qquad \text{insulin-independent utilization} \quad (1)$$

$$U_u = \frac{I \times F_i \times G}{G + K_M} \qquad \text{insulin-dependent utilization} \quad (2)$$

$$U_r = (GFR \times G) - 375 \quad \text{and} \quad 0 \text{ for } U_r < 0 \quad \text{renal glucose excretion} \quad (3)$$

wherein:
G,I glucose and insulin concentration; F_i tissue sensitivity for insulin.
K_M saturation constant for utilization;   GFR glomerular filtration rate

Hepatic glucose production G_{prod}:[8,24,26-28]

G_{prod} is both insulin- and glucose dependent. The output from the liver is suppressed by either increasing glucose or insulin concentrations (or both). In the case of a diabetic state, it can only be suppressed to a basal production rate P_{NS} . The hyperbolic tangens functions are expedient, because a increase of a function can be adjusted in the predefined range.

$$G_{prod} = 150 \times fkt_1 \times fkt_2 + P_{NS}$$
$$fkt_1 = 0.5 \times [1 - \text{tanhyp } 0.08 \times (I - 4)]$$
$$fkt_2 = 0.5 \times [1 - \text{tanhyp } 0.3 \times (G - 77)] \qquad (4)$$

wherein:
I= serum insulin concentration, G= glucose concentration ,P_{NS} = Non-suppressible part. fkt_1 and fkt_2 are funktions representing the insulin- and glucose dependent part.

Exogeneous glucose appearance rate G_{ex}:[24,25]

$$G_{ex} = \frac{CH}{(k_1 + k_2)} \times e^{-k_1 t} - e^{-k_2 t} \tag{5}$$

wherein the time constants k_1 and k_2 were obtained by least-squares fit to already published appearance rates, for an oral glucose tolerance test [24] and a mixed meal [25]. CH the amount of ingested carbohydrates.

Insulin concentration in equilibrium:[13,29].

$$I = \frac{k_{sens} \times G}{G + k_i} \tag{6}$$

wherein:
K_{sens} ß-cell sensitivity for glucose, and k_i is a saturation constant.

Dynamic insulin secretion (feedback control): [1,12,29]

The biphasic secretion pattern of insulin was modeled by use of a *PID*-control-law. The first phase can be described by the differential part and the second phase is described by the proportional and integral parts. The lack of the differential action or of the first phase is considered to be typical for type II diabetes [7,22].
   The following control algorithm is used in the modelling of the pancreatic function :

$$I_{inf} = K_{sens} \times (G + \frac{1}{T_N} \times \int G \, dt \times + T_V \times \frac{dG}{dt}) \tag{7}$$

wherein:
T_N/T_V is the integration/differentiation constant,
K_{sens} is the ß-cell sensitivity for glucose (= gain of the feedback loop)
The 'first phase' can be switched off, by setting $T_V=0$.

Feedback controlled glucose infusion (external glucose-clamp-algorithm) [1,15]

The control law in a discrete-time system is a dead-beat controller of 2nd order (for details see [1])

$$G_{inf}(k) = G_{inf}(k-1) - [G_{set} - (1+a) \times G(k) + a \times G(k-1)] \times \frac{b}{1-a}$$

$$a = e^{-b \times dt} \tag{8}$$

wherein:

G_{inf} glucose infusion rate, G_K, G_{k-1} glucose concentration at sampling events

G_{set} set point for glucose, a and b parameters resulting from the model,

Balance equation for glucose:

$$\frac{dG}{dt} = G_{prod}(G,I) + G_{ex}(CH,t) - U_i(G) - U_u(G,I) - U_r(G) \tag{9}$$

B. Mathematical solution, hardware and software requirements:

The differential equations are transformed into difference equations and integrated using the Runge-Kutta method of second order. The programm is written in BASIC, runs on an IBM PC/AT with DOS and an EGA/VGA graphics monitor. The program occupies about 140 kbyte of memory. A mathematical coprocessor is helpfull in some cases.

References

[1] K.J.ASTRÖM, B.WITTENMARK: Computer controlled systems. Prentice Hall, Englewood Cliffs (1984)

[2] BAILEY C.J.: Biguanides and NIDDM;   *Diabetes Care 15* (1992) 755-772

[3] M.BERGER, H.J.CÜPPERS, H.HEGNER, V.JÖRGENS, P.BERCHTOLD: Absorption Kinetics of Subcutaneous Injected Insulin Preparations ,   *Diabetes Care 5* (1982) 77-91

[4] BIERMANN E.,MEHNERT H., DIABLOG: A computer program for simulation of insulin glucose dynamics for education of diabetics, *Computer Meth.Progr.Biomed.* 32 (1990) 311-318

[5] BIERMANN E.: DIACATOR: simulation of metabolic abnormalities of type II diabetes mellitus by use of a personal computer.   *Computer Meth.Progr.Biomed.* 41 (1994) 217-229

[6] C.BOGARDUS,S.LILLIOJA,B.HOWARD,G.REAVEN,D.MOTT: Relationship between insulin secretion, insulin action, and fasting plasma glucose concentration in nondiabetic and noninsulin-dependent diabetic subjects ;   *J.Clin.Invest.* 74 (1984) 1238-46

[7] BRUCE D.L., D.J.CHISHOLM, L.H.STORLIEN AND E.W.KRAEGEN: Physiological Importance of Deficiency in Early Prandial Insulin Secretion in Non- Insulin-Dependent Diabetes mellitus, *Diabetes* 37 (1988) 735-44

[8] P.CAMPBELL,L.MANDARINO,J.GERICH: Quantification of Relative Impairment in Actions of Insulin on Hepatic Glucose Production and Peripheral Glucose Uptake, *Metabolism* 37 (1988) 15-22

[9] CARSON A.,C.COBELLI, R.FINKELSTEIN, The mathematical Modelling of Metabolism and Endocrinology Wiley New York (1983)

[10] C.COBELLI,G.FEDERSPIL, G.PACINI, A.SALVAN, C.SCANDARELLI: An Integrated Mathematical Model of the Dynamics of Blood Glucose and Its Hormonal Regulation,   *Math. Biosci.* 58 (1982) 27-60

[11] EDELMAN S.V, M.LAAKSO, P.WALLACE, G.BRECHTEL, J.OLEFSKI AND A.D.BARON: Kinetics of Insulin-Mediated and Non-Insulin-Mediated Glucose Uptake in Humans, *Diabetes* 39 (1990) 955-64

[12] U.FISCHER,E.JUTZI,H.BOMBOR,E.-J.FREYSE,E.SALZSIEDER,G.ALBRECHT, W.BESCH,B.BRUNS: Assessment of an Algorithm for the Artificial B-Cell Using the Normal Insulin-Glucose Relationship in Diabetic Dogs and Men,   *Diabetologia* 18 (1980) 97-107

[13] R.A.DEFRONZO, E.FERRANINI: Influence of Plasma Glucose and Insulin Concentration on Plasma Glucose Clearance in Man,   *Diabetes* 31(1982) 683-688

[14] R.A.DEFRONZO ,R.C.BONADONNA, E.FERRANNINI: Pathogenesis of NIDDM, A Balanced Overview, *Diabetes Care* 15 (1992) 318-68

[15] R.A.DEFRONZO,J.D.TOBIN,R.ANDRES, The glucose clamp technique: A method for quantifying insulin secretion and resistance,   *Am.J.Physiol.* 6(1979) E214-E223

[16] H.GINSBERG, E.RAYFIELD: Effect of Insulin Therapy on Insulin Resistance in Type II Diabetic Subjects, Evidence for Heterogeneity   *Diabetes* 30 (1981) 739-45

[17] GOTTESMAN I.,L.MANDARINO ,J.GERICH; Estimation and Kinetic Analysis of Insulin-Dependent Glucose Uptake in Human Subjects ,   *Am.J.Physiol.* 244:(1983)E632-E635

[18] L.C.GROOP: Sulfonylureas in NIDDM ; *Diabetes Care* 15 (1992) 737-754

[19] M.HANEFELD, S.FISCHER, J.SCHULZE, M.SPENGLER, M.WARGENAU,K.SCHOLLBERG, K.FÜCKLER: Therapeutic Potentials of Acarbose as First-Line Drug in NIDDM Insufficiently Treated with Diet Alone, *Diabetes Care* 14 (1991) 732-37

[20] A.HASMAN; Use of authoring systems for constructing medical teachware , in M.P.Baur,J.Michaelis ;*Computer in der Ärzteausbildung* , Oldenburg Verlag München, Wien 1990 p.147-164

[21] J.P.HOSKER, M.A. BURNETT, E.G.DAVIES, E.A.HARRIS AND R.C.TURNER: Sulphonylurea therapy doubles B-cell response to glucose in Type 2 diabetic patients , *Diabetologia* 28 (1985)

[22] H.H.KLEIN UND H.U.HÄRING: Pathogenese des Diabetes mellitus Typ II; *Internist* 31 (1990) 168-79

[23] M.MATTEI, G.ERMINI, P.MALAGUTI,E.RUFFILLI,D.MAZOTTI,E.SARTI: Comparison of glucose metabolism in subjects with different liver diseases and chemical diabetes by digital computer model. *VI Congres of Gastroenterology,* Madrid (1978)

[24] A.MITRAKOU, D.KELLEY,TH.VENEMAN,T.JENSSEN ,TH.PANGBURN, J.REILLY, AND J.GERICH: Contribution of Abnormal Muscle and Liver Glucose Metabolism to Prandial Hyperglycaemia in NIDDM, *Diabetes* 39 (1990) 1381-90

[25] G.PEHLING, P.TESSARI, J.E.GERICH, M.W.HAYMOND, F.J.SERVICE, R.A.RIZZA: Abnormal Meal Carbohydrate Disposition in Insulin-dependent Diabetes , *J.Clin.Invest.* 74(1984) 985-991

[26] R.PRAGER P.WALLACE, J.M.OLEFSKY: In Vivo Kinetics of Insulin Action on Peripheral Glucose Disposal and Hepatic Glucose Output in Normal and Obese Subjects, *J.Clin.Invest.* 78 (1983) 472-81

[27] R.R.REVERS, R.FINK, J.GRIFFIN, J.M.OLEFSKY, O.G.KOLTERMAN: Influence of hyperglycaemia on Insulin's in Vivo Effects in Typ II Diabetes; *J.Clin.Invest.* 73 (1984) 664-672

[28] R.RIZZA ,L.J.MANDARINO, J.E.GERICH: Dose-response Characteristics for Effects of Insulin on Production and Utilisation of Glucose in Man; *Am.J.Physiol.* 240 (1981) E630-639

[29] R.C.TURNER ,R.R.HOLMAN, D.MATHEWS, D.R.HOCKADAY AND J.PETO: Insulin Deficiency and Insulin Resistence Interaction in Diabetes: Estimation of Their Relative Contribution by Feedback Analysis From Basal Plasma Insulin and Glucose Concentration, *Metabolism* 28:(1979)1086-96

[30] H.YKI-JÄRVINEN, A.YOUNG , C.LAMKIN, J.E. FOLEY : Kinetics of Glucose Disposal in Whole Body and across the Forearm in Man , *J.Clin.Investigation* 70 (1987) 1713-19

Rechargeable Glucose Biosensors

E. Wilkins and P. Atanasov

1   Introduction

There are several studies by numerous investigators in the area of glucose sensing, aimed at fulfilling the need for an implantable glucose sensor to close the loop for an insulin pump, and obtain a complete artificial pancreas[1-3.] However, a continuously functioning implantable glucose sensor with long-term stability has not yet been achieved. Research is being carried out on several types of potentially implantable glucose sensors: electrochemical sensors based on direct electrooxidation of glucose on noble metal electrodes, and biosensors combining the selectivity of substrate-specific (in this case glucose-specific) enzymes with the versatility and simplicity of electrochemical electrode systems.

Glucose biosensors are generally based on the enzyme Glucose Oxidase (GOD). This enzyme catalyzes the oxidation of β-D-glucose by molecular oxygen producing gluconolactone and hydrogen peroxide:

$$\beta\text{-D-glucose} + O_2 \xrightarrow{\text{Glucose Oxidase}} \text{gluconolactone} + H_2O_2$$

Either the increase in hydrogen peroxide concentration or the decrease in oxygen concentration due to this reaction can be detected electrochemically, both being proportional to the glucose concentration. Hydrogen peroxide is detected amperometrically during its anodic oxidation on catalytic electrode. In the case of oxygen detection, the amperometric signal is a result of the electrochemical reduction of oxygen carried out usually on a platinum electrode. A linear dependence of the amperometric signal is obtained when the mass transfer of both electrochemically active species (hydrogen peroxide or oxygen) and glucose are the limiting processes. This is achieved by use of a variety of diffusion membranes in the biosensor construction.

In the development of an implantable sensor, the most important problem is that of long-term operational stability. Since the glucose biosensor will be a part of the implantable insulin delivery system, its lifetime must be at least as long as that of the other parts of the system. The lifetime of the biosensor is limited mainly by the processes of enzyme deactivation. The immobilization of the enzyme in gels, on membranes, or on inert dispersed carriers (usually carbon materials) can significantly increase its stability. However, the lifetime of the biosensors with immobilized enzymes is still limited.

2   Rechargeable Glucose Biosensors Design

A new approach, developed in our laboratory, makes it possible to extend the sensor lifetime by *in situ* sensor refilling - replacing spent immobilized enzyme with fresh enzyme[4,5]. The enzyme Glucose Oxidase is immobilized on fine dispersed carbon powder, which is then held in a liquid suspension. The construction of the biosensor is such that the spent immobilized enzyme can be removed from the sensor body and fresh enzyme suspension injected *via* a septum, without sensor disassembly. This concept facilitates recharging of the implanted sensor without surgical removal from the patient.

Fig. 1 shows a crossectional schematic of the rechargeable glucose biosensors based on hydrogen peroxide (Fig. 1.a) and oxygen measuring principle (Fig. 1.b).

The glucose biosensors consist of two parts: an amperometric electrode system and an enzyme micro-bioreactor. A three electrode amperometric scheme is used in both types of the biosensors: a platinum wire (diameter 0.25 mm, length 4 mm) as a working electrode (1), a silver/silver chloride reference electrode (2) and another platinum wire as a counter electrode (3). The working electrode is polarized +0.6 V for hydrogen peroxide oxidation, or -0.6 V for oxygen reduction *vs.* the reference electrode.

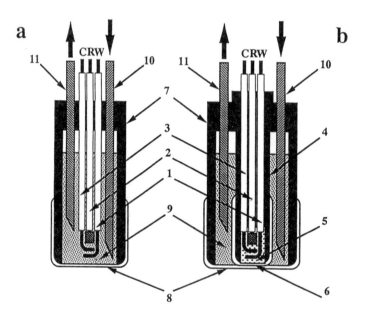

Figure 1. Rechargeable Glucose Biosensor Based on Hydrogen Peroxide (**a.**) and Oxygen (**b.**) Measuring Principle

In the oxygen measurement based sensor (Fig. 1.b) these three electrodes are symmetrically assembled and housed in a separate glass tube (4) and cemented by epoxy resin. At the face-end of the housing, a cavity is formed and filled with gelled electrolyte: agarose[7] or poly-hydroxyethylmethacrylate (HEMA)[8-10] matrix (5) soaked in phosphate buffer solution containing potassium chloride. An oxygen-permeable hydrophobic membrane (6) is used to cover the three electrode system separating it from the enzyme micro-bioreactor. Mechanically attached Teflon membranes or membranes made by dip-coating of the oxygen electrode with silastic latexes are used[7].

In hydrogen peroxide based biosensor (Fig. 1.a) the three electrode amperometric system is directly inserted in the sensor housing (7), a plastic tube, face-side closed by a glucose diffusion membrane (8).

The glucose diffusion membrane is used to limit and control diffusion of the substrate by allowing a small amount of glucose to enter the sensor and react with the enzyme. Particular membranes utilized in the rechargeable glucose biosensor construction are: polycarbonate membranes with different pore size[3,8,11], cellulose acetate membranes[12], Teflon membranes[4,13], membranes formed from silicone elastomers on a polyethylene support[7], as well as membranes constructed from acrylate. These membranes are used also being coated (modified) with a polymer (Nafion, silastic, hydrogel) layer to obtain the necessary diffusion properties. The optimization of the preparation of this external glucose membrane has been the subject of our recent report[8].

The enzyme (Glucose Oxidase) is immobilized on fine carbon powder in a liquid suspension. The liquid state of the enzyme material allows withdrawal of inactive enzyme and injection of fresh enzyme *via* a septum.

A technique for immobilization of the enzyme Glucose Oxidase on carbon materials carbodiimide linking, has been developed[3,4]. Several disperse carbon materials: graphite powder, carbon blacks, and synthetic carbons has been screened[4]. The highest activity of the enzyme and the most reproducible results has been obtain when Ultra Low Temperature Isotropic carbon powder (fraction less than 325 mesh) was used as a carrier for Glucose Oxidase immobilization[4,13].

This enzyme-modified dispersed carbon is used to form a liquid suspension for refilling of the enzyme micro-bioreactor (9) - recharging of the sensor. The suspension is prepared by dispersing in an ultrasonic bath 100 mg of the enzyme-modified carbon powder in 1 mL phosphate buffer solution containing 20 mg Glucose Oxidase. Two capillary plastic tubes (diameter *c.a.* 1mm) - inlet recharge tube (10) and exhaust discharge tube (11) - are used for replacing spent enzyme from the micro-bioreactor without sensor disassembly. Refilling of the sensor is achieved using two septums via these tubes: one for injecting a fresh enzyme suspension, another for exhausting the spent enzyme.

Biosensor dimensions are roughly 3 mm in diameter for the hydrogen peroxide based sensor, and 6 mm in diameter when an oxygen-selective electrode is used. Both bigger and smaller sensors have been and are being constructed.

A conventional potentiostat or specially developed electronic device, in which the biosensor is an integrated part (the prototype of the implantable telemetry system) is used to evaluate the glucose sensor performance.

3   Evaluation of the Rechargeable Glucose Biosensors

We have obtained reproducible results using various modifications of the rechargeable biosensors[6-8]. The amperometric output of the biosensor is linearly proportional to glucose concentration in the entire range of physiological and pathophysiological interest. The upper limit of the linear range of the biosensor varied from 220 to 1000 mg/dl, depending on the type of the sensor and particularly of the type of glucose diffusion membrane used[7,8]. Response time of the sensor - time to reach 95% of the steady-state signal value - is from 1 to 10 minutes depending of the glucose concentration step change and the diffusion membrane used[7-9].

A long term stability test was conducted to demonstrate the potential of the rechargeable glucose sensor for implantation. These experiments were carried out with the glucose biosensor in constant operation throughout the duration of the test. Concentration of glucose in the measuring cell was 100 mg/dL, approximating the normal physiological value. During the test time, the calibration curves in the glucose concentration range from 0 to 300 mg/dL were periodically obtained. The sensitivity of the biosensor (the slopes of these calibration curves) was obtained and used as a parameter for the biosensor characterization.

The biosensors are stable while operated continuously in a model phosphate buffer solution (pH 7.4) for a period of 7 months at 25°C, or at least 3 months at 37°C, before refilling with fresh enzyme suspension[6]. The dependence of the steady state sensor signal *vs.* glucose concentration demonstrated significant reproducibility in their main characteristics: linear range and sensitivity. Standard deviation from linearity for any of the measurements does not exceed ±0.002449.

Stability of the sensor may be defined as the period of operation before sensor sensitivity drops out of the 10% range. With this definition, the rechargeable biosensor is stable for 210 days - sensitivity is constant for nearly seven months[6]. At the end of this period, sensor sensitivity decreases through the natural process of enzyme deactivation. However, the response curve of the sensor is still linear, so it can be used as an analytical device at the end of this period for at least two more months, although recalibration of the sensor before further measurements is necessary. The sensitivity of the sensor at 37°C remained essentially constant during the first two months of testing, varying in the acceptable 10% range. However sensitivity significantly decreased after this period.

Recharging of the sensor with fresh enzyme (refilling the sensor body with new glucose oxidase immobilized on the carbon powder after replacement of the old suspension) can successfully increase the sensor lifetime[6,7,9,10].

The sensitivity of a biosensor (based on the oxygen electrode[10]) *vs.* time of continuous operation through several recharge cycles is presented on Fig. 2. It should be noted that during the first days of operation sensor sensitivity decreases from its initial value, probably due to the loss of activity of the Glucose Oxidase in the solution (not chemically immobilized on the carbon carrier). After this initial period the sensitivity of the biosensors is essentially constant for a period of about 14 days. At the end of this period the sensitivity begins to decrease. When the biosensor sensitivity drops to more than 10% of the previous constant value, the sensor is refilled with fresh immobilized enzyme suspension, using septums (without sensor disassembly). After refilling and initial stabilization the biosensor continued to demonstrate the same constant value of sensitivity as during the first recharge cycle.

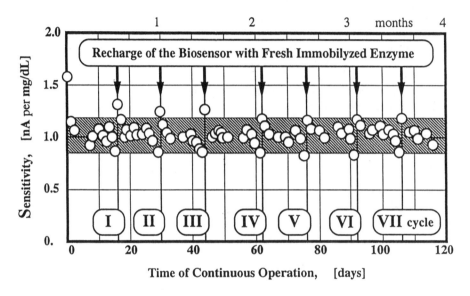

Figure 2. Long-Term Operational Stability of the Glucose Biosensor (dependence of the sensitivity *vs.* time of continuous operation)

The biosensor has been refilled seven times with the fluid enzyme-modified carbon suspension, running over eight recharge cycles of about two weeks each, during the four month continuous test (Fig. 2). The standard deviations of all measured points (including initial stabilization periods and final periods of the decrease in sensitivity for every cycle) from the average value of the biosensor sensitivity is less than 10% over the entire period of continuous operation of the sensors.

By using the combination of two enzymes (in oxygen electrode based biosensors) - Glucose Oxidase and Catalase - the recharge cycle (time between biosensor refilling) can be extended up to one month due to the elimination of the hydrogen peroxide as an enzyme deactivator[7,9].

The effects of possible interference by chemicals in body fluids - ascorbic acid, bilirubin, creatinine, L-cystine, glycine, uric acid and urea - on the amperometric signal of the biosensors were studied. Charged hydrogel coatings over the polycarbonate[11], cellulose acetate[12] and Teflon[13] membranes has been used in order to lower or prevent diffusion of these species into the biosensor based on measuring hydrogen peroxide. It was found that membranes coated with the negatively charged hydrogel layer provided good protection for the biosensor, especially from anionic interferences.

In the glucose biosensor based on oxygen electrode principles interference effects are generally avoided by separation of the amperometric system from the enzyme micro-bioreactor, using a hydrophobic oxygen permeable membrane. The rechargeable biosensor is unaffected by interferences usually present in body fluids[10].

The amperometric response of the biosensor to the glucose levels in sera has been obtained and the concentration of glucose estimated using the sensor calibration curve obtained in the model buffer solutions. The correlation between the glucose concentration values measured by the sensor, and the data for the same samples measured by conventional method, is characterized by a slope of 0.950 with a regression coefficient = 0.993. This is a demonstration of the accuracy of the sera glucose level measurements by the rechargeable biosensor[9].

Glucose concentration in sera samples has been monitored by the biosensor for 16 hours (*in vitro*) with no noticeable change in the sensor response. After this test a calibration curve of this sensor in phosphate buffer glucose solution is obtained and the results (sensor sensitivity and linearity of the response) coincide with the calibration curve obtained before sera measurements.

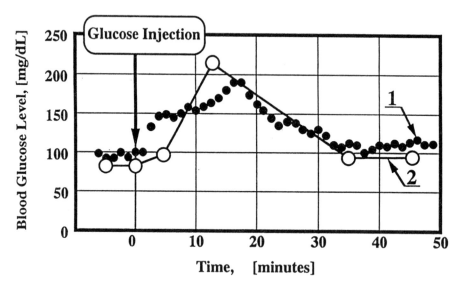

Figure 3. Correlation Between the Values of Glucose Concentration Measured by the Implanted Sensor In Vivo (1) and Blood Glucose Level Estimated by a Standard Method (2)

The glucose biosensor has been implanted and tested *in vivo* in a sheep. Glucose concentration in the animal interstitial fluids has been monitored by the biosensor and juxtaposed with the blood glucose levels estimated by external conventional method in the periodically obtained samples (Fig. 3). The test includes a step change in glucose concentration by injection of a glucose dose into the animal. The data shows good correlation between the profiles of the glucose concentration evolution in body fluids obtained by these two different methods.

4   Conclusions

The rechargeable glucose biosensor has several features different from existing ones:
1. The use of very fine biomaterial particles as carriers (carbon, graphite), with the enzyme chemically immobilized or crosslinked on the particle surface. This is accomplished by first preparing the surface of the powder to be active so as to accept the enzyme bonding, usually by the use of the carbodiimide technique.Glutaraldehyde and bovine serum albumin are used for crosslinking.
2. The fine powder in bulk with immobilized or crosslinked enzyme can be replaced *in situ* for rejuvenation of the sensor without removal from the implantation site. The use of bulk amounts of immobilized or crosslinked enzyme greatly extends both the life of the system and the time between refills.
3. The biosensor is constructed so as to have a large surface area for flux of glucose and oxygen into the sensor by using of variety of hydrophobic and/or hydrophilic diffusion membranes.
4. A reference electrode and counter electrode are incorporated into the biosensor body forming an integrated device

In view of the above characteristics, our glucose sensor is a significant improvement over all other types of sensors currently available, due to the capability of recharging it by refilling the sensor while implanted, the ability to control the oxygen and glucose fluxes, and to control passage of undesirable body chemicals into the sensor.

The capability to refill by external means an implanted glucose sensor without surgical removal would be a major advance in the development of a practical glucose sensor for diabetes control. Our use of enzyme in the form of a liquid suspension of carbon powder with immobilized Glucose Oxidase on the powder allows enzyme replacement without surgery by injection *via* a subcutaneous septum using a procedure similar to that developed for refilling of implantable insulin pumps[14]. The potential of this approach opens the possibility of constructing sensors suitable for implantation with lifetimes greater than several months in the body.

References

1. Turner, A. P. F., Pickup, J. C., *Biosensors*, 1, 85-115, 1985
2. Reach, G., Wilson, G.S., *Anal.Chem.*, 64, 381A-386A, 1992
3. Wilkins, E., *J.Biomedical Engineering*, 11, 353-361, 1989
4. Xie S.L., Wilkins, E., *J. Biomedical Engineering*, 13, 375-378, 1991
5. Xie, S.L., Wilkins, E., *Biomed.Instrum.&Technol.*, 25, 393-39, 1991
6. Wilkins, E., *Boomed.Instrum.& Technol.*, 27, 325-333, 1993
7. Xie, S.L., Wilkins, E., and Atanasov, P., *Sens.& Actuat.B*, 133-142, 1993,
8. Atanasov, P., Wilkins, E., *Analytical Letters*, 26, 1587-1612, 1993
9. Atanasov, P., Wilkins, E., *Analytical Letters*, 26, 2079-2094, 1993
10. Atanasov, P., Wilkins, E., *Biotechol.& Bioengin.*, 43, 262-266, 1994
11. Vaidya, R., Wilkins, E., *Boomed.Instrum.& Technol.*, 27, 486-494, 1993
12. Vaidya, R., Wilkins,E., *Electroanalysis*, in press, 1994
13. Vaidya, R., Wilkins,E., *J. Biomedical Engineering*, in press 1994
14. Wilkins, E., Radford, W., *Biosensors & Bioelectronics*, 5, 167-213, 1990

The Influence of Temperature on pH-Responsive Chemical Valve Systems: Thermodynamic Aspects for Drug Delivery

M. Casolaro

1 Introduction

With the continued development of controlled-release technology, the need has arisen for materials with more specific drug-delivery properties. These materials include ionic or water-soluble polymers, with both hydrophilic and hydrophobic characteristics, that respond to external physical (pH, redox, heat, etc) and chemical (glucose) signals [1]. In the latter case the response is obtained with biological components, such as enzymes (e.g. glucose oxidase, GOD), that act as signal transducers in glucose-sensitive insulin releasing systems. GOD catalyzes the conversion of glucose to gluconic acid and hydrogen peroxide, liberating protons and electrons. The combination of the enzyme with a pH or redox-sensitive system gives a glucose-sensitive system. A self-regulated insulin releasing system responding to a range of glucose concentrations would have important applications. To achieve such a system many hydrogels have been studied in the past to control insulin delivery from a reservoir [2]. An elegant system, based on conformational states of polyelectrolytes grafted on porous membranes was recently developed to control water and solute permeability [3,4,5]. This system, having *chemical valve* function, consists of a porous membrane onto which stimuli-responsive polymers can be graft-copolymerized. The graft polymer changes shape in response to external physical stimuli. Regulation of the gating or valving function depends on the expansion and contraction of the graft chains. The enzyme GOD has been used as a sensor of glucose. Covalent bonding of GOD to the carboxyl group of polymers grafted on cellulose membranes [3] gave a glucose-sensitive membrane (*Figure 1*).

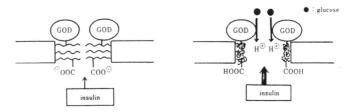

Figure 1. Principle of controlled release of insulin (GOD: glucose oxidase)

As glucose diffuses into membrane pores obstructed by graft chains in the extended state, GOD catalyzes its conversion to gluconic acid, thereby lowering the pH in the microenvironment of the pH-sensitive polymer. Low pH brings the polymer chains back to the conformation of a collapsed coil. The composite membrane thus exhibits high insulin permeability [3,6]. Many polymer systems, that are responsive to a single type of stimulus, such as pH [poly(acrylic acid), poly(N-acryloyl-glycine)] or redox reactions [poly(3-carbamoyl-1-(p-vinylbenzyl)pyridinium chloride)], have recently been investigated for a variety of biotechnological uses [3-6]. We have been studying polymers sensitive to combinations of environmental stimuli, especially when the different responses are competitive. We synthesized a pH-sensitive polymer, poly(N-acryloyl-*L*-valine) (polyAVA), carrying a carboxyl group and other groups (amido and isopropyl) present in the well known poly(N-isopropylacrylamide) (PNIPAAm), a temperature-sensitive non ionic polymer with an LCST (Lower Critical Solution Temperature) of 32°C [7]. The latter polymer undergoes a volume phase transition in water as the temperature is raised above the LCST. The mutual influence of pH and temperature was investigated for water permeability through a porous cellulose membrane (0.2 μm pore size) grafted with polyAVA.

We examined the thermodynamic aspects of this multiple-responsive system with the aim of achieving a reversible and faster response to body needs.

2 Thermodynamics

The protonation of carboxylate groups in polymers grafted on hydrophilic cellulose (CL) was studied in aqueous media and in heterogeneous phases at 25°C. The same polymers were studied when grafted on cast polyurethane (PU) film to compare the hydrophilic quality imparted by the support. Potentiometry, solution calorimetry, and FT-IR spectroscopy were the main techniques used for the thermodynamic analysis of graft systems [5,6,8]. Information from potentiometric titrations was limited to the amount and type of titrable carboxyl groups on the surface. The protonation mechanisms revealed a slow kinetics, since the thermodynamic equilibrium was longer than for the protonation of COO^- in the corresponding free polymer analogues [6,8]. Large hysteresis loops were displayed between forward and backward titrations with H^+/OH^-, at shorter titrant addition times. In agreement with previous studies [5,6,8], the hydrophilicity of the support played an important role in the magnitude of the polymer-substrate interaction. Generally, the weaker the interaction, the shorter the time necessary to reach a condition of thermodynamic equilibrium, reflected by closer hysteresis curves. Equilibrium configuration of the chains takes longer to attain because intra- and inter-molecular interactions hinder the movement of polymer segments.

A similar hydrophilic or hydrophobic nature always leads to a stronger interaction between polymer and substrate, thus increasing the time lag, i.e. the time necessary to reach equilibrium pH, at each titration point. *Figure 2* shows a typical potentiometric titration plot with different hydrophilic components.

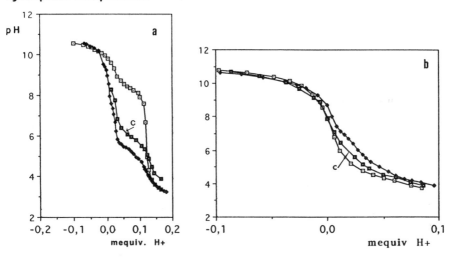

Figure 2.   Potentiometric titration curves at 25°C in 0.1M NaCl for: **(a)** methacryloyl-L-leucine grafted on PU and **(b)** acryloyl-L-valine grafted on cellulose membranes. Stabilization time: 600 sec for each titration point. Equilibrium curves **(c)** is enclosed: **(a)** 7-14 hrs; **(b)** 3-6 hrs.

A hydrophobic substrate (polyurethane) grafted with a hydrophobic polymer [poly(N-methacryloyl-*L*-leucine)] displayed larger potentiometric hysteresis loops than the hydrophilic cellulose substrate grafted with polyAVA (hydrophobic). The latter showed closer curves and the equilibrium conditions were reached faster (3-6 hours) than in the former system (7-14 hrs). The basicity constant (logK = 4.6) for AVA-g-CL was in line with that of the free polyAVA (logK = 4.59). While calorimetric data provided information on the energy of protonation of chain ends fixed to the surface [5,8], with relative freedom of rotation, IR spectral analysis [6] revealed two kinds of titrable carboxyl groups on the hydrophilic surface grafted with hydrophilic polymer.

3 Permeability studies

Water permeation.   Water filtration rates can be controlled by changing the conformation of the grafted polymer chain by suitable stimulation. Redox action, pH and temperature were effective in changing graft chains from extended to coiled states, allowing the water to permeate reversibly through the pores [4,5]. Polymers carrying charges or ionized and hydrated functional groups have extended chains that obstruct the membrane pores, limiting water permeation. As external physical stimuli bring the polymer chains back to collapsed coil conformation, water permeation increases. *Figure 3* shows a multiple-responsive system made of porous cellulose (0.2 μm) grafted with polyAVA.

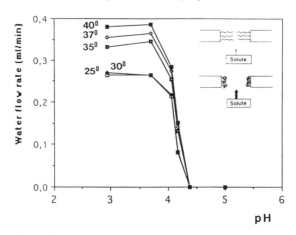

Figure 3. Effect of pH on the water flow rate through acryloyl-L-valine grafted on porous cellulose membranes (0.2 μm pore size) at different temperatures. The inset show the *chemical valve* mechanism.

At pHs > 4.4 the carboxyl groups on the polymer chain are mostly ionized and the macromolecule assumes an extended conformation, due to electrostatic forces. The water flow rate is zero. For pHs < 4.4, the graft chains are in a collapsed state. Water permeation increases with increasing charge neutralization till a plateau is reached in the completely protonated state. The shape of the water flow rate/pH plot is almost unchanged when temperature is varied from 25° to 40°C. Increased water permeability is observed in the range of low pH, where further shrinkage of collapsed polymer chains probably occurs as a result of the LCST phenomenon [7].

The neutral form of polyAVA at pH 3, being closer to that of PNIPAAm, showed a sharp increase in the water flow rate at 31°C. This increase cannot be attributed to ionization processes due to variations in logK values with temperature, because in the range of temperature investigated the basicity constants of polyAVA were almost unchanged.

Insulin permeation. The permeation of solutes of different molecular weight (POE, polyoxyethylene, 1000 to 80000) was studied through cellulose grafted with poly(carboxyl acid) with different graft contents [3]. The higher the graft content, the lower the permeation, and more solute permeated at low pH. Permeation was also controlled by pore size. POE (with a molecular weight in the 6000-20000 range) was found to be the most suitable solute for permeation studies.

The insulin permeability always increased linearly with time, after a short time lag, through grafted and non-grafted porous cellulose membranes [3,6,9]. Different permeation of insulin/time slopes were observed at the two chosen pHs for grafted membranes. The acidity-enhanced permeation may be related to the conformational change of the graft chains acting as gates. The graft polymer chains are mostly ionized at pHs higher than the logK value of poly(carboxyl acid) [5,6]. This causes repulsion between charged chains, obstructing the membrane pores and preventing insulin permeation. On the other hand, low pH enhanced collapse of the polymer onto the substrate, increasing pore size and enabling faster permeation. The pH-sensitive *chemical valve* with the enzyme GOD immobilized on the graft polymer chains was also studied as a self-regulating insulin delivery device [3,6,9]. The enzyme was coupled to the carboxyl group by the carbodiimide method. The total amount of immobilized GOD was determined by spectroscopic methods [3,6]. Its activity was much lower than that of the free enzyme in solution. This was tested by following the decrease in pH with time on addition of a bolus of glucose to a solution containing a sample of grafted and GOD-immobilized membrane. The control of insulin permeation was studied at physiological pH in 0.01M Tris-HCl buffered solution. Permeation data for insulin diffusion through poly(N-acryloyl-glycine)-grafted and GOD-immobilized cellulose membrane is reported in *Figure 4*.

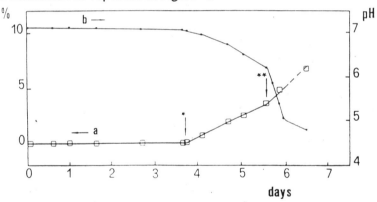

Figure 4. a: percentage of insulin permeating through poly(acryloyl glycine)-grafted and GOD-immobilized cellulose membrane; **b:** pH-change profile with glucose addition (* 50 mg/100ml; ** 400 mg/100ml).

The release rate was negligible for several days. When glucose was added it diffused into the porous cavity and was converted to gluconic acid, catalyzed by GOD. The pH in the microenvironment dropped, allowing protonation of the COO⁻ groups on the chains. These groups, having a higher logK value than gluconic acid [10] (logK = 3.6), protonated, causing the rod-like polymer chains to coil. The pores opened allowing passage of the insulin molecules. On addition of glucose, a faster response was obtained since the insulin permeation/time and pH/time plots changed slope [3,6]. Changes in bulk pH were observed and with the further addition of glucose to higher concentration, they were larger, causing a drop to pH < 5, where the aggregation of insulin was enhanced [6]. Thus, a higher buffer strength with more carboxylate groups is required to avoid critical phenomena due to pH decrease in composite systems, prototypes of an *artificial pancreas*.

4 Conclusions

In the drug delivery technology the devices are becoming more sophisticated for delivering the drug with precise zero-order release or biofeedback mechanisms. We have reported a composite system made of porous membrane onto which stimuli-sensitive polymers have been graft-copolymerized. The grafted polyelectrolytes changed their conformation, behaving as a *chemical valve* to control the water flow rate or solute permeability through the membrane pores. The nature of either the substrate and the polymer influences the reversible ionization process that is reflected in the kinetics of protonation. The experimental results indicate that weak polymer-substrate interactions have a faster response in the protonation thermodynamics. To realize a self-regulating system in which the synthetic macromolecules involved are sensitive to changes in the environment (such as pH, temperature, or glucose concentration) and undergo subsequent conformational changes it is necessary to introduce at least one feedback system. Therefore, the glucose oxidase immobilized on the carboxyl group of the polymer has been used as a sensor of glucose. The system behaves as an artificial pancreas that allows the delivery of insulin at rates dependent on the glucose concentration.

Acknowledgements. *This research was partially supported by a Grant to the International Joint Research Project from NEDO, Japan. Partial support from the Italian MURST (60% funds) is gratefully acknowledged.*

References

[1] Proceedings of the *First International Conference on Intelligent Materials*. Takagi,T. *et al.*, Eds, (1993) Technomic Publ.Co., Inc., Basel

[2] *Polyelectrolyte Gels*. Harland,R.S. and Prud'homme,R.K., Eds, (1992) ACS Symp. Series 480

[3] Ito,Y., Casolaro,M., Kono,K. and Imanishi,Y. *J.Controlled Release* (1989) **10**,195

[4] Ito,Y., Inaba,M., Chung,D.J. and Imanishi,Y. *Macromolecules* (1992) **25**,7313

[5] Casolaro,M. in *Chemistry and Properties of Biomolecular Systems*. N.Russo *et al.*, Eds, Kluwer Academic Publishers (1994) **2**,127

[6] Barbucci,R., Casolaro,M. and Magnani,A. *J.Controlled Release* (1991) **17**,79

[7] Feil,H., Bae,Y.H., Feijen,J. and Kim,S.W. *Macromolecules* (1993) **26**,2496

[8] Casolaro,M. and Barbucci,R. *Coll.Surf. A: Phys.Eng.Asp.*, (1993) **77**,81

[9] Casolaro,M. and Barbucci,R. *Int.J.Art.Organs* (1991) **14**,732

[10] Martell,A.E. and Smith,R.M. in *Critical Stability Constants*, (1974) N.Y., London: Plenum Press

Control Algorithms for a Wearable Artificial Pancreas

P.G. Fabietti, L. Tega, and S. Allegrezza

I. Introduction

Past studies in insulin-dependent diabetes mellitus have shown that the duration of disease and the level of hyperglycaemia are correlated with the onset of micro- and macroangiopathic complications (1,2,3). This observation indicated that patients should be treated with intensive insulin therapy to obtain blood glucose control which mimics the physiological model.

The basis of intensive insulin therapy (4) is multiple daily insulin injections before meals and bedtime, or the use of programmable micro-infusors which are able to administer continuous subcutaneous doses of insulin at a variable rate. This type of feedforward control has led to satisfactory results in terms of daily blood glucose and glycosylated haemoglobin levels, but has also demonstrated the impossibility of reaching a physiological model of blood glucose regulation.

The only artificial organ able to approach a physiological response which is close to the natural model, and can induce metabolic stability, is the artificial pancreas (5,6,7). There are many prototypes, but the Biostator produced by Miles in 1976 was the first on the international market. This instrument functions on the basis of continuous blood glucose monitoring of venous blood drawn through a double-lumen catheter and diluted with an anticoagulant (figure 1).

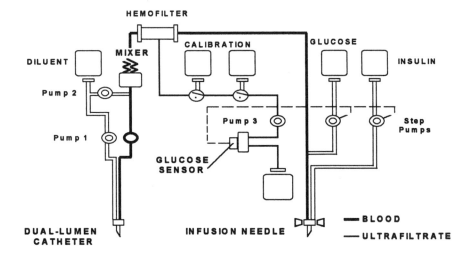

Fig. 1
Diagram of the artificial pancreas

A hydraulic circuit transports the mixture to an extracorporeal enzymatic-amperometric sensor for blood glucose measurement. A minicomputer controls the administration of insulin and glucose by elaborating a proportional derivative algorithm. In 1985, the Esaote Biomedica produced the Betalike, an instrument whose weight (19 kg) and size are reduced in comparison to Biostator (60 kg).

To date, the types of artificial pancreas that have been produced have had limited application in clinics and research because of their cumbersome size and the impossibility of allowing the patients who were connected to them to carry out normal daily activities. Recent technological developments have prompted several researchers (8,9,10,11) to dedicate their energy towards the realisation of a portable artificial pancreas.

Our studies, aimed towards this ambitious goal, have led to the development of a portable unit whose weight and size are reduced (250 g: 8.5x14.5x3.5 cm). It can feed an amperometric sensor working in the range of 10 to 200 nA full scale, and perceive and control the signal that it produces. (12). This unit is based on an Intel 80C196 KB digital microcontroller with an internal 16-bit CPU running at 5 MHz, which processes the sensor output and generates a control signal to drive an infusion pump.

II. Auto - Tuning Controller

To solve proportional-derivative traditional controller limits where the parameters in the control algorithm are fixed, we have designed an auto-tuning controller.

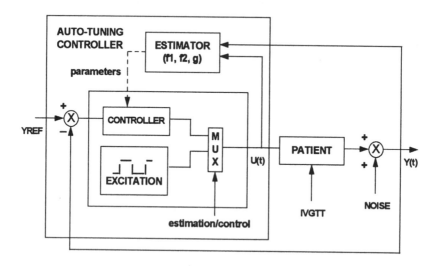

Fig. 2
Auto-tuning controller diagram block. (IVGTT = intravenous glucose tolerance test)

The auto tuning methodology (13,14) is used in biological applications, essentially for two reasons:
1) The artificial pancreas must be used by several patients, so the clinician cannot easily change the parameters for each new patient.
2) The response of the patient to insulin infusion changes during the day, so fixed parameters do not optimize insulin infusion.
Figure 2 shows the auto-tuning controller diagram block.

Estimation algorithm

The estimation algorithm is based on a patient mathematical model. Here Cobelli's non-linear compartmental model of glucose metabolism is used. (15,16,17) The model describes the relationship between insulin infused in venous blood and glucose concentration. It can, however, be modified to take into account different administration routes and sensor locations which are more acceptable for a portable device.

Cobelli's model is linearized around an appropriate working point obtaining a sixth order: the continuous time linear model (see preceding chapter entitled "Recent development and open problems in feedback control of the glucose system in diabetes" Eugenio Sarti et al.). Using a common discretizing algorithm, the discrete time model of the patient is obtained: the input is the insulin infusion rate, at sample k, ins(k), and the output is the blood glucose, at sample k, gli(k). The simplified patient model contained in the estimation block is therefore obtained.

After many attempts, the reduction of the order was obtained through computer simulations. The transfer function is a second-order function with no zeroes and with a finite delay fixed at 6 min. This value was chosen in accordance with experimental tests in glucose metabolism as reported in the medical literature. The discrete expression of the reduced continuous system can be put in the following difference equation:

$$\Delta gli(k+1) = f_1 \Delta gli(k) + f_2 \Delta gli(k-1) + g[\Delta ins(k) + 2\Delta ins(k-1) + \Delta ins(k-2)] \qquad [1]$$

in which:

$$\Delta gli(k) = gli(k) - gli(set\_point) \qquad [2]$$

$$\Delta ins(k) = ins(k) - ins(set\_point) \qquad [3]$$

wherein the k sample is taken as equal to the finite delay of the system because it is an acceptable compromise between the length of time calculation and the patient's behaviour.

At the end of the estimation stage, the parameters f1,f2 and g are introduced into the control algorithm.

Estimation of parameters

The control strategy adopted is sometimes called *auto-tuning*: i.e., as the parameters vary slowly with respect to the dynamics of the system, their evaluation takes place at rather long intervals of time, while the control loop is open and the insulin infusion rate is kept independent from blood glucose variations.

Therefore, the estimation stage must be carried out when the patient is between meals and when his physiological values are steady. Moreover, it should be performed once a day, perhaps during the night.

Estimation algorithm RLS RW

From various algorithms in the literature, a very simple one known as the *recursive least square random walk* (*RLS RW*) was chosen: the small amount of computational work is suitable for the calculation capabilities of the microcontroller unit.

However, there is a solution to the many problems that can arise, such as low speed computation, low accuracy, difficulties of convergence. In order to obtain the parameter values of the controller, which minimize the covariance matrix of estimation error, sufficiently long input and output sequences must be observed.

Determination of the excitation wave

When the high level of logical Multiplexer in figure 2 is selected, the feedback control is active with the parameter borne in the previous estimation. In figure 2 yref is the basal value of blood glucose (95.5 mg/decilitre), i.e. the normal value in a healthy patient between meals. In agreement with control system theory, the noise that affects the blood glucose measurements can be placed in the output of the controlled system.

Between meals, for example during the night, the feedback control is locked and the estimation stage begins, which is necessary to update f1, f2, g, according to the changes in the patient's physiological condition. Therefore, the low level of logical Multiplexer is selected and consequently at the controlled system input a signal performed to stimulate the estimation is applied.

This stage involves three fixed conditions:

1) This stage must be done when the blood glucose values are steady.
2) The time duration must be short, because the patient is not under control.
3) A blood glucose value above 200 mg/decilitre cannot be tolerated.

Based on numerous attempts, a *pulse wide modulation* signal (*PWM*) has been chosen for the excitation signal of parameter estimation. The best shape of the signal is showed in figure 3.

The features are:

1) Time duration of single edge: 2 sampling periods
2) High edge level: 1.6 * (basal rate infusion of insulin)
3) Low edge level: 0

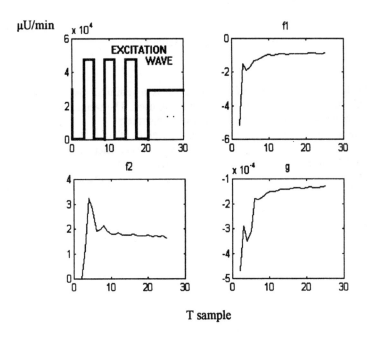

Fig. 3
Parameters of the trend

Figure 3 also shows the parameters of the trend.

The fast convergence of the parameters allows the process to be stopped at sample 6: after many attempts, this time period is the best value that has been obtained for a compromise between the hyperglycaemia, due to the excitation signal, and the convergence of the parameters. Therefore, the objective has been achieved because 6 samples are equivalent to 36 minutes and the patient is not under control for a short length of time during the estimation stage. It must be noted that the same simulations, performed in absence of noise, lead to a much slower parameter convergence. In other words, sensing noise helps to estimate parameters.

Control algorithm

Since a small computation power is available in the portable unit, the number of computations needed to calculate gli(k) must be very small. In the minimum variance regulators, the structure and the finite delay of the model are set previously.

Having obtained the parameters of the model by a least-square method, the controller parameters are calculated according to an optimal control strategy in order to minimize a performance index.

The performance index is:

$$Jk = \Delta gli(k+h)^\wedge 2 + Q0 * \Delta ins(k)^\wedge 2 \tag{4}$$

where h is the finite time delay of the model, Δ is a reminder that a model linearized around a set point and Qo is the weight factor being used. The weighting coefficient Qo is very important because it determines the therapeutic strategy.

In fact, raising its value means increasing the weight of infused insulin in the index J, and therefore reducing the amount of administered substance, at the price of higher tolerance in the blood glucose variations above the basal value.

The parameter Qo can also be used to compensate behaviour changes in the controlled system due, for instance, to changes of feedback control signal.

Thus, the control law obtained is

$$\Delta ins(k) = -(1/(g+Q0/g))[f1\Delta gli(k) + f2\Delta gli(k-1) + 2g\Delta ins(k-1) + \Delta ins(k-2)] \tag{5}$$

III. Simulations And Discussion

Simulations in a non-deterministic environment

Mathematical models were implemented in Matlab 4.0 for Windows and the simulations were performed as described below.
   T= 0: estimation of the parameters
   T=25: intravenous (IVGTT) infusion of 20g of glucose
   T=50: end of simulation

Since blood glucose sensors, based on enzymatic oxidation, introduce non-negligible noise, it was necessary to introduce an output noise in our simulations.

This noise was modelled upon real data sequences recorded from patients connected to Biostator, as the Biostator sensor has been widely tested by clinicians.

The statistical features of the noise have been found by evaluating the error between the values of a interpolation polynomial and the effective values recorded, and as a result an *additive white gaussian noise (AWGN)* has been obtained, with standard deviation approximately equal to 1.

Discussion on simulations

Figures 4 and 5 show the results of simulating the above test (parameter estimation followed by an IVGTT after 60 min.) on an "average"diabetic patient. It shows fast recovery to normoglycaemia.

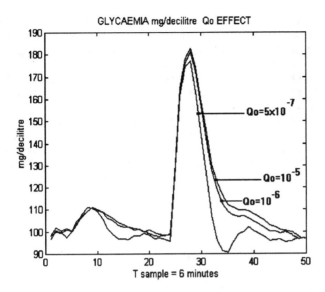

Fig. 4
Glycaemia, Qo effect

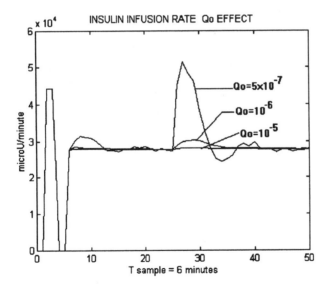

Fig. 5
Insulin infusion rate, Qo effect

The weight factor Qo present in the performance index J(k) plays a fundamental role in the control performance, ruling the balance between the amount of infused insulin and hyperglycaemia. Clinicians can choose a reduced insulin infusion or a reduced post-IVGTT hyperglycaemia by increasing or decreasing Qo, respectively.

Moreover, an increase of Qo decreases the controller gain, and therefore increases the stability on the steady-state error in basic conditions, whilst reducing the incidence of problems that typically arise in an auto tuning controller.

The parameter estimation is, in fact, sensitive to non-measurable disturbances, such as intestinal glucose uptake. In this case, estimation takes place by comparison between measured input and output of the controlled system. Since the output increase caused by such a disturbance is not being correlated with the input, it cannot be interpreted by the estimator but rather, it is the effect of a change in the parameters of the controlled system, mainly in its gain.

IV. Conclusions

The marketing of the first artificial pancreas (Biostator, 1976) created the illusion that within a short space of time a self-regulating insulin treatment system for clinical practice would be available. However, the completely automatic handling of a feedback-controlled metabolic system, such as blood glucose, is still beyond our reach.

Recent technological advancements such as the availability of rapidly absorbed insulin, inexpensive hand-held computers capable of elaborate calculation and the availability of biocompatible materials, have allowed many researchers to come closer to reaching this ambitious goal.

Nevertheless, present knowledge is still far from implantable substitute or portable organs that can carry out insulin-replacement therapy according to physiological needs.

To reach this goal, there are still many problems that must be solved, such as the lack of reliable biocompatible and long-lasting sensors, although recent technological advancements indicate that the portable devices for blood glucose control and other metabolites will be available in the not-too-distant future. Therefore, we have developed a control strategy that can be set easily in a low-cost wearable unit and be used easily and safely by clinicians.

The next step is to conduct *in vivo* tests on the auto-tuning controller and, above all, on the mathematical model of the patient.

V. References

1   Pirart J: Diabetes mellitus and its degenerative complications: a prospective study of 4400 patients observed between 1947 and 1973. Diabetes Care 1, 168-188, 1978

2.   Lauritzen T, Frost-Larsen K, Larsen HW, Deckert T, The Steno Group: Two years experience with continuous subcutaneous insulin infusion in relation to retinopathy and nephropathy. Diabetes 35, 74-79, 1985

3.   Brinchmann-Hansen O, Dhal-Jorgensen K, Hanssen K, Sandvik L and The Oslo Study Group: Effects of intensified insulin treatment on various lesions of diabetic retinopathy. Am J Ophthalmol 100, 644-653, 1985

4.   Gerich JE: Treatment of diabetes mellitus. In: De Groot LJ (Ed) Endocrinology, 2nd Edition. WB Saunders Company, Philadelphia, 1989, p.1424

5.   Pfeiffer Ef, Thum C, Clemens AH: The artificial beta cell. A continuous control of blood sugar by external regulation of insulin (Glucose Controlled Insulin Infusion System) Horm Metab Res 6, 339-342, 1974

6.   Kraegen EW, Campbell LV, Chia YO, Meler H, Lazarus L: Control of blood glucose in diabetics using an artificial pancreas. Aust NZ J Med 7, 280-286, 1977

7.   Slama G, Keiin JC, Tardieu MC, Tchobroutsky G: Normalisation de la glycemie par pancreas artificial non miniaturisé. Application pendant 24 jeures chez 7 diabetiques insulino-dependants. Nouv Presse Med 6, 2309-2312, 1977

8.   Selam JL, Charles MA.: Devices for insulin administration. Diab Care 1990; 13: 955-79

9.   Shichiri M, Kawamori R, Yamasaki Y, Hakui N, Abe H.: Wearable-type artificial endocrine pancreas with needletype glucose sensor. Lancet 1982; II: 1129-31

10. Shichiri M, Fukushima H, Sakakida M, et al.: Clinical application of wearable artificial endocrine pancreas. Problems awaiting solutions for long term use (Abstract). Med & Biol Eng & Comput 1991; 29: 358

11. Fischer U, Detschew W, Yutzi E, et al.: *In vivo* comparison of different algorithms for the artificial beta-cell. Int J Artif Organs 1984; 4: 794-800

12. Fabietti PG, Massi Benedetti M, Bronzo F, Reboldi GP, Sarti E, Brunetti P.: Wearable system for acquisition, processing and storage of the signal from amperometric glucose sensors. Int J Artif Organs 1991; 14: 175-8

13. Wellstead PE, Zarrop MB.: Self-tuning systems control and signal processing.Chichester: John Wiley, 1991: 147

14. Brunetti P, Cobelli C, Cruciani P, Fabietti PG, Filippucci F , Sarti E: A self-tuning portable feedback glycaemia controller. Med Biol Eng Comput 29 (Suppl 1), 892, 1991

15. Cobelli C, Mari A.: Validation of the mathematical models of complex endocrine-metabolic systems. A case study on a model of glucose regulation. Med & Biol Eng & Comp 1983; 21: 390-9

16. Cobelli C, Toffolo G, Ferrannini E.: A model of glucose kinetics and their control by insulin, compartmental and noncompartmental approaches. Math Biosc 1984; 72: 291-315

17. Carson E R, Cobelli C, Finkelstein L.: Mathematical modelling of metabolic and endocrine systems. New York, Wiley, 1983

Adaptive Glycaemia Control Using Neural Networks

R. Bafunno, C. Coltelli, P. Ciaccia, and Y. Takahashi

1. Introduction

1.1 Overview

The aim of our research was to study an adaptive glycaemia controller based on a Neural Network. Adaptivity means that the controller must be able to adjust its coefficients to follow the alterations in the dynamics of the diabetic patient during the whole day (e.g. after meals), without having to retune the device frequently, letting the controller behave in the most similar way to the human pancreas.

Moreover, since this controller is implemented into a wearable unit, the patient can live an almost normal life, not having to recourse to the usual insulin injections or to day-hospitals.

The adaptivity is here obtained using a Neural Network (NN).

1.2 Why using a Neural Network?

The NN major property, here used, is the capability of reproducing non-linear mappings after an appropriate training phase; and particularly the capability of reproducing the functions that could be found in the state equations of the controlled non-linear system, such as the diabetic patient model in our study.

The NN, suitably trained, is inserted in a feedback loop, in order to achieve a kind of pole-zero cancelation technique.

A hard difficulty related to this technique is the *training* of the NN, i.e. a phase in which the coefficients of NN interconnections are adjusted in order to minimize an error function obtained from outputs of the network and a reference model, when they are subject to the same input signal. In fact, this procedure can occasionally require much time, and sometimes it never reaches acceptable error values.

2. A mathematical model of the controlled system

First we have to analyze the mathematical model of the diabetic patient, which is represented in equations 1(a,b,c):

$$\begin{cases} \dot{G}(t) = -(p_1 + X(t)) \cdot G(t) + p_4 \\ \dot{X}(t) = -p_2 X(t) + p_3 I(t) \qquad (1a) \\ \dot{I}(t) = -n I(t) + u(t) \end{cases}$$

$$\begin{aligned} G(0) &= G_0 \\ X(0) &= I(0) = 0 \end{aligned} \qquad (1b)$$

$$\begin{aligned} p_1 &= 0.0054 \ \text{min}^{-1} \\ p_2 &= 0.0329 \ \text{min}^{-1} \\ p_3 &= 0.0178 \ \text{min}^{-2} \cdot \text{U}^{-1} \cdot \ell \qquad (1c) \\ p_4 &= 1.7 \ \text{mg} \cdot d\ell^{-1} \cdot \text{min}^{-1} \\ n &= 0.077 \text{min}^{-1} \end{aligned}$$

We developed this non-linear, 3rd order model on Cobelli-Bergman minimal model that appears in the first two equations of (1a). The third equation we added represents the kinetics of exogenous insulin infusion (i.e. subcutaneous injection).

This model can be schematized as a cascade of three first order blocks, the last of which is the non-linear block, as seen in fig. 1.

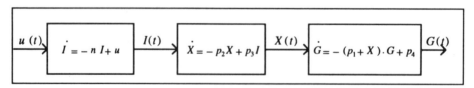

Fig. 1: The model seen as a cascade of three first order blocks, last of which is non-linear.

The input for the overall system is therefore the insulin infusion rate, while its output is the blood glucose concentration.

The physical meaning of parameters and variables of such model is the following:

G is blood glucose concentration (glycaemia);

X is insulin concentration in a compartment remote from plasma;
I is blood insulin concentration;
u is insulin infusion rate;
p_1 is insulin-independent glycaemia decay;
p_4 is hepatic glucose production;
p_2 is coefficient of insulin disappearance from X;
p_3 is coefficient of insulin exchange between X and blood;
n is coefficient of blood insulin decay;

2.1 Discrete-time model

In order to build a digital controller, we always need an equivalent discrete-time model of the controlled system. Hence we performed a separate discretization of linear and non-linear blocks.

To discretize the linear part we followed four standard steps that are
1. Differential equations;
2. Laplace transform;
3. z-transform;
4. Difference equation;

The result was a linear, 2nd order, discrete-time equation:

$$X(k+1) = \alpha X(k) + \beta X(k-1) + \gamma u(k)$$
$$\alpha = e^{-p_2 T_s} + e^{-n T_s}$$
$$\beta = -e^{-(p_2+n)T_s} \tag{2}$$
$$\gamma = \frac{p_3}{p_2 - n}\left(e^{-n T_s} - e^{-p_2 T_s}\right)$$

There are not standard procedures to perform the non-linear block discretization, therefore we proceeded first with an integration of the non-linear equation; after that we sampled the resulting time function; then we did some approximation to finally obtain this non-linear, 1st order, difference equation:

$$G(k+1) = e^{-p_1 T_s} e^{-T_s \frac{X(k+1)+X(k)}{2}}\left[G(k) + p_4 T_s\right] \tag{3}$$

Now we want to combine the two previous results in order to obtain an overall discrete-time 3rd order equation. To do this we must introduce another approximation. As a matter of fact, it is not right to consider the discretization of a two blocks cascaded system as the cascade of the corresponding discretized blocks, if sample time is not sufficiently small with respect to the dynamics of signal between the two blocks. If we choose a sample time equal to 30 seconds, this approximation results to be quite good. The subsequent equation is:

$$G(k+1) = \left[G(k) + p_4 T_s\right]\left[\frac{G(k)}{G(k-1) + p_4 T_s}\right]^{\alpha} \cdot$$

$$\left[\frac{G(k-1)}{G(k-2) + p_4 T_s}\right]^{\beta} e^{-p_1 T_s(1-\alpha-\beta)} e^{-\gamma T_s \frac{u(k)+u(k-1)}{2}}$$

(5)

This means that the overall system is ruled by an equation like:

$$G(k+1) = f\left[G(k), G(k-1), G(k-2); u(k), u(k-1)\right]$$

in which the future output is a non-linear function of 3 samples of the output itself and 2 samples of the input signal.

With a simple mathematical trick, we can simplify the discrete-time equation. As a matter of fact, if we consider the logarithm of normalized glucose concentration as plant output, we can obtain the following equation:

$$y(k) = \ln\left[\frac{G(k)}{G_0}\right]$$

$$y(k+1) = \ln[e^{y(k)} + \frac{p_4 T_s}{G_0}] + \alpha y(k) - \alpha \ln[e^{y(k-1)} + \frac{p_4 T_s}{G_0}] + \beta y(k-1) +$$

$$-\beta \ln[e^{y(k-2)} + \frac{p_4 T_s}{G_0}] - p_1 T_s(1 - \alpha - \beta) - \frac{\gamma T_s}{2} u(k) - \frac{\gamma T_s}{2} u(k-1)$$

(6)

in which the future value of output is the sum of a non-linear function of past values of output and a linear function of past values of input variable. In this form, we found a quite simple way to identify and control the plant.

$$y(k+1) = f\left[y(k), y(k-1), y(k-2)\right] + \lambda_1 u(k) + \lambda_2 u(k-1)$$

3. Identification Steps

For the identification, we used a model based upon a structure-oriented neural network, that is a network whose structure is not fully connected; therefore it is not a general purpose one, but a special purpose one, that somehow follows the plant equation structure.

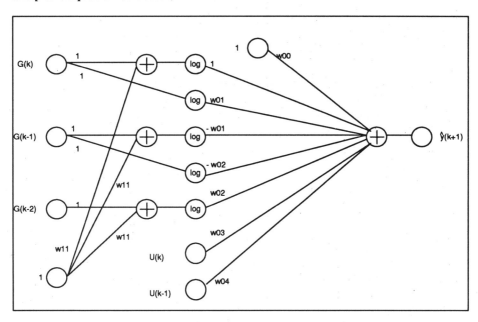

Fig. 2: The structure-oriented Neural Network.

Here we see its output equation

$$\hat{y}(k+1) = \ln\left[\frac{G(k)}{G_0} + w_{11}\right] + w_{01}\ln\left[\frac{G(k)}{G_0}\right] - w_{01}\ln\left[\frac{G(k-1)}{G_0} + w_{11}\right] +$$

$$-w_{02}\ln\left[\frac{G(k-1)}{G_0}\right] + w_{02}\ln\left[\frac{G(k-2)}{G_0} + w_{11}\right] + w_{00} + w_{03}u(k) + w_{04}u(k-1) \tag{7}$$

As it can be seen from a comparison between network output (7) and plant output equations (6), there is a clear correspondence between the weights of the network and the coefficients of plant equation, as shown in table 1, so the identification process tries to adjust network weights until they well approximate the expected values.

Table 1

Weight	Corresponding value
w_{00}	$-p_1 T_s(1-\alpha-\beta)$
w_{01}	α
w_{02}	β
w_{03}	$-\gamma T_s/2$
w_{04}	$-\gamma T_s/2$
w_{11}	$p_4 T_s/G_0$

The scheme we used for the identification is shown in fig. 3. The input signal is a random pulse wave corresponding to random insulin infusions. Such a signal cannot be used directly on-line, that is on the patient, because it can be somehow harmful for him.

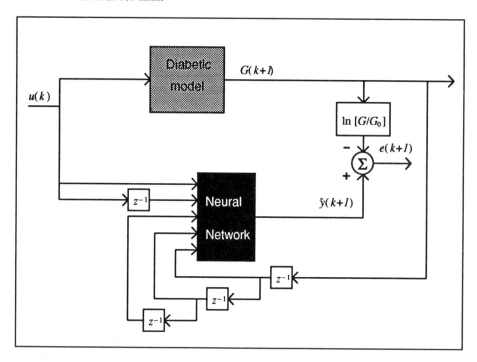

Fig. 3: Plant identification scheme using the neural Network.

This process can be considered complete after a short training period of about five hundred steps, and we can see that the waveforms of the output of the plant and of the identification model are perfectly overlapped. Moreover, the identification error falls under 0.001%, as you can see from figs. 4a and 4b.

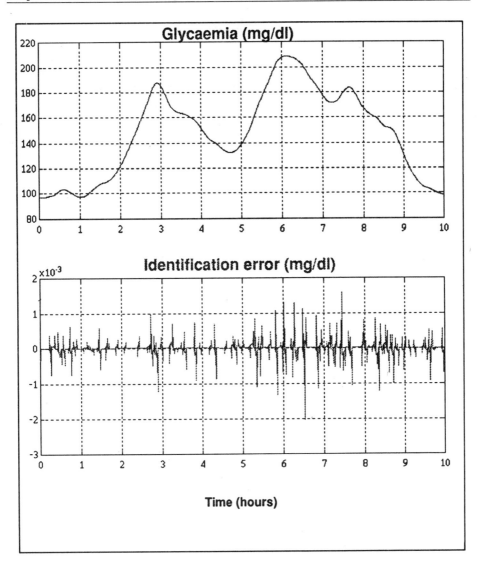

Fig. 4a : Time courses of plant output and NN output are completely overlapped.
Fig. 4b: Time course of identification error.

4. Control

Before explaining how our control system works, it is better to introduce the technique called *feedback linearization*. If the plant equation has the form

$$y(k+1) = f\big[y(k), y(k-1), y(k-2)\big] + \lambda_1 u(k) + \lambda_2 u(k-1),$$

and the corresponding neural network identification model is ruled by the following equation:

$$\hat{y}(k+1) = \hat{f}[y(k),y(k-1),y(k-2)] + \hat{\lambda}_1 u(k) + \hat{\lambda}_2 u(k-1)$$

where $\hat{f}(.)$, $\hat{\lambda}_1$ and $\hat{\lambda}_2$ are respectively the estimations of the function $f(.)$ and the coefficients λ_1 and λ_2, then we can build a control signal

$$u(k) = \frac{r(k) - \left\{\hat{f}[y(k),y(k-1),y(k-2)] + \hat{\lambda}_2 u(k-1)\right\}}{\hat{\lambda}_1},$$

where $r(k)$ is a suitable reference input. In this way, the plant output will be

$$y(k+1) = \frac{\lambda_1}{\hat{\lambda}_1} r(k) + \frac{\lambda_1}{\hat{\lambda}_1} \left\{ f[y(k),y(k-1),y(k-2)] + \right.$$
$$\left. -\hat{f}[y(k),y(k-1),y(k-2)] + (\lambda_2 - \hat{\lambda}_2) u(k-1) \right\}$$
$$\Rightarrow y(k+1) \approx r(k)$$

and, therefore, if the identification process has reached a good level of approximation, we can say that the future sample of the output approximates the actual value of the reference input signal. Hence the term feedback linearization.

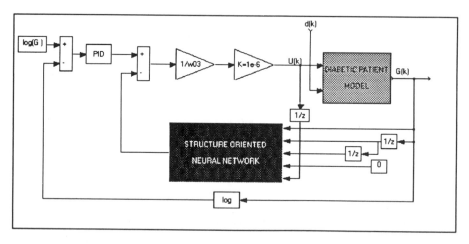

Fig. 5: Control scheme using Feedback Linearization.

Now, let us examine the way we build an adaptive control system. First, we had an inner feedback linearization loop, as it can be seen in fig. 5, but unfortunately it resulted unstable. Therefore we must insert a gain block with value K set to 10^{-6}. On the other side, this block caused an imperfect linearization of the system: consequently, we closed the overall system in a PID loop that completes the control action.

Such loop is needed in order to keep the logarithm of glycaemia close to the logarithm of the basic glycaemia (that is 97 milligram per decilitre), even in presence of disturbances.

5. Simulation results

The disturbance we used in the simulations is an intra-venous glucose injection, analogous to IVGTT (intra-venous glucose tolerance test).

In order to achieve a valid comparison term, let us examine the disturbance response of the system (fig. 6), without control action, i.e. with only basic insulin infusion (that means the infusion needed to keep the glycaemia at the constant basic value, in absence of disturbance).

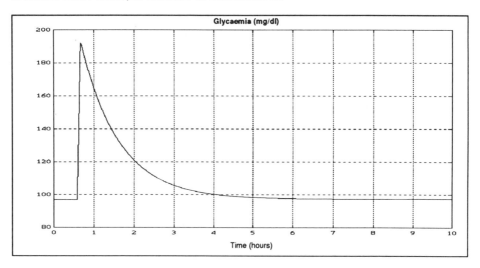

Fig. 6: Disturbance response with only basic insulin infusion.

In about 5 hours, we can notice a slow return to the basic value, due to the lack of a system able to release the insulin necessary to metabolize the excess of glucose.

On the contrary, when the system is under control the response is considerably quicker than the previous case, and the recovery time falls under 1 hour and a half, as it is shown in fig. 7. We can also notice a little undershooting that, however, never reaches hypoglycaemia.

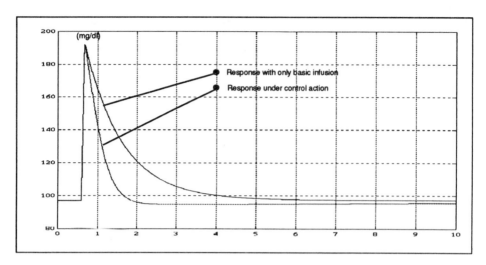

Fig. 7: Disturbance response of the controlled system.

To proof the adaptivity of our control system, we changed a parameter in the
diabetic patient model (for example we could reduce its gain to half) and then
performed a short retraining of the neural network on the new plant model.
The major result is that the behaviour of controlled system does not change, as
far as the controlled variable is concerned, if and only if the neural network is
retuned over the new set of plant parameters.
In the last picture we can see what happens if we change the plant gain but do
not retrain the network: the corresponding disturbance response is different
from the previous case and it is about 40 minutes longer.
Whereas, if we retrain the network, the subsequent disturbance response
overlaps again with the first case.

6. Conclusions

At this point we can draw some important conclusions:
First, we can affirm that the neural network works well in our adaptive control
scheme. The presence of a little undershooting in the disturbance response, due
to the imperfect nonlinearity elimination, is not a serious problem.
What is worth to emphasize is that the parameters of PID controller and the
value of gain K can be changed by the physician in order to follow some
suitable therapeutical strategy on the diabetic patient.
There still remains some open problems such as the search for a training signal
which should be harmless for the patient and, at the same time, really effective
for the neural network training.
Furthermore, we will have to implement the control algorithms into a wearable
unit, that is just the ultimate goal of the research we started here.

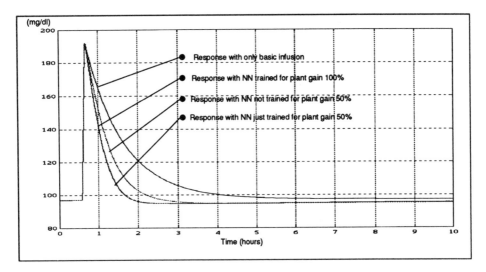

Fig. 8: Disturbance response of the controlled system in all the four cases.

7. Bibliography

P.G. FABIETTI, F. FILIPPUCCI, P. CRUCIANI, E. SARTI, A Computer study on feedback control of glucose concentration through subcutaneous sensors, *Proc. VI Mediterranean Conf. on Medical and Biological Engineering*, Bracale-Denoth Eds., p. 231-234, CNR, Pisa 1992.

K. J. HUNT, D. SBARBARO, Neural Networks for Control Systems – a Survey, *Automatica*, **vol. 28**, no. 6, p. 1083-1112

K. S. NARENDRA, K. PARTHASARATHY, Identification and Control of Dynamical Systems using Neural Networks, *IEEE Trans. on Neural Networks*, 1990.

K. WARWICK, G. IRWIN, K. J. HUNT, *Neural Networks for Control and Systems*, Peter Peregrinus, London 1992.

Y. TAKAHASHI, Adaptive Predictive Control of Nonlinear and Time-varying Dynamical Systems using Neural Networks, Bologna 1992 (priv. not. n. p.).

Neuromuscular Remodelling During Development and Aging

M.A. Fahim

During an animal's life span, the neuromuscular junction undergoes remodelling from early development through maturity, where hormones and activity may be important until aging introduced new perturbations. In this section, we have grouped together 4 chapters dealing with different aspects of nerve and muscle remodelling.

In the first chapter, structural and functional remodelling that occurs during aging is discussed. The chapter's conclusions indicate that acute disuse, exercise and aging affect the morphology and physiology of the neuromuscular junction. Despite these findings, synaptic transmission seems well preserved at aged-neuromuscular junctions.

In the second chapter, experimental data is presented to demonstrate that small and short-lasting variations in the locomotor activity of the normal adult rat within a physiologic range can result in increased remodelling changes at the neuromuscular junction. Specifically, the complexity finally attained by motor nerve endings can be inversely related to the level of usage of the junction. Main changes affect the linear extension of the distally placed free-end segments, whereas the segments between branch points are stable. In animals trained to walk, motor nerve ending adaptation occurs along with a rearrangement of the pre- and postsynaptic membrane structural differentiations.

In the third chapter, experiments dealing with reinnervation after denervation are presented to demonstrate the specificity of innervation of the tibialis anterior muscle of the cat after a complete lesion of a peripheral nerve. Additional evidence is also presented to show that the ability of a nerve to change the properties of a muscle fiber may relate to the state of the fiber at the time of the reinnervation.

In the final chapter there is an extensive overview of the fast-progressing field of molecular biology related to steroid hormones and nerve-muscle remodelling. It is known that anabolic-androgenic steroids, such as testosterone, can alter the molecular, biochemical and physiological characteristics of cells that express androgen receptors. Both skeletal muscle fibers and motoneurons express androgen receptors. Most studies examining the effects of anabolic-androgenic steroids on mammalian neuromuscular systems have focused on skeletal muscle. In this chapter we examine the effect of anabolic-androgenic steroids on the expression of molecules involved in intercellular communication in lumbar spinal cord motoneurons. *In situ* hybridization histochemistry was used to investigate how changes in the circulating levels of testosterne alter the expression of the mRNAs which encode α-calcitonin gene-related peptide (α–CGRP) and choline acetyl transferase (ChAT) in motoneurons. α–CGRP is an important effector molecule with both modulatory and trophic actions at the neuromuscular junction; and ChAT is the synthetic enzyme for the impulse-carrying neurotransmitter acetylcholine. The castration-induced decrease in serum testosterone levels increase α–CGRP mRNA levels and decrease ChAT mRNA levels in motoneurons of the spinal nucleus of the bulbocavernosus. In other lumbar motoneurons castration decreased ChAT mRNA but not α–CGRP mRNA levels. Supraphysiological levels of plasma testosterone did not alter significantly the expression α–CGRP mRNA levels in any motoneurons. In contrast, ChAT mRNA levels increased dramatically in motoneurons throughout the lumbar spinal cord. These experiments indicate that the anabolic-androgenic steroid regulation of motoneuronal α–CGRP mRNA levels is mediated by muscle-nerve interactions, while the regulation of motoneuronal ChAT mRNA levels appears to be mediated through the interaction of anabolic-androgenic steroids with motoneuronal androgen receptors.

Synaptic Plasticity During Aging of the Neuromuscular Junction

M.A. Fahim

1.   Introduction

Of the many physiologic events that occur with senescence, perhaps the most apparent change is the universal decrement of strength and locomotor co-ordination. What was once a simple joy of lifting a child has now become a major, if not unrealistic task to accomplish.   Such an age change is a multi-faceted phenomenon in origin, that involves not only muscle but also central and peripheral nervous systems.

Remodeling of the neuromuscular junction (NMJ) appears to be a lifelong process, as indicated by indirect evidence of continual degeneration and regeneration. Several investigators have suggested that these dynamic changes in the morphology and physiology of the aging neuromuscular junction may be reflected in age-related variations in physical activity [2, 3, 4, 5, 27]. Indeed, hypokinetic studies involving inactivity or experimental disuse in young animals produces degeneration and regeneration of the neuromuscular junction [10, 21, 42] similar to those observed during aging [5, 6, 7, 8, 11, 15, 18, 20, 44, 48, 49, 51, 52]. However, the net morphological and physiological consequences of this dynamic remodeling throughout life have received little attention.

It is therefore of paramount importance to isolate the aging changes from those of the pathological ones in the hope of finding the cellular basis for the above-mentioned dynamic remodeling that occurs at the neuromuscular junction.   The research on mechanisms of age-related changes in rodent neuromuscular junction may shed light on the observed ongoing plasticity, and present a possibility of how they can be enhanced and, perhaps, applied to human aging.

2.   Scope

It was previously suggested that age changes of the neuromuscular junction may reflect altered physical activity levels rather than the unique effects of aging.

Both qualitative and quantitative analyses of presynaptic nerve terminals of old animals show decrease in area, mitochondria and synaptic vesicle density. On the post-synaptic side there are increases in complexity of junctional folds and occasional denervated regions. Physiologically, an increase in evoked transmitter release has been observed.

A decreased number of fast motor units has been reported together with significant decrease in conduction velocity of most motor units. Changes in muscle excitation-contraction coupling are also reported but the changes in animal activity level rather than from aging are thought to be the primary cause. Moreover, the usual laboratory rat, grows obese and inactive with age, due to age-related degenerative, inflammatory and neoplastic lesions.   However, successful compensation in old humans and animals does take place to maintain the integrity of the neuromuscular junction.

3.   Animal Model

Selection of the animal is an important step to the research model. Choosing a species should be the first priority.   This usually relies heavily on how widely the species is used, the type of background information available, the cost, convenience of use and above all, the replicability of results.   Also, of equal importance is the genetic background of the chosen strain to minimize the sample size, decreased individual variability and maximize the power to detect the real age-changes.

One would hope to avoid an animal model which would grow obese and inactive with age since this would contribute to degenerative, inflammatory and neoplastic lesions. The optimum is to characterize the age-related changes in any animal which maintains locomotor activity throughout the lifespan and particularly free of age-related organ pathology. The choice of the F.1., hybrid offers superior animal model because of its less variability and because it also fulfills the above requirements.

4.   Techniques Used

Electrophysiological techniques occupy a strategic position in the evaluation of central or peripheral nervous system alterations during aging. Levels of analysis range from evaluation of specific ion channels through gross potential phenomena very closely related to cognitive function. Even with relatively gross recordings, a temporal resolution of function is possible that extends far beyond that addressable by neurochemical or anatomical techniques. Despite this, the data base available using neurochemical and neuroanatomical techniques is far larger.

The most productive areas at the present time are those neuronal systems that are well understood with respect to their anatomical organization and synaptology, and have a favorable architecture for interpretation of the electrophysiological recordings. Therefore, this review will discuss the plasticity of NMJ and skeletal muscles during aging, studied by both morphological and physiological techniques.   Although work in these areas, by a relatively small number of investigators, has already produced a coherent database in which data can be interrelated.

5.   Morphological Changes

While aging affects the axonal conduction of the nerve impulse signal to muscle, there are also morphological changes that occur at the point where the nerve innervates the muscle fibers - the neuromuscular junction (NMJ). Human endplates of external intercostal muscle, stained with a silver-cholinesterase, maintained the same size and number of nerve endings of the axon and showed no sprouting of terminal axons with age.   However, at the ultrastructural level complex, secondary synaptic clefts, with increased irregular shape of nerve terminals and Schwann cell processes intruding into the primary synaptic cleft,

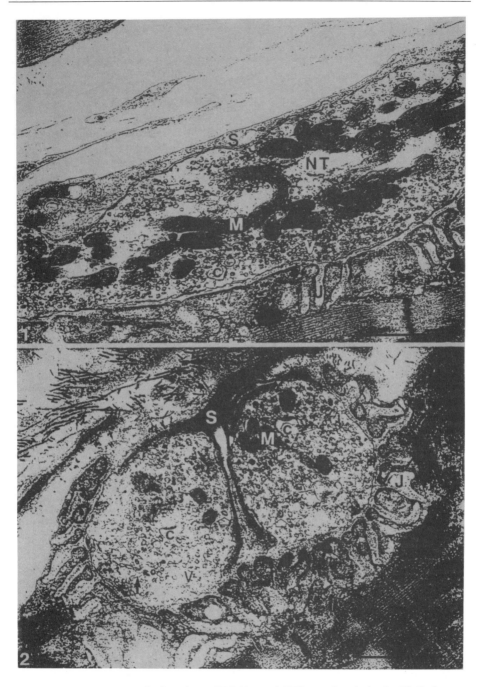

Figures 1 & 2 Neuromuscular junctions of 7 & 29 month EDL muscle. A single elliptical nerve terminal (NT) lies in a depression of the sarcoplasmic surface and is covered by a thin lid of Schwann cell cytoplasm (S).   The nerve terminal is filled with synaptic vesicles (V) and mitochondria (M), with only a single cistern (c). J, functional folds. The aged nerve terminal contains fewer synaptic vesicles and mitochondria (compare with Fig. 1).   There is an increase in the cisternae and of profiles of smooth endoplasmic reticulum (black arrows). The Schwann cell protrudes between two nerve terminals.   There is also increased branching of the junctional folds (compare with Fig. 1). Open arrows point to coated vesicles, which are increased. Scale bar: 1 μm.

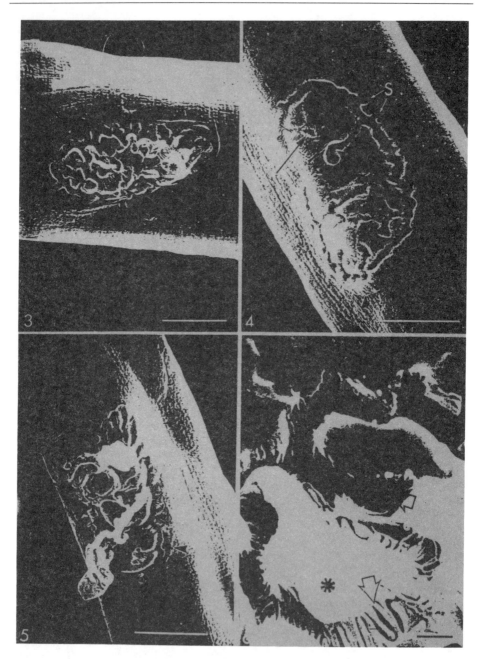

Figures 3 -6. Scanning electron micrographs of endplates of 7-month EDL muscles. Note smooth-surfaced 'raised area' (borders marked with arrowheads in these and subsequent figures); side-branches (s) of primary clefts; insular bulges in raised areas (asterisks); and isolated oval primary cleft region (arrow). Note remnants of intrasynaptic nerve terminals (clear curved arrows in Fig. 5). Scale bar: 10 µm. A higher magnification (Fig.6), showing the arrangement of secondary folds (clear arrows) and insular bulge in raised area (asterisk). Scale bar: 1 µm.

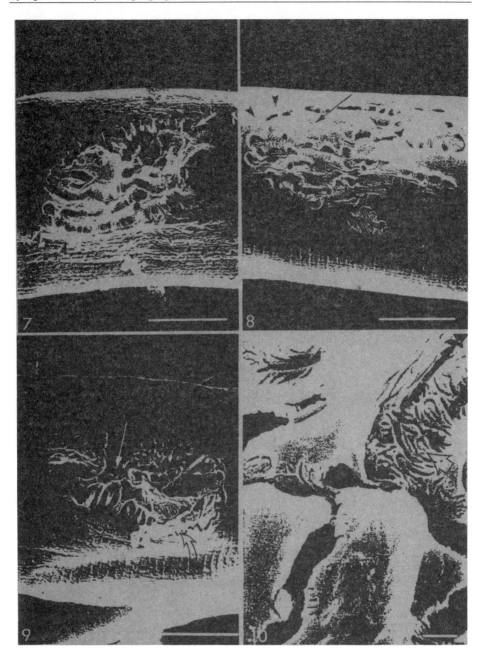

Figures 7 - 10. Scanning electron micrographs of 29-month EDL muscles illustrating U-shaped (short arrow) and concave (long arrow) primary cleft regions (Fig. 7). A scanning electron micrograph showing nerve terminals attached to the outpouching regions of primary cleft (clear curved arrow) which characterized the aged neuromuscular junction (Fig. 8). The short black arrow denotes the u-shaped primary cleft concavity (Fig. 9). Scale bar: 10 um. Scanning electron micrograph of 29-month EDL muscle at higher magnification (Fig. 10), showing an increased randomness of secondary fold orientation (clear arrow) compared to that in Fig. 6. Scale bar: 1 μm.

were evidenced in aged NMJ.

Secondary synaptic clefts in elderly subjects increased branching of junctional folds, degenerating folds and frequently denuded folds. A minor degree of muscle fiber type grouping was observed, which could be explained by a slow process of denervation followed by reinnervation rather than a loss of motor neurons with age. In NMJ's of old mice, compared to young mice, there were fewer mitochondria, fewer synaptic vesicles and more smooth endoplasmic reticulum (SER), coated vesicles, cisternae , neurotubules and neurofilaments (Figs. 1 and 2) [20].   In general, the changes were similar but more pronounced in soleus (SOL) than in extensor digitorum longus (EDL). The diaphragm, on the other hand, showed considerably less change with age [11].   Therefore, aging at NMJ's varies in the same animal with the muscle type but not necessarily with activity patterns [43].

6.   Physiological Changes

The profound ultrastructural changes contrast with the physiological findings that nerve terminals at old EDL and SOL neuromuscular junctions release more transmitter than in young animals [2, 3, 11].   Two hypotheses are plausible:

1.   Increased transmitter release leads to depletion of synaptic vesicles and other hallmarks of rapidly driven synapses (increased cisternae, SER and coated vesicles reflecting a higher turnover of vesicles, [18]) or

2.   Fewer synaptic vesicles may be produced and maintained in old terminal.

As a compensatory mechanism, there is faster recycling of the fewer vesicles present - hence the relative abundance of coated vesicles, SER and cisternae linked with vesicle recycling (Fig. 2).   However, this process would be strained by the capacity of the reduced number of mitochondria to supply adequate energy, and it is not obvious why transmitter release should in fact be increased. Therefore, an additional hypothesis is suggested. Calcium in old nerve terminal may be stored in the increased numbers of SER, cisternae and coated vesicles as well as in the remaining mitochondria. If this compensatory storage were not adequate, so that free Ca^{2+} would be sequestered less rapidly in old nerve terminals, higher levels of intracellular free Ca^{2+} would be present, especially following stimulation. This would lead to increased transmitter release as observed.   At the very least, the quantitative correlation of age-related changes in morphology and physiology points to some remarkably effective compensatory mechanisms for further investigation.

Further changes at the NMJ level also include an augmented proliferation of post-synaptic receptors i.e., acetylcholine receptors (AchR), in aging animals. The remodelling of the NMJ by increasing the number of myelinated axon branches entering a junction (collateral sprouting) and the rise of AchR with age suggests a possible attempt to re-innervate inactive muscle fibers. [44, 45, 46, 48]. Such NMJ plasticity and AchR proliferation is characteristic not only of aging rats but of disuse [21] and post-natal development as well.   This suggests a possible compensatory mechanism by re-innervation for a proportional loss of fast-twitch fibers and partial denervation through inactivity and indicates that aging (like disuse) similarly effects neuronal activity which may subsequently affect fiber type distribution and ultimately the dynamic properties of muscle [15, 20]. At the nerve terminal level, quantal content of transmitter release increases with sedentary aging [2, 3, 6, 48, 49].

Ultimately a single or a combination of pre-synaptic and/or post-synaptic process(es) must be responsible for the observed increase in quantal content at old terminals. First, the quantal content of the end-plate potential is consistently increased, regardless of the quantum size [2, 3, 6, 48]. Second, the greater quantal content during aging is observed, whether or not muscle input resistant (R) is increased and whether or not acetylcholinesterase activity is reduced [2, 3, 6, 48, 49, 51]. Third, post synaptic junction AchR's in mice are not increased in number despite the increased quantal content during aging [11]. However, it is unknown whether properties of EDL and SOL AchR's change during aging. For instance, in the rat diaphragm(phasic fast-twitch, vital muscle), low-affinity binding of AchR's increases during aging [48]. Fourth, nerve terminal expansion (Figs. 3-10) does not account for the increased quantal content during aging [18, 19, 20]. Indeed, a comprehensive study of EDL and soleus nerve terminals at eight different ages (4-32) has revealed that remodelling of nerve terminals is a dynamic but not a progressive process, with no net trend towards larger nerve terminals with advancing age [5]. Fifth, after tetanic nerve stimulation, free intraterminal Ca^{2+} is higher in old than in young mouse soleus nerve terminals [3]. It is believed that this difference is mainly due to an increase in the voltage dependent Ca^{2+} clearance uptake by organelles and/or extrusion by transport system may or may not be involved.

Increased Ca^{2+} influx has been suggested to be responsible for the observed increase of quantal content during aging [2]. Increased re-cycling of synaptic vesicles may also be involved (Fahim unpublished results). Morphometric analysis of aged nerve terminals has revealed an increased number of coated vesicles (49% for EDL, 135% for soleus) which are thought to be involved in transmitter recycling [18]. Moreover, acetylcholine efflux has been reported to increase with high choline uptake at aged nerve terminals [49].

7. Skeletal Muscles

Thus far, no normal skeletal muscle demonstrates universal, morphological, histochemical and dynamic changes that correlate with senescence rather than disease. However, the rate of such alterations varies widely, and is most likely a function of the activity levels of an individual.

Morphologically, muscle atrophy is the most conspicuous change to impair motor performance with age [9]. Muscle mass is usually associated with isometric tension development and an attenuation of mass would then suggest a decrement of strength [24, 38, 39, 40]. Previously, it was believed that such senile atrophy was the result of some myopathic or neuropathic disorder. However, more recent research suggests that such atrophy may be the consequence of reduced neural activity with progressive disuse or increased immobility [26, 27]. Further, other factors affecting atrophy may include functional denervation concomitant with a chronic decrement of neurotransmitter release and, possibly, neurotrophic agents which are, as yet, unidentified. Along with a diminution of mass, a loss of elasticity and an increase of connective tissue within the muscle may also contribute to locomotor and strength loss with age [23].

Senile atrophy, or the decrement of muscle mass during aging, may be accounted for by three factors - a decline in the number of muscle fibers, a decrease in fiber size or a combination of both. Animal studies have indicated a range from 25-40 percent in the reduction in muscle fiber number. In human studies, a decline in the number of functioning motor units has been determined electrophysiologically, suggesting a similar decline of muscle fiber number with age [13, 23, 39]. However, due to severe methodological limitations, direct measurement in humans is not practical, leaving only indirect approaches available, i.e., tomography and muscle biopsies.

With regard to size, fiber diameter has actually been observed to increase with senescence in animals [9]. This may be a response to compensate for loss of fibers. In human studies, however, it has been demonstrated that muscle fibers reach their maximal size by the third or fourth decade and then subsequently decline. Of particular interest is that fiber atrophy tends to be manifested in type II or fast twitch fibers, while type I fibers are relatively unaffected [24]. In fact, by the seventh decade, type II fibers, which are normally larger power-generating fibers twice the diameter of their type I counterparts, are nearly identical in size. By the middle of the eight decade, the area of type II fibers may be as little as 50 percent that of type.

Histochemical alterations that occur with aging are much more subtle than morphological ones but they present profound implications for the dynamic properties of muscle [9]. Originally based on myoglobin content, muscle fibers are classified as either red or white. Additional electrophysiologic analysis of the dynamic properties of these fibers has revealed that red fibers contract more slowly and exert less force than white fibers. Further, they have a lower threshold for activation. However, red fibers are more fatigue-resistant or, in other words, have a high endurance capacity. With more advanced histochemical techniques and biochemical analyses, enzymatic differentiation between fibers has been determined. Slow muscle fibers have high levels of oxidative enzymes such as succinate dehydrogenase (SDH) while fast (type II) have high levels of glycolytic activity. Myofibrillar ATPase, and enzyme responsible for the hydrolysis of ATP to produce energy, can also be differentiated between type I and II fibers. Type I fibers also have a high density of mitochondria.

Post-natally, much of skeletal muscle is composed of slow or intermediate fibers, which have characteristics of both type I and II. With growth and development these fibers differentiate more specifically to type II [1]. With advancing age however, the metabolic differences between fibers becomes less distinct in laboratory animals [38]. The glycolytic-to-oxidative enzyme ratio, for example, declines in fast twitch muscle. Histochemical staining for SDH or ATPase is also less differentiated between fibers [47] In terms of dynamic properties, fast muscle appears to get slower while slow muscle tends to become faster. As mentioned in the morphological studies, type II fibers appear to decrease selectively in size and percentage. Type I fibers tend to increase in number but not in area. Human muscle biopsy studies have been consistent with this trend. The percentage of fast fibers diminishes linearly from the mid-thirties to the seventies in man. It is interesting to note that this selective atrophy of type II fibers with aging is also observed with chachexia, inactivity and denervation, all conditions which have some degree of resemblance to senescence.

Electrophysiologically, membrane resistance and capacitance increase with aging in EDL rat muscles. The change in membrane resistance is due to a decrease of chloride conductance which increased the latency of action potential and consequently prolongation of the action potential duration [13, 16, 17]. However, age changes in fast-twitch muscle is not universal phenomenon since no alterations are reported in contractile, metabolic and histochemical properties of fast flexor digitorum longus muscle of Fishcer 344 rat [50] nor any change in the proportion of myosin type in relation to age. Variation of the fast and slow myosin light chains results mainly from the changes in activity level rather than from age [41]. Finally, changes in excitation-contraction coupling parallel the prolongation of contractile times observed during aging in mammalian skeletal muscle which is consistent with the known reduction in rate and extent of Ca^{2+} uptake by sarcoplasmic reticulum in aged rats [17].

8.   Exercise Modulation of Age-Related Changes

It is important to distinguish between aging and functional disuse. Muscle undergoes a reduction in size, and consequently strength, with aging, which is related to a selective loss of muscle fibers. These changes result in decreased functional capacity and quality of life. A substantial portion of this decrease is the result of sedentary life style and not aging per se. An appropriate exercise program can result in increased strength and endurance [22, 32].

Human volunteers (aged 60-72 years old) have increased their muscle strength and protein turnover after exercise [22]. Interestingly, following strenuous exercise, the amount of damage and capacity for repair were the same among young male 24 and 67 years old volunteers [33].

In rodents, endurance exercise prevents the age-related decline of creatine phosphokinase and other enzyme related to mitochondrial energy generation [29, 30, 31, 32, 33, 34]. The positive effect of exercise to meager muscle glycogen during sustained exercise is not lost with age and, in fact resistance exercise initiated in old rats has resulted in hypertrophy of wrist flexor muscles [32].

Adding endurance exercise to the aging condition has prevented the age-related increase of quantal content in the EDL but not in the soleus [2]. The different physiological responses of these muscles to exercise parallel their morphological responses, recently reported from the same laboratory for exercised 24-month C57BL/6NNia mice [4]. In the EDL, nerve terminal area, nerve terminal perimeter and nerve terminal branching all decrease after exercise [4]. By contrast, in the soleus, there is no change in nerve terminal area of perimeter while nerve terminal branching decrease after exercise [4].

Factors such as hormones, trophic substances or blood supply might play a role in the exercise-induced reduction of quantal content. However, the role of nerve activity has been clearly demonstrated; quantal content in frog cutaneous pectoris muscle was reduced after prolonged *in vivo* nerve stimulation [14, 25]. Hence, the different effects of exercise on EDL and soleus muscles could be due to their characteristic daily activities *in vivo*. EDL motor units are active only 0.04-5.00% of the total time, while soleus motor units are active 22-35% of the total time [8, 13]. Perhaps, the difference in soleus nerve impulse activity between exercised and old animals was not sufficient to induce a modification which would reduce quantal content. This study demonstrates that the mouse NMJ undergoes a process of physiological remodelling during aging. Conceivably, such plasticity might rely on a complex milieu of genetic and hormonal factors that could be modulated by extrinsic elements such as exercise.

9.   Conclusion

Researchers have suggested that morphological, histochemical and electrophysiologic (as manifested by the dynamic properties) changes that occur with advancing years may be the consequence of "functional denervation", resulting in a loss of trophic influence by the nervous system and subsequent reinnervation. Functional denervation may be characterized by a decline of neuronal conduction velocity or impulse activity, a reduction of neurotransmitter release and the possible loss of motor neurons. It has been demonstrated that neuronal stimulation is essential for muscle development and maintenance. In fact, it has been determined that the characteristic of the neuronal activity influences the phenotypic expression of skeletal muscle. For example, stimulation of type II fibers brings about histochemical and electrophysiologic transformation of fast and slow-twitch fibers. It may be suggested also that neural stimulation not only influences the development of young skeletal muscle but also affects the development of aging muscle. Thus, inactivity or hypokinesia, which is prominent in senescence, may result in muscle fiber transformation or redistribution and reduce the proportion of type II fiber.

A number of conclusions can be drawn about the aged NMJ by combining results from our laboratory and other investigators. First, the previously reported 55% diminution in nerve terminal cross sectional area (Figs. 1 and 2) combined with the reported 58% increase in total length of nerve terminal branches (Fig. 3-10) indicate that nerve terminal volume (area x length) was approximately preserved in aged NMJ's. Second, histologic studies in the same strain of aged mice revealed no deficit in transmitter release, despite substantial reduction in numbers of synaptic vesicles and mitochondria. Thus, the growth of functional nerve terminals within the junctional region may serve to compensate for possibly reduced transmitter release per site by providing more release sites. Alternatively, given that presynaptic 'active zones' are in register with secondary clefts the increased complexity and branching of such clefts in aged muscles may indicate that more presynaptic active zones are also present. The growth in synaptic area reported here resembles that occurring in neuromuscular synaptic development, in which there is coordinated enlargement and expansion of the presynaptic nerve terminal, increasing transmitter release as the muscle fiber enlarges.

10.   References

[1]   Alnaqeeb, M.A. and Goldspink, G.   Changes in fiber type, number and diameter in developing and aging skeletal muscle. J. Anat. 153: 31-45, 1987.

[2]   Alshuaib, W.G. and Fahim, M.A. Effects of exercise on physiological age-related change at mouse neuromuscular junctions. Neurobiol. Aging, II: 555-561, 1990.

[3]   Alshuaib, W.B. and Fahim, M.A.   Depolarization reverses age-related decrease of spontaneous transmitter release. J. Appl. Physiol., 70: 2066-2071, 1991.

[4]   Andonian, M.H. and Fahim, M.A.   Effects of endurance exercise on the morphology of mouse neuromuscular junctions during aging. J. Neurocytol. 16: 589-599, 1987.

[5]   Andonian, M.H. and Fahim, M.A.   Nerve terminal morphology in C57BL/6NNia mice at different ages. J. Gerontol. 44(2): B43-51, 1989.

[6]   Anis, N.A. and Robbins, N.   General and strain-specific age changes at mouse limb neuromuscular junctions. Neurobiol. Aging. 8: 309-318, 1987.

[7]   Ansved, T. and Larsson, L.   Quantitative and qualitative morphological properties of the soleus motor nerve and the L5 ventral root in young and old rats.   Relation to the number of soleus muscle fibers. J. Neurol. Sci., 96: 269-282, 1990.

[8]   Ansved, T., Wallner, P. and Larsson, L.   Spatial distribution of motor unit fibers in fast and slow-twitch rat muscles with specific reference to age. Acta Physiological Scandinavica, 143(3): 345-354, 1991.

[9]   Arabadjis, P.G., Heffner, R.R. Jr., and Prendergast, D.R.   Morphologic and functional alterations in aging rat muscle. Journal of Neuropathology and Experimental Neurology, 49(6): 600-9, 1990.

[10] Atwood, H.L.   Age-dependent alterations of synaptic performance and plasticity in crustacean motor systems. Experimental Gerontology, 27(1): 51-61, 1992.

[11] Banker, B.Q., Kelly, S.S. and Robbins, N.   Neuromuscular transmission and correlative morphology in young and old mice. J. Physiol. (Lond.), 339: 355-375, 1983.

[12] Barker, P.C.H. The aging neuromuscular system. Seminars in Neuro 9: 50-59, 1989.

[13] Bischoff, C., Machetanz, J. and Conrad, B.   Is there an age-dependent continuous increase in the duration of the motor unit action potential? Electroencephalography and Clinical Neurophysiology, 81(4): 304-311, 1991.

[14] Brown, M.   Resistance exercise effects on aging skeletal muscle in rats. Physical Therapy. 69(1): 46-53, 1989.

[15] Cardasis, C.A.   Ultrastructural evidence of continued reorganization at the aging (11-26 months) rat soleus neuromuscular junction. Anat. Rec., 207: 399-415, 1983.

[16] DeLuca, C., Mambrini, M. and Conte Camerino, D.   Changes in membrane ionic conductances and excitability characteristics of rat skeletal muscle during aging. <u>Pflugers Archiv - European Journal of Physiology</u>, <u>415(5)</u>: 642-644, 1990.

[17] DeLuca, A. and Conte Camerino, D.   Effects of aging on the mechanical threshold of rat skeletal muscle fibers. <u>Pflugers Archiv - European Journal of Physiology</u>. <u>420(3-4)</u>: 407-409, 1992.

[18] Fahim, M.A. and Robbins, N.   Ultrastructural studies of young and old mouse neuromuscular junction. <u>J. Neurocytol.</u> <u>11</u>: 641-656, 1982.

[19] Fahim, M.A., Holley, J.A. and Robbins, N.   Scanning and light microscopic study of age changes at a neuromuscular junction. <u>J. Cell Biol.</u>, <u>12</u>: 13-25, 1983.

[20] Fahim, M.A., Robbins, N. and Price, R.   Fixation effects of synaptic vesicle density in neuromuscular junctions of young and old mice. <u>Neurobiol. Aging.</u> <u>8</u>: 71-75, 1987.

[21] Fahim, M.A.   Rapid neuromuscular remodelling following limb immobilization. <u>Anat. Rec.</u> <u>224</u>: 102-109, 1989.

[22] Frontera, W.R., Meredith, C.N., O'Reilly, K.P., Knuttgen, H.G. and Evans, W.J.   Strength conditioning in older men: skeletal muscle hypertrophy and improved function. <u>J. Appl. Physiol.</u>, <u>64</u>: 1038-1044, 1988.

[23] Grimby, G. and Saltin, B.   The aging muscle. <u>Clinical Physiol.</u>, <u>3</u>: 209-218, 1983.

[24] Grimby, G., Aniansson, A., Zetterberg., D. and Satin, B.   Is there a change in relative muscle fiber composition with age? <u>Clin. Physiol.</u>, <u>4</u>: 189-194, 1984.

[25] Grinnell, A. D. and Herrera, A.A.   Specific and plasticity of neuromuscular connections; long-term regulation of motoneuron function. Pro. <u>Neurobiol.</u>, <u>17</u>: 263-282, 1981.

[26] Hartung, V. and Asmussen, G.   The influence of life span on extra and intrafusal muscle fibres in the soleus muscle of the rat. <u>Z. Mikrosk. Anat. Forsch.</u>, <u>102</u>: 677-693, 1988.

[27] Hashizume, K., Kanda, K. and Burke, R.E.   Medial gastrocnemius motor nucleus in the rat: age-related changes in the number and size of motoneurons. <u>J. Comp. Neurol.</u>, <u>269</u>: 425-430, 1988.

[28] Hennig, R. and Lomo, T.   Firing patterns of motor units in normal rats. <u>Nature</u>. <u>314</u>: 164-166, 1985.

[29] Herscovich, S. and Gershon, D.   Effects of aging and physical training on the

neuromuscular junction of the mouse. Gerontology. 33: 7-13, 1987.

[30] Heuser, J.E. and Reese, T.   Evidence for recycling of synaptic vesicle membrane during transmitter release at the frog neuromuscular junction. J. Cell Biol., 57: 315-344, 1973.

[31] Hinz, I. and Wernig, A.   Prolonged nerve stimulation causes changes in transmitter release at the frog neuromuscular junction. Journal of Physiol. (Lond.). 401: 557-565, 1988.

[32] Holloszy, J.O. Aging and exercise: physiological interactions. Federation Proc., 46: 1823, 1987.

[33] Howald, H., Hoppeler, H., Claasen, H., Mathieu, O. and Straub, R. Influences of endurance training on the ultrastructural composition of the different muscle fiber type in humans. Pflugers Arch., 403: 369-376, 1985.

[34] Irintchev, A. and Wernig, A. Muscle damage and repair in voluntary running mice: strain and muscle differences. Cell Tissue Res., 249: 509-521, 1987.

[35] Kanda, K., Hashizume, K., Nomoto, E. and Askaki, S.   Effects of aging on physiological properties of fast and slow motor units in rat gastrocnemius muscle. Neurosci. Res., 3: 242-246, 1986.

[36] Kanda, K. and Hashizume, K.   Changes in properties of medial gastrocnemius motor units in aging rats. J. Physiol., 61: 737-746, 1989.

[37] Kobayashi, H., Robbins, N. and Rutishauser, U.   Neural cell adhesion molecule in aged mouse muscle. Neuroscience. 48(1): 237-248, 1992.

[38] Larsson, L. and Edstrom, L. Effects of age on enzyme - histochemical fiber spectra and contractile properties of fast and slow-twitch skeletal muscles in the rat. J. Neurol. Sci., 76: 69-89, 1986.

[39] Larsson, L. and Ansved, T. Effect of age on the motor unit. A study on single motor units in the rat. In: Central determinants of age-related declines in motor function. Ed. Joseph, J.A. Ann. N.Y. Acad. Sci., 515: 303-313, 1988.

[40] Larsson, L., Ansved, T., Edstrom, L., Gorza, L. and Schiaffino, S.   Effects of age on physiological, immunohistochemical and biochemical properties of fast-twitch single motor units in the rat. Journal of Physiology., 443: 257-275, 1991.

[41] Oudet, C., Petrovic, A., Champy, M. and Kahn, J.L.   Is the myosin type altered in the aging platysma? Journal of Cranio-Maxillo-Facial Surgery. 17(4): 190-194, 1989.

[42] Pachter, B.R. and Eberstein, A.   Neuromuscular plasticity following limb immobilization. J. Neurocyt. 13: 1013-1025, 1984.

[43] Robbins, N. and Fahim, M.A. Progression of age changes in mature mouse motor nerve terminals and its relation to locomotor activity. <u>J. Neurocytol.,</u> <u>14</u>: 1019-1036, 1985.

[44] Rosenheimer, J.L.   Effects of chronic stress and exercise on age-related changes in end-plate architecture. <u>J. Neurophysiol., 53(6)</u>: 1582-1589, 1985.

[45] Rosenheimer, J.L. and Smith, D.O.   Differential changes in the end-plate architecture of functionally diverse muscles during aging. <u>J. Neurophysiol.,</u> <u>53</u>: 1567-1581, 1985.

[46] Rosenheimer, J.L.   Factors affecting denervation-like changes at the neuromuscular junction during aging. <u>Int. J. Devel. Neuroscience.</u>, <u>8</u>: 643-654, 1990.

[47] Silbermann, M., Finkelbrand, S., Weiss, A., Gershon, D. and Reznick, A. Morphometric analysis of aging skeletal muscle following endurance training. <u>Muscle Nerve, 6</u>: 136-142, 1983.

[48] Smith, D.O. and Chapman, M.R. Acetylcholine receptor binding properties at the rat neuromuscular junction during aging. <u>J. Neurochem.,</u> <u>48</u>: 1834-1841, 1987.

[49] Smith, D.O. and Weiler, M.H.   Acetylcholine metabolism and choline availability at the neuromuscular junction of mature adult and aged rats. <u>J. Physiol. (Lond.).</u>, <u>383</u>: 693-709, 1987.

[50] Walters, T.J., Sweeny, H.L. and Farrar, R.P.   Aging does not affect contractile properties of the Type IIb FDL muscle in F344 rats. <u>Am. J. Physiol.</u>, <u>258</u>: 1031-1035, 1990.

[51] Washio, H., Imazato-Tanaka, C., Kanda, K. and Nomoto, S.   Choline acetyltransferase and acetylcholinesterase activities in muscles of aged mice. <u>Brain Res.</u>, <u>416</u>: 69-74, 1987.

[52] Wokke, J.H., Jennekens, F.G., Van den oord, C.J.   Morphological changes in the human end plate with age. <u>J. Neurol. Sci.</u>, <u>95(3)</u>: 291-310, 1990.

Neuromuscular Remodelling in the Adult Induced by Small Physiologic Changes in the Locomotor Activity

M.R. Fenoll-Brunet, J. Tomas, M. Santafé, and M.A. Lanuza

1 Introduction

Neuromuscular junction morphology itself is the result of a developmental process that probably is going on during the whole life span of an animal. This process is based on several mechanisms: those that require neural activity and those that do not , see Goodman and Shatz [13] for a recent review). In general the initial steps of axonal guidance and target recognition can occur before neurons become functionally active and so, diffusible gradients and extracellular matrix molecules play a significant role [27,30,31,34,35]. As development progresses and synaptic terminals display adult characteristics, miniature endplate potential frequency and endplate potential quantal content increase gradually achieving their adult values in rat [9] and amphibian muscle fibers [5]. So, thereafter neural activity-related mechanisms can induce plastic modifications tending to remodelate circuits and terminate the molecular organization of the connections [21,32,33,40,41]. This process of activity-dependent synaptic plasticity does not stop at birth but seems to continue throughout the lifetime of the organisms. In the adult neuromuscular junctions there are many situations indicating that synapse or muscle cell activity-depending mechanisms govern plasticity, growth responses and the final size that can be attained by motor nerve endings in a particular situation [3,43,52]. For instance, the activity-dependent seasonal remodelling changes in frog neuromuscular junctions [50].

During the last years, we have performed experiments designed to test the hypothesis that small physiologic variations in the locomotor activity of normal adult rats during a short period can induce remodelling changes in the motor nerve terminals of the fast-twitch muscle *Extensor digitorum longus* (EDL muscle). Some of these results have been previously reported [43,44].

2 Material and Methods

We moderately reduced (serie 2) the locomotor activity of 12 two-month-old male Sprague-Dawley rats during four weeks by maintaining them in metabolic cages (Panlab Ref. 13.2016). This housing procedure reduced by 75 % the surface of the animal's habitat with respect to controls (12 conventionally housed rats during three months, serie 1). This method diminished in a consistent and reproducible way the normal cage activity of the animals without any abnormal postural or locomotor behavior. The locomotor activity was increased in another set of 12 two-month-old animals (serie 3), by a training walk program applied during four weeks (27 m/min.; 55 min./day; 5 days/week). Daily training was divided in 11 walking periods of 5 minutes each separated by resting intervals of 8 minutes. Twenty animals initiated the training but only 60 % of them adapted spontaneously to the walk exercise and so, were allowed to complete the program, see Tomas et al. [43].

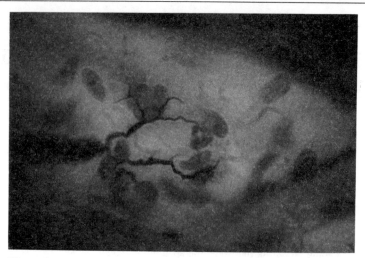

Figure 1
EDL motor endplate stained with the Gros-Bielschowsky silver method
viewed under an immersion oil planapochromatic objective (X100).

Animals were killed under ether anesthesia and both, right and left, EDL muscles were processed to study the morphology of the nerve terminal arborizations, with the Bielschowsky-Gros silver impregnation method performed in 25 micrometer sections (*Figure 1*). Motor nerve terminal arborizations (492 synapses from each serie) were digitized with an automatic image analysis system (VICOM-VME). The images of the terminal branches were processed with a specific program for evaluating many geometric characteristics of the nerve endings.

For the morphometric evaluation, the considered parameters can be grouped as: 1. Terminal length parameters; 2. Angular parameters and 3. Other parameters.

1. *Terminal length parameters.* We analyzed the nerve terminals in segments as can be seen in *Figure 2*. L2 are the trunk segments contained between branch points, whereas L3 are the distal free-end terminal segments. Both L2 and L3 segments can be of first, second or higher order according to their origin. The order is represented by -n (e.g., L3-3 represents a distal free-end segment of third order). Because, in most synapses, more than one L2 or L3 segment of the same order can exist, we added to the parameter notation the character (s) indicating the summed value of these various segments (e.g., L2s-2), or the character (m) indicating the mean value of these segments (e.g., L3m-3). We considered also the summed value and the mean value of all the L2 and L3 segments in each synapse which is noted as: L (2 or 3) (s or m)-t (e.g., L2s-t for the summed length of all L2 segments in a synapse, named also "trunk length", and L3s-t for the summed length of all L3 segments in a junction, named also in this case "distal length"). The parameter L0 represents the length of the initial segment from the end of the myelin sheath to the first branch point. L is the total length of the motor nerve terminal from the first branch point.

2. *Angular parameters.* We measured the intraterminal angles formed between the L2 segments originated in the same branch point (A2). The angles formed between L2 and L3 segments or between two L3 segments originated, in both situations, in the same branch point were also evaluated (A3).

3. *Other parameters.* The number of branch points of the nerve terminal (BP) and the muscle cell diameter (MD).

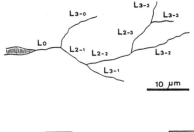

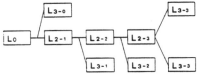

Figure 2.
Camera lucida drawing of a motor nerve terminal arborization to show the initial segment (L0), the trunk segments (L2) and the distal segments (L3). Trunk and distal segments are classified according to their centrifugal order (Tomas et al. [46], with permission).

We also investigated for the ultrastructural correlates of described light microscopy changes. Here, several preliminary data will be described concerning to the ultrastructure of the adaptation rearrangement of the neuromuscular junctions to the physiologic increase of the animal's locomotor activity. In brief, locomotor activity in another set of animals (six two-month-old rats) was increased as previously described. All animals were killed under ether anesthesia and the right and left EDL muscles pinned on Silgard in Petri dishes were processed for conventional electron microscopy. Images of 227 (control) and 357 (increased use) cross sectioned different end plates were directly acquired in magnetic support with a high definition video system and analyzed at a final magnification of X 12.500. Morphometric analysis was performed with a semiautomatic image analysis system (MOP Kontron, Germany). In all cases, we tested the shape of parameter distribution by using the Kolmogorov-Smirnov test and accordingly, a t-test or U-test (Mann-Whitney) was applied for mean values comparison. Values below 0.05 were considered significant.

3 Results

Rabdomyolisis, inflammatory infiltrates, regenerative signs or ultrastructural lesions are absent in the studied muscles so, the locomotor activity variation procedures used in this work did not result in muscle nor nerve damage. However, as showed in *Table 1*, there is a small increase of the EDL muscle weight in trained animals whereas the mean muscle cell diameter increases significantly both in animals trained to walk and in the rats maintained in metabolic cages with respect to the control animals.

In the normal adult animals of various species there are always few motor nerve terminals showing sprouts and/or abnormal branches (short, thick and swollen terminal branches) which are currently interpreted as signs of normally-occurring remodelling of the neuromuscular contacts. We observed in trained animals (*Table 2*) a reduction of the percentage of the motor endings showing spontaneous sprouts, and this finding appears complemented with an increase of terminals having one abnormal branch. On the contrary, in the decreased use serie only a reduction of the normal sprouting was observed.

Table 1. Animals body weight , EDL muscles weight and EDL muscle cells diameter (MD).

	Serie 1	Serie 2	Serie 3	Pm 1 vs 2	1 vs 3
Rat Weight	371.90 ± 3.80	363.90 ± 17.30	388.10 ± 27.70	0.850	0.045
EDL Weight	0.18 ± 0.04	0.21 ± 0.01	0.21 ± 0.01	0.058	0.031
MD	8.20 ± 10.30	62.70 ± 9.67	70.6 ± 8.53	0.000	0.000

12 animals (24 muscles per serie). Weight in grams. Muscle cells diameter (MD) in micrometers. Pm is the P value from U-Test. Values are expressed as Mean ± SD.

Table 2. Percentage of nerve endings showing terminal sprouts and abnormal branches.

	Serie 1	Serie 2	Serie 3	Pm 1 vs 2	1 vs 3
Normal Endings	90.20 ± 4.00	92.64 ± 2.73	89.74 ± 6.11	0.127	0.453
Terminal Sprouts	6.26 ± 1.46	3.65 ± 1.46	4.23 ± 1.37	0.003	0.024
Abnormal Branches	3.54 ± 1.16	3.71 ± 0.91	6.03 ± 2.10	0.699	0.014

Values are expressed as Percentage ± SD. Pm is the P value from U-Test. N (serie 1): 2653 endings; N (serie 2): 1361 endings; N (serie 3): 653 endings. 24 muscles per serie.

Branch points

In the control nerve terminal arborizations the mean number of branch points (BP) was almost three (2.99 ± 1.29, range 1-8). In trained muscles we observed, on the average, a reduction of this mean value (2.61 ± 1.13, range 1-7, P = 0.002 t-test), whereas in the animals with decreased locomotor activity the mean value of BP becomes higher than the control (3.32 ± 1.47, range 1-8, P = 0.029 t-test).

In addition, we observed in most terminals of the control animals that branch points and terminal branches are situated very regularly around the geometric center of the endings. The generality of this tendency is confirmed when we measured in the complete set of control synapses, the inertia momentum of the branch points with respect to the geometric center point of the nerve terminal image (vectorial angles were always referred to the regression line of the nerve terminal image points). In control endings the module of the vector resulting from the summation of the individual vectors joining the branching points with the geometric center of a nerve ending image (the angular value of this vector remains unmodified in the three experimental series considered), amounted 5.68 ± 4.5 micrometers. This small value (less than 5% of the total terminal length, see later) means that the growth and branching of the terminal segments during normal development can proceed tending to place the branch points equidistant between themselves and with respect to the geometric center of the nerve terminal tree. In mammals, this regular disposition would represent probably the best configuration to circumscribe presynaptic branches into a small and almost circular limited synaptic area perhaps to optimize their capacity of depolarizing the sarcolemma.

This nerve terminal pattern becomes clearly altered with the introduced experimental use changes mainly because an asymmetric plastic modulation of the distal branches length (see length parameters). The mean value of the considered module can be 25.49 ± 5.98 in serie 2 and 25.50 ± 6.33 in serie 3 (in both cases P<0.0001, t-test, with respect to the control endings). By only using the value of the aforesaid module it is possible to realize the correct classification of 64.45 % of the synapses, with 97.5 % sensitivity and 96 % specificity to differentiate the control serie from altered use serie (discriminant analysis with SPSS/PC+ advanced).

Although in the rat EDL muscle most nerve terminal branch points are binary, there is a small percentage of instances in which three segments arise from the same branching point (see *Table 3*). Multifurcation points are less frequently observed both in series 2 and 3 when compared with the control serie. In trained animals multifurcation becomes restricted to the first branch point after the initial segment whereas in the decreased use serie 2 we observed a rearrangement and proximo-distal displacement of these points.

Table 3. Multifurcation points. Percentage of motor endings having multifurcation points and their topography (order of the segment originating multifurcation).

	Serie 1	Serie 2	Serie 3	Pm 1 vs 2	Pm 1 vs 3
Total Percentage	15.20 ± 2.3	10.37 ± 1.7	4.88 ± 1.2	0.008	0.001
Initial Segment	64.00 ± 12.0	64.70 ± 10.0	100.00	< 0.01	> 0.01
First Order Trunk Segment	28.00 ± 3.2	17.65 ± 2.4	-	> 0.01	-
Second Order Trunk Segment	8.00 ± 1.3	11.76 ± 1.2	-	> 0.05	-
Third Order Trunk Segment	-	5.88 ± 0.7	-	-	-

Values are expressed as Percentage ± SD. N (serie 1): 2653 endings, N (serie 2): 1361 endings, N (serie 3): 653 endings. 24 muscles per serie. Pm is the P value from U-Test.

Nerve terminal *segments*

Table 4 shows the mean number of trunk, distal and total nerve terminal segments found in the experimental and control situations considered. Data indicate that, on the average, motor arborizations in the animals trained to walk have fewer constitutive segments when compared with the control because a significant reduction in the mean number of both, trunk (19 %) and distal (11 %) segments. On the contrary, in the decreased use serie we found an opposite significant increase of the trunk segments (19 %) and of the distal ones (8 %).

The *Table 5* shows the percentage of the motor nerve terminals having segments of different order. In trained animals results indicate that, on the average, there are fewer synapses having trunk and distal segments of high order (2nd and 3rd) whereas the number of endings with low order distal branches (0 and 1st) was correspondingly higher than in the control. On the other hand, in the decreased use serie, it becomes evident the existence of a greater percentage of endings with high order trunk (specially the order three) and distal segments (2nd and 3rd). In this serie, the growth of trunk and distal branches of higher order (4th and 5th) can be observed in several endings.

Table 4. Mean number of trunk and distal segments.

	Serie 1	Serie 2	Serie 3	1 vs 2 Pm	1 vs 2 Pd	1 vs 3 Pm	1 vs 3 Pd
Trunk Segments	1.96 ± 1.27	2.33 ± 1.47	1.59 ± 1.10	0.015	0.416	0.006	0.006
Distal Segments	4.10 ± 1.31	4.43 ± 1.50	3.65 ± 1.15	0.035	0.415	0.001	0.006
Total Segments	6.06 ± 2.55	6.76 ± 2.95	5.24 ± 2.23	0.022	0.342	0.001	0.002

N: 492 synapses from each serie. Values are expressed as Mean ± SD. Pm: t-test. Pd: Kolmogorov-Smirnov.

Table 5. Mean percentage of motor nerve endings with segments of different order.

	Order	Serie 1	Serie 2	Serie 3	Pm 1 vs 2	Pm 1 vs 3
Trunk Segments	1	86.6 ± 5.6	95.1 ± 1.1	87.8 ± 6.7	**	ns
	2	45.1 ± 4.3	56.1 ± 4.9	35.4 ± 3.7	**	**
	3	9.8 ± 1.6	21.9 ± 2.7	4.9 ± 0.7	*	**
	4	-	3.7 ± 0.0	-	-	-
	5	-	0.6 ± 0.0	-	-	
Distal Segments	0	58.5 ± 6.2	66.4 ± 7.3	75.6 ± 7.4	ns	*
	1	82.9 ± 7.0	89.6 ± 9.7	88.6 ± 6.9	ns	**
	2	45.7 ± 5.4	54.9 ± 6.1	35.4 ± 4.6	**	**
	3	10.4 ± 2.1	22.0 ± 2.8	5.4 ± 1.1		*
	4	-	3.7 ± 0.0	-	-	-
	5	-	0.6 ± 0.0	-	-	-

Values are expressed as Percentage ± SD. Pm: U test. *: $P < 0.01$; **: $P < 0.05$; ns: no significant.

Nerve terminal *length parameters*

In control animals the mean total length of the terminal arborization from the first branch point (L) amounted 117.1 ± 38.0 micrometers. The introduced experimental changes in the locomotor activity can result in significant ($P < 0.001$, t-test) variations of this parameter. In trained animals, L becomes reduced (99.6 ± 44.5 micrometers) whereas in the animals maintained in metabolic cages increased (138.4 ± 42.5 micrometers). *Table 6* shows the corresponding values found with respect to the initial and the individual trunk segments classified according their centrifugal order whereas the same classification for distal segments are showed in *Table 7*.

Table 6. Initial and trunk segments lengths.

	Serie 1	Serie 2	Serie 3	1 vs 2 Pm	1 vs 2 Pd	1 vs 3 Pm	1 vs 3 Pd
L0	12.10± 8.35	13.30± 7.10	13.00± 6.32	0.038	0.043	0.038	0.013
L2s-1	25.10±17.10	21.50±15.70	20.90±14.60	0.070	0.104	0.051	0.051
L2m-1	16.10± 9.46	15.10± 8.02	15.60± 8.13	0.532	0.476	0.814	0.687
L2s-2	17.60±14.40	17.90±10.80	17.70±12.70	0.159	0.118	0.696	0.597
L2m-2	13.10± 7.12	13.50± 5.82	13.80± 6.55	0.351	0.255	0.383	0.392
L2s-3	10.10± 6.09	16.30± 8.41	12.30± 3.41	0.005	0.029	0.075	0.139
L2m-3	10.10± 6.09	14.10± 6.67	12.30± 3.41	0.017	0.029	0.075	0.139
L2s-4	-	8.73± 4.76	-	-	-	-	-
L2m-4	-	8.73± 4.76	-	-	-	-	-
L2s-5	-	14.80± 4.70	-	-	-	-	-
L2m-5	-	14.80± 4.70	-	-	-	-	-
L2s-t	35.40±21.80	36.20±23.80	28.50±19.80	0.954	0.977	0.003	0.023
L2m-t	15.60± 8.61	14.90± 6.35	15.54± 7.40	0.949	0.435	0.694	0.619

N: 492 synapses from each serie. Values are expressed as Mean (mm) ± SD. Pm: t test, Pd: Kolmogorov-Smirnov test.

In muscles from the animals trained to walk both, the trunk and distal portions of the terminal arborizations (parameters L2s-t and L3s-t respectively) become similarly reduced around 20 %. This finding was in accordance with the above mentioned effective diminution of the branch points, the mean number of the trunk and distal segments and with the diminution of the percentage of synapses possessing segments of high order. Nevertheless, whereas the mean length of the trunk segments did not change significantly, we observed that the mean length of distal segments was higher than in the control. Although, topographically, it is interesting to note that the mean length of the distal segments placed proximally to the origin of the terminal arborization (specifically L3m-0) becomes significantly reduced.

In animals with restricted locomotor activity we observed that, on the average, the trunk length and the mean length of their constitutive segments remain unmodified. Nevertheless when analyzing the individual trunk segments according their topographic centrifugal order we observed a significant increase in the mean length of the third order trunk segments. Moreover, only in this case trunk segments of 4th and 5th order have been observed. In this serie, there is a marked increase of the distal length and also the mean length of their segments specially those of the first and second order.

Table 7. Distal segments length.

	Serie 1	Serie 2	Serie 3	1 vs 2		1 vs 3	
				Pm	Pd	Pm	Pd
L3s-0	43.0 ±25.20	40.1 ±22.60	34.9 ±18.80	0.522	0,624	0.018	0.005
L3m-0	32.9 ±14.60	32.2 ±13.60	28.3 ± 8.18	0.292	0.459	0.003	0.011
L3s-1	39.1 ±20.20	49.4 ±27.40	51.5 ±24.20	0.003	0.030	0.000	0.002
L3m-1	16.0 ± 5.77	23.8 ± 8.77	24.8 ± 7.98	0.000	0.000	0.000	0.000
L3s-2	36.6 ±17.90	44.1 ±24.70	43.8 ±21.10	0.031	0.094	0.026	0.031
L3m-2	16.9 ± 8.45	19.9 ± 8.77	20.5 ± 8.50	0.004	0.018	0.003	0.005
L3s-3	28.3 ±14.00	35.1 ±21.60	42.2 ±12.50	0.423	0.242	0.000	0.001
L3m-3	14.1 ± 7.00	15.8 ± 6.47	21.1 ± 6.26	0.290	0.242	0.000	0.001
L3s-4	-	28.9 ±13.10	-	-	-	-	-
L3m-4	-	15.9 ± 6.01	-	-	-	-	-
L3s-5	-	19.7 ± 6.20	-	-	-	-	-
L3m-5	-	9.8 ± 3.20	-	-	-	-	-
L3s-t	86.3 ±26.60	103.9±28.20	71.1 ±30.70	0.000	0.000	0.007	0.032
L3m-t	22.6 ± 9.07	26.06± 9.32	23.9 ± 7.53	0.000	0.000	0.000	0.000

N: 492 synapses from each serie. Values are expressed as Mean (mm) ± SD. Pm: t-test.
Pd: Kolmogorov-Smirnov test.

Angular parameters

The activity-dependent metric and topographic variations previously described can be accompanied by some angular displacements between the terminal segments. Only the mean angle between trunk segments (A2-t) become significantly different in the enlarged endings of the animals maintained in metabolic cages with respect to the control terminals (control serie: 75.9 ± 27.7 degrees; decreased use serie: 62.4 ± 20.9 degrees. P<0.001, t-test).

Ultrastructural approach

   We are specially interested in the study of the adaptative changes in the neural connections induced by an increase of use. We found in animals trained to walk, an increase of the percentage of neuromuscular junctions showing abandoned postsynaptic structures because an effective retraction of terminal branches (see *Table 8*). Neither axonal nor muscle cell ultrastructural lesions have been observed. Some preliminary data related with the synaptic membranes (active zones and postsynaptic membrane densifications) are showed in *Table 9*. At the presynaptic level, physiological increase of use enlarged presynaptic membrane densifications (AZL, active zone mean length). In accordance, we found also an increase in the mean length of the adaxonal postsynaptic membrane densifications in fold crests (PMD) with respect to the control synapses. So, ultrastructural findings of training effects on these neuromuscular junctions were consistent with the existence of a specific modulation of these neurotransmission-related structures.

Table 8. Percentage of synapses showing abandoned postsynaptic gutters.

	NS	MAS	WAS	TAS
Control (n = 227)	35.3 ± 4.2	27.6 ± 5.9	33.3 ± 2.0	3.8 ± 1.7
Increased use (n = 357)	29.5 ± 1.0	18.4 ± 5.4(*)	41.9 ± 3.1	10.2 ± 0.3 (*)

Normal Synapses (NS): no one abandoned gutter present. Minimally Abandoned Synapses (MAS): those showing one gutter unopposed to the axon. Widely Abandoned Synapses (WAS): these having more than one unopposed gutters. Totally Abandoned Synapses (TAS): no terminal axon present. Values are expressed as Percentage ± standard error. (*): indicates $P < 0.05$ with respect to the control, X^2 function.

Table 9. Pre- and postsynaptic membrane densifications.

	Control	Increased use	P
AZL	95.7 ± 38.0	106.9 ± 30.4	0.000
AZN	2.84 ± 2.2	2.58 ± 2.0	0.330
PMD	306.7 ± 115.0	448.3 ± 261.4	0.000

Length of the active zone associated densifications (AZL). Number of densifications per axonal profile (AZN). Length of the postsynaptic membrane densification (PMD). Values are expresed as mean (nm) ± SD. N: control 227 synapses, increased use serie 357 synapses. P is the P value from t test.

4 Discussion

   The described plastic changes developed in absence of cell and tissue lesions. In particular, the walking work of the animals in the treadmill was spontaneously accepted and performed according the physiological principles of exercise training [7].
   Results clearly indicate the existence of an activity-dependent plastic adaptation mechanism in the neuromuscular connections of the normal adult. Furthermore, findings suggest that on the average, the size and complexity of the motor nerve endings can be inversely related with the level of usage of the junctions in a particular moment [43]. Nevertheless, because the existence of a moderate correlation between nerve terminal size and muscle cell diameter in several muscles [4,17,42] we cannot completely discard that the growth of the terminals in the decreased use serie would be related to the small increase of the muscle cell size in this situation. On the other hand, there are some well defined situations in which nerve terminal and muscle cell size are unrelated

[19,28,45]. In the present study, muscle cell fiber diameter increases whereas nerve terminal size becomes reduced in exercised muscles. Probably the postsynaptic size may determine a minimum size for nerve endings whereas further variations in the terminal complexity would be regulated by mechanisms depending on the activity-related remodelling [45].

In trained animals we observed: first, a reduction in the normal remodelling growth capacity of the nerve endings because the existence of fewer synapses having normally-occurring sprouts and also multifurcation points which may be considered the nerve terminal sites manifesting the highest pressure of growth. Second, metric and topographic, light and electron microscopy observations indicate a compactation of the presynaptic nerve terminal arborization because an unequal reduction in the length of the distal branches with a remarkable fixation of the trunk portion of the endings. The most easy explanation for this set of results can be an increased activity-related actual reduction in the length of all distal segments perhaps more effective in branches far of the nerve terminal origin. In a number of endings this reduction would conclude with the complete elimination of a distal branch and a branch point then transforming a set of one trunk and two distal segments in a longer distal branch. In accordance, the light microscopy observations on the decreased use serie can be explained by the existence of a real increase in the linear elongation of both, the distally placed free-end segments and the more distally placed trunk ones. This elongation of the distal branches can result in the generation of a new branch point in some endings.

That the normal adult neuromuscular connections are dynamic structures capable to undergo remodelling changes is well documented [3,14,17,48,50,53], although depending on the methodological approach to the remodelling concept, some results disagree about the magnitude of these growth and retraction phenomena [23]. Structural remodelling is currently related to activity changes and probably nerve and muscle cells can adjust their properties continuously to meat functional demands [3,16]. This could be the case of the growth and retraction seasonal changes observed in nerve terminals of frogs (reviewed by Wernig and Herrera, [51] and Wernig and Dorlöchter, [52], for the increased growth of endings after limb immobilization [10,11], and also for the increased activity dependent delay in the ageing-related neuromuscular changes [1,2].

We found that synaptic rearrangement normally occurs in normal adult EDL muscle and can be modulated by just applying different minimal changes on the animals locomotor activity within a physiological range. Our results also indicate that the activity-dependent remodelling changes mainly affects the distal portions of the nerve endings whilst the most proximal trunk segments undergo minimal angular displacements, see Hill et al. [18]. Some parts of the nerve terminals remain almost unchanged whereas others could be more sensitive to develop structural modifications. In this sense, some morphological and functional proximo-distal gradients have been detected in amphibian neuromuscular junctions [29,47].

We are specially interested to know how the described light microscopy adaptations can develop along with a possible modulation in the morphology of the neurotransmission machinery. In this work we focused attention in the ultrastructural study of the changes observed in the exercised animals. We found that the retraction of the distal terminal branches in trained animals is correlated with an increased differentiation of pre- and postsynaptic membrane structures. In particular both, the pre- and postsynaptic membrane densifications become enlarged.

Although the membrane densities at the active zones are not necessary for quantal release they can facilitate neurotransmission by aligning vesicles [21]. So, their length can be a good indicator of the differentiation grade of the active zones themselves. In amphibian neuromuscular junctions good correlations are found between quantal content and active zone length [6,15,26]. Despite of the small sized motor endings observed in exercised animals the increased length of their active zone-associated densifications would suggest the existence of a compensatory adaptation tending to optimize synaptic efficacy in this situation. The possible existence of such compensation is reinforced by the parallel increase of the membrane densifications at the postsynaptic component, the sites of neurotransmitter receptor accumulation [25,32,33]. This morphologic postsynaptic adaptation may could be associated to an agrin mediated activity dependent action [12,24]. It is unknown at present how synaptic activity controls the mechanisms governing the longitudinal extension and plasticity of the terminal branches and confines the terminal arborization to a limited synaptic area. The main hypothesis relating synaptic activity to growth responses of the nerve terminals suggests that a motor neuron growth factor (MNGF) released by muscle cells in inactive junctions may promote sprouting of the terminals; opposite changes (retractions) would occur in highly active synapses [38]. Also, a more simple hypothesis for the control of junctional remodelling and terminal size proposes that growth is inherent to the nerve terminals but is counterbalanced or even reversed by activity of the junctions through a postsynaptic growth inhibitory substance or simply because the higher calcium influx into the more active terminals may induce the breakdown of the cytoskeletal elements by a calcium-dependent protease system [8,22,36,37,39]. Perhaps the surface-to-volume ratio in the thinnest distal branches [43,49] favors the efficacy of the proposed calcium-activated neutral protease system in controlling the assembly of the cytoskeletal elements. In any case, in addition of the activity level, the activity pattern can play a role in the control of the nerve terminal size and junctional remodelling [46].

The possible functional significance of the activity-dependent variations we observed in the normal adult motor nerve terminals arborization pattern and complexity is not immediately evident. Presynaptic motor endings innervate only one target cell and the safety factor of the neurotransmission is high. Then, it is difficult to argue as a hypothetical finality of the morphological plasticity, eventual variations in synaptic specificity or efficacy. In this context, it can be of interest to know if similar activity-dependent remodelling changes could occur also in the axonal arborizations of the more complex adult polineuronal networks. Probably, the observed phenomenon could suggest the persistence, in the adult neuromuscular junction, of the cellular mechanism controlling growth and retraction of the nerve terminals during synaptogenesis that arises also as a consequence of normal neuromuscular activity [20].

This work is supported by grants from FISSS 91/1082 and 93/0362.

References

[1]   W.B. Alshuaib and M.A. Fahim. Effect of exercise on physiological age-related change at mouse neuromuscular junctions. *Neurobiol. Aging,* 11:555-561, 1990.

[2]   M.H. Andonian and M.A. Fahim. Effects of endurance exercise on the morphology of mouse neuromuscular junctions during ageing. *J. Neurocytol,* 16:589-599, 1987.

[3]   H.L. Atwood, P.V. Nguyen and A.J. Mercier. Activity-dependent adaptation neuromuscular systems: comparative observations. In A. Wernig, editor, *Restorative Neurology (Vol. 5, Motoneuronal Plasticity),* pages 101-114, Amsterdam, 1991, Elsevier.

[4]   R.J. Balice-Gordon and J.W. Lichtman. *In vivo* visualization of the growth of pre- and postsynaptic elements of neuromuscular junctions in the mouse. *J. Neurosci.,* 10:894-908, 1990.

[5]   M.R. Bennett and A.G. Pettigrew. The formation of synapses in amphibian striated muscle during development. *J. Physiol. (Lond.),* 252:203-239, 1975.

[6]   M.R. Bennett, N.A. Lavidis and F. Lavidis-Armson. The probability of quantal secretion at release sites of different length in toad (*Bufo marinus*) muscle. *J. Physiol. (Lond.),* 418:235-249, 1989.

[7]   R. Casabury. Principles of exercise training. *Chest,* 102s:263-276, 1992.

[8]   A.L. Connold, J.V. Evers and G. Vrbová. Effect of low calcium and protease inhibitors on synapse elimination during postnatal development in the rat *soleus* muscle. *Brain Res. Dev. Brain Res.* 28:99-108, 1986.

[9] J. Diamond and R. Miledi. A study of foetal and new-born rat muscle fibres. *J. Physiol. (Lond.)*, 162:393-408, 1962.

[10] M.A. Fahim and N. Robbins. Remodelling of the neuromuscular junction after subtotal disuse. *Brain Res.*, 383:353-356, 1986.

[11] M.A. Fahim. Rapid neuromuscular remodeling following limb immobilization. *Anat. Rec.*, 224:102-109, 1989.

[12] M.J. Ferns and Z.W. Hall. How many agrins does it take to make a synapse? *Cell*, 70:1-3, 1992.

[13] C.S. Goodman and C.J. Shatz. Developmental mechanisms that generate precise patterns of neuronal connectivity. *Cell 72/Neuron 10(Suppl.)*:77-98, 1993.

[14] C.A. Haimann, A. Mallart, J. Tomás i Ferré and N.F. Zilber- Gachelin. Patterns of motor innervation in the pectoral muscle of adult *Xenopus laevis*: evidence for possible synaptic remodelling. *J. Physiol. (Lond.)*, 310:241-256, 1981.

[15] A.A. Herrera, A.D. Grinnell and B. Wolowske. Ultrastructural correlates of naturally occurring differences in transmitter release efficacy in frog motor nerve terminals. *J. Neurocytol.*, 14:193-202, 1985.

[16] A.A. Herrera and M.J. Werle. Mechanisms of elimination, remodeling, and competition at frog neuromuscular junctions. *J. Neurobiol.*, 21:73-98, 1990.

[17] A.A. Herrera, L.R. Banner and N. Nagaya. Repeated, *in vivo* observation of frog neuromuscular junctions: remodelling involves concurrent growth and retraction. *J. Neurocytol.*, 19:85-99, 1990.

[18] R.R. Hill, N. Robbins and Z.P. Fang. Plasticity of presynaptic and postsynaptic elements of neuromuscular junctions repeatedly observed in living adult mice. *J. Neurocytol.*, 20:165-182, 1991.

[19] H. Jans, R. Salzmann and A. Wernig. Sprouting and nerve retraction in frog neuromuscular junction during ontogenesis and environmental changes. *Neuroscience*, 18:773-781, 1986.

[20] J.K.S. Jansen and T. Fladby. The perinatal reorganization of the innervation of skeletal muscle in mammals. *Prog. Neurobiol.*, 34:39-90, 1990.

[21] C.-P. Ko. Formation of the active zone at developing neuromuscular junctions in larval and adult bullfrogs. *J. Neurocytol.*, 14:487-512, 1985.

[22] R.J. Lasek and P.N. Hoffmann. The neuronal cytoskeleton, axonal transport and axonal growth. In R. Goldman, T. Pollard and J. Rosenbaum, editors, *Cell Motility*. pp 1021, 1976, Cold Spring Harbour.

[23] J.W. Lichtman, L. Magrassi and D. Purves. Visualization of neuromuscular junctions over periods of several months in living mice. *J. Neurosci.*, 7:1215-1222, 1987.

[24] R.M. Nitkin, M.A. Smith, C. Magill, J.R. Fallon, Y-M.M. Yao, B.G. Wallace and U.J. McMahan. Identification of agrin, a synaptic organizing protein from Torpedo electric organ. *J. Cell Biol.*, 105:2471-2478, 1987.

[25] K. Peper, R.J. Bradley and F. Dreyer. The acetylcholine receptor at the neuromuscular junction. *Physiol. Rev.*, 62:1271-1340, 1982.

[26] J.W. Propst and C.-P. Ko.Correlations between active zone ultrastructure and synaptic function studied with freeze-fracture of identified frog neuromuscular junctions. *J. Neurosci.* 7:3654-3664, 1987.

[27] C.F. Reichardt and K.S. Tomaselli. Extracellular matrix molecules and their receptors: function in neural development. *Annu. Rev. Neurosci.*, 14:531-570, 1991.

[28] N. Robbins and M.A. Fahim. Progression of age changes in mature mouse motor nerve terminals and its relation to locomotor activity. *J. Neurocytol.* 14:1019-1036, 1985.

[29] R. Robitaille and J.P. Tremblay. Non-uniform release at the frog neuromuscular junction: evidence of morphological and physiological plasticity. *Brain Res.*, 12:95-116, 1987.

[30] U. Rutishauser. N-CAM and its polysialic acid moiety. A mechanism for pull/push regulation of cell interactions during development. *Dev. Suppl.*, 1992:99-194, 1992.

[31] U. Rutishauser. Regulation of cell-cell interactions by N-CAM and its polysialic acid moiety. In *Advances in Neurobiology*, pages 215-227, New York, 1993, Raven Press.

[32]   M.M. Salpeter. Vertebrate neuromuscular junctions: general morphology, molecular organization and functional consequences. In Salpeter, M.M., editor, *The vertebrate neuromuscular junction*, pages 1-55, New York, 1987a, Alan R. Riss Inc.

[33]   M.M. Salpeter (1987b) Development and neural control of the neuromuscular junction and of the junctional acetylcholine receptor. In Salpeter, M.M., editor, *The vertebrate neuromuscular junction*, pages 55-115, New York, 1987b, Alan R. Riss Inc.

[34]   J.R. Sanes. Roles of extracellular matrix in neural development. *Annu. Rev. Physiol.*, 45:581-600, 1983.

[35]   J.R. Sanes. Laminin for axonal guidance?. *Nature*, 315:714-715, 1985.

[36]   W.W. Schlaepfer. The nature of mammalian neurofilaments and their breakdown by calcium. *Progress in Neuropathology*, 4:101-123, 1979.

[37]   W.W. Schlaepfer and S. Micko. Calcium-dependent alterations of neurofilament proteins of rat peripheral nerve. *J. Neurochem.*, 32:211-219, 1979.

[38]   J.R. Slack., W.G. Hopkins and S. Pockett. Evidence for a motor nerve growth factor. *Muscle Nerve*, 6:243-252, 1983.

[39]   G.J. Swanson and G. Vrbová. Effects of low calcium and inhibition of calcium-activated neutral protease (CANP) on mature nerve terminal structure in the rat sternocostalis muscle. *Brain Res. Dev. Brain Res.*, 33:199-203, 1987.

[40]   W.J. Thompson. Synapse elimination in neonatal rat muscle is sensitive to the pattern of muscle use. *Nature*, 302:614-616, 1983.

[41]   W.J. Thompson. Activity and synapse elimination at the neuromuscular junction. Cell Mol. *Neurobiol.*, 5:167-182, 1985.

[42]   J. Tomas i Ferré, E. Mayayo and R. Fenoll i Brunet. Morphometric study of the neuromuscular synapses in the adult rat with special reference to the remodelling concept. *Biol. Cell*, 60:133-144, 1987.

[43]   J. Tomas, R. Fenoll, M. Santafé, J. Batlle and E. Mayayo. Motor nerve terminal morphologic plasticity induced by small changes in the locomotor activity of the adult rat. *Neurosci. Lett.*, 106:137-140, 1989.

[44]   J. Tomas, V. Piera, R. Fenoll, M. Santafé, J. Batlle, M.A. Lanuza and E. Mayayo. Motor nerve terminal morphologic plasticity induced by "physiological" increase in the locomotor activity of the adult rat. *Eur. J. Neurosci.*, 3(suppl):210, 1990a.

[45]   J. Tomas, R. Fenoll, E. Mayayo and M. Santafé. Branching pattern of the motor nerve endings in a skeletal muscle of the adult rat. *J. Anat.*, 168:123-135, 1990b.

[46]   J. Tomas, M. Santafé, R. Fenoll, E. Mayayo, J. Batlle, M.A. Lanuza and V. Piera. Pattern of arborization of the motor nerve terminals in the fast and slow mammalian muscles. *Biol. Cell*, 74:299-305, 1992.

[47]   J.P. Tremblay, R. Robitaille, O. Martineau, C. Labrecque and M.A. Fahim. Proximodistal gradients of the postjunctional folds at the frog neuromuscular junction. *Neuroscience*, 30:535-550, 1989.

[48]   A.R. Tuffery. Growth and degeneration of motor end-plates in normal cat hind limb muscles. *J. Anat.*, 110:221-247, 1971.

[49]   G. Vrbová, M.B. Lowrie M.B. and J. Evers.Reorganization of synaptic inputs to developing skeletal muscle fibres. In *Ciba Foundation Symposium, Vol. 138: Plasticity of the neuromuscular system*. Chichester: Wiley, pages 131-151, Chichester, 1988, Wiley.

[50]   A. Wernig, M. Pécot-Dechavassine and H. Stöver . Sprouting and regression of the nerve at the frog neuromuscular junction in normal conditions and after prolonged paralysis with curare. *J. Neurocytol.*, 9:277-303, 1980.

[51]   A. Wernig and A.A. Herrera. Sprouting and remodelling at the nerve-muscle junction. *Prog. Neurobiol.*, 27:251-291, 1986.

[52]   A. Wernig and M. Dorlöchter. Plasticity of the nerve muscle junction. In H. Rahmann, editor, *Prog. in Zool. (Vol. 37) Fundamentals of Memory Formation: Neuronal Plasticity and Brain Function*, pages 83-99, Stuttgart, 1989, Gustav Fisher Verlag.

[53]   D.J. Wigston. Remodeling of neuromuscular junctions in the adult mouse *soleus*. *J. Neurosci.*, 9:639-647, 1989.

Anabolic-Androgenic Steroid Regulation of Gene Expression in Spinal Motoneurons

P.E. Micevych, P. Popper, and C.E. Blanco

1. Introduction

Circulating levels of anabolic-androgenic steroids influence the physiological and biochemical properties of rat skeletal muscle [31,32,33,37,76]. These steroids increase reliance of muscles on aerobic metabolism of carbohydrates and fatty acids [76,37,31] and increase muscle fatigue resistance[32]. The regulation by anabolic-androgenic steroids the transcription of specific genes is mediated the androgen receptor (AR), a transcription factor which binds to specific DNA sequences located in the promoter region of certain genes known as hormone response elements (HRE/AREs [11,88]). The derived HRE/ARE consensus sequence, GGNACAnnnTGNTCT/C [65, 74], also is recognized by other nuclear receptors, including the glucocorticoid, mineralocorticoid and progesterone receptors [24,38]. The specificity of hormonal regulation of gene expression is insured by: specific amino acid sequence in the NH_2 terminus of each steroid receptors (i.e., the transactivational domain [44]); the selective interaction of steroid receptors with accessory factors [2]; binding to other DNA sequences within the genes promoter region, [1,73]; and the existence of composite regulatory elements with DNA sequences that are recognized by various transcription factors including steroid receptors [29,66]. Since AR is present in both skeletal muscles [30,50,62] and motoneurons [20,79,84], anabolic-androgenic steroids may act on both motoneurons and muscles [19,46,47]. Most of the studies on the effects of anabolic-androgenic steroids on motoneurons have been on the spinal nucleus of the bulbocavernosus (SNB)-bulbocavernosus/levator ani muscle complex (BC/LA), a sexually dimorphic neuromuscular system in the rat [7]. In adult male rats SNB motoneurons are located in the lumbar spinal cord and innervate two sexually dimorphic perineal muscles, the BC/LA and the external anal sphincter muscle [20,58,80]. The BC/LA is composed exclusively of histochemical type IIb muscle fibers with extremely low oxidative capacity, which resemble the metabolic profiles of muscle fibers belonging to fast-twitch fatigable motor units [13]. The BC/LA muscle complex is active only during penile reflexes which are androgen sensitive [40,41,77]. The BC/LA and the SNB neurons bind anabolic-androgenic steroids [20,30,50]. The androgen sensitivity of both the motoneurons and muscles makes this neuromuscular system particularly well-suited for studies on the effects of anabolic-androgenic steroids on neuromuscular systems. In adult, male rats, both the motoneuronal and muscular components of the SNB-BC/LA system respond to changes in circulating levels of testosterone [7].

Castration of adult male rats induces profound changes in the morphology, connections, and gene expression of the SNB-BC\LA neuromuscular system [34,51,55,56,57,67-70]. Castration results in reduced contractile activity and atrophy of the BC/LA muscles [35,78,86,93]. Castration induces ultrastructural changes in the BC and LA motor endplates [9,39] which are similar to those observed in pharmacologically induced models of inactivity [45]. There are also increases in miniature endplate potentials [86], up-regulation of extrajunctional nicotinic acetylcholine receptor (nAChR) synthesis [25] and decreased junctional nAChR levels [17,87]. In the SNB motoneurons, decreasing plasma testosterone

Table 1:

Myosin heavy chain (MHC) isoform composition of the Sol, DIA, EDL, TFL, and the BC/LA muscles.

| | **Muscle** | | | | |
ISOFORM (% of Total)	Sol[a]	DIA[b]	EDL[b]	TFL[a]	BC/LA[c]
MHC-s	80	28	2	-	-
MHC-IIa	20	21	9	-	-
MHC-IIx	-	45	29	-	-
MHC-IIb	-	6	59	100	100

[a] Derived from Tsika et al. [91]; [b] Data from Sugiura et al. [89]; [c] Data from d'Albis et al. [27]. The MHC isoform expressed by the muscle fibers innervated by a single motoneuron is associated with the physiological characteristics of the motor unit. Slow-twitch (type S) motor units are the most commonly activated and are generally composed of muscle fibers which express the slow myosin heavy chain isoform (MHC-S). Among fast-twitch motor units, the fatigue resistant motor units (type FR) are recruited more often than the fatigue intermediate (FInt) and the fatigable (type FF) motor units [23,42]. Muscle fibers belonging to FR, FInt and FF motor units express the MHC-IIa, MHC-IIx and MHC-IIb isoforms, respectively (for review, see [82]).

levels alters the amount of of mRNA coding for connexins [34]; the neuroeffector molecules α-calcitonin gene-related peptide (α-CGRP [69]) and cholecystokinin [67]; and the synthesizing enzyme for the neurotransmitter acetylcholine (ACh), choline acetyltransferase (ChAT [71]) in SNB motoneurons. All these molecules are involved in intercellular communication. We have focussed our efforts on examining the effects of anabolic-androgenic steroids on the motoneuronal expression of α-CGRP and ChAT.

CGRP is a 37 amino acid peptide that exists in an α- and a β- form [5]; α-CGRP is an alternate RNA splicing product encoded by the calcitonin gene [6,75], while β-CGRP is encoded by a separate gene [5]. CGRP synthesis, like that of most neuropeptides, is regulated at the level of transcription [26,68,69]. The expression of α- and β-CGRP is differentially regulated in motor and sensory neurons [64]. It has been postulated that CGRP may be an important effector molecule with both modulatory and trophic actions at the neuromuscular junction [60]. Although motoneurons that are CGRP immunoreactive or that express α-CGRP mRNA have been localized throughout the mammalian spinal cord and brainstem, not all motoneurons express α-CGRP in adulthood [10,15,68,83]. α-CGRP mRNA levels in SNB motoneurons are modulated by one or more factors derived from inactive BC/LA muscles [67,70]. This observation suggests that muscle fibers that are innervated by motoneurons which are infrequently activated may release higher levels of this inactivity-induced muscle factor(s).

The impulse-carrying neurotransmitter at the vertebrate neuromuscular junction is ACh. The rate of ACh synthesis is dependent on intracellular ChAT and choline concentrations.

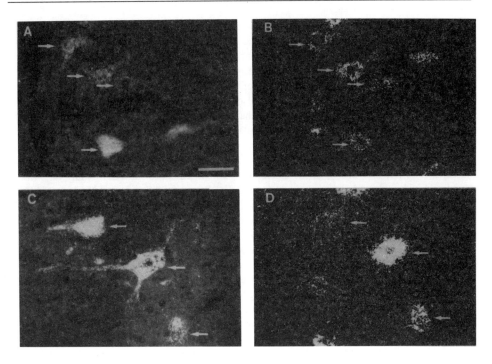

FIGURE 1: Identification of Sol motoneurons using Fluroruby (A) and TFL motoneurons using Fluorogold (C). Darkfield illumination (B and D) reveals that a number of Sol and TFL motoneurons express α-calcitonin gene-related peptide (arrows). Scale bar represents 50 μm.

In motoneurons, ChAT levels are transcriptionally regulated. For example, studies have shown that castration decreases ChAT activity levels in the BC\LA muscle complex [36,92]; this effect may be a consequence of a down-regulation of ChAT gene transcription.

Motoneurons innervating four additional muscles were also examined. The lumbar spinal cord motor pools innervating the soleus (Sol), extensor digitorum longus (EDL) and the tensor fascia latae (TFL) muscles, as well as the cervical spinal cord motor pool innervating the diaphragm (DIA) muscle. These four muscles were selected on the basis of their myosin heavy chain (MHC) isoform composition (Table I).

2. Alterations of α-CGRP and ChAT mRNA Levels in Motoneurons.

α-CGRP: Specific hybridization for α-CGRP in lumbar spinal cord motoneurons, including the SNB, are shown in Figure 3. The α-CGRP mRNA levels in SNB motoneurons were greater in rats castrated for 28 days than in age-matched gonadally intact males [69]. However, in motoneurons of the retrodorsal lateral nucleus (RDLN), which also express AR, α-CGRP mRNA levels were not altered by castration [69]. This suggests that α-CGRP mRNA expression is not regulated directly by testosterone through an AR-dependent mechanism. (For this review, "direct AR action" refers to the interaction of a ligand activated AR with an ARE). Subsequent experiments demonstrated that the increase in α-CGRP mRNA levels in SNB motoneurons was due to a concomitant decrease in BC/LA muscular activity following castration [67,70], since the activity of the BC/LA muscle complex is highly dependent on the circulating levels of testosterone [78,86]. Paralysis of the BC/LA muscle complex, by chronic injections of bipuvicaine to the BC/LA muscles of gonadally intact male rats produces similar increases in the α-CGRP mRNA levels in SNB motoneurons to those that occur following castration. In addition, injection of homogenates prepared from the BC/LA muscle complex of denervated or castrated rats into the BC/LA muscle complex of gonadally intact rats also increased α-CGRP mRNA levels in SNB motoneurons [67,70]. These data imply that an inactivity-induced factor produced by inactivated BC/LA muscle complex stimulates α-CGRP mRNA expression in SNB motoneurons [70].

In situ hybridization histochemistry with a ^{35}S-labelled antisense riboprobe complementary to the 3'-end (exons 5 and 6) of the calcitonin gene, which codes for αCGRP mRNA, was combined with retrograde labelling (intramuscular injection of a 4% solution of Fluorogold™ or fluororuby) of the soleus (Sol), diaphragm (DIA), extensor digitorum longus (EDL), and the tensor fascia latae (TFL) motoneurons to determine if αCGRP mRNA expression was related to contractile activity (Figure 2 [15]). Motoneurons were considered to contain α-CGRP mRNA if the silver grain density over the soma was at least three times greater than the background silver grain density. Using this criterion, the SNB (89.2%±4.8%) and TFL (84.2%±2.2%) motor pools contained the greatest number of α-CGRP mRNA expressing motoneurons, the EDL (74.8%±4.4%), Sol (55.3%±6.8%) and DIA (12.3%±4.8%) motor pools had fewer α-CGRP mRNA positive cells. There was a positive correlation (r=0.984) between the proportion of α-CGRP mRNA expressing cells in the SNB, TFL, EDL and DIA motor pools and the proportional expression of two myosin heavy chain isoforms, MHC-IIx and MHC-IIb (Table I). Muscle fibers which express these two MHC isoforms belong to fast-twitch fatigable and fast-twitch fatigue intermediate motor units and have low to intermediate succinate dehydrogenase activity levels [12,82]. This correlation suggests that motoneurons expressing α-CGRP mRNA in these motor pools may innervate fast-twitch fatigable and fast-twitch fatigue intermediate motor units. The correlation does not explain the significant number of Sol motoneurons expressing α-CGRP even though the Sol muscle does not contain any fast-twitch fatigue intermediate or fatigable motor units.

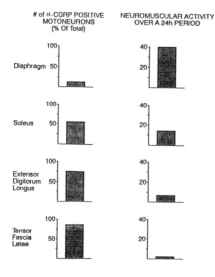

FIGURE 2:

Histographs depicting the inverse relationship between the number of motoneurons expressing α-CGRP mRNA in a motor pool and the average daily activation (neuromuscular activity) of various muscles.

testosterone -insensitive motor pools. These included the SNB motor pool, lumbar motoneurons innervating the Sol, EDL and TFL muscles, and cervical motoneurons innervating the DIA muscle.

α-CGRP: Anabolic-androgenic steroids increase the fatigue resistance and the aerobic capacity of fast-twitch muscles which are primarily composed of type IIb fibers [31,32,76]. Generally, increases in muscle aerobic capacity and fatigue resistance of a muscle are an indication of increased neuromuscular activity [22]. These changes in muscle physiology are consistent with our hypothesis that anabolic-androgenic steroids may indirectly decrease α-CGRP mRNA expression in motoneurons through the production and release of an

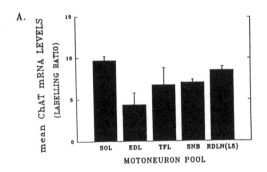

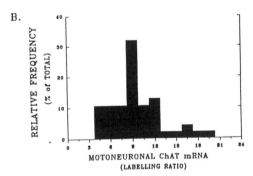

FIGURE 3:
ChAT mRNA levels, expressed as labelling ratio, in specific motor pools of the lumbar spinal cord of gonadally intact male rats. (A) The mean motoneuronal ChAT mRNA level of five motor pools. (B) Distribution of ChAT mRNA levels among Sol motoneurons

inactivity-induced muscle factor [67,70]. Indeed, α-CGRP mRNA levels in SNB motoneurons of TP-treated castrated rats were lower than those which are seen in castrated and intact males [69]. Implantation of a single 45mm TP capsule into castrated rats, which produces physiologically high circulating testosterone levels, depressed levels of α-CGRP mRNA in SNB motoneurons. These results strongly suggests that supraphysiologic doses of anabolic-androgenic steroids increase BC/LA muscle activity and concomitantly depress α-CGRP mRNA expression in SNB motoneurons. In males whose serum testosterone were chronically elevated to five times physiological levels for 28 days (10 ng/ml vs 1.8 ng/ml testosterone, respectively), the number of α-CGRP mRNAexpressing SNB motoneurons was 24% lower than in gonadally intact males. The 10 ng/ml testosterone level also resulted in an increase the weight of the BC/LA muscle complex. After standardizing to body weight, BC/LA muscle complex mass was increased by 34%. Increasing plasma testosterone levels to ten times physiological levels (i.e., 20 ng/ml) did not result in a further increase in BC\LA muscle complex weight, nor a decrease in α-CGRP mRNA levels in SNB motoneurons.

ChAT: In gonadally intact male rats whose serum testosterone levels were elevated to five times physiological (10 ng/ml for 28 days), there were significant increases in ChAT mRNA levels in motoneurons of the SNB and caudal RDLN but not in the Sol, EDL and TFL motor pools (Figure 4). When serum testosterone were elevated to ten times physiological levels (20 ng/ml), ChAT mRNA labeling ratio of SNB, caudal RDLN, Sol, EDL and TFL motoneurons were dramatically elevated. These results indicate that testosterone increases ChAT mRNA levels in specific motor pools in a dose-related manner. The sensitivity of specific motoneurons to alterations in plasma testosterone levels may be related to the AR levels. Studies are currently underway to determine the level of AR expression of SNB, caudal RDLN, Sol, EDL and TFL motoneurons.

4. CONCLUSIONS

The results of these studies have demonstrated that anabolic-androgenic steroids alter the expression of two molecules involved in intracellular communication in motoneurons.

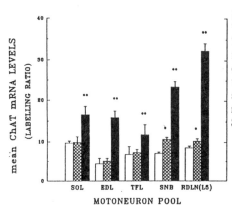

FIGURE 4:
The mean ChAT mRNA levels in the Sol, EDL, TFL, RDLN (L5) and SNB motor pools of intact males (open bars), intact males treated with 1 TP capsule (cross-hatched bars) and intact males treated with 2 TP capsules (filled bars). Error bars are ± sem. * Significantly different from intact males and males treated with 2 TP capsules. (p < 0.05). ** Significantly different from intact males (p < 0.01).

α-CGRP is a neuroeffector molecule that maintains neuromuscular morphology [60] and ChAT mediates the synthesis of ACh, the neurotransmitter at the neuromuscular junction which is responsible for initiating excitation-contraction coupling. There is substantial evidence that the expression of α-CGRP in motoneurons is regulated indirectly by anabolic-androgenic steroids through muscle-derived factor(s) and/or changes in spinal inputs to motoneurons [8,60,64,70]. In anabolic-androgenic sensitive neuromuscular systems, such as the SNB, testosterone-induced changes in CGRP levels indicate that there may be substantial molecular reorganization at the neuromuscular junction. Indeed, in the mouse SNB-BC/LA system, alterations in circulating testosterone levels induce parallel changes in the muscle fiber and neuromuscular junction size [9]. Subphysiologic levels of anabolic-androgenic steroids induce changes in the synthesis, release and degradation of ACh, in the BC/LA muscle complex [36,53,85,87,92]. Castration also decreases the density of postsynaptic nAChR in LA muscle fibers [17,87]. Testosterone replacement therapy of castrated male rats restores ChAT and acetylcholinesterase activity levels to the levels observed in gonadally intact rats [36,85,92]. However, the physiological significance of the regulation of ChAT expression by supraphysiologic levels of anabolic-androgenic steroids is unknown. Together with the present data, these results suggest that one of the regulators of ChAT activity at the neuromuscular junction is the level of ChAT expression. An elevation in ChAT levels could increase the rate of ACh synthesis at the presynaptic terminal of the neuromuscular junction increasing the amount of ACh available for release.

The sites and mechanisms through which anabolic-androgenic steroids regulate ChAT mRNA expression remain to be elucidated. Although several regulatory mechanisms may be involved, such as alterations in motoneuronal inputs and in the availability of retrogradely transported muscle-derived factors, the most parsimonious explanation is that ChAT mRNA levels are regulated directly by anabolic-androgenic steroids.   Sequence analysis of the upstream promoter region for the most prevalent ChAT mRNA isoform in the murine spinal cord which also regulates the neuron specific expression of ChAT [63] indicates that it contains two potential half-palindromic AREs. Various studies have shown that homology with the right-half of the HRE/ARE palindrome is sufficient for AR dependent transcriptional regulation [4,38,54,73,90]. Whether there is a similar promoter region in the rat ChAT gene is unknown, however, the organization and sequence of the exons is highly conserved in the mouse and and rat ChAT genes [48]. The presence of potential AREs in a highly active promoter region of the ChAT gene and the expression of AR by motoneurons suggest that AAS, such as testosterone, may directly regulate ChAT gene transcriptional activity through AR dependent mechanisms.

Studies in this laboratory and others (R. Handa, personal communication), suggest that testosterone up-regulates the expression of AR mRNA in the central nervous system. In male rats, testosterone binding in the hypothalamus decreases with castration and is restored to the binding levels of gonadally intact rats by testosterone replacement therapy [49]. Testosterone positively regulates AR concentrations in muscle and in specific regions of the prostate [18,62,72,95]. Moreover, supraphysiologic doses of anabolic-androgenic steroids increase the intensity of AR immunoreactivity in neurons, suggesting that the increase in mRNA levels is translated into functional AR [59]. Our data suggest that the anabolic-androgenic steroid regulation of ChAT mRNA levels in motoneurons is dependent on intracellular AR levels. Thus, we predict that in motor pools designated "anabolic-androgenic steroid-insensitive," such as the Sol, EDL and TFL respond to anabolic-androgenic steroids by increasing ChAT mRNA levels. These changes in ChAT message, we predict, are preceded by increases in AR expression. In cells expressing steroid receptors, the limiting factor in determining the capacity of a cell to respond to steroid levels is the receptor concentration [94].

In summary, these data indicate that skeletal muscle fibers are not the only target for anabolic-androgenic steroid action in the neuromuscular system. The changes in gene expression of motoneurons in response to alterations in the circulating anabolic-androgenic steroid levels may be mediated via muscle-nerve interactions, as suggested by our observations on the regulation of α-CGRP expression in SNB motoneurons, or by direct regulation of gene transcription, as we have hypothesized for ChAT expression. Thus, supraphysiological levels of testosterone alter mRNA levels of molecules involved in intercellular communication which alter motoneuronal gene transcription. Further studies are needed to determine the physiological consequences of these changes and their impact at the neuromuscular junction have not been investigated.

ACKNOWLEDGEMENTS

The authors would like to thank Dr. John Lu for assaying serum testosterone levels, Dr. Catherine Priest for critical commentary of this manuscript. This research was supported by NS-21220 and HD-07228.

LITERATURE CITED

[1] Abelson,L.A. and Micevych, P.E (1991). Distribution of preprocholecystokinin mRNA in motoneurons of the rat brainstem and spinal cord. *Molec. Brain Res.* **10**:327-35.

[2] Adler, A.J., Danielson, M., and Robins, D.M. (1992). Androgen-specific gene activation via a consensus glucocorticoid response element is determined by interaction with nonreceptor factors. *Proc. Natl. Acad Sci. USA* **89**:11660-63.

[3] Adler, A.J., Scheller, A., Hoffman, Y., Robins, D.M. (1991). Multiple components of a complex androgen-dependent enhancer. *Molec. Endocrinol.* **5**:1587-96.

[4] Adler, A.J., Scheller, A., Robins, and D.M. (1993). The stringency and magnitude of androgen-specific gene activation are combinatorial functions of receptor and nonreceptor binding site sequences. *Molec. Cell. Biol.* **13**:6326-6335.

[5] Amara, S., Arriza, J., Leff, S., Swanson, L., Evans, R. and Rosenfeld, M. (1985). Expression in brain of a messenger RNA encoding a novel neuropeptide homologous to calcitonin gene-related peptide. *Science,* **229**:1094-7.

[6] Amara, S., Jones, V., Rosenfeld, M., Ong, E., and Evans, R. (1982). Alternative RNA processing in calcitonin gene expression generates mRNAs encoding different polypeptide products. *Nature (London),* **298**:240-4.

[7] Arnold, A.P., Matsumoto, A. and Micevych, P.E (1988). Neural plasticity in a hormone-sensitive spinal nucleus. *Bull. TMIN* **16(suppl. 3)**:41-66.

[8] Arvidsson, U., Cullheim, S., Ulfhake, B., Hokfelt, T. and Terenius, L. (1989). Altered levels of calcitonin gene-related peptide (CGRP)-like immunoreactivity of cat lumbar motoneurons after chronic spinal transection. *Brain Res* **489**:387-91.

[9] Balice-Gordon, R.J., Breedlove, S.M., Bernstein, S., and Lichtman, J.W. (1990). Neuromuscular junctions shrink and expand as muscle fiber size is manipulated: *in vivo* observations in the androgen-sensitive bulbocavernosus muscle in mice. *J. Neurosci.* **10**:2660-71.

[10] Batten, T. F. C., Maqbool, A. and MacWilliam, P. N. (1992). CGRP in brain stem motoneurons. Dependent on target innervated? *Annals NY Acad. Sci.* **657**:458-60.

[11] Beato, M., Arnemann, J., Chulepekis, G., Slater, E. and Williams, T. (1987). Gene regulation by steroid hormones. *J. steroid Biochem.* **27**:9-14.

[12] Blanco, C.E., Fournier, M. and Sieck, G.C. (1991). Calcium-activated (Ca^{2+}-) myosin ATPase activity of muscle unit fibers: relation to muscle unit fatigue resistance. *Soc. Neurosci. Abstr.* **17**:650.

[13] Blanco, C.E., Micevych, P.E and Sieck, G.C. (1995). Succinate dehydrogenase activity of the sexually dimorphic muscles of the rat. *J. Appl. Physiol.* (In Press).

[14] Blanco, C.E., Mosconi, T., Kruger, L., and Micevych, P.E (1993a). Alterations in motoneuronal α-CGRP, GAP-43, and ChAT mRNA levels following mononeuropathy produced by a fixed diameter nerve "constriction" in the rat. *Soc. Neurosci. Abstr.* **19**:1508.

[15] Blanco, C.E., Popper, P. and Micevych, P.E (1992). αCGRP mRNA expression in motoneurons is related to myosin heavy chain composition of specific rat hindlimb muscles. *Soc. Neurosci. Abstr.* **18**:858.

[16] Blanco C.E., Popper P. and Micevych P.E (1993b). Anabolic-androgenic steroid regulation of choline acetyltransferase (ChAT) mRNA in spinal motoneurons. *Anat. Rec.* **Suppl. 1**:38.

[17] Bleisch, W.V. and Harrelson, A. (1989). Androgens modulate endplate size and ACh receptor density at synapses in rat levator ani muscle. *J. Neurobiol.* **20**:189-202.

[18] Blok, L.J., Bartlett, J.M.S., Bolt-De Vries, J., Themmen, A.P.N., Brinkmann, A.O., Weinbauer, G.F., Nieschlag, E., and Grootegoed, J.A. (1991). Regulation of androgen receptor mRNA and protein in the rat testis by testosterone. *J. Steroid Biochem. Molec. Biol.* **40**:343-47.

[19] Breedlove, S.M. (1992). Sexual dimorphism in the vertebrate nervous system. *J. Neurosci.* **12**:4133-42.

[20] Breedlove, S.M. and Arnold, A.P. (1980). Hormone accumulation in a sexually dimorphic motor nucleus of the rat. *Science* **210**:564-6.

[21] Brice, A., Berrard, S., Raynaud, B., Amsieau, S., Coppola, T., Weber, M.J., and Mallet, J. (1989) Complete sequence of a cDNA encoding and active rat choline acetyltransferase: a tool to investigate the plasticity of cholinergic phenotype expression. *J. Neurosci. Res.* **23**: 266-73.

[22] Burke, R.E. (1980). The stability of motor types in response to altered functional demands: hypertrophy, atrophy, and reinnervation models. In: Mechanisms of Muscle Adaptation to Functional Requirements, Vol. 24, (eds. Guba, F., Marechal, G. and Takacs, O.), Pergamon, NY, pp. 45-56.

[23] Burke, R.E. (1981). Motor units: anatomy, physiology and functional organization. In: Handbook of Physiology: The Nervous System. Motor Control. Section 1, Volume II, Part 1, Bethesda, MD: Am. Physiol. Soc., pp. 345-422.

[24] Cato, A.C., Skroch, P., Weinmann, J., Butkeraitis, P., and Ponta, H. (1988). DNA sequences outside the receptor-binding sites differently modulate the responsiveness of the mouse mammary tumour virus promoter to various steroid hormones. *Embo. J.* **7**:1403-10.

[25] Chin, H. and Almon, R. (1980). Fiber-type effects of castration on the cholinergic receptor population in skeletal muscle. *J. Pharmacol. Exp. Therapeutics* **212**:553-59.

[26] Comb, M., Hyman, S.E. and Goodman, H.M. (1989). Mechanisms of trans-synaptic regulation of gene expression. *TINS* **10**:473-8.

[27] d'Albis, A., Couteaux, R., Janmot, C., and Roulet, A. (1989). Specific programs of myosin expression in the postnatal development of rat muscles. *Eur. J. Biochem.* **183**:583-590.

[28] Damassa, D.A., Smith, E.R., Tennant, B., and Davidson, J.M. (1977). The relationship between circulating testosterone levels and male sexual behavior in rats. *Hormones and Behav.* **8**:275-86.

[29] Diamond, M.I., Miner, J.N., Yoshinaga, S.K., and Yamamoto, K.R. (1990). Transcription factor interactions: Selectors of positive or negative regulation from a single DNA element. *Science* **249**:1266-72.

[30] Dube, J.Y., Lesage, R., Tremblay, R.R. (1976). Androgen and estrogen binding in rat skeletal and perineal muscles. *Can. J. Biochem.* **54**:50-5.

[31] Eggington, S. (1987a). Effects of an anabolic hormone on aerobic capacity of rat striated muscle. *Pflugers Arch.* **410**:356-61.

[32] Eggington, S. (1987b). Effects of an anabolic hormone on striated muscle growth and performance. *Pflugers Arch.* **410**:349-55.

[33] Exner, G.U., Staudte, H.W. and Pette, D. (1973). Isometric training of rats. Effects upon fast and slow muscle and modification by an anabolic hormone (nandrolone decanoate). *Pflugers Arch.* **345**:15-22.

[34] Fisher, R. and Micevych, P. E (1992). Distribution of mRNA coding for connexin 32 and connexin 43 in the central nervous system with special reference to the cerebellar cortex. In: Progress in Cell Research, Vol. 3: Gap Junctions, (J. E. Hall, G. A. Zampighi and R. M. Davis, eds.), Elsevier Science Publishers, New York, pp. 141-8.

[35] Gori, Z., Pelligrino, C and Polkera, M. (1967). The castration atrophy of the dorsal bulbocavernosus muscle of rat: electron microscopic study. *Exp. Mol. Pathol.* **6**:172-98.

[36] Gutmann, E., Tucek, S., and Hanzlikova, V. (1969). Changes in the choline acetyltransferase and cholinesterase activities in the levator ani muscle of rats following castration. *Physiol. Bohem.* **18**:195-203.

[37] Guzman, M., Saborido, A., Castro, J., Molano, F. and Megias, A. (1991). Treatment with anabolic steroids increases the activity of the mitochondrial outer carnitine palmitoyltransferase in rat liver and fast-twitch muscle. *Biochem. Pharmacol.* **41**:833-5.

[38] Ham, J., Thomson, A., Needham, M., Webb, P., and Parker, M. (1988). Characterization of response elements for androgens, glucocorticoids, and progestins in mouse mammary tumour virus. *Nucleic Acids Res.* **16**:5263-76.

[39] Hanzlikova, V. and Gutmann, E. (1978). Effect of castration and testosterone administration on the neuromuscular junction in the levator ani muscle of the rat. *Cell Tiss. Res.* **189**:155-66.

[40] Hart, B. L. (1983). Role of testosterone secretion and penile reflexes in sexual behavior and sperm competition in male rats. A theoretical contribution. *Physiol. Behav.* **31**:823-7.

[41] Hart, B. and Melese d'Hospital, P.Y. (1983). Penile mechanism and the role of striated penile muscles in penile reflexes. *Physiol. Behav.* **31**:807-13.

[42] Henneman, E. and Mendell, L.M. (1981). Functional organization of motor pool and its input. In: Handbook of Physiology: The Nervous System. Motor Control. Section 1, Volume II, Part 1, Bethesda, MD: Am. Physiol. Soc., pp. 423-507.

[43] Hernsbergen, E., and Kernell, D. (1993). Daily duration of activity in ankle muscles of the cat. *IUPS Meeting Abstr.*, Glasgow, Scotland.

[44] Janne, O.A., Palvimo, J.J., Kallio, P., Mehto, M. (1993). Androgen receptor and mechanism of androgen action. *Annals Med.* **25**:83-9.

[45] Jirmanova, I. (1975). Ultrastructure of motor end-plates during pharmacologically-induced degeneration of skeletal muscle. *J. Neurocytol.* **4**:141-55.

[46] Jordan, C.L., Breedlove, S.M., and Arnold, A.P. (1991). Ontogeny of steroid accumulation in spinal motor neurons of the rat: implications for androgenic site of action during synapse elimination. *J. Comp. Neurol.* **313**:441-48.

[47] Jordan, C.L., Pawson, P.A., Arnold, A.P. and Grinnell, A.D. (1992). Hormonal regulation of motor unit size and synaptic strength during synapse elimination in the rat levator ani muscle. *J. Neurosci.* **12**:4447-59.

[48] Kengaku, M., Misawa, H., and Deguchi, T. (1993), Multiple mRNA species of choline acetyltransferase from rat spinal cord. *Molec. Brain Res.* **18**: 71-76.

[49] Krey, L.C. and McGinnis, M.Y. (1990). Time-courses of the appearance/disappearance of nuclear androgen+receptor complexes in the brain and adenohypophysis following testosterone administration/withdrawal to castrated males: relationships with gonadotropin secretion. *J. steroid Biochem.* **35**:403-8.

[50] Krieg, M. (1976). Characterization of the androgen receptor in the skeletal muscle of the rat. *Steroids* **28**:261-74.

[51] Kurz, E.M., Brewer, R.G., and Sengelaub, D.R. (1991). Hormonally mediated plasticity of motoneuron morphology in the adult rat spinal cord: a cholera toxin-HRP study. *J. Neurobiol.* **22**:976-88.

[52] Leslie, M., Forger, N.G., and Breedlove, S.M. (1991). Sexual dimorphism and androgen effects on spinal motoneurons innervating the rat flexor digitorum brevis. *Brain. Res.* **561**: 269-273.

[53] Lima-Landman, M.T., Goncalo, M.C., and Lapa, A.J. (1991). Efflux of acetylcholine in dimorphic skeletal muscle from castrated male rats. *Brazilian J. Med. Biol. Res.* **24**:1137-40.

[54] Lund, S.D., Gallagher, P.M., Wang, B., Porter, S.C., and Ganschow, R.E. (1991) Androgen responsiveness of the murine β-glucuronidase gene is associated with nuclease hypersensitivity, protein binding, and haplotype-specific sequence diversity within intron 9. *Molec. Cell. Biol.* **11**: 5426-34.

[55] Matsumoto, A., Arai, Y. and Hyodo, S. (1993). Androgenic regulation of β-tubulin messenger ribonucleic acid in motoneurons of the spinal nucleus of the bulbocavernosus. *J. Neuroendocrinol.* **5**:357-63.

[56] Matsumoto, A., Arnold, A.P., Zampighi, G.A. and Micevych, P.E (1988a). Androgenic regulation of gap junctions between motoneurons in the rat spinal cord. *J. Neurosci.* **8**:4177-83.

[57] Matsumoto, A., Micevych, P.E and Arnold, A.P. (1988b) Androgen regulates synaptic input to motoneurons of the adult rat spinal cord. *J. Neurosci.* **8**:4168-76.

[58] McKenna, K.E., and Nadelhaft, I. (1986). The organization of the pudendal nerve in the male and female rat. *J. Comp. Neurol.* **248**:532-49.

[59] Menard, C.S. and Harlan, R.E. (1993). Up-regulation of androgen receptor immunoreactivity in the rat brain by androgenic-anabolic steroids. *Brain Res.* **622**: 226-36.

[60] Micevych, P. E and Kruger, L. (1992). Status of CGRP as an effector peptide. *Annals NY Acad. Sci.*, **657**:379-96.

[61] Micevych, P.E, Popper, P., and Blanco, C.E. (1993). Rat motoneuronal choline acetyltransferase (ChAT) mRNA levels are increased by exposure to supraphysiological amounts of an anabolic-androgenic steroid. *Soc. Neurosci. Abstr.* **19**:985.

[62] Michel, G. and Baulieu, E.-E. (1980). Androgen receptor in rat skeletal muscle: characterization and physiological variations. *Endocrinol.* **107**:2088-98.

[63] Misawa, H., Ishii, K., and Deguchi, T. (1992). Gene expression of mouse choline acetyl transferase: alternative splicing and identification of a highly active promoter region. *J. Biol. Chem.* **267**: 20392-99.

[64] Noguchi, K., Senba, E., Morita, Y., Sato, M. and Tohyama, M. (1990). α-CGRP and β-CGRP are differentially regulated in the rat spinal cord and dorsal ganglion. *Mol. Brain Res.* **7**:299-304.

[65] Nordeen, S.K., Suh, B.J., Kuhnel, B., and Hutchison, C.A. (1990). Structural determinants of a glucocorticoid receptor recognition element. *Molec. Endocrinol.* **4**: 1866-73.

[66] Pearce, D. and Yamamoto, K.R. (1993). Mineralocorticoid and glucocorticoid receptor activities distinguished by nonreceptor factors at a composite response element. *Science* **259**:1161-5.

[67] Popper, P., Abelson, L. and Micevych, P.E (1992a). Differential regulation of α-calcitonin gene-related peptide and preprocholecystokinin messenger RNA expression in α-motoneurons: effects of testosterone and inactivity induced factors. *Neuroscience* **51**: 87-96.

[68] Popper, P. and Micevych, P.E. (1989). The effect of castration on calcitonin gene-related peptide in spinal motor neurons. *Neuroendocrinol.* **50**:338-43.

[69] Popper, P. and Micevych, P.E. (1991). Steroid regulation of calcitonin gene-related peptide mRNA expression in motoneurons of the spinal nucleus of the bulbocavernosus. *Mol. Brain Res.* **8**:159-66.

[70] Popper, P., Ullibarri, C. and Micevych, P.E (1992b). The role of target muscles in the expression of calcitonin gene-related mRNA in the spinal nucleus of the bulbocavernosus. *Molec Brain Res* **13**:43-51.

[71] Priest, C., Popper, P. and Micevych, P.E (1992). The neurochemistry of the motoneurons of the spinal nucleus of the bulbocavernosus. *Soc. Neurosci. Abstr.* **18**:858.

[72] Prins, G.S., and Birch, L. (1993). Immunocytochemical analysis of androgen receptor along the ducts of the separate rat prostate lobes after androgen withdrawal and replacement. *Endocrinol.* **132**:169-78.

[73] Rennie, P.S., Bruchovsky, N., Leco, K.J., Sheppard, P.C., McQueen, S.A., Cheng, H., Snoek, R., Hamel, A., Bock, M.E., MacDonald, B.S., Nickel, B.E., Chang, C., Liao, S., Cattini, P.A., and Matusik, R.J. (1993). Characterization of two *cis*-acting DNA elements involved in the regulation of the probasin gene. *Molec. Endocrinol.* **7**:23-36.

[74] Roche, P.J., Hoare, S.A., and Parker, M.G. (1992). A consensus DNA-binding site for androgen receptor. *Molec. Endocrinol.* **6**: 2229-35.

[75] Rosenfeld, M. G., Mermod, J-J., Amara, S. G., Swanson, L. W., Sawchenko, P. E., Rivier, J., Vale, W. W. and Evans, R. M. (1983). Production of a novel neuropeptide encoded by the calcitonin gene via tissue-specific RNA processing. *Nature (London)* **304**:129-35.

[76] Saborido, A., Vila, J., Molano, F. and Megias, A. (1991). Effect of anabolic steroids on mitochondria and sarcotubular system of skeletal muscle. *J Appl Physiol* **70**:1038-43.

[77] Sachs, B. (1982). Role of striated penile muscles in penile reflexes, copulation, and induction of pregnancy in the rat. *J. Reprod. Fertil.* **66**:433-43.

[78] Sachs, B. and Leipheimer, R. (1988). Rapid effect of testosterone on striated muscle activity in rats. *Neuroendocrinol.* **48**:453-59.

[79] Sar, M. and Stumpf, W.E. (1977). Androgen concentration in motor neurons of cranial nerves and spinal cord. *Science* **197**:77-9.

[80] Schroder, H. (1980). Organization of the motoneurons innervating the pelvic muscle of the male rat. *J. Comp. Neurol.* **192**:567-87.

[81] Sieck, G.C. (1995). Physiological effects of diaphragm muscle: Denervation and disuse. IN: <u>Clin. Chest Med.</u>, (editor, L. Fanberg), **IN PRESS**.

[82] Sieck, G.C., and Fournier, M. (1991). Developmental aspects of diaphragm muscle cells: Structural and functional control. IN: <u>Developmental Neurobiology of Breathing</u>, (eds., G.D. Haddad and J.P. Farber), Marcel Dekker, New York, pp. 375-428.

[83] Sieck, G.C., Popper, P., Zhan, W.Z., and Micevych P.E (1991). Absence of CGRP mRNA expression in phrenic motoneurons. *Soc. Neurosci. Abstr.* **17**:467.

[84] Simerly, R.B., Chang, C., Muramatsu, M. and Swanson, L.W. (1990). The distribution of androgen and estrogen receptor mRNA-containing cells in the rat brain: an *in situ* hybridization study. *J. Comp. Neurol.* **294**:76-95.

[85] Souccar, C., Godinho, R.O., Dias, M.A.V., and Lapa, A.J. (1988). Early and late influences of testosterone on acetylcholinesterase activity of skeletal muscles from developing rats. *Brazilian J. Med. Biol. Res.* **21**:263-71.

[86] Souccar, C., Lapa, A. and do Valle, J.R. (1982) Influence of castration on the electrical excitability and contraction properties of the rat levator ani muscle. *Exp. Neurol.* **75**: 576-88.

[87] Souccar, C., Yamamoto, L.A., Goncalo, M.C. and Lapa, A.J. (1991). Androgen regulation of the nicotinic acetylcholine receptor-ionic channel in a hormone-dependent skeletal muscle. *Brazilian J. Med. Biol. Res.* **24**:1051-4.

[88] Spelsberg, T.C., Rories, C., Rejman, J.J., Goldberger, A., Fink, K., Lau, C.K., Colvar, D.S. and Wiseman, G. (1989). Steroid action on gene expression: possible roles of regulatory genes and nuclear acceptor sites. *Biol. of Repro.* **40**:54-69.

[89] Sugiura, T., Morimoto, A., Sakata, Y., Watanabe, T., and Murakami, N. (1990). Myosin heavy chain isoform changes in rat diaphragm are induced by endurance training. *Japanese J. Physiol.* **40**:759-63.

[90] Tan, J.-A., Marschke, K.B., Ho, K.-C., Perry, S.T., Wilson, E.M., and French, F.S. (1992). Response elements of the androgen-regulated C3 gene. *J. Biol. Chem.* **267**: 4456-4466.

[91] Tsika, R.W., Herrick, R.E., and Baldwin, K.M. (1987). Subunit composition of rodent isomyosins and their distribution in hindlimb skeletal muscles. *J. Appl. Physiol.* **63**:2101-10.

[92] Tucek, S., Kostirova, D., and Gutmann, E. (1976). Testosterone-induced changes of choline acetyltransferase and cholinesterase activities in rat levator ani muscle. *J. Neurol. Sci.* **27**:353-62.

[93] Venable, J. H. (1966). Morphology of cells of normal, testosterone-deprived, and testosterone-stimulated levator ani muscles. *Am. J. Anat.* **119**:271-302.

[94] Webb, P., Lopez, G.N., Greene, G.L., Baxter, J.D., and Kushner, P. (1992). The limits of the cellular capacity to mediate an estrogen response. *Molec. Endocrinol.* **6**:157-67.

[95] Wolf, D.A., Herzinger, T., Hermeking, H., Blaschke, D. and Horz, W. (1993). Transcriptional and posttranscriptional regulation of human androgen receptor expression by androgen. *Molec. Endocrinol.* **7**:924-36.